W0042150

# A NEW HISTORY OF GREEK MATHEMATICS

The ancient Greeks played a fundamental role in the history of mathematics. Their innovative ideas were reused and developed in subsequent periods, all the way down to the scientific revolution and beyond. In this, the first complete history for a century, Reviel Netz offers a panoramic view of the rise and influence of Greek mathematics and its significance in world history. He explores the Near Eastern antecedents and the social and intellectual developments underlying the subject's beginnings in Greece in the fifth century BCE. He likewise leads the reader through the proofs and arguments of key figures, such as Archytas, Euclid, and Archimedes, while considering the totality of the Greek mathematical achievement, which includes – as well as pure mathematics – such applied fields as optics, music, mechanics, and above all, astronomy. This is the gripping story of not only a major historical development but also some of the finest mathematics ever created.

REVIEL NETZ is Patrick Suppes Professor of Greek Mathematics and Astronomy at Stanford University. He is the author of many books on Greek mathematics and culture, including *The Shaping of Deduction in Greek Mathematics* (Cambridge, 1999); *The Archimedes Codex* (coauthored with William Noel; Da Capo Press, 2007); *Ludic Proof: Greek Mathematics and the Alexandrian Aesthetic* (Cambridge, 2009); and *Scale, Space and Canon in Ancient Literary Culture* (Cambridge, 2020).

# A NEW HISTORY OF GREEK MATHEMATICS

REVIEL NETZ
*Stanford University, California*

Shaftesbury Road, Cambridge CB2 8EA, United Kingdom

One Liberty Plaza, 20th Floor, New York, NY 10006, USA

477 Williamstown Road, Port Melbourne, VIC 3207, Australia

314–321, 3rd Floor, Plot 3, Splendor Forum, Jasola District Centre, New Delhi – 110025, India

103 Penang Road, #05–06/07, Visioncrest Commercial, Singapore 238467

Cambridge University Press is part of Cambridge University Press & Assessment,
a department of the University of Cambridge.

We share the University's mission to contribute to society through the pursuit of education, learning and research at the highest international levels of excellence.

www.cambridge.org
Information on this title: www.cambridge.org/9781108833844

DOI: 10.1017/9781108982801



First published 2022

*A catalogue record for this publication is available from the British Library*

*Library of Congress Cataloging-in-Publication data*
Names: Netz, Reviel, author.
TITLE: A new history of Greek mathematics / Reviel Netz, Stanford University, California.
DESCRIPTION: Cambridge, United Kingdom ; New York, NY : Cambridge University Press, 2022. | Includes bibliographical references and index.
IDENTIFIERS: LCCN 2022022814 (print) | LCCN 2022022815 (ebook) | ISBN 9781108833844 (hardback) | ISBN 9781108987202 (paperback) | ISBN 9781108982801 (epub)
SUBJECTS: LCSH: Mathematics, Greek–History. | Mathematics, Ancient. | BISAC: PHILOSOPHY / History & Surveys / Ancient & Classical
CLASSIFICATION: LCC QA22 .N282 2022 (print) | LCC QA22 (ebook) | DDC 510.9–dc23/eng/20220513
LC record available at https://lccn.loc.gov/2022022814
LC ebook record available at https://lccn.loc.gov/2022022815

ISBN 978-1-108-83384-4 Hardback

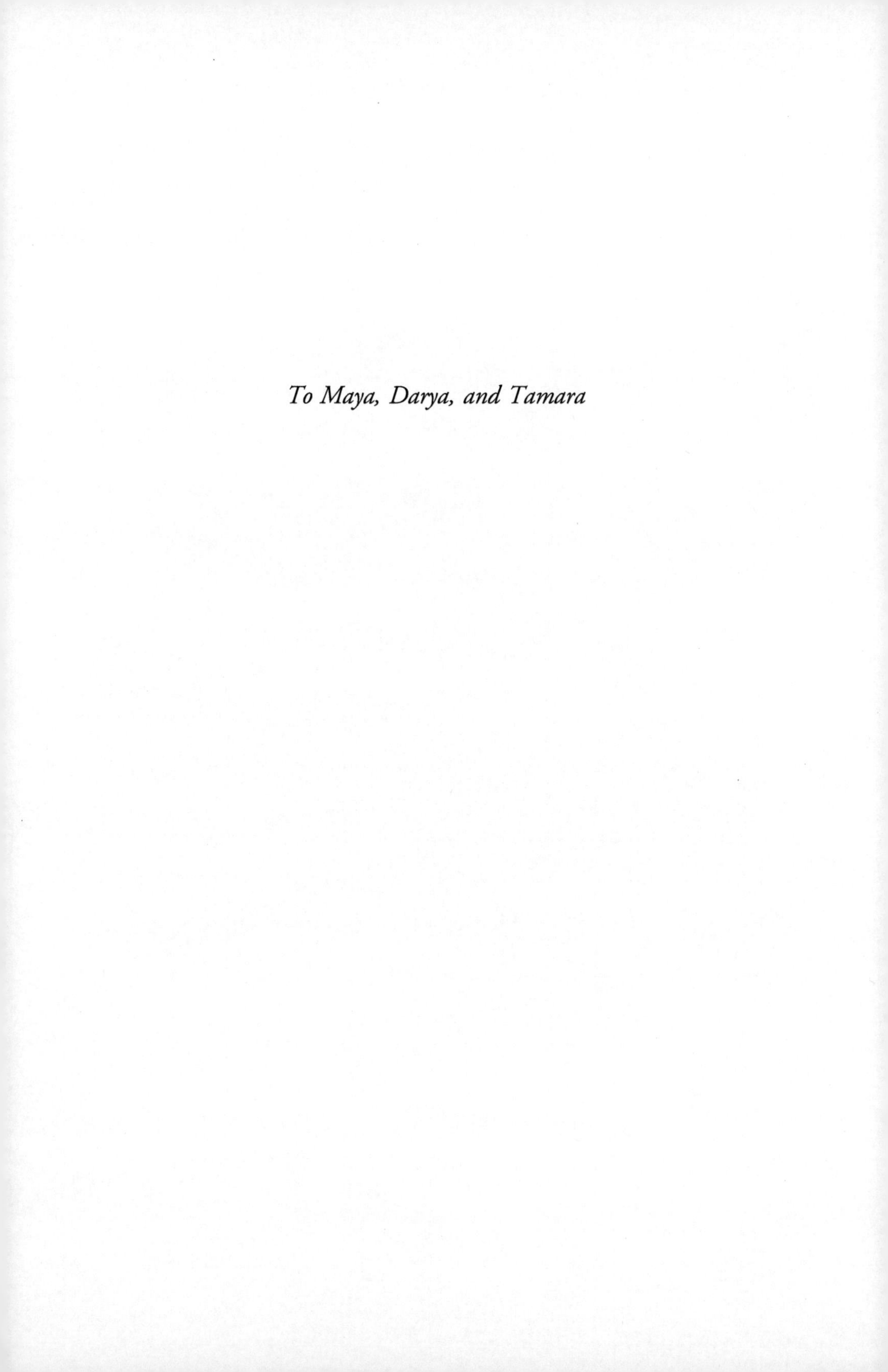

*To Maya, Darya, and Tamara*

# *Contents*



*A plate section will be found between pages 272 and 273.*

# *Plates*

1 Pages from Euclid's *Elements* (Book XIII.16, showing the construction of the regular icosahedron). MS D'Orville 301, Bodleian Library, Oxford University. Image courtesy of the Clay Mathematics Institute.

2 P. Oxy. 5299. With brief statements and rough figures from the first book of Euclid's *Elements*. Courtesy of the Egypt Exploration Society and the University of Oxford Imaging Papyri Project.

3 Doryphoros (450–400 BC). Designed by sculptor Polykleitos. Roman marble copy of Herculaneum. Naples National Archaeological Museum. Italy. Photo: PHAS/Universal Images Group/Getty Images.

4 Pleiades or Seven Sisters or the Messier 45 star cluster rising above the mountains of Leh, Himalayas. © Sukanya Ramanujan. (https://sukanyaramanujan.wordpress.com).

5 Late Babylonian clay tablet with the mul.apin. Purchased from Messrs Mann & Bishop, 1889. Museum number 86378 © The Trustees of the British Museum.

6 Mosaic floor from Boscotrecase, Pompeii, showing Plato's Academy at Athens. The philosopher – sitting in the middle – teaches a group of disciples. National Archaeological Museum, Naples, Italy. Photo by Leemage/Corbis/Getty Images.

7 Exploded computer model of Antikythera mechanism. © 2020 Tony Freeth.

8 The Archimedes Palimpsest, folia 102r–98v: a diagram of a spiral, with an initial from the prayer book laid over it. Image produced by the Rochester Institute of Technology and Johns Hopkins University. Copyright courtesy of the owner of the Archimedes Palimpsest.

# *Preface*

In 1586, Galileo Galilei was twenty-two years old and ready to do great things. He authored his first work, the *Bilancetta*, or *Small Balance*, a gem of a treatise. As so often with Galileo, it brings much together: literary erudition, mathematical sophistication, experimental precision.

Galileo begins with Vitruvius's story of Archimedes's solution to the problem of the crown. We all remember this story: The king's goldsmith was provided the materials to make a crown of pure gold – but could he have returned, instead, a crown made of an alloy containing silver? Could he have pocketed the difference? Stepping into the bath – so Vitruvius continued his tale – Archimedes suddenly realized that his body displaced a volume of liquid equal to that of his own body; excitedly, he ran home naked, shouting, "*Eureka*!" ("I have found it!"). It is possible to measure how much of a liquid is displaced by a given body and thus to measure the volume of any given body. And if so, we can also weigh the body and, by dividing the two quantities – weight divided by volume – we can find the body's (in modern terminology) "specific density" or, effectively, "specific gravity." It is well understood that gold is rather heavier than silver (has a higher specific gravity). So, the specific gravity of the crown would tell us whether it is or is not made of pure gold. It was not, we are told by Vitruvius. Archimedes – the mathematician detective – confronted the goldsmith, who confessed. A triumph for science!

This Galileo retells – and refutes. Of course Vitruvius must have been wrong. Why? Because Galileo read Archimedes himself (made newly available in Commandino's edition from 1565). The observation reported by Vitruvius – that the surface of a liquid rises as you step into it – is trivial; a really deep mathematical fact, however, which Archimedes did discover, is the law of buoyancy. A body immersed in a liquid *loses a weight equal to the weight of the liquid it displaces.* The fraction of the original weight lost as a body is immersed in a liquid can therefore be used directly to measure specific gravities – no need for messy (and inherently imprecise) measures of volume displaced. Instead, use the "small balance" of Galileo's treatise, in which a counterweight is moved so as to balance an arm immersed in water. Galileo shows the mathematical principles underlying the instrument and their grounds in actual Archimedean science, and he crowns this all with concrete, technical suggestions on how the small balance can be made truly precise.

And so, for Galileo, a beginning. Not yet an original contribution, but immediately following that, Galileo set his eyes on his own path in science. Aristotle had argued that the motion of heavy bodies was simply downward, but Archimedes (so Galileo learned, from Commandino's edition) had shown the mathematical reason why objects, when immersed in a liquid heavier than themselves, are pushed upward. A thought then suggested itself to the young Galileo: Would it not be possible, in general, to account for the motions of bodies through the relative weights of the moving object and the medium through which it moved? Could one extend Archimedes's theory of floating bodies into a theory of free fall? This was Galileo's first major theoretical attempt, *De Motu* (*On Motion*), embarked upon when he was just twenty-five. Like the *Bilancetta* itself, it was never published in his lifetime, and the quest for a theory of free fall ultimately would lead Galileo in many other, different directions. The fruit of all this labor would wait for many years, until the *Dialogues Concerning Two New Sciences*, Galileo's mature contribution, published in 1638. A major book, now – it set out the foundations of modern classical mechanics. As a seventy-two-year-old, Galileo still reflected on where it all started from: "It was Archimedes' own books – which I had already read and studied with infinite astonishment – that rendered credible to me all the miracles described by various writers."[1]

Galileo set himself, from the beginning, as a critic of Aristotle, and this critique, indeed, would loom large in his career. We imagine him (probably a mere legend, though) hauling light and heavy balls up the Leaning Tower of Pisa (light and heavy, they fall at the same speed, thus refuting Aristotle!). More historically, we can spy him squinting at the sky with his homemade telescope, so crude to our eyes and yet so powerful in Galileo's hands. He detects moons circling Jupiter, mountains and seas on the surface of the moon. In other words, the heavens were not the ethereal, unchanging realm of pure circular motions all centered around the earth, as envisaged by Aristotle. They were a load of coarse matter pressing upon

[1] G. Galilei, *Dialogues Concerning Two New Sciences*, trans. H. Crew and A. de Salvio (New York: Macmillan, 1638/1914), p. 41.

coarse matter, just like the earth – Copernicus, after all, must have been right! The church, defending Aristotle's orthodoxy, found objection to all of this, and Galileo's ensuing trial set out a powerful image. Historians came to think of this era in such terms, breaking away, decisively, from the constricted dogmas of Aristotle. In 1957, Koyré called this the transition "from the closed world to the infinite universe"; Thomas Kuhn, a few years later, saw this transition as a *paradigm shift*, the prime example of his 1962 classic, *The Structure of Scientific Revolutions*.

More recently, historians have become wary of such sweeping and even teleological categories as the "scientific revolution." Steven Shapin, famously, began his study from 1996, titled simply *The Scientific Revolution*, with the following words: "There was no such thing as the Scientific Revolution, and this is a book about it." I am not here to beat a dead horse. But this should be emphasized. The narrative of a scientific revolution is mistaken, first of all, in that it hinges on encounters such as that of *Galileo with Aristotle*. But these, in fact, were not the decisive moments leading to the rise of modern science. Far more crucial were encounters such as that of *Galileo with Archimedes*. What gave rise to modern science was a new appreciation of the science of antiquity and the attempt, finally, to emulate and outdo it. To a large extent, modern science came not from a scientific revolution but a *scientific renaissance*.

Modern science is firmly rooted in the science of antiquity. To be clear: this ancient science was not *only* mathematical. William Harvey's revival of Galen's medicine would ultimately be as important, in its own way. And yet, the main line of development of modern science does begin with Copernicus, passing through Galileo, Fermat, Kepler, and Descartes, and leading on to Leibniz and Newton. This line of development, throughout, can be characterized as a revival of the ancient tradition of the exact sciences. And in this book, I set out to provide a new history of this ancient tradition.

A topic as important as this – the soil from which grew modern science – might be expected to attract significant scholarly attention. Astonishingly, this book is the first such contribution in precisely a century. Thomas Little Heath's *History of Greek Mathematics*, published in 1921, has served as a reliable guide to many generations of scholars and curious readers. Historiographies went in and out of fashion, but Heath still stands, providing a clear and readable survey of the contents of most of the works of pure mathematics attested from Greek antiquity. To a modern reader, used to more critical, analytical historical approaches, Heath's work reads most like an encyclopedia, arranged by chronological principles. One may turn to the entry on Apollonius of Perga, for instance, and find seventy highly informative pages summing up the contents of the Conics (forty-one pages), followed by smaller surveys of the contents of the minor, indirectly attested works. I keep Heath by my side, and I urge you to do so as well. This new history does not aim to replace Heath's, and I do not aim at his encyclopedic coverage. My goal, instead, is to provide a historical *account*. Something quite deep – indeed, transformational – took place in the ancient Greek world. A new kind of science emerged, ultimately providing the tools for modernity. Why and how did that happen? What we need is to understand the conditions and scope of this achievement.

This Greek achievement belongs, in my view, to the history of science. *Science* itself is not a word the Greeks used, although I think it is useful for our purposes. The modern word *mathematics* is indeed ultimately Greek, but what the Greeks called *ta mathēmata* was usually wider in meaning than the modern term implies (and wider than what Heath understood as the scope of his own history). Besides pure geometry and stereometry (as well as the much less central field of pure arithmetic), the ancient Greeks always included within mathematics fields such as astronomy and theoretical music, and they often added optics and mechanics as well. That there was a difference between "pure" and "mixed" mathematics was often acknowledged, but that the two belonged together was also clear. How could it not be? As I will point out throughout this book, the identity of Greek mathematics was, above all, that of a literary genre. And at this level – the way they were written about – the different fields were not all that far apart. Whether in geometry or astronomy, optics or mechanics, one would encounter nearly the same diagrams, nearly the same formulaic language. It would be ahistorical as well as misleading to produce a history of Greek mathematics and leave out the more applied fields. If you will, you may think of it as a history of the Greek exact sciences. But I prefer to keep the word *mathematics*: it is, in fact, closer to the way the Greeks, themselves, thought about it. Even in the more applied fields, it was a theoretical study, organized around the idea of proof, not around the idea of experiment. But I am getting ahead of myself. I shall argue for all of this throughout the book.

One final word. Heath required two volumes, even while excluding all the applied mathematical sciences. What I wish to produce is different: a single narrative account, of use for the general interested public, as well as for undergraduate classes and for those graduate students and scholars looking for some entry point into the historical foundations of science.

A wider public readership demands a slimmer apparatus. At the end of each chapter, I provide a list of additional readings – those that directly further the main topic – and offer only a handful of footnotes suggesting sources for more particular claims. My goal is simple: to make the story interesting enough so that my readers, indeed, look further.

## Plan of the Book

The seven chapters of the book are mostly – but not entirely – arranged in chronological order. The first three chapters provide a chronological survey of (mostly) pure geometry as it developed from the beginnings to the era of Archimedes (so, roughly until the year 200 BCE). Chapter 1, "To the Threshold of Greek Mathematics," provides a comparative context, zooms in on the Babylonian antecedents to Greek mathematics, and then argues that the beginnings of Greek mathematics are to be found in the second half of the fifth century BCE. I then argue that much of the creative activity in Greek pure mathematics took place in two generational events. The first occurred early in the fourth century, and it is the subject of Chapter 2, "The Generation of Archytas." The second occurred late in the third century, and it is the subject of Chapter 3, "The Generation of Archimedes" (which also surveys the developments in the era in between the generations, including the important figure of Euclid). I argue that there is a significant difference between the two generations: in the generation of Archytas, mathematics was in dialogue with philosophy; in the generation of Archimedes, it was much more autonomous. Chapter 3 has the most mathematics: many authors of this era are extant, and their contributions are extremely important; as a consequence, it is also a longer chapter.

The two chapters that follow take up the more applied mathematical sciences: Chapter 4, "Mathematics in the World," looks at various "mechanical" and similar applications, and Chapter 5, "Mathematics of the Stars," looks at astronomy. Although these chapters break away from the chronological sequence, many of the developments in those more applied fields took place in the late Hellenistic era and then in the Roman imperial period, so those two chapters, taken together, extend the survey all the way down to around 200 CE. Astronomy is a huge field (which is why Heath put it aside); thus, Chapter 5, once again, is a longer chapter.

From 200 CE onward, the legacy of Greek mathematics was formed and carried forward by many civilizations. This is the subject of the final two chapters. Chapter 6, "The Canonization of Greek Mathematics," discusses the absorption of mathematics into Neoplatonist philosophy, and specifically into the practice of philosophical commentary, in Late Antiquity. The final chapter, Chapter 7, "Into Modern Science: The Legacy of Greek Mathematics," considers both the survival of Greek mathematics, through the transmission of the works in manuscript in Byzantium, and its impact in later scientific civilizations, such as the Islamic world and, finally, early modern Europe – with the renaissance of Greek mathematics giving rise, ultimately, to the rise of modern science.

# *Acknowledgments*

Back in 1986, the Tel Aviv University's Bulletin was still an actual printed book. Flipping through it, I chanced upon an intriguing class: "Euclid's Elements," taught by Professor Sabetai Unguru. I went in, very soon realized I had much more to learn; I still do. Thank you, Sabetai.

To list the people to whom this book owes a debt would require a fully-fledged memoir. Let me just say that I had plenty of good luck in my encounters since. Avoiding a long apparatus, this book provides instead suggestions for further reading. Looking again through them all, I notice how many of the names I cite are those of my teachers or of my peers. This book emphasizes the role of the scientific network: the people you talk to shape the scholarship you produce. I was formed as a scholar in the 1990s, and the network organized around the study of early science was, back then, bursting with creative energy, with Geoffrey Lloyd as its philosopher-king. I just list those of my friends that I cited in the Suggestions for Further Reading (apologies to the many of you I have learned from in other ways): Len Berggren, Alain Bernard, Alan Bowen, Christián Carman, Karine Chemla, Jean Christianidis, Leo Corry, Serafina Cuomo, Daniella Dueck, Michael Fried, Jens Høyrup, Carl Huffman, Alexander Jones, Henry Mendell, William Noel, Josiah Ober, Eleanor Robson, Courtney Roby, Francesca Rochberg, Ken Saito, David Sedley, Michalis Sialaros, Nathan Sidoli, Natalie Tchernetska, Nigel Wilson, Ido Yavetz, and Leonid Zhmud. A particular thank-you to Christian Carman and to Alexander Jones, who read and vastly improved an early version of Chapter 5, and to Geoffrey Lloyd and Yuval Wigderson, who commented insightfully on the entire manuscript. A particular sorrow is that one's new work can no longer be shared with Andrew Barker, Myles Burnyeat, and Ian Mueller; I constantly remember David Fowler, the loveliest of friends.

I never met Wilbur Knorr. This book owes everything to his research. In 1999 I stepped into his giant shoes in the department of Classics at Stanford. This book would have been impossible without the generosity of my department and of my deans. I wrote it during a sabbatical year at the Center for Advanced Studies in Behavioral Sciences (CASBS), where I benefited from the idyllic site and from the delightful company of my co-scholars of the last academic year before the pandemic, 2018–2019. Sixty years before, in the same place – a few studies away from my own – Thomas Kuhn wrote his *The Structure of Scientific Revolutions*. As I was writing this book, I became more and more convinced that Kuhn was completely wrong about ancient science. But then again, I came to wonder, how would my own book fare? Scholars come and go; places such as CASBS remain; scholarship progresses from one wrong book to the next.

For this, we have publishers to thank. The profession of classics is lucky to have Michael Sharp as the editor of Classics at Cambridge University Press, bringing into being, with his assured touch, so much of the literature on which we rely. Bethany Johnson, as content manager, deserves special honor for steering so many books through the ravages and backlogs of the pandemic. Special thanks to Kirsten Balayti, my copy-editor, and Amy Carlow, my research assistant who helped with preparation of the index and also discovered typos I am too ashamed to reveal. And now, thanks to all of you, it is finally ready, a printed book. I can now hope for chance encounters: future readers, chancing upon the book, flipping through its pages. Maybe, some of my readers might even decide to embark upon the study of Greek mathematics. . . . Thank you, my readers.

CHAPTER I

# *To the Threshold of Greek Mathematics*

## Plan of the Chapter

This chapter is introductory. I first survey, in a quick sweep, mathematics before Greece. This is followed by the historical context for the rise of mathematics in Greece itself (a discussion heavy with historiographical problems because so much is speculative). Finally, I conclude with a picture of the earliest known Greek mathematics.

I start with a section titled "Before Greece" – indeed, before any organized science at all. What are the universally shared bits of mathematics known even to simple societies? We find considerable but shallow knowledge. Familiarity with numbers and shapes is nearly universal – but does it amount to mathematics? "Empire and the Invention of Mathematics" brings in the rise of the state and with it, I argue, mathematical knowledge; "Beginning in Babylon" zooms in on the most important antecedent to the Greeks: the mathematics of Mesopotamia.

This, then, provides one kind of introduction. Another has to do with the Greeks themselves. The section titled "The Greeks: Standing Apart?" brings in the basic historical context: the unique characteristics of early Greek civilization. But where and how does mathematics emerge in Greece? "Greek Mathematical Myths" argues against some traditional narratives (most important: Pythagoras the mathematician was, I argue, indeed a myth). Another problematic context is that of Mesopotamian mathematics, and the following section, "Greeks and the Near East: A Historiographical Detour," tries to delineate a possible account of the debt owed by the Greeks to their predecessors.

With all of this in place, we may finally get to "The Threshold of Mathematics," which I identify as the mathematics attested to Hippocrates of Chios, and I conclude with "Assessing the Threshold": the historical meaning of this new Greek invention of mathematics.

## Before Greece

Throughout this book, I will argue that Greek mathematicians had achieved something quite unprecedented. But of course, people everywhere know some mathematics, and the Greeks specifically must have owed something to past cultures. They did not start from scratch!

All of this sounds nearly obvious. In fact, we've merely started – and have already entered a minefield. The question of the cognitive universals underlying mathematics is invested with political meaning.

The issue can be stated quite simply, and it should be stated right as we begin. Students from disadvantaged backgrounds do much worse in mathematical tests. The response to this fact varies. Some take comfort. (They see the results of mathematical tests as proof of their belief in their group's superiority over others.) Most, aware of the enormous difference that social conditions make to cognitive growth, are less surprised that the underprivileged are also the mathematical underachievers. The explicit racist position is, frankly, preposterous, but it is stated by some and perhaps harbored by many. And so it is right that I should address it, head-on, right at the beginning. Consider the following two statements: (A) "The Greeks invented mathematics because they were white," and (B) "John is good at math because he is a white boy." If A strikes you as implausible, so should B. And if A does not strike you as implausible. . . . Well, this is, in part, why I've written this book.

So, what to do with mathematical tests? Some would say that they should not matter. Do math for the intellectual satisfaction it brings, not to get a good grade! But mathematical educators do not have the luxury of retreating into such fantasies. They have to go and teach in a world where mathematical tests do matter, and so the urgent task is this: How can we make mathematics more accessible to underprivileged students?

Now, this brings us back to the history of mathematics and to the question of universals. This question – how to make mathematics more accessible to the underprivileged – became especially acute in the global scene in the aftermath of decolonization in the 1960s and 1970s. New states in the Third World aimed to make education universal; however, this newly available education, more often than not, did not empower students but instead instilled in them a sense of helplessness and dependency. The mathematics was alien and forbidding, and so the best educators looked for ways to make it grow directly out of the students' own culture. Paulus Gerdes, for instance, as a young mathematics teacher in Mozambique, noticed that fishermen prepare their haul for sale by drying their fish near a fire built in the sand by the seashore. To make sure all the fish become dry at the same time, they follow a certain procedure. First, plant a stick in the ground, then attach a rope, and with a second stick attached at the other side of the rope, draw a circle in the sand. At this point, place all the caught fish along this drawn line, and finally, build the fire at the center. Gerdes's idea was revolutionary – and straightforward: Instead of starting with some abstract definitions, would it not make more sense to teach the children of those fishermen the concepts of "circle," "center," and "circumference" based on this procedure?[1]

Multiply this kind of example hundreds of times, and you have the discipline of ethnomathematics. Anthropologists, even apart from any application to the education of mathematics, came to be interested in the mathematical ideas available to preliterate societies; cognitive psychologists soon came to appreciate the significance of this research for the study of the universal human mind.

Thanks to the work of the ethnomathematicians, several observations emerged. First, numbers are pretty much universal. To be clear: it has been observed that the Pirahã tribe in the Amazon has no words for numerals. (There is some scholarly debate over this: Do the Pirahã words *hói* and *hoí* mean "one" and "two," respectively, or do they mean – as the best experts now seem to believe – merely something like "small" and "larger"?) It is extremely interesting to cognitive psychologists if, indeed, even a single language could fail to develop numerical terms – and so, perhaps, number is not directly hardwired into the human brain.[2] However, from the point of view of the anthropologist or of the historian, the example of the Pirahã is striking primarily for its freakish rarity. Everywhere you go around the globe, languages possess varied systems of counting. A few might be more impoverished (in particular, the Amazon has a number of less numerical societies, of which the Pirahã are an extreme and relatively well-studied case). But more often, simple societies have highly sophisticated numerical systems, with addition, multiplication, and iteration encoded into language itself. (Only one among these is the base-ten numerical systems now used by nearly all humans; it is nearly universal, perhaps, because it is, if anything, mathematically simpler than many of its alternatives.)

[1] This example and more like it are detailed in P. Gerdes, "Conditions and Strategies for Emancipatory Mathematics Education in Undeveloped Countries," *For the Learning of Mathematics* 5 (1985): 15–20.

[2] For a fascinating account, see M. C. Frank, D. L. Everett, E. Fedorenko, and E. Gibson, "Number as a Cognitive Technology: Evidence from Pirahã Language and Cognition," *Cognition* 108 (2008): 819–824.

Second, geometrical terms are not as universally verbalized, but once again, one of the most persistent features of almost all cultures is some kind of attention to patterns – molded, painted, tattooed, drawn in the sand. Those patterns often display symmetries and occasionally involve more precisely drawn geometrical shapes. Does this amount, in and of itself, to geometry? Is any of this *mathematics*?

Authors in the tradition of ethnomathematics often elide this question, and one sometimes has the impression that they try to impute to indigenous cultures geometrical knowledge concerning figures, where in fact, all that those cultures have is the habit of producing those figures. Some ethnomathematicians probably are overenthusiastic in this sense, but mostly this is a misleading framing. Once again, let us take an example from Paulus Gerdes. He describes the following pattern in Mozambique weaving baskets:

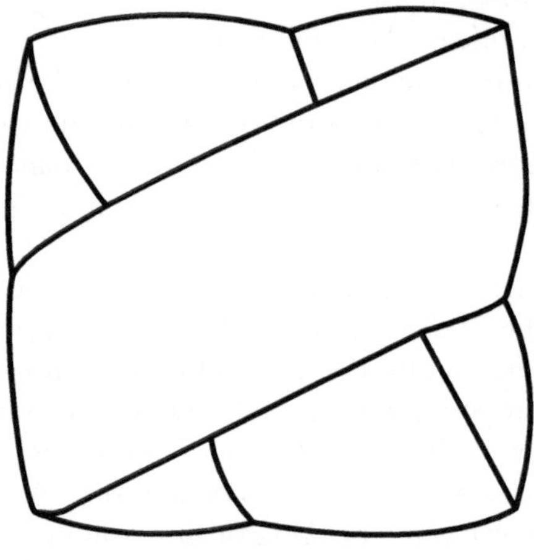

Figure 1

A nice geometrical pattern! But more than this, Gerdes observes, we may share this pattern in class and then proceed to discuss, with our students, how we may find here a relation between the various areas. In fact, with a little manipulation, we can derive, from this pattern, Pythagoras's theorem itself! (The main idea is that we see a big square – composed of four identical right triangles – and a smaller square enclosed in the middle. It is likely, I believe, that Pythagoras's theorem was indeed discovered around such drawings – by Babylonian teachers, working in a very different milieu. We shall return to see this in the discussion that follows.) Now, Gerdes does not mean that African basket-weavers are aware of Pythagoras's theorem; but it is nonetheless likely that the near-universal presence of patterns, of one form or another, is a significant precondition for the rise of geometrical reflection.

However, let us not get carried away. This is not yet reflective of explicit knowledge of geometrical properties, nor is the presence of a numerical vocabulary tantamount to the explicit knowledge of arithmetic. The discipline of ethnomathematics is useful for its scope – as well as for its limits. All humans, everywhere, talk about quantity and operate with shapes. But they almost never reflect on them explicitly, let alone develop a specialized craft of talking about numbers and figures. The discipline of mathematics and the profession of the mathematician are extremely rare.

Ethnomathematics is, of course, part of ethnography, and ethnographers tend to focus on what people *do* – how people interact, form kinship structures, cook, talk, sing. Anthropologists are trained to observe action, and so ethnomathematicians, quite properly, observe actions that are rich in mathematical meaning: counting, calculating, patterning. Those actions are real and form the background for the history we are about to explore in this book. Still, we should try to draw a line between an action that can be explained, *by us*, through our own mathematical understanding and the actors' own mathematical knowledge. Fishermen in Mozambique draw lines in the sand to dry their fish, and it is right and proper that we describe those lines in terms of circle, center, and circumference. It is also important to draw the conclusion that those fishermen have what it takes to create geometry. And finally, it is reasonable to say that the fishermen act in a geometrically intelligent way, without possessing any knowledge of a theoretical field such as geometry.

Many of you would probably agree that drawing a circle in the sand does not display, yet, knowledge of the theoretical field of geometry. I would say that the same is true about drawing a route, from point A to point B, along a straight line. This is a geometrically intelligent practice – but not a display of geometrical theory. I would also say the same about the building of a straight canal of irrigation leading to your fields. If you construct such a canal, then it is still the case that you may, or may not, have some theoretical understanding of "lines." I would also say the same about a straight road, faced by straight walls that form rectangular houses. And I would continue to say the same even if the houses become very imposing and perhaps assume the more complicated forms of various temples and pyramids. A pyramid, in and of itself, implies no more science than a line drawn for drying fish on the sand.

All of this is relevant to the question of the rise of mathematics as a theoretical discipline. We can find extremely sophisticated architecture and town planning around the globe – from the imperial cities of China to those of the Aztecs – and it is often assumed, especially by nonspecialists, that such imposing structures must involve theoretical mathematical knowledge. They certainly could, but the buildings, themselves, are not dispositive. And in fact, when we do find mathematics emerging, the context seems to be somewhat different.

## Empire and the Invention of Mathematics

We can locate several historical moments where mathematics was independently invented. Taking them together, we may form certain conclusions about the natural context of such an invention – which brings us back to politics.

The Inca empire, ruling over a vast region of the Andes in South America, left behind many monuments – but no writing. From the very beginning, Western observers noted a puzzling and rather humble artifact. Known as the "quipu," this is a system of knotted threads (often made of cotton) that can usually be spread out as pendants – one main thread, with many others hanging on the main one; occasionally, this can become a many-layered object. Each of the threads has a pattern of knots attached to it, and throughout the twentieth century, as more of these artifacts were surfaced and analyzed, the system came to be understood as essentially numerical (and base ten). Roughly speaking, the knots on a cord form clusters. To simplify things a little, it works like this: if you have a cluster of three knots, a space, and then a cluster of two, this can stand for "32." Such individual numbers on the hanging cords are summed up as the number recorded on the main cord. This, then, seems like an accounting device. The research leading up to this basic decipherment, based purely on a mathematical analysis of the extant quipus (of which there are now several hundred), can be found in the work of Ascher, *Code of the Quipu* (1981). I mention this because Ascher is also one of the most brilliant scholars in ethnomathematics and the author of the basic monograph in the field, *Ethnomathematics: A Multicultural View of Mathematical Ideas* (1991). For her, quipus are an example of ethnomathematics: an indigenous culture's preliterate display of mathematical sophistication. We should, in fact, note an ambiguity: Is that display, strictly speaking, *pre*literate? Or was the quipu, instead, simply the Inca form of literacy? As more evidence came to light in the last generation, based on more careful excavations, we came to understand better the original function of the quipu. As was often suspected in the past, it seems to represent a tax system based on geographical allocation through subdivisions. We find that several quipus replicated each other (a guarantee of accounting consistency), and some quipus may be identified as summing up the results in other quipus (apparently, this represents lower and higher layers of the geographical subdivision). Most spectacularly – a veritable Rosetta Stone – a very late set of quipus from after the Spanish conquest was seen to match a Spanish written list of tributes from across many villages. It now seems likely that the colors of the threads were also meaningful, perhaps encoding geographical regions – thus, quipus were an even more informative system than we had ever assumed. The upshot of this research is that the Inca empire produced a specialized class of quipu masters whose job was to maintain information on the tribute required from across the empire. Now, as a matter of fact, we cannot really say how much "mathematics" those quipu masters knew, precisely because the Inca produced no writing. Whatever education was involved in the perpetuation of the quipu-master technique was purely oral and is now lost. But some education of this kind certainly existed, and so we can say this: in the Andes, prior to Pizarro, there must have been some mathematics actively produced, with people explicitly discussing rules of calculation and accounting.

And another remarkable observation: numeracy was so central, in this particular civilization, that it completely supplanted literacy. To explain: the tool that the state needed was some kind of numerical record. This was efficiently achieved by the quipu, and this did not give rise to literacy as a spin-off.

In other places, of course, states did rely much more on writing. Once again, it is useful to start from as far away from Greece as possible: let us get a sense of the entire range of possibilities. We may begin with China, where finally, we see a very clear tradition of theoretical mathematics. Here it is useful to focus on a relatively late work, *The Nine Chapters on the Mathematical Art*, a work that may have reached something like its current form under the Han dynasty (perhaps in the second century CE?). The Chinese court always required a large retinue of scholars, the bulk of whom were masters of religious rituals, but many specialized in fields such as astrology or other forms of scientific knowledge. It seems that at the latest by the Middle Ages, but perhaps even in the very earliest times, some were trained, and examined, based on their knowledge of the *Nine Chapters* – which is appropriately, then, seen to concentrate around accounting-like needs.[3] The measurement of fields and of heaps of rice and grain, taxation, and distribution by proportion – all brought under a set of general, well-understood algorithms, which then become a subject of study in their own right. The needs of the state, generalized – and turned into a mathematical art. Once again, our evidence in this case is late, and it is hard to tell how mathematics first emerged in China. But more recent archeological excavations do provide us with more context and push the evidence further back. One dramatic find is that of "The Book on Numbers and Computation," a set of inscribed bamboo strips that a civil servant took to his tomb, sometime early in the second century BCE. Much earlier, then, in Chinese imperial history – but still well after the formation of the first Chinese states – yet we see here the same kind of material as that found in the *Nine Chapters*. Problems that relate to concrete bureaucratic needs – solved with considerable general sophistication.

## Beginning in Babylon

This brings us to the best-documented and most significant emergence of mathematics – and also, much closer to Greece itself. To the extent that the emergence of Greek mathematics was in debt to previous civilizations, it was to Babylonian mathematics.

This begins very early, along the shores of the Tigris and the Euphrates, and especially near their southern marshes.[4] This is one of the origins of urban civilization, and from the beginning, we find a system of accounting – not unlike that of the Quipu, perhaps – based, this time, on clay. (In the steep Andes, transportation is at a premium, and one looks for light tools; in the flat, river-based civilization of Mesopotamia, heavy but durable inscriptions are favored.) Archeologists have noted small, variously shaped pieces of clay found in many sites from the late Neolithic. Schmandt-Besserat was the first to offer a general account of those tools, and although she is not without her critics, very few doubt her basic interpretation (Schmandt-Besserat's critics mostly point out that the

[3] For the relation between mathematics and administration in the early Chinese state, see K. Chemla and B. Ma, "How Do the Earliest Known Mathematical Writings Highlight the State's Management of Grains in Early Imperial China?" *Archive for History of Exact Sciences* 69, no. 1 (2015): 1–53. Chemla and Ma, remarkably, are able to extract detailed information on the working of the administrative state, based on theoretical mathematical writings!

[4] The history of Mesopotamia is complicated: not a single state but a plethora of city-states and kingdoms, whose kaleidoscope kept shifting over millennia. I skip all the details (this is a history of Greek mathematics!), but read, for instance, N. Postgate, *Early Mesopotamia: Society and Economy at the Dawn of History* (New York: Routledge, 1994).

small pieces of clay could have been used for a variety of purposes beyond those she emphasizes; this is a reasonable critique). Most likely, different shapes stood for different commodities – so, for instance, could it have been a particular shape, say, for one head of cattle? Economic obligations – in the form of contracts or even taxation – could have been certified by an archive of such small tokens. This is all still ethnomathematics, a direct reliance on basic calculation and simple tools. And then, Schmandt-Besserat noted, something dramatic happened: it was realized that one could make impressions on clay, whose shape resembled the actual tokens. Late in the fourth millennium BCE, people in Mesopotamia began to use such tracings as economic records. A new idea, then: visual traces to mark numerical quantities. Pretty soon, instead of being tied to particular commodities, symbols emerged to represent *number as such*, and at this point, it took a mere step (or, if you will, a leap of genius) to begin to record other linguistic elements as well – at first, names of the objects counted and, very soon, language itself with its full vocabulary. By the end of the fourth millennium BCE, one of the major Mesopotamian languages – Sumerian – became fully written, the first ever. Literacy emerged, piggy-backing on numeracy.

Skipping many centuries of Mesopotamian history, we may look at the same shores of the Tigris and the Euphrates almost a millennium later. They are now dominated by different people, speaking a different language (Akkadian, a Semitic language that is somewhat similar to Hebrew or Arabic), still using the same script, the same inscriptions on clay. The technical knowledge of the Sumerians was not lost, in this and in other matters. The rivers themselves required constant attention – digging the canals and irrigating the fields. A lot of engineering, planning, and control was necessary, and throughout, Mesopotamia saw the rise of strong central authorities, powerful temple centers, and kings and their retinue. In the late third millennium, we see clear evidence for a specialized bureaucracy. Scribes were trained in writing, keeping accounts, and advising the rulers. What is most important: they did not just use the basic techniques of writing and calculation; they took pride in becoming genuine masters in all of those. Thus, besides simply writing down bureaucratic records in Akkadian, they also transcribed (a much harder task) the old literary legacy in Sumerian. And they did not just calculate, say, how many workers were required to dig a canal or how much tax should be levied on a field – they also invented particular fictional problems of a more abstract character, where one calculated volumes, plane areas, and work rates. In the Chinese *Nine Chapters* (or in the somewhat earlier "The Book on Numbers and

Computation"), we see the end result of, perhaps, a similar trajectory: bureaucratic training becoming its own raison d'être, giving rise to the problem-set version of a mathematics, which, although quite elementary, is already sophisticated. In Mesopotamia, our evidence is much more plentiful (early Chinese writing used a variety of delicate surfaces, such as the bamboo strip; from Mesopotamia, we have the clay tablet, history's most robust writing material). And so we get a closer sense of the entire transition: tokens, then writing, a bureaucracy, and this, finally – sublimated into mathematics. We have massive evidence, from the end of the third millennium to the beginning of the second millennium BCE. The evidence stops quite abruptly a little after 1800 BCE, for reasons we cannot quite fathom (for indeed, we no longer have substantial evidence!). It appears that the same old cities came under different sets of rulers and that the scribal traditions were disrupted. Little is known, then, for over a millennium – but clearly, there was some continuity. Beginning in the eighth century BCE, we find, once again, Mesopotamian palaces – preserving masses of clay tablets and a lot of the ancient culture. There is little mathematics to be found, though, in this later material (but plenty of astronomy; we shall return to this in Chapter 5). The object we study, then, is fantastically distant in time: the mathematics produced early in the second millennium BCE, or roughly four thousand years ago.

Just what is this mathematics? Let me paraphrase a very simple tablet (BM 13901 #1):

> I have it that the surface of the square, and its side, taken together, are three quarters.
>
> [Implicitly, our task becomes to find the numerical values of the side and area of this square. We're no longer just calculating taxes on fields; we're doing clever problems that build off such calculations! I attach Figure 2; notice that here, as in most cases, we do not have a figure on the clay tablet itself.]
>
> Here is what you should do. Make one as a projection to the side.
>
> [We now have in Figure 2 an elongated rectangle, divided into two parts, of which the right one is the original square, and the left one is a rectangle, one of whose sides is the original side of the square, its other side – one. The area of this left rectangle, then, is equal to 1 × the original side of the square, so its area is taken to be equal to the original side of the square. At this point, we can say that the entire elongated rectangle is equal to the original square plus the original side of the square. This is all equal to three-quarters, then.]

Break [the left rectangle] into two equal parts.

[And it is also implicitly understood that the broken left rectangle is now rearranged as in Figure 2, in the shape of a gnomon, a square minus a square. This gnomon, too, is equal to three-quarters].

Multiply half by half [to get a quarter].

[This is the area of the small square "implied" inside the gnomon because its side is the broken-into-two one, that is, 1/2].

Add the quarter to three-quarters, so you get 1.

[This, 1, is the area of the big square we would form from "completing the gnomon" and is also therefore the side of the big square.]

Take away the one-half in the inside, and one-half remains.

[Take away the side of the small square "implied" inside the gnomon, and you have, obviously, the side of the original square that we set out to find.]

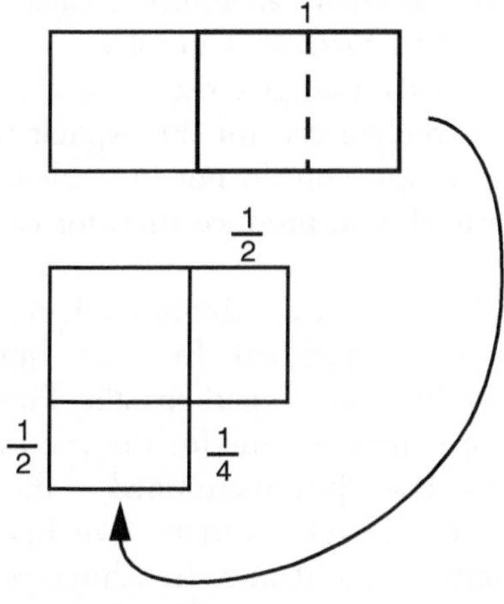


Figure 2

And this is how mathematics first emerges in the historical record: the simple, clever games accompanying the education of bureaucrats.

We have come a long way from the fishermen drawing a circular line on the sand. Here, surely, is mathematics. And it is precisely *here* that mathematics is to be found, in this particular variation on bureaucratic education. The Egyptian builders of pyramids may have been no more than the makers of glorified lines in the sand. Babylonian schoolmasters, however, created theoretical knowledge. And the difference is clear. As one builds a pyramid, one engages, throughout, in a concrete endeavor. There is no occasion to abstract away from the actual slabs of limestone to purely geometrical prisms: the slabs are what you handle throughout. But in the schoolroom, for instance, in the calculation example cited earlier, one no longer deals with actual measured fields. One deals with rules of calculation for the measurement of objects *such as* a square field. The schoolroom is at a remove from the field itself, and so its subject matter is not the concrete objects under calculation but, rather, the terms for calculation itself. Paulus Gerdes's project, setting out on his campaign for ethnomathematics, was to use the concrete knowledge of the fishermen of Mozambique as a starting point for the teaching of theoretical knowledge in the classroom. And in this, he traced the very same movement through which mathematics first emerged, four thousand years before.

## The Greeks: Standing Apart?

This is nearly universal to humans: an ability to calculate with integers, the manipulation of shapes and patterns. Complex states give rise to bureaucracies, and this, occasionally, may give rise to the training of a scribal elite, which, finally, provides the context for the explicit statements of mathematical facts. And so, at last, you do not just calculate or draw patterns without reflection. Instead, you produce rules for calculation and for the measurement of areas.

This is a valid broad outline of the rise of mathematics in many parts of the world, but it has to be qualified. Even the human universals have a great deal of variety in them – perhaps the Pirahã don't even have numbers! – and the same may be true for the rise of state bureaucracies. In fact, even the three cases just mentioned – the Andes, China, and Mesopotamia – show considerable variety. We have no evidence for a more reflective geometry in the Andes. In China, reflective mathematics seems to postdate empire; in Mesopotamia, mathematics emerged almost simultaneously with the market economy itself. It might be argued that all of this is a matter of the different sources of our evidence. In the Andes, one relied less on bulky artifacts, and so we have merely the threads of the quipu to tell our stories; presumably, much more oral lore circulated and is now lost. The Chinese bamboo strip is only slightly more robust than the quipu; we must have lost a lot from the initial stages of Chinese mathematics. Mesopotamian clay, finally, is extremely durable, providing us, in such a way, a much more detailed panorama of early Mesopotamian civilization. All of this is true but perhaps misses the point. The various societies used different media *because* state bureaucracy was not always the same. Mesopotamia really was – at least at times and in certain places – a heavily regimented society, recording the tiniest details of property and labor. It used an abundant, robust form of writing because this is what it required. Writing is not some kind of ornament; it may be built into the very fabric of society, defining its overall practices and achievements.

And so let us approach the Greek evidence in an open-minded way. Did the Greeks have bureaucratic state mathematics? Were they like Mesopotamians? At first, they seem so. Indeed, once again, we find the very same medium already familiar from Mesopotamia, that of the clay tablet.

Since the beginning of the twentieth century, excavations have discovered extremely ancient temples and palaces in Greece and, in particular, in Crete. Dating from the fifteenth to twelfth centuries BCE, several of those sites also yielded written tablets. For a long time, it was not even clear which civilization – or language group – occupied, in such ancient times, the lands now known as "Greek." It was only after the decipherment of Linear B in the 1950s – a triumph of linguistic deduction made by Alice Kober and Michael Ventris[5] – that the language of the tablets was identified as Greek. In some other ways, we can say, the decipherment produced an anticlimax. The contents of the tablets, themselves, were very mundane pieces of accounting. All this linguistic brilliance put in by Kober and Ventris – and then: "two tripod cauldron of Cretan workmanship . . . One tripod cauldron with a single handle . . ."

An anticlimax, perhaps, but also a meaningful result. Early Greek civilization, in the Bronze Age, blended together with that of the ancient Near East as a whole. Not just the tool of the clay tablet – we find an entire cultural practice shared and perhaps transmitted: centers of political authority, where detailed numerical accounts are written down and stored. The implied sociology – the rise of some kind of state power (kings or king-priests?), with its bureaucrats, is clear enough. Did this go together with a more explicit training in numeracy? Was there, then, Bronze Age Greek mathematics? If so, it left no traces in writing. This is not where our story begins, and – the more general point – this is not, really, where the

[5] In the Suggestions for Further Reading, I go on to recommend John Chadwick's book from 1958, detailing the decipherment of Linear B. John Chadwick is a very fair and reliable narrator, and he does give proper credit to Alice Kober, but reception at the time was such that only Ventris's contribution was etched into shared historical memory (this is all very similar to Rosalind Franklin's place in the discovery of DNA). Kober's major contribution was A. E. Kober, "The Minoan Scripts: Fact and Theory," *American Journal of Archaeology* 52 (1948): 82–103.

Greek story begins, in general. What the archeological evidence suggests, instead, is dramatic rupture. At about 1200, early Greek civilization comes nearly to a halt. Palaces fall down; plunderers set in. The writing stops, seems to be forgotten; cities and their civilization shrink and disappear. This rupture is not isolated and is instead seen across the Near East, where many states seem to fail at about the same time: the rare event of a civilizational near-extinction. Well, elsewhere in the Near East, the habits of the state were perhaps more powerful, and the rupture was not as total. But in the Greek-speaking world, the year 1200 could well have been a kind of year zero. The culture of Linear B would have been as puzzling to the Greeks of the year 800 as it would be, many centuries later, to Kober and Ventris.

But then, at about the year 800, something new began to stir. And here, finally, we come to our proper topic. To clarify the contrast: earlier, in the second millennium BCE, we clearly found state formation in the Greek-speaking world, along the familiar lines of kings or king-priests setting up strong centers of power. This early state formation is knocked out by the crisis of the twelfth century, and when the state begins to reemerge, it seems to take a different shape from that of the previous temples and palaces and that of the other early empires noted in this chapter.

What was remarkable about the reemerging Greek state? That it was weak and small. The usual tale of state formation is that as societies become more complex, they become more unequal. A few individuals emerge as the powerful rulers, and if lucky, they manage to consolidate power over ever-larger groups. Eventually, state becomes empire. This was definitely the case in the pre-Columbian Andes and in China, and although Mesopotamian states often controlled no more than a mere region (sometimes, not much more than a single city), control was often centralized, a king, priests, their retinue – and a mass of subjects. Greek cities just did not go along such a route. Emerging from the rubble of the post-1200 collapse, villages gradually grew in size, but they never reached a very large extent. Greece, quite simply, was not a great river civilization. It was always marked by steep mountains, sharply cutting into the sea, isolating small islands and tiny valleys. At its height, centuries later, Athens would have hundreds of thousands of residents; but for now, in the eighth or seventh century, even larger settlements would have no more than a few thousand residents. The people whose voice mattered for any kind of political or cultural life – the more or less self-sufficient male adults – would number much fewer, surely no more than a few hundred.

To start with, those settlements were not very rich, and thus even the rulers could not be very rich themselves. (If you rule over a couple hundred small-scale farmers, how much wealth can you amass?) And so: scale creates habits, which can then become self-sustaining. Instead of relying on smaller armies of noble horsemen and charioteers, one relied on the larger standing army of the entire polis (*polis* is the Greek word for "city," which we will use from now on; the Greek term evokes a distinctive cultural model). It is fitting that those soldiers are armed with the cheaper and more widespread material of *iron* – not the expensive, specialized bronze of predecessor states. Kings in the ancient Near East counted their wealth in heavy ingots of precious metals. Citizens of the Greek polis would begin to use a more manageable currency, the coin. Invented in Asia Minor, or present-day Turkey, in the seventh century, the coin would become a hallmark of the Greek world: small pieces of metal, widely owned across the social strata, easy to transport and manipulate. Also, in Mesopotamia and in Egypt, literacy was a scribal specialty, sometimes monumental, always based on arcane, difficult systems of writing. The Greeks borrowed the cheapest and most portable writing surface available from Egypt – the papyrus roll – and they borrowed the simplest and easiest-to-learn writing system from Phoenicia: the alphabet. Writing, for the Greeks, was never a matter of some rarified scribal elite.[6] Not that it mattered all that much in the eighth or seventh century. Culture, quite simply, did not belong to any inward-looking court, with its established retinue of bureaucrats. There was nothing like that in early Greece. Culture belonged to the open spaces of the polis – perhaps in a festival, the youth singing together as they walk in procession; perhaps in a public square, a professional bard reciting an epic; sometimes, the richer folk, relaxing together and singing in a symposium with their guests from other poleis (because, in such a world of small poleis, there are always many other poleis, not far away, their people coming and going through your own).

The thread running through all of this is *being spread out.* It is the opposite of the concentration of Mesopotamia. The ancient Near East had a few expensive bronze chariots; hard-to-decipher hieroglyphs and cuneiform scripts; and rare, heavy ingots of gold; its culture was reproduced among a tiny group of trained scribes. Greece had plentiful iron spear tips

[6] It seems likely that the alphabet was, to begin with, the invention of subaltern groups that appropriated to themselves, in a much-simplified form, the complex systems of scribal elites. See O.Goldwasser, "How the Alphabet Was Born from Hieroglyphs," *Biblical Archaeology Review* 36 (2010): 40–53.

and relatively accessible alphabetic writing on papyrus, and eventually, it would have plenty of tiny silver coins – and it had the culture of the open polis, which could be shared with almost all members of the community. The political and demographic forces that made for shared culture became entrenched in tools that had a way, in turn, of maintaining the features of this culture, even as states did become, eventually, somewhat more powerful. Silver coins would tend to preserve a market economy, just as iron spear tips would tend to preserve military and political order based on the citizen body, fighting side by side. Public performance – recorded in widely accessible script – would become the most stubbornly entrenched of all those cultural habits. We still read Homer.

Without any specialized bureaucracy, archaic Greeks surely did not develop any specialized scribal schools. It is, in fact, very hard to reconstruct anything about Greek mathematical education in the archaic era. (I shall return to this topic in Chapter 4, which discusses Greek mathematical education in the Hellenistic era and later, where some evidence is available.) All we can say is that the Greeks knew little, and what they knew must have come from elsewhere. But surely, if we want to understand the rise of Greek mathematics, we ought to make some guesses concerning the contents of such knowledge!

To get there will require a double detour, into the historiography of early Greek science and philosophy and into the historiography of the ancient Near East. Why "historiography"? Because in those early mists of history, so little is known, and so much is based on speculation, that it is impossible to discuss the past in separation from the way in which modern historians have interpreted it. And so, Thales and Pythagoras, and then, Babylon.

## Greek Mathematical Myths

When Heath wrote his own history of Greek mathematics a century ago, he thought he could give a very detailed survey on the question of origins, of first steps. Following his initial introductory chapter and a second chapter, "Greek Mathematical Notation and Arithmetical Operations," the titles of his Chapters 3, 4, and 5 are, respectively, "Pythagorean Arithmetic," "The Earliest Greek Geometry. Thales," and "Pythagorean Geometry." Heath's picture was essentially of a few isolated geniuses, such as Thales and Pythagoras, coming up with a new kind of science that eventually led to the great achievements of Greek mathematics.

When I lecture on Greek mathematics, I am often asked about Pythagoras (Thales, not so much), who still occupies a central role in the cultural imagination, and so this is a point I need to emphasize. Thales and Pythagoras were real historical persons, both active in the sixth century BCE (Thales somewhat earlier, Pythagoras somewhat later). However, although different scholars take different views on this question, the standard view is that *Thales and Pythagoras did no mathematics whatsoever.* This is all a myth. To be sure, it was a myth started by the Greeks themselves – who did like to invent stories, projecting their contemporary achievements into the distant past. But those later stories tell us more about the agenda of later Greeks than they do about the Greek culture of the Archaic era.

I have already mentioned that early Greek civilization was all based on public performance, and what comes to mind, first of all, is the public performance of recitation and song. Alongside the *bard*, the early Greeks also had the *sage*: the man respected for his words of wisdom and advice, words that might ring paradoxical and yet impress for their kernel of truth, often touching on the political life of the community but sometimes reaching beyond that, to speculation about the cosmos, about the human condition, about truth itself.[7] Such wise words could sometimes be noted and commemorated, and eventually – in the fifth century – a wise man even could decide, occasionally, to write down a book. But it was all about the public performance of wisdom – as it would still be all the way down to Socrates. Thales and Pythagoras certainly put nothing down in writing, and it is not clear that they had a "something" – a clearly articulated body of doctrines – for which "putting down in writing" would be the appropriate exercise. The sense that they had such a thing is a later construct that we can see emerging very clearly from the works of Aristotle. One of Aristotle's favorite techniques was to go through past views on a particular topic so as to see them as approximations – but no more – of Aristotle's own views. Thus, past philosophers, according to Aristotle, only looked for the material cause, "the stuff out of which things are made," unlike Aristotle, who developed a complex, nuanced understanding based on different kinds of causes. And so past sayings attributed to past figures were shoehorned into the model of material cause, and so also, whatever

[7] The claim here – that Archaic Greek sages should be understood alongside other public performers – is based on R. P. Martin, "The Seven Sages as Performers of Wisdom," in C. Dougherty and L. Kurke (eds.), *Cultural Poetics in Archaic Greece: Cult, Performance, Politics* (Cambridge: Oxford University Press, 1993), pp. 108–128.

sage words were attributed to Thales concerning the cosmos got transformed, in Aristotle's telling, to "Water is the Material Cause," which still survives as the first thing one usually reads in a general history of philosophy: "Thales was the first philosopher, and he said that all is water."[8]

It is reasonable enough that Thales, the wise man, would talk about the cosmos, and so we may perhaps be justified in saying that he probably did like to mention water in whatever it was he was saying. It is not entirely clear if he, or Pythagoras, had anything to say specifically about mathematics. Consider, for instance, the tradition concerning the most famous piece of such early mathematics: Pythagoras's theorem itself. It all goes back to a story of Pythagoras's pride in the finding of this theorem, a story that we find in many sources. The earliest of these, however, is Cicero.[9] A first-century BCE Roman figure, Cicero was active about five hundred years later than Pythagoras. Cicero and others later than him all seem to rely on a piece of verse, written about a century or two earlier than Cicero – so we are getting closer – but this piece of verse, by a certain Apollodorus, was apparently intended to be *funny*. The background is the following (it's an awful thing – I have to explain a joke now; by the time I'm done, there will be nothing funny left). Pythagoras, the wise man, did, in fact, promulgate a code of conduct demanding a certain kind of purification, of which one of the key demands was vegetarianism. Further: Greeks often thought of a remarkable achievement in terms of the religious ritual to celebrate it. The more remarkable the achievement, the more spectacular the ritual. And so the anecdote relayed by Apollodorus in his verse: Pythagoras celebrated the discovery of the most famous theorem by slaughtering one hundred bulls. This, I repeat, is meant to be *funny*. (A vegetarian – slaughtering bulls!) When Apollodorus writes it down, his audience perfectly understands the intention, and no one thinks that any of this is the literal historical truth. It does show two things, indeed.

First, in the time of Apollodorus, what we know as "Pythagoras's theorem" was already a well-known result, perhaps the most well-known mathematical result of all. This is not surprising: this result is widely

[8] The argument that Aristotle is an unreliable narrator of the earlier history of philosophy was made forcefully already by H. F. Cherniss, *Aristotle's Criticism of Pre-Socratic Philosophy* (Baltimore, MD: John Hopkins Press, 1935). Most scholars today believe that Aristotle aimed to be faithful to his sources while being completely shaped by the assumptions and agendas of his own time and place. Is this not true of all historians?

[9] Cicero alludes to the story in *On the Nature of the Gods* III.88. (He comments, pedantically, or perhaps mock-pedantically, that the story is impossible.) The verse itself is cited by several sources, beginning with Plutarch (first century CE), *Moralia* 1094b.

assumed in the educational documents we have, and it is referred to as the "schoolchildren result" by Polybius, a historian writing at about the same time as Apollodorus (I return to all of this in Chapter 4). In the Hellenistic era, and probably even before, Greeks knew "Pythagoras's theorem" as a basic part of their education.

Second, in the time of Apollodorus, it was already widely assumed that Pythagoras made contributions to mathematics. To cut a long and complicated story short: in the late fifth and early fourth centuries, there were authors, such as Philolaus and Archytas, active in the same geographical region that resonated with Pythagoras's fame (we will see much more of those authors in the next chapter). Such authors were all interested in mathematics, and they also shared something of the otherworldly, purity-seeking approach of Pythagoras's original type of wisdom. Aristotle (who, once again, is our main source) thought of Philolaus and Archytas as "Pythagorean"; they probably did so themselves. We shall revisit all of that in the following chapter. The point, for now, is that it is through association with those authors that the image of Pythagoras as a mathematician likely emerged.

We can go, in such a way, through the traditions concerning the various pieces of mathematics attributed to early mathematicians prior to the end of the fifth century. The evidence is extremely late. Very often, our earliest sources are from Late Antiquity (when, indeed, a mathematizing version of Platonism becomes very widespread and is often considered "Pythagorean"; we return to this in Chapter 6). Aristotelian guesses, amplified by later readers, Roman, late ancient – all constructing a myth that remains embedded in the Western historical narrative, all the way down, indeed, to Heath's *History of Greek Mathematics.*

Aristotle had very few sources to work from as regards Thales and Pythagoras, but the point, concerning the nature of the evidence, is not merely methodological. The point has to do with the underlying historical reality itself. The world of Thales and Pythagoras had relatively little use for extensive writing, and so whatever knowledge they uttered would have to be oral. This is why they were "wise men." Preliterate knowledge, preliterate science – wisdom. Thus, the same must be true for mathematical knowledge as well. Whatever the Greeks knew, back then, in the field of mathematics, would have to be understood along the terms of oral knowledge. This is not a matter of scientific doctrine but of shared cultural lore.

But where did this shared cultural lore first emerge? I believe it did, ultimately, in Babylon. But for this, yet another detour is called for.

## Greeks and the Near East: A Historiographical Detour

So far, we have considered the mathematics of the ancient Near East essentially as a contrast to the Greeks (Mesopotamia had centralized bureaucracies; the Greeks did not). But can there be, instead, some kind of a *link*? That we even raise this question puts us, already, ahead of Heath. In 1921, historians of Greek mathematics knew essentially nothing of the mathematics of the earlier civilizations of the ancient Near East. It is significant that this changed – and changed so rapidly.

Otto Neugebauer was born in Innsbruck, Austria, in 1899. In 1922, he became a graduate student in, arguably, the best department of mathematics in the world: Göttingen. It was a time and a place where mathematicians were intensely engaged with the very foundations of their discipline. This concern with foundations brought Neugebauer, as a student, to a concern with *origins*. Finding himself fascinated by Egyptian archeology (a common enough fascination), he realized that there was something deeply significant about the fact that the very early civilizations developed abstract, symbolic systems for the representation of number. Mathematics – always looking for deeper abstractions. Throughout the 1920s, he produced important interpretations of the arithmetical systems in Egyptian papyri, followed by an account of the origins of the base-sixty system in Mesopotamia. Neugebauer stayed on in Göttingen after completing his studies, and he became the first lecturer, ever, in a mathematics department, to specialize in the history of mathematics itself. In Neugebauer's hands, the history – even the most distant one – looked forbiddingly impressive, and relevant, to contemporary mathematicians, in its abstraction. Then, in January 1933, the Nazis came to power. Not Jewish himself, Neugebauer could simply go on as before, but when German university teachers were required to declare loyalty to the regime, he refused – very few had this courage, made this sacrifice – thus giving up his position at the best department in the world and leaving Germany for good.

The University of Copenhagen saw the opportunity and offered him a professorship. (Eventually, he would get, in 1939 – another lucky break – to Brown University, which, for half a century, would become the center for research in the ancient exact sciences.) And so, in the year 1935, Otto Neugebauer, a Copenhagen professor, began to publish his *Mathematische Keilschrift-Texte* (*Mathematical Cuneiform Texts*).[10] This was the first

[10] For more on Neugebauer's fascinating intellectual trajectory – much more complicated than my brief outline suggests – see A. Jones, C. Proust, and J. M. Steele (eds.), *A Mathematician's Journeys: Otto Neugebauer and Modern Transformations of Ancient Science* (Berlin: Springer, 2016).

significant scholarly resource with which one could begin to make sense of the mathematics of the ancient Near East. Bear in mind: it was only in the nineteenth century that even the script of the clay tablets – cuneiform – was deciphered. At first, scholars were busy looking for the literary evidence for the biblical world, and only gradually did they come to consider the great bulk of numbers in cuneiform. Even here, the emphasis, to begin with, was on metrology: What were the units? What do these tell us about the society, the economy, and the history of antiquity? Some tablets did not yield that many interesting economic details and instead were more purely numerical – and these, in turn, were very obscure. I cited earlier a Babylonian text, but I did so through the useful interpretation now produced by Jens Høyrup. Imagine how obscure this was without any interpretation. I cite it again, now trying to preserve the sense of the original difficulty:

> My surface and the-equal-to-itself I added: 45′. 1, the beyond, you posit. Half of 1 you break, 30′ you make hold. 15′ to 45′ you add: 1. 1 is side. 30′, which you made hold, take away inside 1: 30′, the-equal-to-itself.

In this passage, I simplified the fractions to modern form, but in the original, they were sexagesimal, or in base sixty, so that 45′ is three quarters, 30′ is a half, and 1 can be thought of as 60′: we may comprehend this easily enough by thinking of minutes as fractions of an hour. (As an aside, this is not an accident: our minutes and hour reach us from Greek astronomy, which, in turn, depended on its Babylonian antecedents; more in Chapter 5.)

But the sexagesimals are relatively easy! The hard part is to make sense of the very point of the mathematical exercise. Neugebauer realized that the surface in question must be that of a square; "the-equal-to-itself" must be its side. However, there is no geometrical meaning to adding together a square and its side. This, then, so Neugebauer understood, must be a more algebraical formulation, dressed in geometry: *geometrical algebra*. The problem, at its core, takes the following form:

$x^2 + x = a.$ (In this case, the value is $a = 45'$, and so the value of $x$ is 30′.)

And Neugebauer argued that the Babylonians knew how to solve this algebraic problem in general, using particular numerical examples for the display of their general algorithms.

And so, according to Neugebauer, mathematics was always looking for deeper abstractions. The earliest mathematics, of the cuneiform tablets, was also essentially a piece of sophisticated science. Those individuals were, quite simply, professional mathematicians, pursuing the numerical solutions of algebraic equations. It was Göttingen on the Euphrates.

It is extremely rare that a field of scholarship passes so dramatically, so quickly, from obscurity to clarity. Babylonian mathematics barely existed as a field of study before 1935. Following Neugebauer's publication, it was grounded in a substantial number of texts, now thoroughly understood. We can readily imagine the authority that Neugebauer rightly assumed over his field, and indeed, for many years, no one would doubt his interpretation.

Now add this: in the 1930s and for many decades hence, the scholarly consensus was also to accept, at face value, the testimonies, such as those of Proclus and Aristotle, concerning the earliest Greek mathematics. Heath, we recall, had three full chapters on the mathematics of Thales and Pythagoras! The implication of that consensus would be, essentially, that even early on, the Greeks had the equivalent of professionalized mathematics, authors whose goal was to promote theoretical mathematical understanding. Put this side by side with Neugebauer's theoretical mathematicians of Babylon, and you have a continuity.

More than this: an interpretation presents itself almost immediately, indeed, was assumed already by Neugebauer himself, even as he was making his very first proposals concerning Babylonian mathematics. It has been central to later reconstructions of early Pythagoreanism that this philosophy somehow involved an early mysticism of number, as if somehow, integer numbers underlay the very structure of the cosmos. It is evident that such a mysticism would sit awkwardly with the discovery of irrationality. If the side and diagonal cannot both be expressed with rational numbers (for which, see pages 78–81), then it becomes impossible to describe the entire cosmos with such rational numbers. But it is clear that very early on, the Greeks did discover irrationality! So this should be difficult for Pythagoreanism, if indeed we believe in its existence as an early mathematical doctrine based on rational numbers. Such a belief began in antiquity itself – perhaps as early as Aristotle – and by Late Antiquity, some authors speculated that the discovery of irrationality could cause a crisis for Pythagoreanism.

The Göttingen era of Neugebauer's mathematical education was consumed by the crisis of foundations. If you try to base mathematics on set theory, this could easily lead to paradoxes; thus, one required special axiomatic assumptions to make a consistent mathematics even possible. The debate ranged further: Does mathematical existence require explicit construction? If so, are we to allow infinite constructions? Is mathematics purely formal? Is it even theoretically possible, formally, to axiomatize all of mathematics? (In 1931, Kurt Gödel – an Austrian mathematician working in Vienna – proved certain surprising results that made such a goal appear impossible, but throughout the 1920s, while Neugebauer was formed as a mathematician, Göttingen was consumed by the dream of a fully grounded, axiomatic mathematics.)

And so, it was tempting and natural to project a crisis of foundations, with its consequences, on the earliest history of mathematics. Babylonians came up with the algebraical study of numerical relations. But then their Greek followers discovered the phenomenon of irrationality and therefore realized that it is impossible to describe all relations in purely numerical terms. One therefore needed to formulate algebra in a strictly geometrical way – hence the mature geometry of Euclid's *Elements*.[11]

This account, then, in outline, was shared by scholars for decades. It came under pressure only very gradually. This happened in many stages. First, in 1962, the German philologer Walter Burkert published *Weisheit und Wissenschaft: Studien zu Pythagoras, Philoloas und Platon*, which – especially following its English translation from 1972, *Lore and Science in Early Pythagoreanism* – would come to define scholarship into Pythagoras. Here was a more careful, professionalized classical philology, keen to understand the authors we read not as mere parrots, repeating their sources, but instead as thoughtful agents who shape and retell the evidence as suits their agenda. Pythagoras, under such a reading, crumbles to the ground: almost everything – as noted earlier – comes to be seen as the making of later authors from Aristotle on.

Never mind: the historians of mathematics went on as before (and Burkert, with all his skepticism, did not consider himself an expert in the history of mathematics; he did not engage directly with this part of the Pythagorean legend). But there were other considerations, from within the discipline of the history of mathematics itself. The history of science, in general, was becoming more professional. It was no longer acceptable to pursue the history of science as an abstract, disembodied history of ideas, completely at a remove from any sense of historical context. In 1975, in his article "On the Need to Re-Write the History of Greek Mathematics," Sabetai Unguru made a very modest plea: mathematical texts should be understood on their own terms, not in some kind of translation to an abstract, ahistorical symbolism. The Greeks did not study equations; they studied geometrical configurations inside diagrams. The plea was modest, but Unguru framed it as a radical critique – which it was – of the dominant

[11] For this supposed crisis of foundations – the evidence and a spirited discussion – see D. H. F. Fowler, *The Mathematics of Plato's Academy* (Oxford: Clarendon Press, 1987), pp. 294–308.

approach to the history of mathematics. It touched a nerve. He was savagely attacked by some of the most prominent historians of mathematics, even mathematicians, of the time. Well, the discipline paid attention. And in the end, the authority of Unguru's detractors made no difference. The need to historicize, to read texts in context, was much too powerful.

Burkert was circumspect and did not seek to dethrone the early Pythagoreans from their mathematical pedestal, but later historians of mathematics, following his lead, had fewer compunctions. In general, it became clear that the very effort to reconstruct the earliest stages of Greek mathematics was fraught with speculation. In Unguru's wake, this endeavor largely went out of fashion. Indeed, until the 1980s, most scholarship in Greek mathematics went into the reconstruction of the earliest mathematics and its interaction with philosophy. Since the 1980s, almost no one kept writing on this topic. Scholarship moved on to the interpretation of later mathematics, based on the firmer grounds of our extant texts from the fourth century BCE onward, texts that one could fully understand in historical context. The tendency was to be agnostic about any earlier Greek mathematics; through the years, scholars became more comfortable with the idea that Greek mathematics, in fact, did not start till quite late. While no one looked, Pythagoras and Thales were quietly removed from the museum.[12]

Meanwhile, things changed in Babylonian studies, as well. Professionalization would, at the end, overrun the authority even of Neugebauer himself. In a series of publications from the 1980s onward, Høyrup went back to the cuneiform texts, paying closer attention to the fine detail of vocabulary. For instance, why does the tablet quoted earlier talk about multiplication as "make hold"? Why is the subtraction it mentions taking from "inside"? Paying attention to such detail, Høyrup succeeded in reconstructing a concrete reference to all those seemingly abstract manipulations of numbers. In fact, Neugebauer had it upside down. This was not Babylonian algebra, later to be turned, by the Greeks, into geometry. The Babylonians, already, dealt with concrete geometrical configurations. It was all about measurement, calculation, patterns. It is simply that in the Babylonian case, the diagrams (or their more concrete, manipulated equivalent) were lost. The clay tablet contained directions for operation, referring to an external object that was

[12] This transformation was noted already by K. Saito, "Mathematical Reconstructions Out, Textual Studies In: 30 Years in the Historiography of Greek Mathematics," *Revue d'histoire des mathématiques* 4 (1998): 131–142.

operated upon concretely. Høyrup evoked a vivid sense of Babylonian mathematics in action – moving stuff about, tearing pieces of fabric or wood apart. He still wrote, however, strictly as a historian of science (albeit with extraordinarily wide-ranging knowledge and curiosity!). More recently, the study of Babylonian science has become even more professionalized, and scholars such as Eleanor Robson, active since the 1990s, are now fully trained not just as historians of science but also as social historians and archeologists. Robson is now able to put the vivid practice, recovered by Høyrup, in the context of the Babylonian school. And so, we now understand how those tablets came about. These are not at all professional mathematicians à la Göttingen, pursuing theoretical knowledge for its own sake. These are teachers in scribal schools, engaging with problems that are at a certain remove – but never entirely divorced from – their practical application.

But if Babylonian mathematics was always strictly a scribal school practice, how exactly did it even endure? The implicit assumption underlying Neugebauer's account of the ancient exact sciences was that there was something, such as a theoretical understanding of algebra, perhaps written down and since lost but, at any rate, circulating, in some form, through the generations, crossing eventually from Babylon to Greece. But this is clearly wrong and, in fact, directly contradicted by the pattern of our evidence. We do not see a theoretical continuity anywhere. What we see is that in several cities across Mesopotamia, centuries and even millennia apart, we find new foundations of scribal schools with their own flavor of mathematical education. And here is the thing: we can trace a continuity between problems set out in tablets from the twenty-first century BCE and, say, the eighteenth century BCE, and this is of a kind that can be traced even between distant cities. But the schoolmasters of a given city did not learn about their distant predecessors by the reading of tablets (which, by the eighteenth century, were already covered with dust, to be dug up only four millennia later!). Although, of course, tablets did travel around to be copied, this was not their main function. These were local tools of the trade, not at all the same as a book. No: whatever tradition there was had to be *oral.* At the first instance, it was not tablets that traveled, but *people.*

More than this: there is a set of numerical values and problem types that can be traced across an even wider geographical and chronological range. Once again, we owe the following observation to Jens Høyrup. Now, in the simple example of Babylonian mathematics quoted earlier, we are given the sum of a square area and its side. A much more specific type of question is where we are given the sum of the square area and its *four sides, taken together*. Remarkably, Høyrup found the exact same problem asked, and solved, in nearly the exact same manner, in Babylonian, Greek, Arabic, and Italian sources. How do problems travel? Well, the story of Cinderella, too, is told far and wide. When people meet, Høyrup explains, they share stories. A mathematical riddle is just that – a riddle, something to tell and to solve. Apparently, this happened frequently enough, throughout history, so as to preserve a certain minimal layer of shared mathematical knowledge. The likeliest account of mathematical transmission between cultures, then, is nearly the opposite of that implied by Neugebauer's account. It is not that some theoretical understanding was transmitted from the Babylonians to the Greeks. Instead, it was the problems themselves – puzzles and riddles, freed from their original school context – that came to be more widely shared. And the very idea of a "Babylonian influence over Greece" is misleading. Not that it ever made any sense to seek the impact of eighteenth-century BCE clay tablets from southern Mesopotamia on the words of a wise man in southern Italy twelve centuries later. There was a continuity, but it must have been, precisely, *along a continuum*. Somehow – even through the major disruption of Mesopotamia from 1800 BCE onward – some kind of mathematical lore survived. Some teaching must have continued; there must have come about a kind of folk koine shared by practitioners of various crafts. It owed its birth to the scribal schools of past empires. But something did survive their demise, a more humble practice – but one freed, perhaps, from the court?

This is all lost: away from the courtly scribal schools, we just do not have the evidence. But we can surmise the presence of real mathematical progress. Let me outline how this could have come about.

The riddles that Høyrup detects, permeating across cultural and linguistic borders, generally take the form of a challenge: If you know X, can you also know Y? He looks specifically at the following riddle:

> If you know the sum of the square area with the square's four sides, can you tell the side of the square?

And we have earlier cited the following riddle:

> If you know the sum of the square area with one of the square's sides, can you tell the side of the square?

A slightly more complicated puzzle is the following:

> If you know the difference between the two sides of a rectangle, and you also know the area of the rectangle, can you tell what the two sides are?

This riddle was solved by the Babylonians with a technique directly analogous to that cited on pages 10–11: Because you know the difference between the two sides, you can mark the difference on the longer side as a projection beyond a square. And now you have the situation where you know the area of a square plus its projection. For instance, suppose the area is 60, and the difference between the sides is 7 (yes, you know the answer already, but please wait till the end of the example).

Because the difference is 7, I can break it into two equal parts and transpose one of the resulting triangles. I get a gnomon, whose area is *still* 60.

This gnomon has a small gap, a square, whose side is 3.5. The area of this small square, then, is $3.5^2 = 12.25$.

Add this to the gnomon, and we have a larger square whose area is 72.25.

So, its side is 8.5.

This is simply the middle between the greater and smaller sides of the rectangle, when their difference is 7. Add 3.5 and remove 3.5, and you have the two sides: 5 and 12.

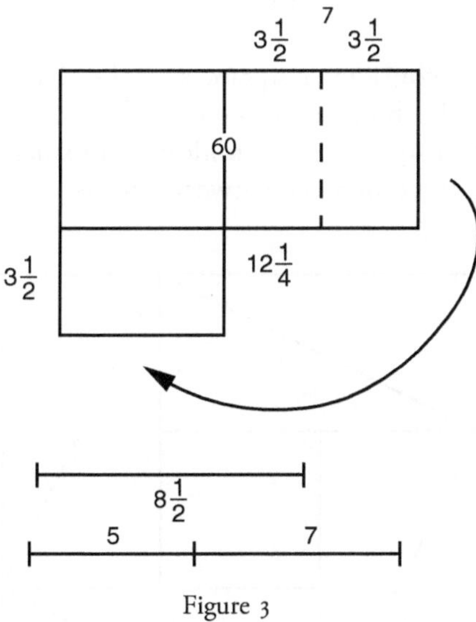


Figure 3

To repeat, what this approach shows us is how we can solve a riddle: given the difference between the two sides of a rectangle, and the area of the rectangle, to find what the two sides are.

At some fairly late stage in the history of Babylonian mathematics, the thought suggested itself to some schoolmaster to improve on this type of riddle by considering the pattern – noted by Gerdes in Mozambique weaving! – such as this:

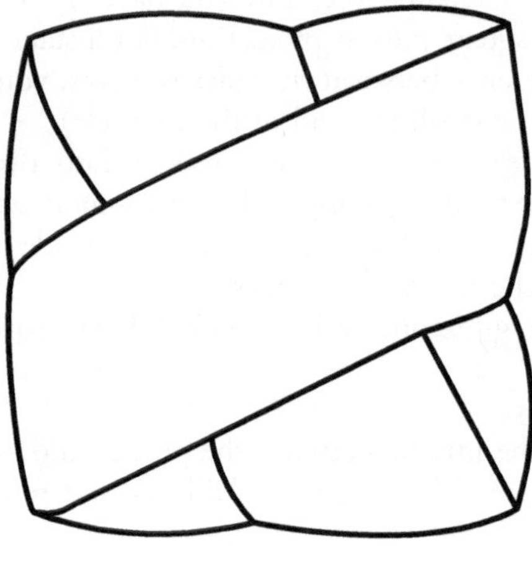

Figure 4

That is, we may consider a square and the diagonals of four equal rectangles drawn on it obliquely. Note that if we draw all four rectangles, we have a small gap, a square, in the middle. This small square stands on a side that is, in fact, the difference between the two sides of the rectangle:

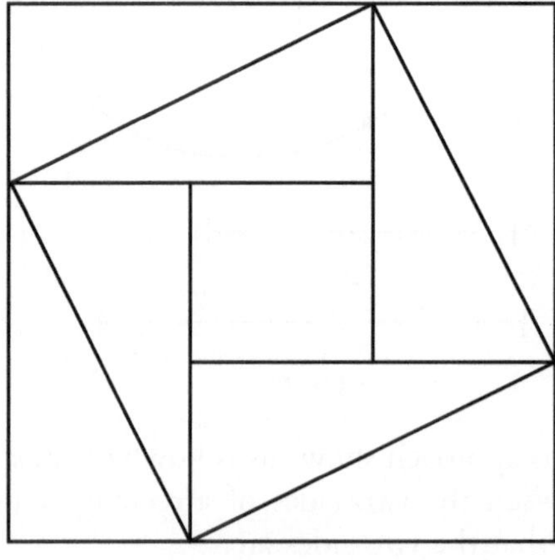

Figure 5

Suppose you have taught your students a hundred times already how, if they are assigned the difference between the two sides, as well as the area of the rectangle, they can find the two sides. It dawns on you that you can have a new way to find this difference because the small gap, the small square, is, in fact, the difference between the slanted square across the figure and *four triangles* – which are the same as *two rectangles.*

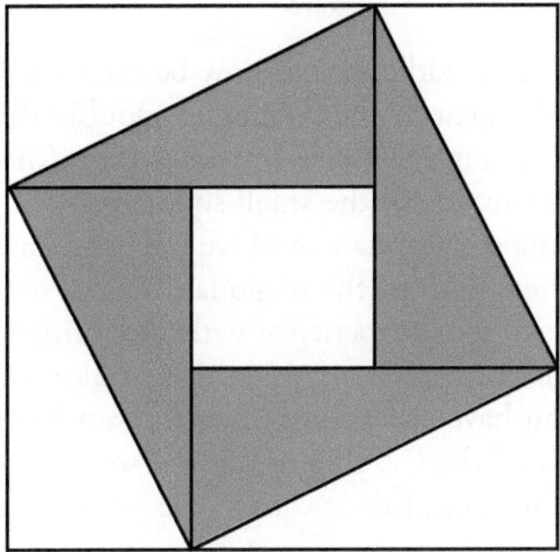

Figure 6

The four triangles are equal to two rectangles, whereas the slanted square is simply the square on the diagonal of the rectangle. Now, in the riddle we are most familiar with, we are given the area of the rectangle and the difference between the two sides of the rectangle. We have now realized that if we are given the area of the rectangle, and we are also given the length of the diagonal, we can find the difference between the two sides!

This will provide us with a much neater, much more impressive riddle:

> If you know the area of the rectangle, and you also know its diagonal, can you tell its two sides?

Wow! And in fact, this is directly doable: take the square on the diagonal, and subtract from that 2 × the area of the rectangle. You come up with the area of the small square in the middle. Find the root of this square. This is the difference between the two sides. So now we've made it to the position where we know the area of the rectangle, and we also know the difference between its two sides – and this, we know already!

This is as far as our evidence leads us. Babylonian geometrical problems are typically related to the measurement of surfaces. They are never entirely detached from, well, the measurement of fields – from which it all began.

However, this riddle did exist, and apparently someone, in the lost mists of time, thought of an extension of the same technique to derive a related riddle, one no longer connected to the measurement of the areas themselves:

> If you know the two sides of the rectangle, can you tell the length of the diagonal?

The approach to this riddle should now be obvious. By multiplying the two sides, we get the area of the rectangle. Double this, and we get two rectangles, or four triangles. Next, by subtracting the smaller from the greater side, we get the side of the small square in the middle. Add up the four triangles and the small square, and we have the slanted square; find its root, and we get the length of the diagonal.

Finally, if you have some experience with the numerical finding of roots and squares (a technique quite central to this schoolmasterish tradition), you are aware that if you have two values, then the result of multiplying them, and doubling that, and then adding to this the square on their difference, is nothing else than the sum of their squares. As we would put this:

$$2ab + (a - b)^2 = a^2 + b^2.$$

And so, this riddle is rather elegantly solved: if you are given the two sides of the rectangle and you want to know the length of the diagonal, simply square each side, add them up, and look for the root of the sum – and there, you have the length of the diagonal. To repeat: this last riddle is a reconstruction. It was probably not offered by the year 1800 BCE. But it appears that it did enter the mathematical koine, the shared template of familiar mathematical riddles, out of which Mediterranean civilizations could – if they wanted to – build up their education. And so, Pythagoras's theorem. Not the fruit of some theoretical reflection by an isolated Greek genius. Rather, the outcome of the school practice and the mathematical folklore of riddling, through the centuries. As we imagine the beginnings of Greek mathematics, we must imagine it against the background of this kind of widely shared, elementary knowledge.

## The Threshold of Mathematics

There is no turning back from the empire to the tribe. In simple societies, there is usually a language for counting and practices of calculation, as well as pattern-making. People *do* things, but they do not produce theoretical reflections on such activities. Empires give rise to bureaucrats, and those (often) give rise to teachers, and teachers, finally, do produce theoretical reflections. And at this point, they begin to do something new. They engage in practices of writing, they impart rules, and they share riddles. And these, finally, will tend to reproduce outside the closed environment of the classroom. Babylon was gone. But the ancient Near East, *even post-Babylon,* would not have the same mathematical practice as that of an isolated tribe in the Amazon. The riddles were already out in the wild, reproducing.

What is described here is oral knowledge that, by definition, can no longer be uncovered. How many people knew just what? All we can do is point at the overall parameters. Certainly, as Greek democracy grew, more Greeks had access to education. From what we have seen in the foregoing discussion, it is likely that such an education included a modicum of mathematics. We may then follow Høyrup's lead and suggest that mathematical knowledge spread through a few orally shared examples/riddles, and we might also suggest, then, that this knowledge, as it spread through the Greek world and became entrenched in elementary mathematical education, included the basic rules of area measurement, up to and including Pythagoras's theorem.

Peering back into their earliest history, Greeks felt that they had to attribute the beginnings of their mathematical knowledge to particular named Greek authors: hence, the invention of Thales and Pythagoras as mathematical authors. But this was a category error, on the behalf of Greek historians as well as their modern followers. The earliest mathematical knowledge shared by the Greeks was neither authored nor Greek. It was, instead, the reflection in oral teaching of past Mesopotamian scribal schools. (Which, to be fair, many Greeks sensed, as well; Herodotus, for instance, was quite explicit that Greek knowledge of mathematics came from the East!)

And yet, the very need to attach cultural achievements to the names of their authors is telling – and specifically Greek. We do not pause to consider this as such because this habit – the reliance on named authors – has become so natural to us. But it is, in fact, a specific historical invention, happening only in some civilizations. We do not have an author for the epic of Gilgamesh, for the Chinese Book of Odes, for the Mesoamerican Popol Vuh. Indeed, Greek bards, too, recited epic stories of the siege of Troy, never attributing them to any named author. As in so many other civilizations, these were just the stories being told.

A Babylonian teacher, imparting the rule whereby one finds the sides of the rectangle from the combination of area and diagonal, did not invoke the "author of the rule." It was just a rule: teachers knew how to apply it. Similarly, a Greek, singing the song of Achilles's wrath, did not invoke the "author of the poem." It was just an epic poem: bards knew how to sing it.

And then, a few Greeks started to *make a name for themselves*. Instead of just reciting a well-known set of epic poems, widely shared among professional bards, some Greeks started to sing more personal songs, produced from a personal angle. Others repeated those poems, still attributing them to their original authors. In this new tradition, what made a particular poem effective was precisely that one could imagine, in concrete detail, a particular person – say, Sappho or Hesiod – in a particular place, say, Lesbos or Boeotia, authoring the song. So much so that soon enough, people began to imagine a concrete person, the author of the epic poems themselves! Perhaps at some point in the sixth century, "Homer" was invented. The Greeks started to believe that there was one particular person who was responsible for just those two epics, the *Iliad* and the *Odyssey*.[13] And now we can see: Thales and Pythagoras became, as it were, the "Homer" of mathematics. They were retrospectively reinvested with a new identity; they were reidentified as the authors of what was, in fact, an unauthored oral tradition.

For indeed, there was no turning back now. All Greek culture, from now on, would have a named author. This was a new, exciting departure, in many ways specific to the Greeks: through literature, one could make a lasting reputation for oneself.

This invention of the author is also the direct background for a new development in the late fifth century: a proliferation of writing. The real – that is, historical – Thales and Pythagoras were sages, performing orally. Sages do not write; they proclaim. Throughout the fifth century, a handful of sage-like figures began to add a book to their résumés. The proclamations, attached to a name, would spread far and wide. Such are the wise, surprising proclamations of Parmenides, Anaxagoras, Philolaus: a single book, distilling a life of wisdom. Right near the end of the fifth century, this practice of book writing becomes an avalanche. Suddenly, many authors try to make their name in prose, and several produce not one book but many. "Author" becomes, in such cases, the key identity. And so, more and more figures of the author emerge. The genres multiply. These

[13] This reconstruction must be speculative, but see M. L. West, "The Invention of Homer," *Classical Quarterly* 49 (1999): 364–382.

were exciting times, and many turned to the writing of history. Political speech mattered a lot: many wrote speeches, and some produced manuals of rhetoric; teachers of speech-making, they wrote about the practice of speech – as many others now did, writing about their craft. Physicians wrote about health and disease; artists wrote about the proportions of their statues; architects, about the proportions of their buildings. And so, a few wrote on mathematics, as well. Our history proper begins here.

Let us, then, lay out the evidence. So far in this chapter, we have measured our narrative in centuries, even millennia. Now, decades matter. It is the invention that we now need to understand, and so we try to isolate a very precise group: authors likely to have circulated mathematical texts throughout the last few decades of the fifth century.

The list of such attested authors is, in fact, very small. A couple of astronomers are mentioned for Athens – Euctemon and Meton – and we will return to discuss them in Chapter 5. As I will explain there, it is at least possible that their brand of astronomy was not the kind we would recognize as mathematical. Meton, at least, is well dated: he is mocked by Aristophanes in a comedy dated to the year 414. Others are much less well dated. Theaetetus was commemorated by Plato, in the dialogue carrying his name, and we get to know the precise historical circumstances of his death in battle. We can independently date this to 369 BCE. In the same dialogue, we find that as a youth, Theaetetus studied with the mathematician Theodorus of Cyrene. Theaetetus, apparently, did not die in old age, so he could not be Theodorus's student that many years before 369. The dialogue does have Socrates conversing with both Theodorus and Theaetetus, a conversation set just as Socrates's trial is about to begin, in 399; but perhaps not too much is to be made of this because Plato did allow himself considerable historical license, and the moment of the trial was one he returned to for dramatic effect. The likelihood remains that Theodorus was active very late in the fifth century or very early in the fourth.

Another name we must mention is that of Democritus: the first genuinely prolific prose author in Greek. Many dozens of works are attributed to him, and ancient catalogues identified an entire set of works on mathematical topics, for instance, *On the Contact of Circles and Spheres*. This, presumably, was a study not of the tangency of a circle with a sphere but, rather, a study of the general question of how straight lines touch curvilinear lines. There is one concrete mathematical result that we can certainly ascribe to Democritus – because we can do so on the authority of Archimedes himself. What we learn from Archimedes is that Democritus asserted – although without proof – that cones are one-third of the cylinder in which they are enclosed, and also that pyramids are one-third of the prism in which they are enclosed.

We can pursue this report a little further. First of all, the key piece of background: the one thing we know for sure about Democritus is that he was the central figure in early atomism. One of the main topics for paradoxical, wise statements about nature, among early authors, was the very makeup of the universe: What are its ultimate constituents? Atomists contended that the world is made of microscopic, unbreakable pieces.

Now, there is a well-known passage where Plutarch, a philosopher and belletrist of the imperial era, quotes Chrysippus, a Hellenistic Stoic philosopher. Chrysippus cites a puzzle raised by Democritus: if you cut a cone by a plane parallel to the base, you get two circular faces, one at the top of the lower cut, the other at the bottom of the upper cut. The interpretation of this passage is contested, but my own understanding is that Chrysippus implies that Democritus said as follows: surely the two faces are not equal to each other – for otherwise, we would have not a cone, but a cylinder. Hence, what appears to us as the smooth surface of a cone must, in reality, be a terraced, jagged surface, made by microscopic steps. This, then, is consistent with atomism. The world is not a continuum but is instead, so to speak, rough at the (invisible) edges: our impressions of smoothness are no more than an illusion. We now see how discussions of the surface of the cone, or of the sphere, can easily belong to such philosophical debates.[14] It seems that Democritus did assert an intuition – which is, in fact, mathematically correct: that the cone in the cylinder is essentially the same as the pyramid in a prism. He probably meant, however, not some kind of theoretical mathematical observation, but instead a philosophical one: his argument was that the cone in the cylinder simply *was* a pyramid in the prism. It is possible, then, that the evidence for Democritus does not imply any original mathematical activity, although it does imply familiarity with a mathematical lexicon and probably also some mathematical results (it seems that Democritus did know, already, that a pyramid is one-third of the prism containing it – an elementary, although not a trivial, result). But this does not add much to our knowledge of the relevant chronology. Democritus was probably born around 460, but he certainly had a very long and very productive life. The implication, then, is that if we consider

[14] For an account of Democritus's discussion of the cone – and of Democritus's atomism as a whole – see N. D. Sedley, "Atomism's Eleatic Roots," in P. Curd and D. W. Graham (eds.), *The Oxford Handbook of Presocratic Philosophy* (Oxford: Oxford University Press, 2008), pp. 305–332.

an arbitrary work by Democritus, it likely originates in the late years of the fifth century or even the beginning of the fourth. Can we not push the evidence any earlier?

We may, in all likelihood – with just two outstanding names. These are Oenopides of Chios and Hippocrates of Chios (not to be confused with that other Hippocrates – of Cos! – the father of medicine). That the two come from the same island may or not may be significant: our evidence hangs on a thin thread. But could mathematical literature – at the moment of its very inception – perhaps be a local phenomenon?

The evidence for Oenopides is tantalizing, meager – and significant. Three observations stand out.

First, Oenopides is consistently credited with some astronomical discovery having to do with the ecliptic. As I will return to explain in Chapter 5, the motion of the sun, moon, and planets takes place on a thin strip, a circle located on the sphere of the fixed stars, called the "ecliptic," and it happens to be set at a particular oblique angle to the equator: roughly twenty-three degrees. It seems likely that Oenopides made some statement concerning the obliquity of the ecliptic. This suggests an interest in some kind of explicit geometrical model of the sky.

Second, this interpretation can be supported by a couple of very late citations produced by Proclus, the fifth-century author mentioned earlier concerning his testimony for Thales and Pythagoras.[15] What Proclus claims, in his commentary to Euclid's *Elements* 1, is that Oenopides first discovered the construction of a perpendicular to a given line (*Elements* 1.12), as well as the construction of an angle equal to a given angle (*Elements* 1.23). As usual, everything has to be taken with a grain of salt, but the report seems significant for several reasons. Concerning 1.23, Proclus claims that his testimony is based on Eudemus – a follower of Aristotle who composed a history of geometry (which is now, unfortunately, lost; it is our best source for early Greek mathematics, but we rely on a few scattered quotations by authors later than Eudemus). Concerning

[15] I will frequently refer, in this book and especially in this chapter, to Proclus's commentary to Euclid's *Elements* (discussed in the "Proclus and the Philosophical Schools" section in Chapter 6). This has an excellent English translation: G. R. Morrow, *Proclus: A Commentary on the First Boom of Euclid's Elements* (Princeton, NJ: Princeton University Press, 1970/1992). Note that it is customary to make references inside the work not according to the page numbers in Morrow's translation but according to the scientific edition from the nineteenth century. Pages 65–68 are especially useful because they contain a brief history of early Greek mathematics, which Proclus apparently had on the authority of the early historian Eudemus.

1.12, Proclus claims that this result was used by Oenopides for the sake of astronomy, and he also adds – which provides the report with extra credibility – that Oenopides used a particular archaic term for "perpendicular" ("at a gnomon," which in itself may or may not carry a specific astronomical implication). The reference to the precise language used suggests quite clearly that Proclus, ultimately, relies on a source who could read Oenopides's own writing. Oenopides, then, was a writer! And in all likelihood, this report, too, ultimately went back to Eudemus.

To bring all this into an appropriate context, it should be noted that scholars have long concluded that Eudemus may have based his history, in part, on his own rational reconstructions. That is, Eudemus would ascribe to past mathematicians knowledge of such things that was required for the sake of what he knew they actually did know. If result A is logically demanded by result B, and Eudemus found evidence that a mathematician asserted B, Eudemus would then claim also that the mathematician knew A. Eudemus, further, seems to have structured his own work primarily as a survey of "first discoveries"; that is, he looked for evidence, however indirect, concerning the identity of the first authors who knew about particular results.[16] Add to this Proclus's own purpose. He does not write a summary of the history of Eudemus; he takes from it such tidbits that are relevant for Proclus's own commentary to Euclid's *Elements* 1.

Bring all of the evidence together, and a likely account emerges: Oenopides may have been the first author to produce a geometrically motivated account of the sky, for which – quite naturally – he relied on various assumptions concerning the construction of angles. All of this is important, and we shall return to it in Chapter 5.

But why do I even trust any of this? Why should I not dismiss the evidence, as I did for Thales and Pythagoras? Why assume that Oenopides was a fully fledged author and not just a wise man, proclaiming his thoughts concerning the stars and the earth?

The main reason has to do with dates. Oenopides is said, by Proclus, to have been somewhat younger than Anaxagoras. This is in the context of a quasi-chronological list of early Greek geometry, and Oenopides is followed by Hippocrates of Chios, who is in turn followed by Theodorus of

[16] I won't go into this historiographical detail at length, but Eudemus is indeed a central pillar to our history. See L. Zhmud, "Eudemus' History of Mathematics," in I. Bodnár and W. W. Fortenbaugh (eds.), *Eudemus of Rhodes* (New Brunswick, NJ: Routledge, 2002), pp. 263–306.

Cyrene. Nothing here inspires huge confidence (the pairing of Anaxagoras and Oenopides seems to depend on a forged dialogue by Plato, *The Lovers*, where the two are mentioned side by side). But we recall that Theodorus was likely active near the very end of the fifth century; Anaxagoras was a philosopher reported to have been close to Pericles, and thus he was active not long after the middle of the fifth century. With Oenopides and Hippocrates of Chios, then, we likely find authors active in the last few decades of the fifth century: two authors, close to each other in time and in place. And so, to the extent that we find that Hippocrates of Chios was likely the fully fledged author of a mathematical book, we are probably justified to assume the same for Oenopides.

And there is, in fact, very little room for doubt: Hippocrates of Chios must have been a fully fledged author because – finally! – we have a very substantial fragment of his original writing still extant. Well, "extant" is – as you would expect by now – a relative term. We have a report from a late commentator to Aristotle, called Simplicius (very late, from the sixth century CE!). He tries to account for a passage in Aristotle where Hippocrates of Chios is briefly mentioned. And to do so, Simplicius makes a lengthy quotation from the very same Eudemus, the early historian of geometry. The quotation certainly includes many interpolations by Simplicius himself, but past scholars have identified certain linguistic hints that allow us to separate Eudemus's original text from that of Simplicius. The result is that we can read, at the very least, a nearly unadulterated citation from Eudemus, which seems to hew closely, at least in parts, to Hippocrates's own words.

This, then, is our first substantial glance at Greek mathematics. It is our first extended close-up, and it is a pivotal moment; we should linger here for a while.

In Eudemus's own presentation, Hippocrates produced four separate results. I will cite the first two in full and then only briefly mention the third and the fourth. In what follows, the assumption is that we translate a text by Eudemus. The word *he* refers to Hippocrates of Chios:[17]

> He first proved by what method a quadrature was possible, of a lunule having a semicircle as its outer circumference. He did this after he circumscribed a semicircle about a right-angled isosceles triangle and, about the

[17] What follows is my own translation. The entire passage may be read in I. Thomas, *Greek Mathematical Works* (Cambridge, MA: Harvard University Press, 1939), pp. I.234–I.252. (This is a translation of selected passages of Greek mathematics, designed to complement Heath's history.)

base, [he drew] a segment of a circle, similar to those taken away by the joined [lines]. And, the segment about the base being equal to both [segments] about the other [sides], and adding as common the part of the triangle which is above the segment about the base, the lunule shall be equal to the triangle. So the lunule, having been proved equal to the triangle, could be squared. In this way, taking the outer circumference of a semicircle as the [outer circumference] of the lunule, he readily squared the lunule.

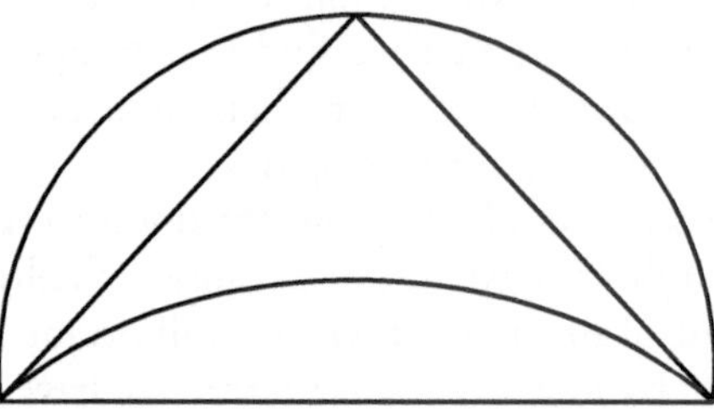

Figure 7

Then, following that, he postulates a [lunule] greater than a semicircle, having constructed a trapezium having the three sides equal to each other, while one, the greater of the parallel [sides], being three times each of them [= the smaller sides], in square, and comprehending the trapezium by a circle and having circumscribed about its greatest side a segment similar to the [segments] cut off from the circle by the three equal [sides]. And that the said segment is greater than a semicircle, is clear when the diameter is drawn in the trapezium. For, of necessity, this [diameter], subtending two sides of the trapezium, is greater than double the remaining one in square; and therefore the greatest of the sides of the trapezium will necessarily be smaller [in square] than: the diameter, and that of the other sides, which the said [= the greatest side] subtends (together with the diameter). Therefore the angle standing on the greatest side of the trapezium is acute. Therefore the segment in which it is (which is the outwards circumference of the lunule) is greater than a semicircle.

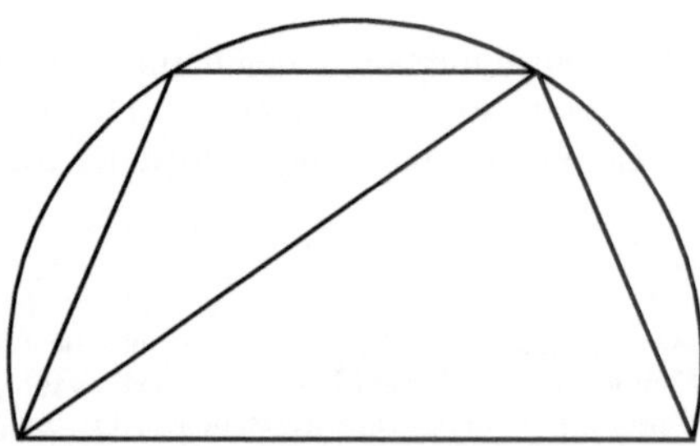

Figure 8

As a teacher, I know that it is never a good idea to describe a certain piece of mathematics as "easy." The students who understand it will feel cheated of their due accolades; those who do not will rightly be upset. But it has to be said that although the first argument is already very ingenious, it is also really quite straightforward. We require three assumptions:

i. Pythagoras's theorem.
ii. The angle on the circle's diameter is right.
iii. Similar segments of a circle take up a similar fraction of the square on their base.

By now, we know how Hippocrates (and his audience) could have been familiar with (i). As for (ii), it is clear that Hippocrates uses this assumption in some generality, but let us note that for what is required in the very first application, he can directly see his result through a simple construction:

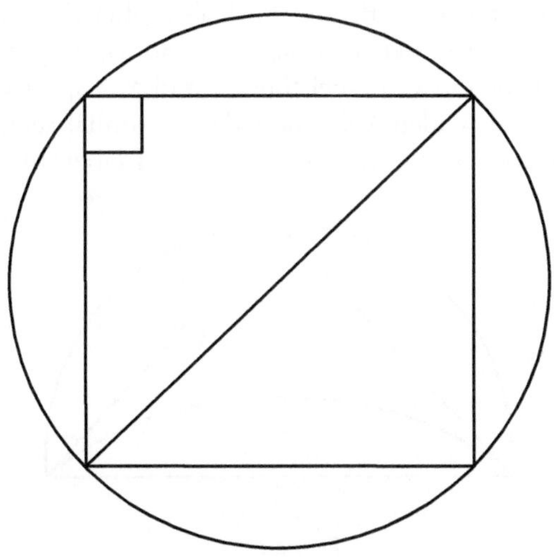

Figure 9

If we draw a square in a circle, and then draw the diagonal of the square, then it becomes obvious that this diagonal is also the diameter of the circle, and the angle on it is indeed right. For the first of Hippocrates's arguments, indeed, this is precisely the required construction.

As for (iii), as soon as we draw two similar segments of a circle, with the squares on their bases, it becomes clear that the two constructions are indeed identical other than their scale; obviously, whatever fraction the left segment is of its square, the right segment will also be of its own square:

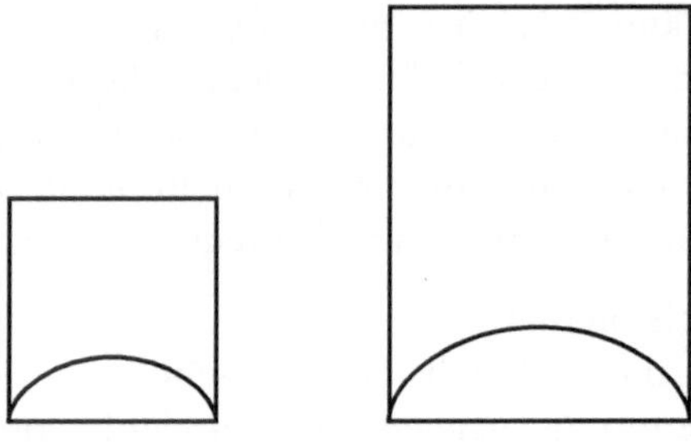

Figure 10

This is so evident that, in all likelihood, Hippocrates need not have formulated this assumption explicitly. It is part of what is *meant*, for him, in the expression, "similar segments."

Now let us look more closely at his construction: We have here an isosceles, right-angled triangle within a semicircle. I label the three sides of the right-angled triangle A, B, and C. I also label three segments: Two segments are contained between sides A and B, respectively, and the circumference of the circle. I label those 1 and 2, respectively. Those two segments are obviously identical. I now draw a similar segment (so, bigger but similar in shape) whose base is the side C. I label this one 3:

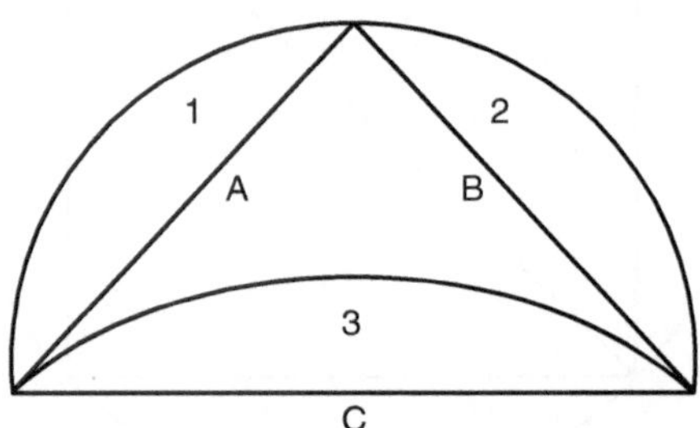


Figure 11

The squares on A and B, taken together, are equal to the square on C ($A^2 + B^2 = C^2$). The segments of circle 1, 2, and 3 take up equal fractions of those squares. If 1 were to be a third of the square on A, so would 2 be of the square on B, and so would 3 be of the square on C. If 2 were to be a tenth of the square on A, so would 2 be of the square on B, and so would 3 be of the square on C. We do not know this fraction, but we know that it holds constantly. Hence, if the squares on A, B, and C are to each other in a certain relation, so would be the segments of circle 1, 2, and 3. These are all one and the same thing – scaled by some mystery fraction.

And so, because $A^2 + B^2 = C^2$, we also know that $1 + 2 = 3$. The two top segments, taken together, are equal to the bottom segment.

But then, equality has consequences! If $1 + 2 = 3$, it means that removing 3 and adding $1 + 2$ should cancel out.

So, let me take the right-angled triangle, remove 3, and add $1 + 2$. I remove the segment at the bottom and add the two segments at the top.

But what did I do? I turned the right-angled triangle into the lunule shape, the thing shaped like a crescent. And so I proved something: the lunule is equal to a right-angled triangle. I can actually measure this crescent shape directly, even though it is circular! Crazy, right?

But this was so simple to achieve! In fact, almost no special assumptions were required beyond what would be obvious to a Babylonian. And indeed, even the basic technique – tearing up parts of a geometrical configuration and adding them up elsewhere – is precisely what we have seen on Babylonian tablets (thanks to Høyrup's interpretation) and, in all likelihood, what pupils were taught, in math classes, for millennia. But this is not your ordinary math class or indeed your ordinary clay tablet, and not just because this is all about a counterintuitive result concerning circular shapes. The most obvious difference from anything seen so far in this book is the absence of numerical values. In my own brief explanation, I did insert numbers ("if this is a third of that"). This is what one does when explaining mathematics. This is pedagogy. But Hippocrates's text is not pedagogical. Instead, it is a tour de force of pure argument for argument's sake.

And a tour de force it is. From the simple right-angled triangle, we move to the trapezium in the second construction. Here the same idea is expanded. It has become clear, from the preceding argument, that we can have other constructions where the squares on the top sides, taken together, are equal to the square on the bottom side. So, let us draw a trapezium, with the three top sides equal to each other. I label the three top sides A, A, and A, and I label the bottom as B.

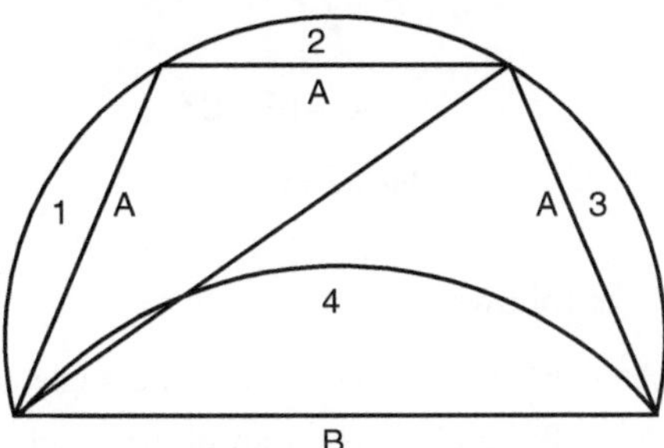


Figure 12

Clearly, the squares on the three top sides are equal to $3 \times A^2$. And so, I simply stipulate that the length of B is such that $B^2 = 3 \times A^2$. This is a simple stipulation concerning the shape of the trapezium (I do not need to know how to find such a B, and there is no trace of an argument that such a trapezium is possible, although this is obvious upon reflection because B can be anything bigger than A but smaller than $3 \times A$). Now, obviously, if I draw a circle around this trapezium, each of the top A lines cuts off an identical segment of the circle. I now draw a similar segment on top of B, and I label the segments as 1, 2, 3, and 4, respectively.

Clearly, now, $1 + 2 + 3 = 4$. (These are similar segments taken out of the squares, and we have established that this relation holds with the squares.) By this token, if I take the trapezium, remove the bottom segment 4, and add the top segments 1, 2, and 3, I end up not changing anything. Hence, we have found a lunule equal to a trapezium. Again, a crescent shape equal to a rectilinear shape! This, by now, is much clearer, and in fact, at least in the mediated text we now possess, calls for no explanation at all. Instead, it seems that Hippocrates dedicated some effort to showing a more complicated observation: a trapezium produced according to this recipe will have its base, so to speak, beneath the diameter of the circle; or in other words, it will occupy more than a semicircle.

How to show this? We need some kind of expansion of Pythagoras's theorem. Indeed, it is not hard to see that if, in a right-angled triangle, $A^2 + B^2 = C^2$, then if I squeeze the right angle smaller, I get C smaller, accordingly; if I expand it wider, I get C greater:

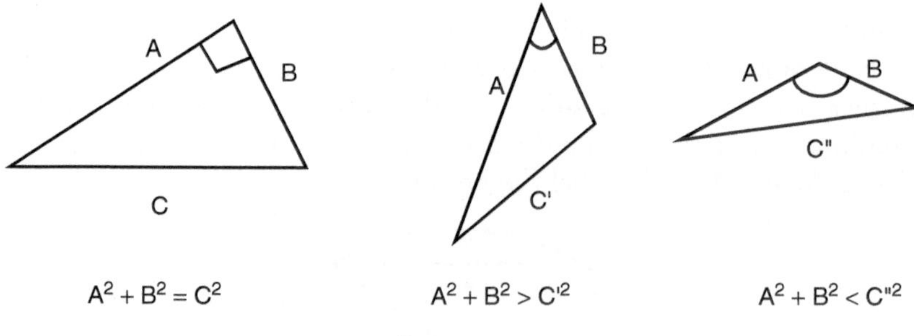


Figure 13

Thus, the rule is as follows:

$$A^2 + B^2 > C^2 \quad \leftrightarrow \quad \text{an acute angle;}$$
$$A^2 + B^2 < C^2 \quad \leftrightarrow \quad \text{an obtuse angle.}$$

By somewhat similar reasoning, it becomes clear that if the right angle rests on the diameter of the circle, then if we squeeze this right angle smaller, we have to push the base lower, and we get more than a semicircle; if we expand this right angle bigger, we have to push the base higher, and we get less than a semicircle:

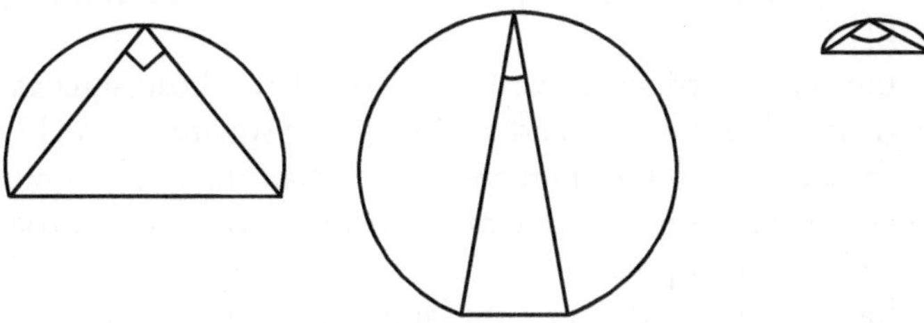

Figure 14

It is easy, at this point, to reflect on the trapezium. Recall that I labeled the three sides on the top with "A" and the bottom with "B." By stipulation, $B^2 = 3 \times A^2$. Now, I also draw a diagonal, which I label D:

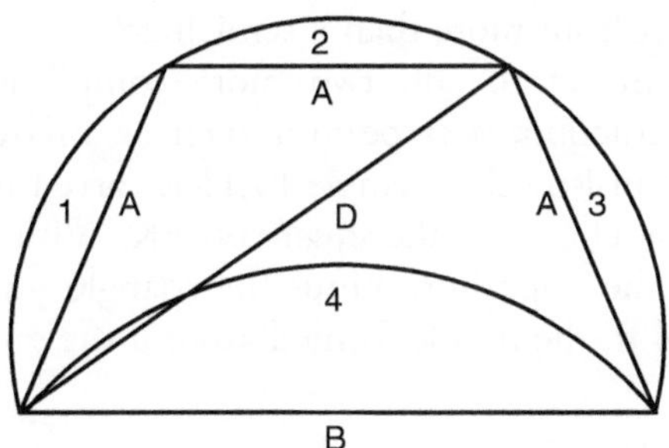


Figure 15

The top triangle, with two As as its sides and D as its base, has an obtuse angle (we see this immediately because this is a trapezium, with the bottom greater than the top). By our expansion of Pythagoras's theorem, we have the following:

$$D^2 > A^2 + A^2.$$

Or

$$D^2 > 2 \times A^2.$$

We now consider the bottom triangle, with D and A as its two sides and B as its base. We do not know the precise length of D, but we do know that $D^2 > 2 \times A^2$. We also know, of course, that the square on A is $A^2$. This means that we know this: the sum of the squares on the two sides of this triangle – the sum of the squares on D and A – is greater than $3 \times A^2$:

$$D^2 + A^2 > 3 \times A^2.$$

But $3 \times A^2$ is precisely what we stipulated to be the value of the square on the base, B!

So, in the bottom triangle, we have two sides, whose squares add up to a sum greater than $3 \times A^2$, and a base, whose square is exactly equal to $3 \times A^2$. The squares on the sides are greater than the square on the base, and so, by our expansion of Pythagoras's theorem, the bottom triangle has an acute angle at the top.

Now, this is important because this angle, at the top, is the angle in the circle, resting on the base. And by our expansion of the relation "right angle lies on the diameter," we can now draw a further conclusion. Acute angles lie on a base that is lower than the diameter, or on more than a semicircle.

And thus, we conclude that the trapezium takes up more than a semicircle. If the first lunule was carved out from a semicircle, the second lunule is carved out from more than a semicircle!

Hippocrates went on to add two more complicated constructions. In one such construction, a trapezium such as EKBH is drawn, with EK = KB = BH; an isosceles triangle EZH is carved inside it, such that the square on EZ is 1½ times the square on EK. With this arranged, it is easy to show that the trapezium minus the triangle – the rectilinear area EKBHZ – is equal to the lunule formed around the same points:

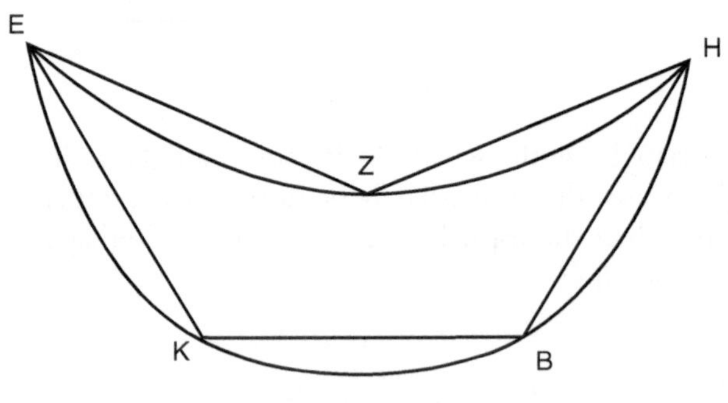


Figure 16

Once again, Hippocrates does not even bother to show this equality and instead develops an argument for how the construction is possible – and also that it now rests on less than a semicircle. For reasons analogous to those mentioned before, the angle EKH has to be obtuse; the line EH has to be above the diameter of the circle.

The final construction sets up two concentric circles, such that the square on the diameter of the greater is six times the square on the diameter of the smaller one. We also set up a regular hexagon within the smaller circle and, in the greater circle, a triangle such as HQI, produced along the projection of two lines from within the hexagon. Under this arrangement, it can be shown that the sum of the hexagon and the triangle is equal to the sum of the smaller circle and the lunule on the triangle:

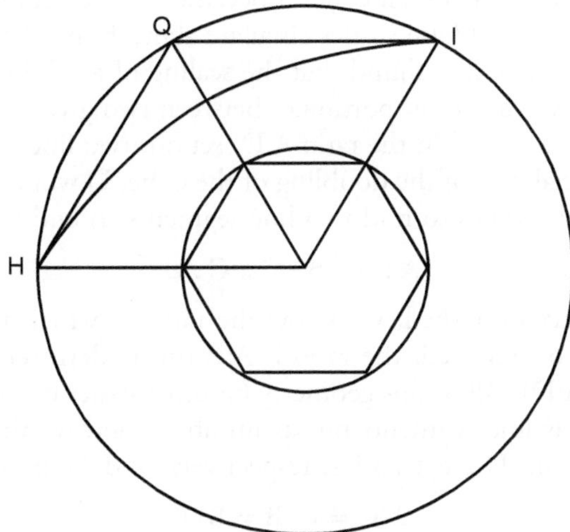


Figure 17

With this, our evidence for Hippocrates's lunules ends.

This is enough. There is nothing remotely like this in the records of all preceding human history. This, then, has been worth our prolonged attention. Let us linger a little further on this moment: we should extract all the history we can get out of it. What else is known about the mathematics of Hippocrates of Chios?

There is no other extended fragment of his works. But we do possess a few tidbits of information.

First, there is a problem that, eventually, would be known as the "doubling of the cube." Suppose you are given one cube, whose side is $S$ and whose volume is $V$, and you are told to find another with double the volume. It quickly transpires that this is a difficult task. If you simply double the side from $S$ to $2 \times S$ – which is perhaps one's immediate impulse – you clearly will have not twice the original volume but *eight times* that. In fact, there is no easy solution at all. (Arguably, the most significant tradition of all of Greek geometry would revolve precisely around this problem; more on this, then, in the following chapters.) We are told that Hippocrates made the first stab at this problem. This is on the authority of Eratosthenes, who worked at the court of the kings in Alexandria in the third century BCE; he could still have had access to Hippocrates's original writings or, at the very least, to some indirect but reliable reports. (And I believe Eratosthenes: Hippocrates of Chios, although well deserving of fame, never became the subject of much ancient notoriety. Why would false stories circulate about him?[18]) Specifically, we are told that Hippocrates found that the scaling of a solid is equivalent to the finding of two mean proportionals between two given lines. That is, if you wish to scale a solid in the ratio A:D, set out two line segments A and D. (In the special case of the doubling of the cube, D would be double A). The task then becomes to find two line segments, B and C, such that

$$A : B :: B : C :: C : D.$$

When this is accomplished, we know the ratio in which the side of the solid needs to be adjusted: the ratio is A:B (in modern terms, A:B is the cubic root of A:D). All of this geometry becomes easier to us if we think of the terms not as line segments but as numbers (and we think of ratio as division). If A and D are 1 and 2, respectively, and we have

$$2/C = C/B = B/1,$$

it is clear that $B^2 = C$ and that $B^3 = 2$, or that B is the cubic root of 2. All of this, however, is the modern way of thinking about it: to Hippocrates, these were simply lines, in proportion.

Second, we are told by Aristotle himself that Hippocrates, as well as his follower Aeschylus (obviously, not the playwright and, otherwise, completely unknown) promoted a particular theory of the nature of comets. The details are difficult to secure, but this much is clear: Hippocrates and Aeschylus thought that comets were like planets.

[18] We have the text from Eratosthenes on the authority of the late ancient commentator Eutocius, and so, of course, there is always room for doubt as to how reliable our report is. I return to discuss Eutocius's commentary in the section "Eutocius – and a Coda" in Chapter 6, and the text may be read in R. Netz, *The Works of Archimedes: Translation and Commentary*, vol. I (Cambridge: Cambridge University Press, 2004), pp. 294–298.

This was unlike Aristotle himself, who thought comets were essentially meteorological phenomena (naturally occurring fireworks, if you will). It should be noted that Hippocrates and Aeschylus are, in fact, correct. Although the details are unclear, it does seem that the Hippocratic theory, in this case, involved some geometrical reasoning concerning planets, and if so, it could well have fitted within the broad tradition of Oenopides.

Finally, Proclus tells us that Hippocrates was the first to write in the tradition of Euclid's *Elements*. This is a much later report, and Proclus's agenda makes it somewhat less credible: Proclus writes this in the introduction to his own commentary to Euclid's *Elements*, and part of the ground he is expected to cover is information on past authors in the field, preceding Euclid himself. He is thus under some compulsion to provide information, even where his sources are thin, and it seems possible that, from Proclus's sources, Hippocrates simply appeared as a likely candidate to have been such an early author of *Elements*. It may be that this is all that Proclus's report can tell us. Or perhaps he did read in some reliable source that Hippocrates was the first author of *Elements* – would that mean that Hippocrates was the first author of a book in the nature of Euclid's *Elements*? Or simply the first author of a book whose contents partially overlap with those of Euclid's *Elements*? (In the latter case, this really provides us with no information beyond what we know already from the lunules themselves!) The most generous interpretation of Proclus's evidence is that Hippocrates wrote a book that collected some elementary results in geometry; so, besides doing the spectacular fireworks of the lunules, he might have also produced a book that contained proofs of fairly basic but still useful elementary results. My own guess is that Proclus's evidence in this case is wrong, but the more generous interpretation is definitely possible.

We find that Hippocrates produced a spectacular result, measuring lunules in terms of rectilinear figures; that he found an interesting approach to another seemingly intractable problem, "doubling the cube"; that he had produced some (geometrical, perhaps?) claims concerning the sky; and that he may or may not have written an elementary work in the *Elements*. We find a strong sense of identity: everything has to do with the exact sciences. This is no mere philosopher, dabbling in some claims that could interest the historian of mathematics. No – we have here the professional identity of the mathematician. We also find an emphasis on strong, surprising results based on rigorous argument. This, indeed, will remain to define ancient mathematicians. The tradition, begun.

## Assessing the Threshold

I suggest here that Hippocrates's works were among the earliest pieces of Greek mathematics ever to be written. And this might seem surprising. Could mathematics emerge like that – springing forth from Zeus's head? Would we not expect mathematics to emerge in a more rudimentary form?

In fact, I think this is precisely how we should expect mathematics to emerge: from Zeus's head, fully armed. What would be the alternative?

For this, too, let us again consider the context for Greek literature pursued in this chapter. It was a unique type of literature, based on the figure of the author – seeking fame for oneself through the making of books, emerging from a culture of public, face-to-face debate. *Of course* the very first mathematical works in circulation would contain remarkable, surprising results. Why else would you even bother to circulate them? I suspect that the counterfactual is sometimes not sufficiently carefully thought through here. Just what would a more rudimentary piece of mathematics look like? Would it prove some truly elementary results, such as, say, the equality of the angles at the base of an isosceles triangle? (This is Euclid's *Elements* proposition 1.5, in some ways the first substantial result achieved by Euclid.) Why would anyone care about such a treatise, proving such a result? Take just a few examples. Parmenides circulated a work arguing that there is no change and that only one thing exists. The author of the medical text *On Diet* circulated a work arguing that everything is due only to fire and water. Gorgias circulated a work arguing that Helen – whose adultery was the cause of the Trojan War – was worthy of *praise*. This seems to have been the more typical purpose for putting a work into circulation in the fifth century: to argue for something *surprising*. Why circulate anything dull?

And indeed, it is clear that Hippocrates's lunules would be considered surprising. Recall the little that is known on Democritus: at some point, he wrote on the tangencies of circles and spheres, obviously reflecting on the paradoxical nature of a straight line touching a curved one. Protagoras, an author active at about the same time, is recorded as explicitly arguing that a straight line does *not* touch a circle at a single point. And Antiphon, another philosopher active at about the same time, is known to have argued – almost certainly, against the received opinion (or else, how would this end up recorded as his, Antiphon's, opinion?) – that the area of the circle can be measured directly by finding a rectilinear figure to which it is equal. (He suggested doing so by inscribing ever-greater polygons within a circle.) Greeks, in general, knew enough geometry, in the late fifth century, so as to know that there was a difference between a curved and a straight line; knew, even as schoolchildren, how to measure a rectilinear area; and knew that measuring a curved area is very different.

Although the mathematical koine implied by Hippocrates's results is certainly continuous with the knowledge generated in Babylon, it is also subtly different. As well it should be, and not only because across more than a millennium, things will change but, more to the point, because the context of mathematical education in Babylon and in Greece was quite different. We can reconstruct in full archeological and historical detail the Babylonian scribal school, ultimately embedded in the state/religion apparatus. We must be much more speculative as we reconstruct early Greek education, but the best guess involves a more widespread practice, distributed across many households, embedded in a greater variety of social settings. Indeed, the mathematical koine implied as a background to Hippocrates's works stands apart, somewhat, from that of the Babylonians. Especially noteworthy is the emphasis on the measurement of *angles*, and in general on rather precise shapes, such as the triangle and (in particular) the circle. Instead of an emphasis on the needs of, say, irrigation, land measurement, and taxation, we see an interest in particular constructions: as it were, less bureaucracy, more carpentry. Is this, then, perhaps, understandable through the specific social setting of Greek education? A practice of education spread among many different social settings can be expected to become more of a bricolage, expressing the various ways in which this society engages with calculation and measurement. And so, through the centuries separating the earliest, indirect Greek contacts with Babylon and the emergence of Greek mathematical literature, could Greek mathematical educators have imperceptibly, gradually, expanded the reach of geometry to emphasize a greater variety of shapes and angles? Some of the credit for the making of Greek mathematics should then go to those unnamed educators.

But there is more than what was given to Hippocrates from the educational koine. Perhaps what I need to defend, first of all, is my overall strategy of ascribing to Hippocrates – when I can no longer depend on his direct reliance on the educational tradition – the ability to *innovate.*

Change, in general, is often slow. Bit by bit, the sea erodes the rocks on the shore; generation by generation, the giraffe's neckbones got ever so slightly longer; animal populations, in their habitat, steadily ebb and flow. Much the same megafauna roamed the Americas for tens of millions of years: giant beavers, giant sloths, and mammoths, the same species, in roughly the same numbers, for untold generations. Change was slow, almost invisible. In this way, indeed, over centuries, Babylonian educational traditions (mostly dealing, in geometry, with basic calculations concerning rectangles) could have expanded, within a Greek setting, with its wider points of contact and with ordinary practice, into a more varied study of shapes and their angles.

And then, I suggest, things can happen, occasionally, quite rapidly. Indeed, go back to the American megafauna, the mammoths and the giant beavers and the giant sloths. Nearly all disappeared, *nearly instantaneously*, at around the year 11,000 BCE. Change, usually, is slow – except for the rare occasions when it is fast. And this is how modern biologists and geologists understand historical processes: often, these are series of rapid transformations between otherwise stable equilibria. Things do tend to emerge – even at the timescales of biology and of geology – fully armed, out of Zeus's head!

And indeed, why so? Why did the American megafauna all disappear, all together? Now, there are many competing theories, and it would be rash of me to rely on any one in particular. Perhaps it was the climate, perhaps some other environmental event. But it is certainly a logical possibility – maintained by serious scholars in the field – that it was the humans coming in. You see, humans will have no trouble at all killing mammoths, giant sloths, and giant beavers. Indeed, it would be absurd to expect those hunters to cross the Bering Strait and then to spend the first millennium of their presence there, say, just killing mammoths, then the next millennium just killing giant beavers, finally turning, in the third millennium of their stay, to the giant sloths. Of course not. The hunters would kill them *all*, as soon as the possibility arose.

And as Greek mathematicians started writing and thinking actively of new, surprising things to find, they had crossed the Bering Strait. They had reached the threshold of mathematics. And all of a sudden, an entire range of results became easily accessible, and so we should simply assume that whatever was achieved would have been achieved quickly – which is what we will find in the following two chapters, each dedicated to a single generation of Greek innovation in the exact sciences. Through the course of these two generational events, several dozens of active mathematicians had, essentially, cleaned up a particular branch of geometry. For two thousand years, there were only pickings left, and when mathematics did make a new push further – finally, in the seventeenth century – this was, you can say, by using the Greek continent as a launching pad to new continents altogether.

For now, we may finish this chapter with these two observations: First, we certainly have lost a lot of what took place in the last few decades of the fifth century. Even from Hippocrates, we probably have no more than a sample; other mathematicians must have been lost entirely from memory. It is perfectly possible that Hippocrates was not literally the first; he could have been the fifth, the tenth. What we can say with great likelihood is that mathematics emerged in the manner in which we see it with Hippocrates's lunules: with a bang.

And second, we can now see that there were many reasons why this happened then, with the Greeks: Because this was a culturally dispersed society, with education reaching fairly widely. Because it was a culture of face-to-face, public debate, favoring radical, surprising argument, powerfully put forward. Because it created the idea of the author making a name for himself, and now made it a habit to produce such names by the writing of prose dealing with a great variety of subject matter. Innovation makes little sense to the mathematics teacher, but it is essential to the mathematical *author*. Bring the figure of the author into mathematics, and innovation will cascade. Soon, indeed, more Greeks would pay attention, and an entirely new discipline would emerge.

## Suggestions for Further Reading

M. Ascher, *Ethnomathematics: A Multicultural View of Mathematical Ideas* (New York: Routledge, 1994) remains the best starting point for mathematics in simple societies. Gary Urton, especially, has contributed considerably since to our knowledge of Andean record-keeping. There is no single monograph (and the field keeps growing), but a brief and readable entry point is G. Urton and C. J. Brezine, "Khipu Accounting in Ancient Peru," *Science* 309 (2005): 1065–1067. An account of the role of the state in East Asian mathematics can be found in A. Volkov, "Argumentation for State Examinations: Demonstration in Traditional Chinese and Vietnamese Mathematics," in K. Chemla (ed.), *The History of Mathematical Proof in Ancient Traditions* (Cambridge: Cambridge University Press, 2012), pp. 509–551.

Essential readings on Babylonian mathematics are as follows: (1) D. Schmandt-Besserat, *Before Writing. Vol. 1: From Counting to Cuneiform* (Austin: University of Texas Press, 1992); (2) J. Høyrup, *Lengths, Widths, Surfaces: A Portrait of Old Babylonian Algebra and Its Kin* (Berlin: Springer, 2013); and (3) E. Robson, *Mesopotamian*

*Mathematics, 2100–1600 BC: Technical Constants in Bureaucracy and Education* (Oxford: Clarendon Press, 1999).

It is obviously hard to select just a few works on the background to Greek civilization. A classic on the earliest Greek civilization is J. Chadwick, *The Decipherment of Linear B* (Cambridge: Cambridge University Press, 1958). For a general background on classical Greece, the best book now is J. Ober, *The Rise and Fall of Classical Greece* (Princeton, NJ: Princeton University Press, 2015). For the interface between social and intellectual history in antiquity, the best is G. E. R. Lloyd, *Demystifying Mentalities* (Cambridge: Cambridge University Press, 1990).

For the specific question of the veracity of reports concerning Pythagoras, the standard view is represented by W.Burkert, *Lore and Science in Ancient Pythagoreanism* (Cambridge, MA: Harvard University Press, 1972); L. Zhmud, *Pythagoras and the Early Pythagoreans* (Oxford: Oxford University Press, 2012) states the case that Pythagoras was, in fact, a mathematician. For the overall historiographical question of the Greek debt to Babylonian mathematics, the key reference will always remain S. Unguru, "On the Need to Rewrite the History of Greek Mathematics," *Archive for History of Exact Sciences* 15 (1975): 67–114.

Readers who wish to read even more on Hippocrates of Chios can turn to an article of mine: R. Netz, "Eudemus of Rhodes, Hippocrates of Chios and the Earliest Form of a Greek Mathematical Text," *Centaurus* 46 (2004): 243–286. For a thorough sense of the context for the very earliest Greek mathematical writings, an excellent resource is I. Bodnár, *Oenopides of Chius: A Survey of the Modern Literature with a Collection of the Ancient Testimonia* (MPIWG Preprint 327, 2007).

CHAPTER 2

# *The Generation of Archytas*

## Plan of the Chapter

We will survey the early history of Greek mathematics through two generational events: this chapter, "The Generation of Archytas," followed by the next chapter, "The Generation of Archimedes." This is a substantive claim: Greek cultural life was generally organized by such isolated, generational events. This is the claim of the first section, "The Hypothesis of Generational Events." Following this general historical statement, the section "What Little We Know" surveys the evidence for Greek mathematics in the first half of the fourth century.

The mathematical achievement of the generation is surveyed through its three key figures, with a section on each. First, Archytas: we follow his spectacular solution to the problem of finding four lines in continuous proportion – and we delve, briefly, into Greek mathematical musical theory or "harmonics." Then, Theaetetus, for a brief glimpse of the Greek treatment of irrationality, followed by Eudoxus, perhaps the most important mathematician of this generation.

To close the survey, the section "Assessing the Landscape" considers the entire group of mathematicians active in this period; the section "The Making of Mathematics" brings forward the nature of the common achievement of the generation as a whole – mathematics, as a genre. Finally, "Coda: The Final Contribution of the First Generation" concludes on one particular result, achieved right at the end of this generation and destined to shape much of the future of mathematics: the discovery of conic sections.

## The Hypothesis of Generational Events

The social history of science is annoying. If I explain the rise of Greek mathematics in terms of "the motivation to project one's identity as an author," "the rise of literacy," or "the proliferation of genres," there's a strong sense that I'm missing the simplest and most important point. Why did people do mathematics in the late fifth through the early fourth century? Because, to them, it was fun. This is the intuition of many readers today – most important, of many working mathematicians.

I hope it's clear that I agree. If, as social historians of science, we lose sight of this point, we are not merely disrespectful but simply wrong. In fact, we miss an important *social* observation. Absent special incentives, you would expect people to do pleasurable things; this determines, then, what mathematics gets done. They didn't need to worry about tenure, and so Greek mathematics is tilted to such mathematical questions that provide – to the mathematically inclined – direct intellectual satisfaction. More to the point, they were not schoolteachers. They were authors, seeking fame – and so, reaching for the more remarkable results. This, after all, is what made people of the following centuries go back to Greek mathematics and try to extend it. The intellectual attractiveness of Greek mathematics – forced upon it, as it were, through the absence of institutional support – is an important part of its significance as a historical presence.

And so, we need a social history. Of course, people do the things they like doing – *among the options they have.* Social history tells us what the options are, and by the late fifth century, those have dramatically changed. Born around the year 430 BCE, you would be surrounded by books. They come with their authors marked prominently, names from far abroad, suddenly gaining the fame once allowed only to successful public performers (are you not tempted to imitate them?). They touch on all subjects. Here is a transcribed speech on the human body (the author says that all men are made of fire and water! It must be against that other book I read last month, which said all men are made of four humors!) Here is the history of the recent war, written by an eyewitness. Here is a treatise on political theory, arguing against Athens' democracy. And here is a papyrus roll dedicated to the measurement of the circle (would you believe this is possible?) – but wait, of course this is wrong; I could do better than that author!

And so, many more try. I imagine that, in the final few decades of the fifth century, no more than a handful of mathematical works got written. Through the first few decades of the fourth, there were probably dozens of mathematical authors who, taken together, produced many dozens of works between them (possibly, even a few hundred): a very substantial mathematical library. This is the subject of this chapter.

This chapter is concentrated on a few decades. The first few decades of a new development are always significant, setting the stage for what follows.

But once again, I think there is a more specific story to be told. Because ancient science (and, with few exceptions, ancient culture as a whole) did not develop stable, impersonal institutions, it was hard for any cultural trend to persist. Many ancient cultural events can be seen as *generational*. More precisely, what we often find is perhaps a generation and a half – a passing fashion or, perhaps, a response to some powerful presence that, after a while, loses its power. We recall: Thales of Miletus is said to have argued, very early on (around the beginning of the sixth century?), something that posterity remembered as *all is water*. In the ensuing generation, Anaximander of Miletus would argue (as transcribed by posterity) that *all is the unlimited*; Anaximenes of Miletus, that *all is air*. This is the gist of the information we have based on Aristotle, and it is probably false in many ways, but one thing is clear: here was a powerful voice – that of Thales – and a generation of local, immediate responses. And then – silence. For a long while, we hear of no more philosophers in Miletus, and elsewhere, wise people seem to be engaged with rather different concerns. A powerful voice – a set of echoes – and then, more or less, silence.[1] This pattern seems to be repeated very often in ancient cultural life. A prominent master is followed by a prominent disciple but with no obvious continuation into a third generation. Parmenides of Elea, and then Zeno of Elea. Leucippus, and then Democritus. In the early third century in Alexandria, there was a flourishing of scientific pursuits of all kinds (which will, of course, greatly interest us in the following chapter). Perhaps chief among this is medicine, where the foundations of modern anatomy and physiology are laid by Erasistratus and Herophilus. They seem to dissect, even vivisect, not only animals but also humans. A deeply disturbing scientific breakthrough! Modern historians of medicine are often puzzled by the fact that this flourishing is so brief. If so much new knowledge is gained, why stop cutting? For in fact, immediately thereafter and then for several centuries, apparently no one else in the Greek world pursued the practice of dissection and vivisection. But perhaps this is not so surprising. This is simply how ancient science worked. A handful of leading practitioners, competing against each other and influenced by each other, working simultaneously in the same place, contributing to the same field. For a while, this was the thing to do. But then, those leading practitioners got

[1] I'm using a familiar example, but a new study suggests that perhaps Anaximander and Anaximenes were not so proximate in time to Thales after all (in which case Thales is more like Pythagoras, a distant inspiration and no more to later "Thalenas" . . .); see P. Thibodeau, "The Chronology of the Early Greek Natural Philosophers," preprint, 2019, www.researchgate.net/publication/333561564_The_Chronology_of_the_Early_Greek_Natural_Philosophers_-_Part_1_of_5.

old, died. And so, quite naturally, things changed. You see, Herophilus and Erasistratus did not set up a research center with funds dedicated to anatomical research. And so: new people would make their names in different, new ways.

There are, of course, exceptions to this rule. Tragedy and comedy persisted, especially in the city of Athens, original and vibrant for at least two centuries. Later on, the same city of Athens had four continuous philosophical traditions – Stoic, Aristotelian, Academic, and Epicurean – once again stable for about two centuries. In both cases, this continuity was based on a social institution. The Athenian state supported dramatic competitions beginning at about the years 500 (for tragedy) and 450 (for comedy). Eventually, starting at around the year 300, there were four schools of philosophy, actual institutions with their own properties and rules (each school had a fixed place; each school had a leader – scholarch – serving for life, a new one elected when the old one died). This kind of institutional stability is the exception in ancient culture and was never approximated in mathematics.

And so, there is nothing too surprising about the hypothesis proposed here. At around the year 400, a whole generation of authors started to produce mathematical works. They were all impressed by a few original contributions; they were probably aware of each other and aware of the significance of the moment as a new development. It must have been exciting. And then, the first significant authors – and even their immediate followers – got old; something of the original excitement was gone. By midcentury, new work continued to be produced, but the pace could have slowed down, and the sense of a meaningful moment was lost. For a while, at least, fewer new authors would be attracted to this field; other pursuits became more fashionable instead. This, then, is the first generation of Greek mathematics.

## What Little We Know

What have we got in support of this hypothesis? This, too, will be a question we will keep returning to throughout the chapter. The problem of evidence is, in this case, especially acute. Not a lot remains, and what we have to go by is mostly names. Of those, however, we have quite a few.

In the following table, I list the attested individuals who I think were born between 430 and 380 and are likely to have authored works in the exact sciences. For each individual, I provide an extremely speculative range of dates for activity, based on the assumption that, unless we are told that the individual was active at an especially young or old age, he would have been active at around the age of thirty to sixty. This operation is indeed very speculative, but its overall contours are relatively solid, and for one reason primarily: most individuals in this list can be related, in various ways, to the securely dated (429–347) Plato. "Born between 430 and 380" could also mean "having the living Plato as a presence in your life." Perhaps this is relevant.

An added word on the evidence: as noted on pages 35–36, in his commentary to the first book of Euclid's *Elements* (fifth century CE), the Neoplatonist philosopher Proclus wrote a very lengthy introduction that, among other things, contained a brief history of early geometry up to the time of Euclid. Scholars usually think that this is a résumé of the history of geometry written by Eudemus, the younger member of the school of Aristotle, whom we have met already for his testimony on Hippocrates of Chios. If so, this is excellent, almost contemporary evidence, even if mediated. In the accompanying table, I put the mark "Eud." next to those individuals who are attested in the list from Proclus. I underline the notation – "Eud." – in the cases where our *only* information for this individual derives from that list. I add two more marks. "Pl." means that we have reason to believe that the individual was in some contact with Plato, and this reason does not depend on Proclus's authority. "Pl.-Pr." means that we believe this individual was in some contact with Plato, based on Proclus's (biased) authority alone.

| **Author** | **Decades** | **Eud.** | **Pl./Pl.-Pr.** |
|---|---|---|---|
| Archytas | 400–360 | Eud. | Pl. |
| Leodamas of Thasos | 400–370 | Eud. | Pl.-Pr. |
| Neoclides | 390–360 | Eud. | |
| Theaetetus | 380–370 | Eud. | Pl. |
| Amphinomus | 380–350 | | Pl. |
| Bryson | 380–350 | | |
| Silanion | 380–350 | | Pl. |
| Zenodotus | 380–350 | | |
| Leon | 370–340 | Eud. | |
| Eudoxus | 360–340 | Eud. | Pl. |
| Amyntas of Heraclea | 360–330 | Eud. | Pl.-Pr. |
| Euphranor | 360–330 | | |
| Helicon of Cyzicus | 360–330 | | Pl. |
| Hermodorus of Syracuse | 360–330 | | Pl. |
| Polemarchus | 360–330 | | |
| Philip of Opus | 360–320 | Eud. | Pl. |
| Xenophilus | 360–320 | | |

| Athenaeus of Cyzicus | 350–320 | Eud. | Pl.-Pr. |
|---|---|---|---|
| Callippus | 350–320 | | |
| Dinostratus | 350–320 | Eud. | Pl.-Pr. |
| Hermotimus of Colophon | 350–320 | Eud. | Pl.-Pr. |
| Menaechmus | 350–320 | Eud. | Pl.-Pr. |
| Theudius | 350–320 | Eud. | Pl.-Pr. |

The table lists twenty-three names. In general, even in extremely well-attested fields, such as Greek tragedy, no more than a third to a half of the original authors are now still attested.[2] Now, later Greek writers did have a significant interest in the history of mathematics (Eudemus wrote a history! Proclus cited it!) But we are looking at a very early moment in history: much has been lost from memory before our sources got down to transcribing what they knew. In fact, not a single mathematical work survives from this era: everything still depends on small fragments. It would be surprising if we were to know even one-fifth of all the mathematicians active in this time, and so there were probably more than a hundred.

All that remains of the mathematics of the era is fragments – and we will look more closely at some of those. But a larger amount of mathematics survives, so to speak, *regurgitated.* We will get to know Euclid in the following chapter, but we should note here that approximately eighteen books are now extant under his name. (By the word *book*, I mean here the ancient sense, which could be either a single work or a *segment* of a work, roughly the equivalent of a single papyrus roll. Euclid's book *The Elements* has thirteen such "books"; there are, by one count, five other single-book works ascribed to Euclid.) There are also two extant books by Autolycus. Both Euclid and Autolycus were active, I believe, right at the end of the fourth century BCE: all of a sudden, our evidence begins to explode just then. But we can say more. Those twenty extant books by Euclid and Autolycus are all, evidently, not so much original works as compilations. They survey and arrange past knowledge, and this past knowledge was probably produced by the authors listed in the previous table.

There were, I argue, more than a hundred individuals writing mathematics, from Archytas to Theudius. But the great bulk ought to have been very minor (indeed, we know *literally nothing* about most of them!), likely the authors of no more than a tiny mathematical contribution. If so, it

[2] For a detailed argument for such claims, see R. Netz, *Scale, Space and Canon in Ancient Literary Culture* (Cambridge: Cambridge University Press, 2020), Chapter 5. (Tragedy is discussed on 552–555, 561–564.)

might be that each mathematician could have written, on average, say, not much more than the equivalent of two books – let us say, then, not much more than two hundred books produced throughout the entire period. Furthermore, there ought to have been considerable overlap in the contents of those books (many of them would introduce the very same material, treated in some variation). If so, as an order of magnitude, the extant works of Euclid and Autolycus should represent something like 10 percent of the original mathematical contributions made throughout the era. There is a lot that we do know, then, and a lot that we don't. As we survey the mathematical achievement of the era, we will look at the fragments of the original authors; we will consider the evidence of Euclid and Autolycus; and we should try to keep our minds open when considering the possibilities for the dark matter, now completely lost. History should account for that, too.

The first impression is that there were many mathematicians at the time – and that many of them had something to do with Plato. The pattern of the evidence is reassuring. Although a lot of our names depend on one lucky find – the survival of a testimony from Eudemus's early history – we would have had many of those names even without this piece of fortune. And although Proclus, the Neoplatonist philosopher, had all the motivation to ascribe bogus connections to Plato, it is reassuring that many of those connections are, in fact, independently verifiable. Surely, there were many other mathematicians, further away from Plato, of whom we know less. There is a Plato-centric, Athens-centric bias in our evidence for the period. Elsewhere, for instance, Proclus quotes a small testimony about Andron and his student Zenodotus, who, in some sense, are said to have belonged to the "school of Oenopides."[3] Now, Oenopides, together with his compatriot Hippocrates of Chios, was, as we recall, one of the earliest pioneers of mathematics. I find it extremely unlikely that there would have been another, otherwise unattested, later mathematician by the same name, and so the likelihood is that we look at two individuals, one perhaps from the late fifth century, the other from the early fourth, both from the same place. Chios – where mathematics perhaps was born! – could have remained, for several decades, as its own mathematical center, now attested by a single passage.

It seems much clearer that later in the fourth century, there was a center of astronomical practice in Cyzicus, further to the north of the Aegean world. This center was probably inspired (led?) by Eudoxus, and if so, it

[3] In Proclus's commentary on Euclid, 80.15–80.18.

was at least indirectly connected with Plato and with Athens; we revisit Cyzicus in Chapter 5.

In our extant sources, we see many in touch with Plato in Athens, but we also see Chios and Cyzicus, and there ought to have been rather smaller centers such as these (all of them, now, among the invisible, dark matter of history). But perhaps many of those were rather like Cyzicus? That is, even if not directly attached to Plato, would they still be indirectly tied to the main figures who do seem to have Athenian connections? If it was a network, surely Plato was not far from its center.

We can add a little more. The tiny piece of information we have attested for the mathematicians of Chios is a meta-mathematical, or philosophical, position concerning the nature of "theorems" and "problems." As for the school of Cyzicus, one of the best-attested facts about it was a criticism leveled against it, from a philosophical point of view, by Epicurus (who stayed, at the time, at Lampsacus, not far from the city of Cyzicus; we return to this in Chapter 5). On the other hand, the school had a lasting influence through its impact on Aristotle, who adopted the model of homocentric spheres and gave it a prominent role within his own teaching. And so: even away from Plato, mathematicians, in this era, seem to be in dialogue with philosophy. Indeed, three of the mathematicians of the era seem to stand out (and we shall look more closely at their contributions): Archytas, Theaetetus, and Eudoxus. Both Archytas and Eudoxus were known, in antiquity, not just for their mathematics but also for their philosophy. Although Theaetetus did not leave a mark as a philosopher (did he die too young?), he is now remembered, above all, as a character in Plato's dialogues, one of them named after him!

One of the important questions for us, in this chapter, is whether mathematics, at this point, is even distinct from philosophy. But what we see already is that those who engage in mathematics, in this era, are as likely to contribute to philosophy and certainly as likely to communicate with philosophers – so much so that the entire network of authors could have had, at its center, not a mathematician but rather the mathematically curious philosopher, Plato.

And is this, perhaps, yet again, a function of the numbers? I suggested that, compared with the fifth century, numbers could have exploded. But even so, we find – even including the dark matter! – perhaps somewhat more than a hundred mathematical authors emerging through half a century. Two per year? Three? Although this adds up to a substantial community, at each given moment, each individual mathematician would have known no more than a handful of colleagues – especially at first. It is likely that when Archytas started writing, there were not many other mathematical authors. And so, his imagined audience was composed of those of his coevals who, although not mathematicians themselves, were at least potentially curious about the kind of arguments a mathematician had on offer. The imagined audience, then, would be made of what we call "philosophers" (a very significant group in numerical terms, even in this early stage). Some of those philosophical authors would indeed be riveted by the new writings in the genre of mathematics. One of those – Plato – will reach prominence. As this nexus fastens, early on in the fourth century, the character of the generational event is formed. Even as there are more mathematical authors, they already emerge into a particular dialogue, set in place by authors and readers such as Archytas and Plato, and the mathematics produced in this generational event would often continue to be written with a philosophical audience in mind.

This is already a hypothesis, and an interpretation, for which the broad statistics are no longer sufficient. Time to bring in a more detailed sense of the mathematics of the fourth century. Let us follow the three main names mentioned previously, in order: Archytas, Theaetetus, Eudoxus.

## Archytas

Archytas was born in Tarentum, a city in southern Italy (a region that – along its coast – was, in this era, predominantly Greek). Like Athens, Tarentum was a democracy, although possibly of a somewhat more "aristocratic" cast. For sure, Archytas was a member of the Tarentine elite, in fact, something like the de facto leader of the city (what comes to mind is the role of Pericles in Athens). More than this: the second quarter of the fourth century was the heyday of Tarentum. It became the focal point of all the cities of southern Italy (what comes to mind, again, is the control Athens had, in the fifth century, over much of the Aegean Sea). Throughout his lifetime, then, Archytas was one of the most powerful individuals in all of the Greek world. Members of this topmost layer of the Greek elite always enjoyed relations of friendship – in the Greek context, a formal social relation – that cut across the borders of individual city-states.[4] It seems clear that Archytas had this kind of formal friendship relation, in particular, with Plato. Quite a figure, then! Remarkably, until recently, there was no monograph dedicated to Archytas, but this lacuna has been admirably filled by Huffman, whose book from 2005 is titled *Archytas of*

[4] This phenomenon of friendship is well outside the scope of this book but worth pursuing for anyone interested in ancient Greek culture. See G. Herman, *Ritualized Friendship and the Greek City* (Cambridge: Cambridge University Press, 1987).

*Tarentum: Pythagorean, Philosopher and Mathematical King*. And for once, the long academic subtitle is apt. Archytas cannot be summed up briefly, and wide ambition – political, cultural, intellectual – seems to be essential to his career.

It is clear that Archytas wrote multiple books (indeed, later grammarians saw him as the model of prose of his native dialect, Doric). None survives, and the testimonies are tiny. What we can now say with certainty is that in the fields of the exact sciences, Archytas made important contributions to music theory and geometry. A probable case can be made that he made similarly important contributions to number theory and optics. Because so little is extant, and even in that small evidence, we see traces of such a wide footprint, it is most likely that Archytas wrote across many mathematical fields. It is also clear that he was a superb mathematician. Only one testimony is at all substantial – but this is spectacular.

We need to go back. Recall Hippocrates of Chios: he had discovered that the problem of scaling a solid is reducible to the problem of finding four lines in continuous proportion. To restate from the earlier discussion (p. 46): if you wish to scale a solid in the ratio A:D, set out A and D as line segments, and then you need to find two more, B and C, such that

$$A : B :: B : C :: C : D.$$

When this is accomplished, we know the ratio in which the side of the solid needs to be adjusted: the ratio is A:B (in modern terms, A:B is the cubic root of A:D).

This is where we left off, several decades back, with Hippocrates of Chios. Let us now start to think about how such a problem might be tackled. Someone (Hippocrates himself?) certainly reduced, early on, the simpler problem of how to scale a *plane* area into a problem in proportion. Think of it as doubling the square. Suppose you are given a square whose side is A. What is the side B, such that the square on B is twice the square on A? Even before we solve this scaling problem, we can reduce it to a problem in proportion, as follows. To double the square whose side is A, find the line C such that it is double A. Then find B that satisfies the following:

$$A : B :: B : C.$$

And B is the side you require.

So, to scale a plane, you need to find *one* mean proportional. I imagine that Hippocrates of Chios first realized this and then went on to realize the extension of the same idea: that the scaling of *solids* can be reduced to the finding of *two* mean proportionals.

Further, it is perhaps meaningful that the very first occurrence of mathematics in a Platonic dialogue, in the *Meno*, has Socrates teaching Meno's slave how to double a square. In the dialogue, this is done in a very elementary – indeed, Babylonian – way that could not be generalized to scaling a plane figure by any factor. In fact, the elementary way of teaching is intentional – this is perhaps the paradigmatic example of "Socratic" teaching, Socrates gradually extracting the result out of the slave's own mouth. In all likelihood, this elementary, Babylonian manner – adding up the pieces of which the square is made – was indeed not far from the way in which Greek schoolteachers would tackle such a question.

This is a significant dialogue, the one that establishes a key idea for Plato's philosophy: knowledge of mathematics is at least a metaphor for true knowledge – that of the *forms*. It is quite likely that Plato, writing the *Meno* in about, say, 390 BCE, was at least somewhat influenced by Archytas's philosophy, and he was also likely familiar with Archytas's finding of two mean proportionals – thus, the lesson could be a subtle Platonic allusion to a great mathematical achievement.

Let us first look, then, at how the Greeks (once again, already Hippocrates?) would have gone about finding the *single* mean proportional, that is, how they solved Meno's problem.

This involves perhaps the key insight of Greek geometry, namely: there is a tight connection between the two separate fields of *quantity* and *geometrical form*. Specifically: "proportion," a quantitative relation, is almost the same as "similarity between figures," a statement of geometrical form. In practice, similar triangles are easiest to find, and as usual, it is best to deal, specifically, with right-angled triangles. So, let's have two similar right-angled triangles, such as the following, with some of the sides marked as A, B, C, D:

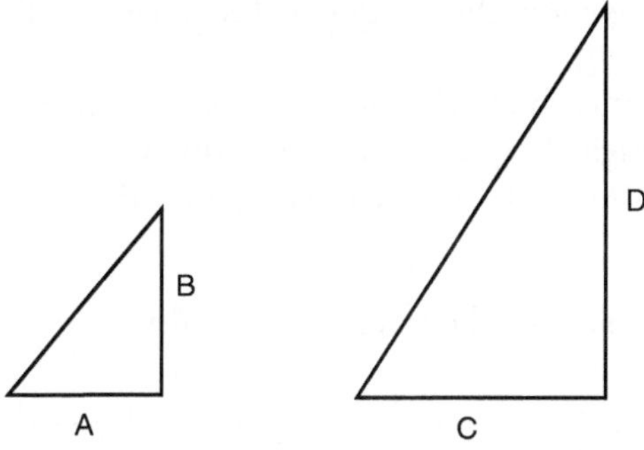


Figure 18

When we say that they are “similar,” we mean that they have exactly the same shape, simply scaled differently. If so, it is also true that

A : B :: C : D.

Shape, turned into quantity; quantity, turned into shape.

Suppose now we want to have

A : B :: B : C.

The way to do that would be to “glue” the two triangles: to have, say, the same side appear twice, as the greater side in the smaller triangle but also as the smaller side in the greater triangle:

We know something about this configuration! If we glue the two triangles like this, we get a new, third triangle. We also know that the sum of interior angles in a triangle is fixed. (As an aside: anyone interested – as Hippocrates of Chios was – in the relation between the angle and the arc of a circle it subtends would come to recognize this fixed sum.) It is thus easy to note the various angles, all equal to each other:

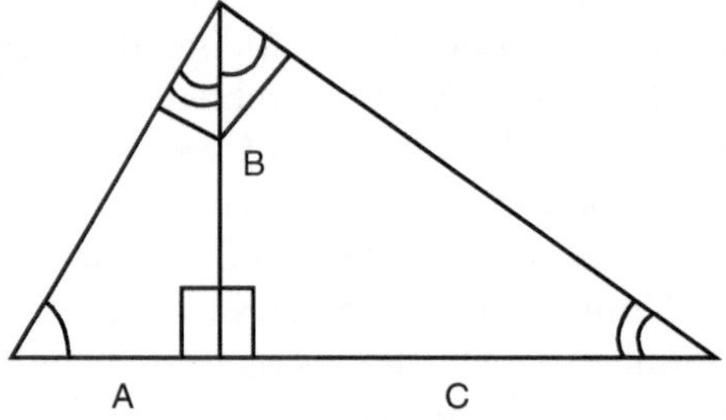


Figure 19

Note in particular that the top angle in the new, glued-up triangle must, once again, be a right angle.

And what do we know about right angles? That they lie on the diameter of the circle!

So now, remember our original task. We were given the line segment A (side of the original square) and the ratio A:C (the ratio in the area between the two squares). We are looking for a B that would satisfy

A : B :: B : C.

And now it is obvious how to achieve this through similar right-angled triangles.

Put A and C together, along a single line; draw a circle with that single line as diameter; raise a perpendicular from the point where A and C meet;

this will hit the circle to produce our B. The perpendicular within the semicircle is the mean proportional.

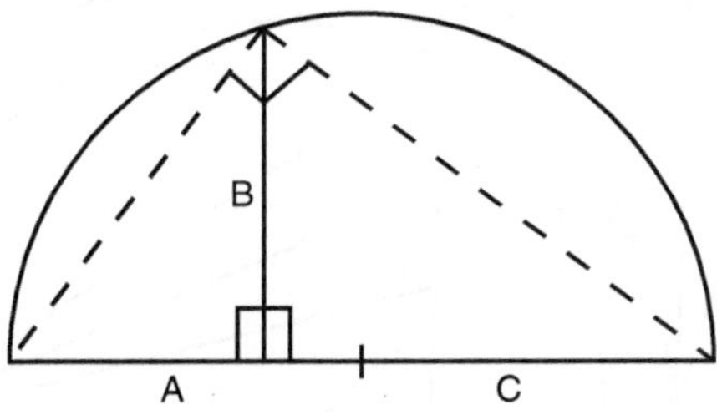


Figure 20

I imagine Hippocrates of Chios did all that, or, if not him, someone very close to him in time. It's one of those results that quickly falls prey to the hunters, who are moving into the new territory of elementary geometrical proofs. (As noted earlier, many of those results are now extant, reused as pieces of the mosaic that is the works of Euclid: in this case, this has become *Elements*, Book VI, proposition 13.)

The next one, however, had to wait for a really skillful master, Archytas.

As usual, our evidence is extremely mediated. In the sixth century CE, Eutocius of Ascalon set out to write commentaries on authors such as Archimedes and Apollonius. As part of those commentaries – on Archimedes's *Second Book on the Sphere and the Cylinder* – Eutocius included an extended excursus with many solutions to the problem of scaling a solid (more on Eutocius in Chapter 6). One of these solutions is said to be "by Archytas, as is told in Eudemus's history." It is provided in full mathematical detail. Good! However, this solution does not have some of the features that we would have expected from Archytas's own writing. He was the model of Doric prose, but it's not in Doric. He must have used some archaic phrases that we find in Hippocrates in Chios (and even in Aristotle), but the language we see is essentially the same as what we find from Euclid onward. Someone – whether Eudemus or an intermediary source used by Eutocius – must have modernized Archytas's proof. And yet, all scholars agree that, in outline at least, this is indeed by Archytas. Why? This will be clearer once we've described the idea for the solution. This is, in fact, extremely complicated.[5]

[5] Archytas's solution may be read in the catalogue of solutions provided by Eutocius, which is part of my translation of Eutocius's commentary. R. Netz, *The Works of Archimedes: Translation and Commentary*, vol. 1 (Cambridge: Cambridge University Press, 2004), pp. 290–293.

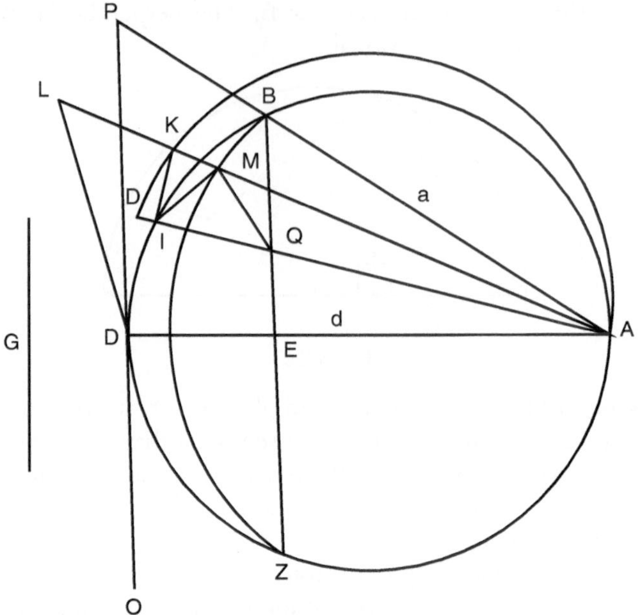


Figure 21

Even our diagram becomes much more congested than anything we have seen so far. Indeed, I will change letters referring to the proportion to lowercase, so as to accommodate Archytas's figure! So: remember that we are assigned a ratio between two lines, such as a:d, and that we seek b, c that satisfy

a : b :: b : c :: c : d.

In this diagram, d – "the greater given line segment" – is the diameter of the circle, AD.

In this diagram, a – "the smaller given line segment" – is an arbitrary smaller line fitted from point A to meet the circumference of the circle, as the line segment AB.

Drop a perpendicular down from B, and you get the points E, Z (I will explain where Q comes from in the following discussion).

Now I ask you to do some mental gymnastics. The line AB is extended to point P so that we have a right-angled triangle APD.

And now we gain liftoff: the figure begins to project into space. This triangle APD is rotated around the axis AD to produce a cone. (You can't really see that cone in Figure 21.) Note, in particular, that through this rotation, point B draws a semicircle in space: BMZ.

Further, we imagine a cylinder set up orthogonally on the circle ABDZ. (You can't see that cylinder either.) Note that point K is understood to be on the surface of that cylinder.

Finally, a copy of the semicircle ABD is lifted up so that it becomes orthogonal to the original circle. It is allowed to rotate in such a way that point A remains fixed, but point D is allowed to move, as long as it remains on the original plane of the original circle – the plane of the paper, so to speak. (It is because of this rotation that the capital letter *D* appears twice in the diagram.) In this diagram, this appears as the semicircle on which are the points A, Q, I, D, K.

And so, we have three objects in space. One is the cylinder set up on the circle ADBZ. Another is the cone made from the rotating triangle APD. And finally, we have the rotating semicircle, such as AKD.

What is point L? It is crucial to the argument. You remember that the cone was produced by rotating the triangle APD around the axis AD. We essentially rotate the side of the triangle AP, keeping point A fixed, keeping the side at the same fixed distance from the axis AD. Well, at some point, this rotating line AP becomes the line AL.

How do we pick this position in the rotation of the line AP so that it becomes AL?

Indeed, how do we pick the position in the rotation of the upstanding semicircle?

We pick both simultaneously:

You will notice that point K is shared by the rotating semicircle AKD and by the rotating side AP. In fact, point K is also on the surface of the cylinder that stands above the circle. This is how we decide how to pick the point where to stop the rotations of the semicircle and of the cone. There is precisely one combination where the two meet so that they share a point *on the surface of the cylinder*, and this is the position chosen for this diagram. To restate this: point K is the servant of *three* masters:

It is on the surface of the cylinder on the circle ADBZ.
It is on the arc of the rotating semicircle AD.
It is on the surface of the cone produced by the rotation of APD.

(The argument for why such a point exists, and exists uniquely on a given side of the diameter, is subtle but also quite simple, and in the proof as it is now handed, it is just taken for granted; I leave it as an exercise.)

Once we have point K, it is not difficult to understand what point I means: it is simply the point on the original circle, immediately beneath point K. This, in turn, clarifies what Q is: it is simply the intersection of the two lines IA, BE. And now it is also clear what M means: it is simply the point in the triangle ALD (which is APD rotated – so, a point on the surface of the cone), immediately above the point Q.

Everything is in place, and it is time to assert the conclusion:

AB : AI :: AI : AK :: AK : AD.

The two line segments AI, AK are the two mean proportionals between AB, AD. In the original terms of our task, they are the b and c required between the two given lines a and d; AI, specifically, solves the problem (and can be described as the "cubic root" of AD).

The formal proof is difficult, but we can see, as follows, why this is true. The key argument is that AMQ, AMI, AKI, and AKD are all right-angled triangles.

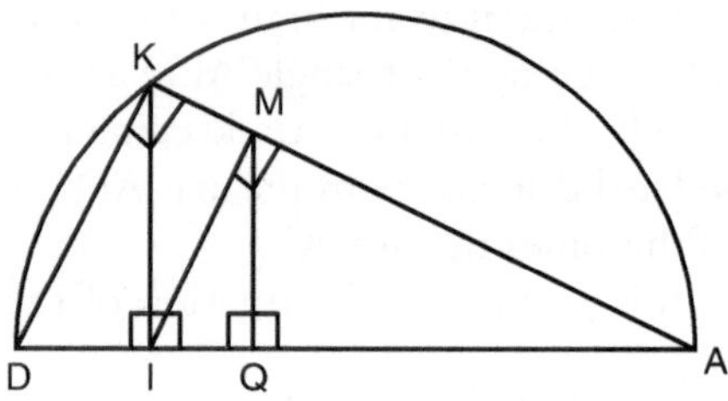


Figure 22

AQM and AIK are right-angled for a rather obvious reason – MQ, KI were raised up as perpendiculars. AKD is right-angled because it lies on a diameter.

The proof is difficult because of one thing and one thing only: we need to show that AMI, too, as constructed, is a right-angled triangle. I linger on it because, although complicated, it is still reminiscent of the mathematics of Hippocrates of Chios. This is historically telling. And so, consider just this arrangement from the basic circle:

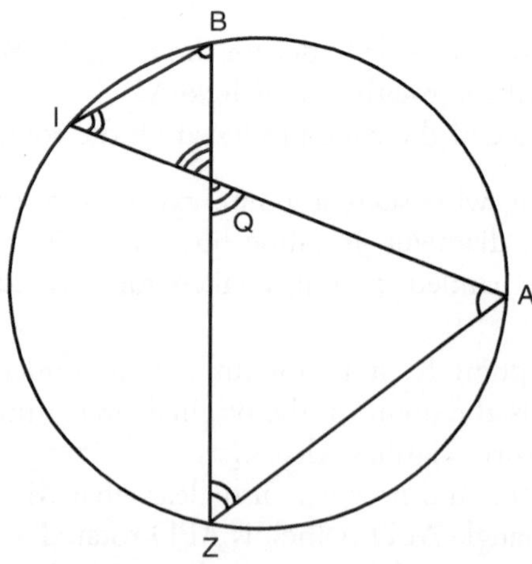


Figure 23

Simple considerations from angles show that the two triangles AQZ, BQI are similar. This will be true for any triangles produced by chords cutting each other in a circle. Hence, we can derive a proportion:

> AQ : BQ :: QZ : QI, which can translate into an equality between two rectangles:

$$AQ \times QI = BQ \times QZ.$$

(This is, once again, true in a general sense: the rectangles produced by two chords in a circle cutting each other are equal to each other. This result is proved – in a more roundabout way, the reasons for which will be made apparent in the following chapter – in Euclid's *Elements* III.35. You recall that Hippocrates of Chios is said to have contributed to the *Elements* – was it with results such as these?)

Now, the basic idea of the solution to the problem of finding a single mean proportional is that a right-angled triangle, encased in a semicircle, gives rise to a rectangle, equal to a square. For instance, in this case, BMZ is a semicircle produced by the rotation of the cone, and it encases a right-angled triangle:

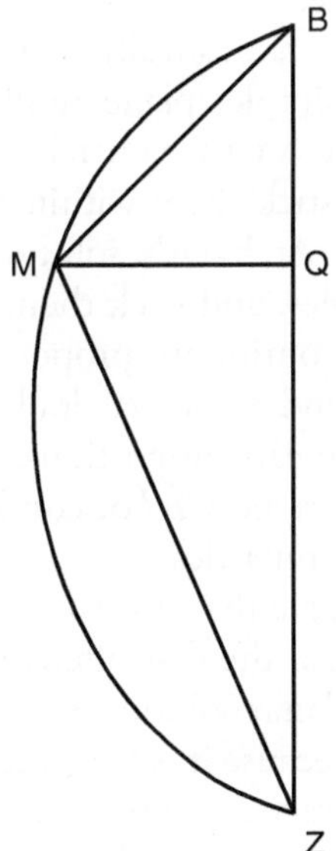


Figure 24

Specifically, we know that $BQ \times QZ = MQ^2$. However, we have just learned that $BQ \times QZ = AQ \times QI$. So that:

$$AQ \times QI = MQ^2.$$

And now, from the converse of our solution for finding a single mean proportional, we know that, because of the previous equality, in a triangle marked by the labels AQIM, the angle AMI must be right-angled:

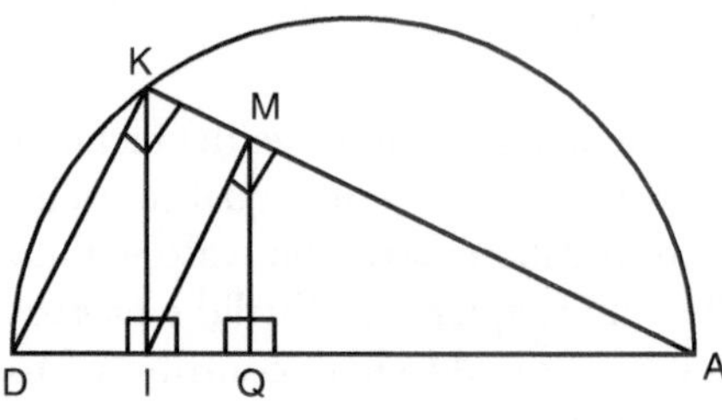


Figure 25

Now, indeed, we have established four triangles in sequence, all right-angled, all similar: AQM, AMI, AIK, AKD. This was the entire point of the exercise, construction, and proof taken together: fit those right-angled triangles, somehow, in space.

Archytas's solution is a fantastically complicated three-dimensional exercise in extending the simple, plane result, which is now known as *Elements* VI.13. In Elements VI.13, you take the two sides, produce two right-angled triangles, and stick them within a third, to get three lines in continuous proportion. In Archytas's solution, you take the two sides, produce right-angled triangles, and stick them together with a third and a fourth, to get four lines in continuous proportion. It is a reasonable story: Archytas's solution was found as he was looking to extend a solution he already knew for a single mean proportional – finding how one can fit triangles in space to get the same *kind* of configuration. It is, in fact, hard to do – hence the spatial gymnastics.

This is one reason to believe that this solution, handed down under the name of Archytas, was indeed the first offered: it is the most *direct* way of extending the idea of the planar solution into space.

I lingered on this proof because it is fantastic and because it is important to see how easy it is to envisage it, coming early on: it is the most difficult combination of the most basic ideas. We can see this because so much survives. We should probably assume the same for the remainder of Archytas's oeuvre – where almost nothing is extant.

So, is it possible that something similar takes place in Archytas's music? Once again, our evidence is highly mediated. Boethius, a Latin author of the fifth century CE, wrote an introduction to music, and in it, he cited a single proposition, stating that it was found by Archytas. Now, centuries after his life, many spurious works circulated in antiquity under the name of Archytas – a part of the overall invention of "Pythagoreanism" (more on this in Chapter 6). In a lot of this, music does play a role: Pythagoreanism, for reasons that we will explore in the following discussion, was closely related to the scientific study of music. However, such spurious works tend to be more metaphysical or, indeed, almost mystical, and the very dry, technical character of this proof inspires confidence. This, then, is another significant piece of mathematics we may quite confidently ascribe to Archytas:[6] "A superparticular proportion cannot be divided into equal parts by the interpolation of a mean proportional number."

This is strange, so let us go through this slowly. A "superparticular ratio" is, we would say, a ratio whose lowest form in integers is $(n+1)/n$, that is, the ratio of two consecutive integers, for instance, 2:1, 3:2, 4:3. The claim is that there is no integer $q$ such that

$$n+1 : q :: q : n.$$

This might seem obvious: after all, there is no integer between 2 and 1, so obviously, there is no integer $q$ such that 2:$q$::$q$:1. But remember that 2:1 is simply the expression, in smallest numbers, of a ratio that can be expressed in bigger multiples as well. Is it no longer obvious, then, that no integer $q$ can be fitted between, say, 120 and 60:

$$120 : q :: q : 60.$$

This, in fact, was my first intuition: the fact there is no integer $q$ such that 2:$q$::$q$:1 does not preclude the possibility that there would be some integer $q$ such that 120:$q$::$q$:60.

This intuition is false, and a fairly sophisticated argument, now extant as Euclid's *Elements* VIII.8, proves that if there are mean proportional integers between any given two integers, there would also have to be similarly located mean proportional integers between the smallest version of the same ratio. In other words, if we have 120:$q$::$q$:60, it must follow that there is some other integer $p$ satisfying 2:$p$::$p$:1, which proves that there is no

6 Boethius, *De Inst. Mus.* III.11.

such $q$ integer that can be positioned between 120 and 60 and, in general, there can be no mean proportional between integer numbers that are in a superparticular ratio. Archytas almost certainly proved such a result, though, to be fair, in this case, I am less sure that we have his original proof. I will not say much more about this line of development here, but it should be clear that, anticipating such results directly related to music theory, authors such as Archytas must have developed a certain amount of pure number theory: this is indeed now known mostly through Euclid's *Elements*, Books VII–IX (there is very little Greek number theory otherwise, perhaps because the ancient Greeks hardly ever pursued it, in fact, other than in the context of music).

Of all of this, we have no more than the slightest indications. Still, the very fact that Archytas provided a proof of this kind is meaningful. Surely, he belongs, then, to a group of authors who founded *music theory*. This indeed goes to the heart of "Pythagoreanism," to Archytas's influence on Plato – and to the nature of the mathematics emerging in this generation.

I mentioned a key insight of Greek geometry: quantitative ratios and the similarity of figures are related. Ratio – and shape. Somehow, Greeks stumbled across another, much more surprising relation: ratio – and *sound*. This involves, specifically, not the ratios between line segments but the ratios between numbers.

Certain musical pitches combine well; others do not. If we play in quick sequence notes that are an octave, a fifth, or a fourth apart (that is, a middle C followed by either a high C, a middle G, or a middle F), the result is pleasing to the ear. If we play in quick sequence a middle C, followed by a middle sharp G, this is less pleasant. The octave, the fifth, and the fourth are harmonies, in a way in which other sound combinations are not.

Apparently, someone in the years in which Archytas grew up came up with the following identification of such harmonies: the octave is like the ratio 2:1; the fifth is like the ratio 3:2; the fourth is like the ratio 4:3 (this is useful, then, in explaining why, as a fifth and a fourth are composed, the result is an octave.) Pleasing sound relations, many Greeks came to believe, are defined by ratios of the form $n + 1{:}n$, with small values for $n$.

It is hard to know what was meant by this identification to begin with, although later on, the ancient Greeks would prove it experimentally by taking a string and stopping it at a bridge. (A bridge placed at the bisection of the string yields a note that is higher by an octave; placed two-thirds of the way, and the note goes up by a fifth; three-quarters, and it goes up by a fourth. We now explain all of this as a function of wavelengths: a pitch higher by an octave is, in fact, of half the wavelength.[7] Of course, the ancient Greeks never came up with this explanation, so the most fundamental observation, for them, was arithmetical rather than physical.) This is more evident to us on an instrument such as a violin or a guitar, on which one plays different pitches by stopping the string at different positions, but the most common type of string instrument in antiquity was the lyre – more like a harp. The *tautness* of each string was adjusted separately to produce its own fixed sound. The discovery of the numerical basis of musical harmony is thus something of a mystery.

We should bring in another name, that of Philolaus: we have not discussed him yet because, in all likelihood, he was not a mathematician. But Philolaus might be considered one of the most influential philosophers in antiquity. I explain: to the extent that we cease believing that there was a specific branch of metaphysics, promulgated by Pythagoras in the sixth century BCE – which is what most scholars today suggest we should, and which is what I have argued in the preceding chapter – then we have to find how this metaphysics became promulgated by someone *else*. In all likelihood, then, Philolaus was the creator of what we understand as Pythagoreanism.

Philolaus was perhaps roughly the coeval of Socrates (and so, a generation older than Plato and Archytas). He was still more of a sage than a writer; he seems to have written only one book that pertained to many sides of his wisdom. Part of the very point of such books was to make outrageous claims – Parmenides, for instance, argued that reality is single and unchanging. Philolaus claimed that everything is made by *being limited*, and that this is, specifically, provided by *number*. Perhaps what he had in mind can be stated as follows. We might think we see some concrete, material things in front of us, but to the extent that we see anything intelligible, it is because it has some definition and structure (*being limited*), and this is, ultimately, some kind of arithmetical relation. For all of this, the strongest example may well have been that of music: what makes intelligible sounds – intelligible song! – is the numerical structure organizing musical harmonies.

7 The physics of musical sound is truly fascinating. Speaking as a nonspecialist, I personally enjoyed A. H. Benade, *Horns, Strings and Harmony* (New York: Dover, 1960).

Whether or not Philolaus was the discoverer of the ratios underlying musical harmony, he is the earliest author known to refer to them. He described more of the structure of the octave in numerical terms, correctly noting that because the fifth is 3:2, whereas the fourth is 4:3, their gap (what we call a "tone") is 9:8. This is as far as any mathematics can be squeezed out of the extant fragments (though, admittedly, not much is extant), and so the likely account is that his treatise evoked and extolled the power of numbers to structure reality, taking music and harmonics as the key example, but did not contain anything we need to call "mathematics." If so, it is quite possible that Archytas's innovation was to turn harmonics into its own science.

Suppose you wish to know how the octave can be divided into integer ratios; it is very important to know, then, that there are *no* integer ratios that divide the octave evenly. No $q$ such that $2:q::q:1$! Nor, for that matter, are there any integers that divide the fifth, the fourth, or the tone (no $3:q:q:2$, $4:q:q:3$, $9:q:q:8$, either). It may well be even that Philolaus was unaware of this and so implied a naive structure of the octave: was Archytas correcting an oversight by Philolaus?

Indeed, the task of the Greek harmonicist was especially difficult. We are accustomed to a nearly equal-spaced octave. The Greeks, like many other civilizations, enjoyed more complex harmonic structures. Indeed, they typically tuned their instruments (and trained their voices) to hit the key harmonies of the fourth, the fifth, and the octave (to repeat: this is like middle C, middle F, middle G, higher C), but they were simply not interested in dividing the remaining gaps as evenly as possible. There would be two notes inside the fourth – but these were not expected to divide evenly the ratio 4:3. The Greeks often had their "D" about a quartertone above C (flatter than a C-sharp!), their "E" about a quartertone above their "D" (so, about a C-sharp!) – leaving a gap of about two tones, unoccupied by any musical pitch, between their "E" (a little above our C-sharp) and their (and our) F. More than this: they had a variety of such tunings – completely different ways of occupying positions in the space of possible pitches, chosen by the musician prior to a performance. A good musical theory was supposed to account not for a single tuning but for the totality of all possible systems.

Once 2:1, 3:2, and 4:3 have been established, the question remains how to account for the remaining fine-grained pitch relations. The task, then, is to find several series of three integer ratios that, multiplied together, produce the ratio of the fourth or 4:3. Two of these should be very near unity; the last one should be closer to 4:3. These four notes between (what we call) C and F are four "chords" (the key model is always that of a stringed instrument), hence the name of the entire exercise: "the division of the tetrachord." It is clear now that this task is not trivial. An extended passage by Porphyry – an ancient philosopher from the third century CE (more on him in Chapter 6) – quotes the Archytean system in detail. Porphyry quotes Archytas's physical theories of sound and also the numbers (but no proofs) for the division of the tetrachord. Archytas, it turns out, provided three possible tunings for the pitches within the fourth:[8]

$$28 : 27, 36 : 35, 5 : 4$$

$$28 : 27, 243 : 224, 32 : 27$$

$$28 : 27, 8 : 7, 9 : 8$$

We should combine the report from Boethius together with the one from Porphyry. They are both mediated, but they support each other to provide a coherent picture. Archytas must have produced a complicated mathematical edifice, with both abstract arguments concerning numerical ratios as such and concrete numerical calculations, all in the service of a physical or even a cosmological theory of music. And it is indeed intriguing that this all ends up with the division of the tetrachord – quite literally, the division of one bigger ratio (4:3) by the insertion of two numerical values in between. The division of the tetrachord is not at all the same as the finding of two mean proportionals between two given lines. The motivation and the mathematical tools are entirely different. One gives rise to sophisticated geometry; the other gives rise to sophisticated number theory. But the two do resonate, quite directly, in fact – they are the discrete and continuous solutions to the same problem of finding four terms in proportion. Both also combine sophistication with conceptual simplicity: this is what one can accomplish, starting with very little – other than enormous ingenuity.

We have seen geometry in some detail, and we have good indications for harmonics, even for some number theory. Our evidence, from here on, runs nearly dry. We do have a few brief reports that ascribe to Archytas certain views on optics. There are also two Euclidean books on optical themes, one on the general field of perspective and the other on the formation of images in mirrors. Most likely, then, some fourth-century authors must have written on such questions, and so the hints we have in the literature could imply that Archytas made mathematical contributions to

[8] For a full survey of Archytas's system, see A. Barker, *Greek Musical Writings, Vol. II: Harmonic and Acoustic Theory* (Cambridge: Cambridge University Press, 1990), pp. 46–52.

this field, as well. If so, Archytas would likely *originate* the field.[9] The essence of the treatment of optics, in the Euclidean works as in subsequent Greek mathematical works, is the reduction of the phenomena of vision to the properties of rays of light (or rays of vision), conceived strictly as geometrical, straight lines. If so, the hypothesis that this field is due to Archytas is an attractive one, obviously on par with that for music. Past philosophers had considered the relation between sounds and numbers; Archytas turned this into a mathematical science. Similarly, past philosophers had considered the relation between vision and rays passing into the eye or emanating from the eye; Archytas's originality consisted, once again, in turning this into a fully fledged mathematical science, the rays becoming geometrical lines.

We have a few fragments and testimonies from Archytas of a more purely philosophical character. But even in discussing questions of, let us say, epistemology, mathematics seems to be front and center. Porphyry, once again, quotes from Archytas:[10]

> Those concerned with the mathēmata seem to me take distinctions well, and it is not at all surprising that they have correct understanding about individual things as they are. For having made good distinctions concerning the nature of wholes they were likely also to see well how things are in their parts. Indeed concerning the speed of the stars and their risings and settings as well as concerning geometry and numbers and not least concerning music, they handed down to us a clear set of distinctions.

Or another quotation, from Stobaeus:

> Logistic [~calculation] seems to be far superior indeed to the other arts in regard to wisdom and in particular to deal with what it wishes in a more manifest way than geometry.

With both quotations, we see Archytas engaged with the ranking of the sciences; mathematics, especially the mathematics of number, turns out to be the top science. We are reminded of Philolaus, again, and of Plato. Philolaus started from Pythagoras: this archaic sage had promulgated a way of life, leading, through purity, toward the otherworldly. Philolaus took this way of life and made it into a metaphysical claim – that underlying apparent reality there is that other, purer version, made of the abstract structures of numbers: a metaphysical claim supported by the supreme

9 The argument was first developed in M. F. Burnyeat, "Archytas and Optics," *Science in Context* 18 (2005): 35–53. This article is among the Suggestions for Further Reading in this chapter; it is a rare example of a mostly speculative argument that appears to be almost certainly true.

10 The two following quotations are from what is known as Archytas's Fragments 1, 4. See C. Huffman, *Archytas of Tarentum: Pythagorean, Philosopher and Mathematician King* (Cambridge: Cambridge University Press, 2005), pp. 103–161, 225–252.

example of music. Archytas, in turn, took this claim to its natural conclusion: to properly have knowledge, then, one ought to become a mathematician and study the mathematical structures underlying reality. Impressed, Plato took over from Philolaus the idea of an otherworldly reality, and he took over from Archytas the idea that mathematicians, indeed, have an insight into that reality. Plato, however, reserved the ultimate insight into the highest of all realities – the forms – to philosophy alone. Archytas was a mathematician, Plato a philosopher: each ranked his own discipline higher. The emerging division between the disciplines becomes clear – as does their proximity in this vision, shared by the philosopher and the mathematician. Many mathematicians, as well as philosophers, took notice.

## Theaetetus

*Theaetetus* is one of Plato's greatest dialogues. It is framed as a eulogy following Theaetetus's recent death in battle. Two of his old friends, Terpsion and Euclid (not the same as the mathematician), recall a conversation Socrates, when very old, had with Theaetetus, when just a child. It is not really clear in which battle Theaetetus died, but the date usually taken by scholars – 369 – makes sense because this dialogue otherwise appears, for various stylistic and philosophical reasons, to belong to Plato's later work. Because Athenians could easily soldier on well into their forties – and Theaetetus, for once, did not merely have Athenian contacts but was a citizen – and because the protagonist of the *Theaetetus* is clearly a bright kid, we seem to get a firm grasp of Theaetetus's chronology. Born in around 415–410, dead in 369? But perhaps such historiographical weight should not be put on this work of fiction.

Past historians of mathematics did not have such scruples, and the dialogue has been squeezed for the last drop of historical implications. In particular, to start off a conversation, Socrates asks Theaetetus what he did today in school. Theaetetus is very glad to answer – he is, in fact, glowing with satisfaction. Theodorus, his teacher, presented the kids with proofs that, effectively (I will immediately qualify this), the square roots of many numbers from 3 to 17 are irrational. Theaetetus then worked out for himself that, in fact, this is true for any nonsquare integer.

There is now a considerable literature trying to reconstruct Theodorus's mathematical achievement (somehow capable of showing that the square roots of 3, 5, . . . 17 are irrational, but not further) and the manner in which Theaetetus generalized it. This literature culminates with Knorr's first book, from 1975, *The Evolution of the Euclidean Elements*, which is a classic – well worth reading even today. But notice that this assumes that Plato's point, in this passage, was to record the recent history of mathematics. That's evidently not what Plato is doing here. He is, generally speaking, the author of philosophical narratives, and the exchange should be understood in terms of its narrative and philosophical functions. In narrative terms, it should create in our mind the image of a bright, enthusiastic kid who is yet charmingly modest. (For instance, Theaetetus insists on sharing the credit for his discovery with a friend of his, although the clear impression is that he himself put in all the work.) In philosophical terms, the topic emphasizes the need for knowledge, generalization, and classification. (Theaetetus's breakthrough was to identify various *classes* of numbers.) The dialogue then picks up such topics directly and seeks to generalize a statement – by way of various classifications – concerning the nature of knowledge.

It would make sense, in narrative terms, for all of this to occur against the background of Theaetetus's real, historical mathematical contribution. Once again, we can now bring in the evidence of a late ancient scholar. Pappus, active in the fourth century CE, wrote a commentary to the Book X of Euclid's *Elements*, which is a study of irrationals. (Pappus's commentary is now extant only in Arabic – the first example of this kind of transmission we encounter in this book; it is, in fact, quite common.[11]) Pappus ascribes to Theaetetus – once again, on Eudemus's authority! – a very specific, otherwise-unattested classification of irrationals. Had it not been for the authority of Eudemus, this would have been highly suspect. (Perhaps later authors have foisted upon Theaetetus a certain classification of irrationals, just because this somehow echoed the famous dialogue?) But Eudemus was too early; the classification described by Pappus is too specific. It is still possible that Pappus failed to recognize a complete fabrication, but more likely than not, he reports a historical truth. This, then, is the historical kernel we can more or less extract from the combination of Plato's and Pappus's evidence: sometime between, say, 390 and 370, Theaetetus of Athens classified irrationals. Nothing less; nothing more.

But that's not nothing! Theaetetus classified irrationals – this is an extremely valuable kernel of historical knowledge!

What does this even mean? Now, when we use the word *irrationals*, these are, for us, a certain class of numbers. This is because the word *number* has expanded its meaning considerably since ancient times. For the Greeks, "number" meant what we think of as a positive integer (usually – not always – the Greeks excluded the number 1 itself, which was not so much a number as the "unit" by which numbers are measured).

[11] For the English translation, we still use: W. Thomson, *The Commentary of Pappus on Book X of Euclid's Elements* (Cambridge, MA: Harvard University Press, 1930).

So, the Greeks surely did not have "irrational numbers." What is an irrational, to them, then? This is a category not of number but of *ratios between line segments*.

Suppose I have two line segments, whose ratio is the same as 4:3. This means that there is some smaller line segment that fits exactly three times into the smaller ("measures it three times") and exactly four times into the greater ("measures it four times"). Because the two lines standing in the ratio 4:3 possess in common such a line – that measures both of them – we call them *commensurable* (co-measure-able). It is also immediately clear that two lines are commensurable if and only if they stand in the ratio of two integers.

If we take a certain fixed line as our basic measure, then we can define lines that are incommensurable with it as "irrational."

Perhaps one's first intuition is that if you look hard enough and are willing to go for really small common measures, you will find one that measures any two lines – perhaps the immediate intuition is that all lines should be commensurable with each other. This is, in fact, wrong, and the Greeks probably knew this early on, based on considerations very close to those we have just pursued, concerning Archytas. Remarkably, this will bring together the two pieces of mathematics we have surveyed for him: the geometry of the duplication of the cube and the harmonics of finding single means between integers.

To begin with, recall that Plato's *Meno* asked how to double the square (perhaps in an oblique reference to Archytas's doubling of the cube?). We have discussed the scaling of a square by an arbitrary ratio on pages 62–63, which the ancient Greeks solved by similar triangles. But Plato considers the simple, school-like, basically Babylonian case of, specifically, *doubling* a square. This is very easy to achieve by a particular configuration (which is the content of the lesson packed into the *Meno*):

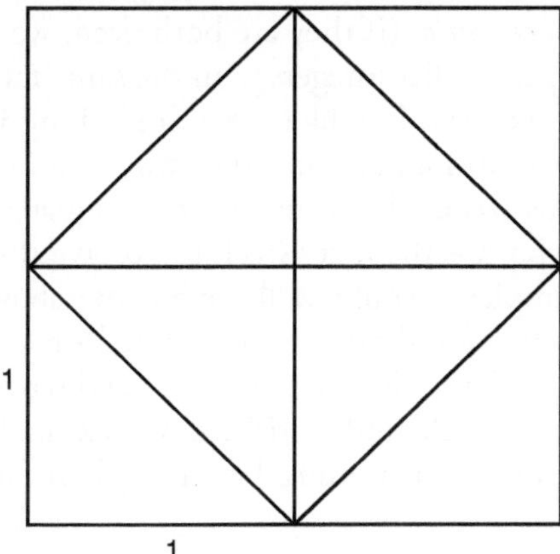


Figure 26

The sides of the original square are marked by the length 1. The area of the original square is 1; the square on the double of the original side has the area 4; and the square on the diagonal, quite obviously, has the area 2, or twice the original square.

If so – already going beyond the scope of Plato's *Meno* – the diagonal of the square is what we would call $\sqrt{2}$ of the side of the square. It appears that Greeks were aware, perhaps very early on, that with the side of the square taken as the basis of measurement, the diagonal is irrational.

How do we know that? This now becomes a little like Archytas's music theory.

Assume the opposite – that the side and the diagonal are commensurable. This is tantamount to the claim that they are in the ratio of two integer numbers, say, $m{:}n$. Let us also assume that $m{:}n$ are the smallest such integers standing in this ratio – the key concern for Archytas's arithmetic! So, $m{:}n$ are not like 6:8, but like 3:4. They are $m{:}n$ – the smallest such numbers.

We also know, by Pythagoras's theorem (always the very same basic tool!), that $2m^2 = n^2$. Now, because $n^2$ is the double of something (it is the double of $m^2$), it is an even number. There is no way for a square number to be even without its root being even as well (odd times odd is odd, not even). So, $n$ itself is an even number. Indeed, because $n$ is an even number, it follows that $n^2$ is divisible by 4. (An even number is twice something; its square is therefore four times the something squared.) But we have $2m^2 = n^2$ or $m^2 = n^2/2$. Now, because $n^2$ is divisible by 4, it follows that $n^2/2$ is divisible by 2 or, in other words, is even.

$$m^2 = n^2/2, \text{and } n^2/2 \text{ is even.}$$

We have learned, then, that $m^2$ is even. By the logic applied before, we've also learned that $m$ itself is even. We have therefore established that both $m$ and $n$ are even, *contradicting our preliminary hypothesis that* m *and* n *are the smallest integers of their ratio.* (If they are both even, we just need to divide them both by 2 to get smaller integers with the same ratio. If they are both even, they are not like 3:4, but like 6:8.) Something is incurably wrong about our initial assumption, and in other words, no two integers can be found to display this ratio. The diagonal of the square is irrational.

I do not know who was the first Greek to observe this. The observation is ingenious but simple; it could well have come about in the very first years of Greek mathematical writing. (As noted on pages 22–23, it was once suggested by modern scholars that this, the discovery of irrationality, could have come as a shock, and a crisis, to Greek mathematicians; this is almost certainly wrong and nothing but a modern fantasy. The Greeks *preferred* surprising results!) We only know about this result now because it is tucked in as a kind of appendix at the end of Book x of Euclid's *Elements*. This is a huge book – 115 propositions! – that does not discuss any particular irrationals at all. Instead, it is all about classifying *types* of irrationals. The most basic category is that of "lines commensurable in square only." (So, to take an example – but note that Euclid does *not* provide any examples – whereas the side and the diagonal of the square are incommensurable, the *squares* set up on the side and the diagonal are commensurable – they are, in fact, in the ratio 1:2. The side and the diagonal, then, are *commensurable in square only*.) Now, take two lines that are commensurable in square only (my example, in modern terms: with lengths 1, $\sqrt{2}$). Set up the rectangle contained by the two lines (the area of this rectangle is obviously, in our terms, $1 \times \sqrt{2}$, or simply $\sqrt{2}$), then find the square equal to that rectangle, and finally, find the side of this square (so, in our terms, the root of $\sqrt{2}$ or $\sqrt[4]{2}$). So, is this new side of a square commensurable with the original side? (Is $\sqrt[4]{2}$ rational?) No, it is not! In fact, it belongs to a category of irrationals, called *medial*, defined in *Elements* x.21 in the manner described previously (the side of the square equal to a rectangle contained by two lines commensurable in square only). Such medial lines have their own properties. For instance (*Elements* x.22, somewhat rephrased): if you take the square on a medial line, and you find a rectangle equal to that square, then, if one side of that rectangle is rational, so would be the other. Throughout Book x, Euclid finds a rich classification of irrationals, with complicated interrelations between the categories – only to admit, in the final proposition, that the classification is not exhaustive. The purpose of all of this is unclear to us (and this area of mathematics is not taken up again, later than Euclid, by any extant work; it appears that there were no more than a few extensions of this theory – all now lost – even in antiquity itself; a dead end). Why pursue this at all? Perhaps, if one is inclined to bring order into seemingly unstructured fields, and one is inclined to do so by way of classification, one would engage with the project of Euclid's *Elements*, Book x.

Euclid himself does apply Book x once, inside of his Book xiii. This book is a study of the five regular solids (the pyramid, the cube, the regular octahedron, the regular dodecahedron, the regular icosahedron). Medieval scholia now found within the text of Book xiii, as well as Byzantine sources, tell us that Theaetetus was the original author of the type of material now in Book xiii. Weak and late evidence, but taken together with the nexus between Book x and Book xiii, it begins to carry some significance. Indeed, Plato's Timaeus makes the choice to construct the cosmos out of those regular solids – and it is tempting to see in this yet another tribute to Theaetetus.

How does Book x feed into Book xiii, a classification of irrationals into a study of regular solids? This comes about because in Book xiii, the focus of attention is on finding the ratio between a given regular solid's edge (obviously, in a regular solid, all the edges are equal to each other) and the radius of the sphere circumscribing it. This ratio is then classified as a type of irrational. There is no obvious reason to engage in this classification or to be interested in this particular ratio, but it should perhaps be borne in mind that a rather natural way of trying to square the circle is to consider the ratio of the side of a regular polygon and the radius of the circle. Considering the same for regular solids and the sphere is, then, a very natural extension. The result can be conceptually and visually striking: see Plate 1.

There is considerable preparation that has to go into the study of such ratios within the regular solids, and this part of Book xiii is among the most beautiful pieces of geometry in all of the *Elements* (so, quite the opposite of the dryness of Book x) – for instance, *Elements* xiii.8, a study of the proportions inside the regular pentagon (the building block of the regular dodecahedron). I paraphrase one part of the argument.

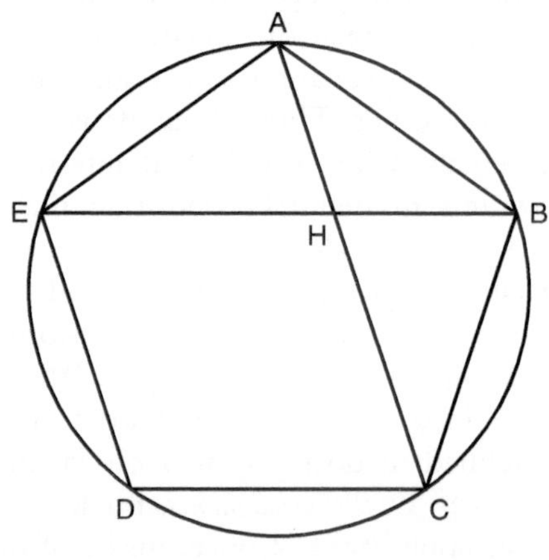


Figure 27

Suppose you have a pentagon ABCDE inscribed in a circle, and two lines drawn in, such as AC, BE, cutting each other at H. Then:

EB : EH :: EH : BH.

(This is what is known, to modern readers – not a Greek term! – as the "golden ratio.")

This is proved as follows. Because the pentagon is regular, all sides are equal, and all angles lying on equal arcs are equal as well. We have, then, (angle BAH) = (angle ABH), and therefore (through basic techniques related to the sum of angles in a triangle):

$$(\text{angle AHE}) = 2 \times (\text{angle BAH}).$$

But also (because of the arcs):

$$(\text{angle EAC}) = 2 \times (\text{angle BAC} = \text{angle BAH}).$$

We thus find

$$(\text{angle AHE}) = (\text{angle EAC}).$$

And so, HE = AE (an elementary result: in triangles, equal angles go with equal sides, and vice versa).

Furthermore, because

$$(\text{angle ABE}) = (\text{angle AEB}),$$

but also,

$$(\text{angle ABE}) = (\text{angle BAH}),$$

it can be shown that if we take the two triangles, ABE, ABH, they both share the same equal two angles (angles ABE, AEB in the triangle ABE; angles ABE, BAH in the triangle ABH). The remaining third angle must be equal, and the two triangles must be similar. Therefore:

$$\text{EB} : \text{BA} :: \text{AB} : \text{BH}.$$

But we have established

AB(obviously, the same as EA) = EH, and so we may rephrase the proportion:

$$\text{EB} : \text{EH} : \text{EH} : \text{BH}.$$

The golden ratio, in a pentagon, found so easily. And once again, the very same techniques: triangles inscribed in a circle, angles adjusted to determine similar shapes, similar shapes turned into proportions. All of this, then, fantastically built as the scaffolds of three-dimensional solids. It is easy to imagine this done by a young mathematician, impressed with the work of Archytas, belonging, once again, to the circle of Plato. We glimpse very little of Theaetetus's mathematics. We see that it could have been brilliant. And we also see that it sought to find, within a seemingly intractable and even inexpressible mess, a rational classification system.

## Eudoxus

Diogenes Laertius tells us that Eudoxus died at age fifty-three[12] – and although Laertius likes to repeat fanciful stories, he is often reliable for his bare facts. Indications from Eudoxus's fragments imply that he died later than Plato – so, later than 347 – but evidently not much later. The parameters are therefore, once again, surprisingly clear. Eudoxus was born in Cnidus (across the Aegean from Athens) in around 395 or 390. This was, perhaps, as Archytas's works began to circulate and as Plato wrote the *Meno*. By the time Eudoxus became an adult, mathematics was already an established genre; Archytas's – and Theaetetus's – work mostly done; Plato at the height of his fame. Among Laertius's stories, there is one where the young Eudoxus comes to study with Plato in around 370. Shall we trust Diogenes Laertius? Why not? Plato was already the most famous philosopher in the Greek world. Indeed, Aristotle – roughly Eudoxus's coeval and certainly a member of Plato's circle – on a couple of occasions, runs into *philosophical* debate with Eudoxus. Just like Archytas, then, Eudoxus was known as a philosopher, and most probably, he did belong to Plato's circle.

It is clear that his writings in the exact sciences – and beyond – were extensive. In the sheer terms of his mathematical achievement, Eudoxus could well have been as significant as Archytas, if not more so. If I survey him more briefly, it is only because we shall return to him later in this book: in Chapter 3, as an inspiration for Archimedes, and in Chapter 5, as a founder of mathematical astronomy.

Indeed, relatively much is known. We have significant fragments from a major contribution to geography, although nothing suggests that this had any specifically *mathematical* geography. More likely, this was a more descriptive exercise. If so, it was somewhat comparable to a couple of reported works in descriptive *astronomy*, providing a geographical tour of the night's sky. These works are well attested: Aratus, a later poet, took this tour and turned it into verse, and his poem is still extant; Hipparchus, a later mathematician and astronomer (on whom, more in Chapter 5), criticized Aratus's version, in part so as to criticize Eudoxus, along the way providing more information on Eudoxus's project.[13] Clearly, Eudoxus

[12] Diogenes Laërtius, *Lives of the Philosophers* VIII.8.90.

[13] There is no English translation of Hipparchus's commentary to Aratus (one of the major remaining lacunae in the translations of ancient Greek works). Aratus's "Phaenomena" was one of the most successful poems in antiquity: for translation, I use G. R. Mair, *Callimachus: Hymns and Epigrams. Lycophron. Aratus* (Cambridge, MA: Harvard University Press, 1921).

observed the sky, and clearly, also, his descriptive astronomy – whatever we may say for his descriptive geography – involved at least some numerical values and calculations. In fact, he certainly made at least one other, and more theoretical, contribution to astronomy: a treatise known as *On Speeds* tried to account for the aberrant motions of the planets. We shall revisit this material again in Chapter 5, but a few words are in order here. It is clear that Eudoxus presented a geometrical model that was meant to account for the motions of the planets (a category that, for the Greeks, included the sun and the moon as well). The key idea was that although the fixed stars were carried by a single rotating sphere, the planets were each carried by several rotating spheres, nested within each other, each moving at its own speed, around its own pole. The combination of those different poles and speeds could give rise to various patterns, approximating the aberrant motions of the planets. There is some debate concerning the intended numerical precision of such models. What is clear is that however numerical, Eudoxus – and his followers in the circle of astronomers established in Cyzicus – produced *a geometrical model* of astronomy: the trajectory of stars, as the trace produced by points, drawn by the rotation of spheres. We are reminded of Archytas, likely producing a geometrical model of optics (vision – the result of lines of vision, or light, converging at the point of the eye). At an even more abstract level, we are reminded of Archytas and his arithmetical model of music. Reality – and underneath it, and somehow explanatory of it, a more abstract structure. The comparison between music and astronomy might seem fanciful, but it was embraced by Plato and by many later philosophers, and it is quite possible that Plato, in his late years, was influenced, in this regard, by Eudoxus. If so, Eudoxus – whatever his actual intentions – deepened the commitment of a certain group of philosophers to a metaphysics where an underlying abstract and fixed mathematical structure explains the apparent material and aberrant structure of the universe.

The interest in more abstract structures is perhaps to be seen in Eudoxus's contributions to pure mathematics, as well. Eudoxus was the first to prove that the cone was one-third of the cylinder containing it. This achievement – now extant as a series of propositions in Book XII of Euclid's *Elements* – would loom large, in particular, for the work of Archimedes. We shall revisit it in detail as we bring in Archimedes's project of measurement. I therefore just note now that we will see how Eudoxus's measurement of the cone seems to have been motivated by an interest in rigorous proof: seeking a way to bypass, so to speak, infinity. It is likely that we can see the trace of Eudoxus's engagement with abstract logical structures involving infinity in another of Euclid's books, namely, Book v. This is what I will concentrate on in this chapter.

The evidence is weak – once again, merely a scholium in the transmitted text of Euclid, stating that this book was the discovery of Eudoxus – but there is no special reason to deny it. Aristotle seems to be aware of something like this book – and seems to know it as a fairly recent accomplishment. In general, it is the kind of book that should come relatively late, not because it takes so much for granted but, to the contrary, because it assumes so little. This is a study in foundations, taking terms that one could easily use without reflection – but then, insisting on defining them and demonstrating their properties. This is what mathematicians do, after much else has been done already. And this is, to the Greeks, the most foundational topic of all: proportions, in full generality.

We have seen proportions applied in music, geometry, and arithmetic. Book v defines proportion and shows that all those properties – which previous mathematicians took for granted – were in fact correctly assumed. The key definition (usually given as number 5) is astonishing:

> magnitudes are said to be in the same ratio, the first to the second and the third to the fourth, when, if any equimultiples whatever be taken of the first and third, and any equimultiples whatever of the second and fourth, the former equimultiples alike exceed, are alike equal to, or alike fall short of, the latter equimultiples respectively taken in corresponding order.

This is so abstract that it can only be legible to a modern reader with the aid of notation.[14] The claim is that for four magnitudes, a, b, c, and d, they are in proportion

$$\mathrm{a} : \mathrm{b} :: \mathrm{c} : \mathrm{d}$$

if and only if for *any* integers *m*, *n*:

$$m\mathrm{a} > n\mathrm{b} \leftrightarrow m\mathrm{c} > n\mathrm{d}$$
$$m\mathrm{a} < n\mathrm{b} \leftrightarrow m\mathrm{c} < n\mathrm{d}$$
$$m\mathrm{a} = n\mathrm{b} \leftrightarrow m\mathrm{c} = n\mathrm{d}.$$

[14] This is also in some ways (but not in all) equivalent to a sophisticated nineteenth-century definition of real numbers: Dedekind's account of real numbers as cuts in the set of rational numbers. (The set of rational numbers is, precisely, the set of all ratios between integers; Eudoxus's definition of a general ratio can be used to define the general ratio – equivalent in a sense to our "real number" – in terms of divisions of this set.) See Appendix 10.1 of L. Corry, *A Brief History of Numbers* (Oxford: Oxford University Press, 2015). (I recommend this book as a whole in the Suggestions for Further Reading.)

And even this is opaque without a concrete example. Suppose I say a:b::c:d. This means, for instance, that if I take a:b, they stand in a certain relation: a is greater than, equal to, or smaller than b. (To take a specific case: suppose my four-term relation is between 5, 4, 10, and 8. a is 5, b is 4, and a > b.) Now, change this by a certain combination of two factors (to take another specific case: say, by multiplying a by 2 and b by 3; so, the factor $m = 2$, the factor $n = 3$). Perhaps this relation has changed; perhaps not. Perhaps they are now equal; perhaps the order of size has flipped. (In fact, in the example provided, the 5 becomes a 10, and the 4 becomes a 12 on the other, so the relation did, in fact, flip.)

But now I apply the same multiplication to c and d. (In this specific case, then, these are c = 10 and d = 8, and I multiply c by 2 and d by 3.) Once again, I do not know what the new relationship is. But suppose that they follow the same order as a and b (that is, suppose that following the new multiplication, d will become greater than c – and indeed 24 is greater than 20!). This is good! If this will be true, *no matter which factors* m *and* n *I pick*, I am allowed to say, based on Eudoxus's definition, that the four terms are in proportion.

You might be inclined to say: of course 5:4 and 10:8 will have the same pattern of relations of size under multiplications by various factors. This is because they have the same ratio! Eudoxus perhaps will not dispute this (it is not clear that his definition is taken ontologically, as the identification of the essence of what makes a proportion). The point is that he has provided a mathematically determined equivalent to the pre-theoretical expression "being in proportion." This term has been analyzed into the simple terms of multiplication, equality, greater than, and smaller than. At some abstract level, "proportion" and a certain set of simpler terms are equivalent.

Because the terms used in the definition have clear mathematical significance – we know what it means to multiply and to state equalities and inequalities between mathematical magnitudes – the definition can, in fact, be applied. For instance, to state that four terms are *not* in proportion is to state that there is at least one couple of factors where the order of size between the terms in the two ratios is *not* the same: we know what this means, then. Book v quickly unpacks the definition, first through some truly elementary results – for instance, v.7, equal magnitudes have, to the same magnitude, the same ratio, and then to more complicated results, for instance, v.16: from a:b::c:d, derive a:c::b:d. It is possible that no one has proved any of those results in the past, and it is possible that no one has even noticed the statements such as v.7. It is also possible, of course, that Eudoxus has joined a more robust field of research into the foundations of proportion, and that his contribution is simply the only one that is now known in detail. There is a passage in Aristotle where he states that proportion was once defined – before Eudoxus, then? – through a different mechanism. The earlier contribution implied by Aristotle seems to imply a narrower, purely geometrical study of proportion. (There is a large literature on this topic, which I mention in the Suggestions for Further Reading.) We see the peak alone of an ancient range – but we notice, once again, the proximity of mathematics to philosophy.

The logical, almost philosophical motivation is palpable. One is reminded of Book X. Theaetetus's classification of irrationals, implied by Book X, is so cumbersome, so little motivated – why would a mathematician, capable of finding charming results such as the golden section in the pentagon, devote so much effort to such a dead end? Perhaps because of the sheer metaphysical fascination of bringing order to the seemingly unordered. Book V is less tedious, but its results are even less surprising. Do we even care about the results? Are they there because, ultimately, the working mathematician needed them?

I am not so sure that there was even an urgent sense of need to prove such basic assumptions. This is because in the practice of Greek mathematics, an important set of results involves ratios' *in*equality: for instance, that from a:b > c:d, one can deduce a:c > b:d. This is harder than the results concerning ratios' equality or proportion. Once we get to this complexity, it does indeed become crucial to distinguish the valid results. And yet, Book V, as it now stands, is almost entirely confined to the very basic, the level that requires proof only if one is truly pedantic, and it does not engage with ratios' inequalities at all. And yet, future mathematicians seem to have been content to take the results concerning proportion inequality for granted! After Eudoxus, mathematicians are content to take for granted that if a:b > c:d, then a:c > b:d. I therefore suspect that prior to Eudoxus, mathematicians were content to take for granted that if a:b::c:d, then a:c::b:d.

The definition is its own point, and its point is philosophical. In fact, the definition plays with philosophical fire. It is nonconstructive. That is, there is no finite, doable procedure that can establish that four terms are in proportion. A single counterexample will establish that the four terms are not in proportion, but to verify that they are, one needs to check *all* equimultiples – all the possible *m*s and *n*s, all infinitely many of them, twice over – in order to conclude that a, b, c, and d are, in fact, in proportion. This does not mean that the definition cannot be applied by the mathematician – but the very willingness suggests an explicit assumption of the existence of a mathematical reality, preceding human construction. I do not believe that a mathematician conversing with Plato would fail to notice this, and therefore I suspect that Eudoxus, in fact, meant this. The definition is not the tool to derive the results (we trust the results regardless). To the contrary: the definition is the point. The function of the results is *to verify the definition.* (We find that the definition allows us to prove the basic results we expect to be true for proportion, and so we conclude that the definition is indeed correct.)

The definition seems to carve out, with supreme analytical precision, something deep about the nature of proportion – proportion, as it were, is a kind of extension of the concept of equality. But it is, precisely, an infinite extension, one that, for better or worse, escapes verification – which is necessary because the definition is required to cover finite as well as infinite cases, rationals as well as irrationals. The quest is for supreme generality and abstraction. As far as we can tell, later mathematicians did not extend or attempt to revise Book V. Just like Book X, it is, for the Greeks, something of a mathematical dead end. But philosophers paid attention immediately. Aristotle seems to refer to Eudoxus's definition, and he does this not as an aside but to make a fundamental claim: definitions should be produced in their full generality. As far as Aristotle was concerned, Eudoxus made a contribution to the *theory of definition as such*. Was this a correct interpretation? Was this Eudoxus's intention? At a more basic level: Was Aristotle, in a sense, the intended audience?

## Assessing the Landscape

Our portraits of the better-known figures are already speculative, based, as they are, on tiny, mediated fragments. But we must become even more speculative than that. Our task is to understand the shape of mathematics, *as a whole*, through the fourth century BCE – and so, we ought to try to guess what it is that we know even less or do not know at all. The bias of the sources is our enemy: it preserves so little. But if we understand this bias, it can become our friend – that which the bias tended to obscure is what we have lost the most.

The main bias of our sources is clear. The single most important source of information is Eudemus's history of geometry, in the brief summary preserved by Proclus and in the various other sources that repeat fragments from that work. As far as the brief summary goes, this has a very particular bias. Proclus writes a commentary to Euclid's *Elements*, and his entire point, in the introduction, is to put this work into context. His manner of abbreviating Eudemus, in this introduction, is to list a few mathematicians and then state that they had contributed to the *Elements* or compiled certain parts of the *Elements*. It is an open question whether this means that those authors actually compiled works in the manner of Euclid's *Elements*, but one thing is clear: this emphasis on Euclid has nothing to do with the focus of the work of early mathematicians. Indeed, how could it? We have evidence that connects the three major figures surveyed previously with the contents of the *Elements*: Books v (Eudoxus), vii–viii (Archytas?), x (Theaetetus), xii (Eudoxus again), xiii (Theaetetus again). The remaining books are the more obvious "tools," the kind of mathematical results that are proved because they are, in fact, required for future applications. The only way they could have been written down at all, as their own separate books, is if someone had, indeed, already prior to Euclid, produced a compilation of elementary results. Perhaps they did; perhaps not. But this could not have been the focus of mathematical activity, which instead ought to have been dedicated to various mathematical applications. More squaring of the circle, more duplication of the cube? The latter, especially, will be mathematically important; I shall return to it in the final section of this chapter.

An even more important bias is not Proclus's but, already, that of Eudemus. Everything suggests that his history was a sober survey of pure geometry. This is useful: thanks to him, we get a very precise sense of how Hippocrates squared his lunules, how Archytas duplicated his cube. But a much larger amount of early Greek mathematics than we can now see must have been less pure and less technical. This is only glimpsed – but it is glimpsed often enough to suggest a lost pattern. Silanion was a sculptor who was famous for his portrait of Plato; apparently, he was also the author of a book on symmetries and proportion in the human body. This was not a new kind of writing – already, the sculptor Polyclitus, in the fifth century, wrote a book explaining the measurements of one of his statues. This is an entire genre that is now almost entirely submerged. Bryson and Antiphon both produced quadratures of the circle that Aristotle (undoubtedly correctly) considered badly false. Apparently, one or both were also philosophers (everything is obscure about such episodes: more than one person going by the same name, some more mathematical; others more philosophical[15]). How likely are we to know about the bad or simply false mathematics produced by less professional authors? Xenophilus of

15 There is little recent scholarship on such authors, who can only be approached with a high dose of speculation. The most sympathetic treatment of Bryson and Antiphon (too sympathetic, I suspect)

Chalcidice is said to have been a Pythagorean and a teacher of the greatest (and nonmathematical) musical theorist of antiquity, Aristoxenus. This Aristoxenus – a follower of Aristotle, although not much younger than him – vehemently opposed what he saw as the Pythagorean effort to mathematize music (we will mention him again briefly in Chapter 5 as we survey Ptolemy's astronomy; we will revisit harmonic theory, as well). Surely, Xenophilus was a mathematical author, but would we even know about him had it not been for his chance association with Aristoxenus? Indeed, if pushed to hazard a guess, it would be that the field of mathematical music is among the least represented in our attestation. With rare exceptions, Hellenistic authors were no longer interested in it, and what came to be remembered from the fourth century was the contributions by two outstanding individuals – Archytas and Aristoxenus. But Archytas was very early and so simply could not report on the many musical theorists who offered their own models, whereas Aristoxenus had no interest in reporting the details of minor authors he despised.

Sculptors, and others, playing with proportions; philosophers, reflecting upon mathematics and offering their own contributions; musical theorists, dividing the tetrachord – there ought to have been dozens of authors active in the penumbra of mathematics. Our evidence now has, on the one side, well-attested authors who clearly had a philosophical audience in mind, but yet their identity was that of "mathematicians"; on the other side, well-attested authors who clearly were interested in mathematics, and yet their identity was that of "philosophers." But the less attested, and so likely also the unattested, could well have fallen, in many cases, in between. "Mathematics" was real enough as a category (I shall return to discuss this later in the chapter), but it subsisted on one end of a well-populated spectrum.

If so, there is no reason why the mathematicians we do know could not have at least dabbled in philosophy. Perhaps it was common for the same person to contribute to both genres. Archytas and Eudoxus certainly did; among the more minor, we may mention Amphinomus, Hermodorus, Philip of Opus, and Zenodotus, for all of whom we have at least *some* report of a philosophical allegiance. And if so, the evidence concerning the contents of the works of the main mathematicians should be considered, as well, in terms of its potential continuity with philosophy.

remains A. Wasserstein, "Some Early Greek Attempts to Square the Circle," *Phronesis* 4 (1959): 92–100.

We see evidence for the mathematization of at least three physical domains: musical harmony, vision, the stars. For each, there is considerable evidence of philosophical speculation preceding (and subsequent to) the efforts of mathematization. But it is not just that mathematicians enter a philosophical domain. The very act of mathematization could be seen, in and of itself, as a philosophical claim: Perhaps that the underlying structure of the universe is to be understood on mathematical, more abstract terms?

Over and above such specific pushes toward abstraction, we see an interest in generalization, definition, and classification. Is this not the point of Archytas's general theory of numbers, of Theaetetus's general theory of irrationals, of Eudoxus's general theory of proportion? Even more specifically, the one theme that binds the entire spectrum together – the results in pure mathematics together with the more philosophical reflections – is that of harmony and proportion. We imagine Plato sitting down to his portrait by Silanion; in between sittings, Silanion is busy with his own work, finding the symmetries and numbers in the human body – perhaps as Plato was busy writing his great work, the *Republic*, whose starting point is the claim that there is a fundamental analogy (so, a proportion) between the individual soul and the city. Their conversation could touch upon the apparent limits of proportion – in the phenomenon of irrationality. Plato would recall his transformative meetings with Archytas; the rush of excitement, reading Archytas's finding of four lines in proportion; above all, the realization that music, deep down, was made of numbers in proportion.

We have started out this chapter from sociology. Of course, the ancient Greeks did mathematics because they liked the contents of the discipline. But people do not simply do what they like; they choose what to do from the available options, and starting with the final years of the fifth century BCE, mathematical authorship became a new option. This is part of the story underlying the flourishing of mathematics in the fourth century. But we have gradually seen a more detailed sociological account. It involves, specifically, networks of actors. People act together: learning from each other, writing to each other. The network to which you belong determines what you can do. Archytas was a pivotal member of Greek elite networks – if nothing else, because he was the most respected citizen in a major city. He belonged to a specific southern Italian network of Pythagoreans – followers of an otherworldly way of life. Several of them, recently, turned to a rather specific metaphysics, identifying the otherworldly with numbers and proportion, all based on the model of proportion, newly identified as the ground for musical harmony.

Archytas went further. With his far-reaching networks, he was aware of new ways of writing about mathematics and was in touch with new philosophical authors. His work pursued the project of musical Pythagoreanism as rigorous, technical mathematics. As such, it had an enormous influence on Plato, another member of the network of the Greek elite but also a central actor in, perhaps, the most significant cultural network of the time – those Greeks who decided to pursue philosophy in the tradition inspired by Socrates.

The nexus of Archytas and Plato – bringing together the previously unrelated traditions of musical Pythagoreanism, on the one hand, and Athenian philosophizing, on the other hand – came to be the center of a new network that endured for decades, perhaps sustained primarily by Plato's authority, persevered for the duration of the lifetime of Plato's direct circle.

The point is not that mathematicians, at this time, were *instructed* by philosophers – as if Plato told Archytas what to believe. Why should Archytas follow? The point, rather, is the continuity of audiences. And so, because philosophers from Plato onward were so interested in what mathematicians had to say, it is only natural that as mathematicians decided what to write, they would consider that particular audience. Plato influenced Archytas, I suggest, not by what he told him but simply by listening.

## The Making of Mathematics

The generation would end. And yet, mathematical continuity did not. Mathematics, from now on, exists as a clearly defined identity.

Now, I did argue that throughout this period, mathematics and philosophy were placed on a single spectrum, the one merging into the other. This would seem to imply that there was no clearly demarcated "mathematics." But the continuum of social interaction coexisted with sharply demarcated *genres*. This is typical for the survival of culture in antiquity as a whole. There were few institutions in antiquity and so little continuity based on them. But the Greeks were habituated – from their performative literature – to the idea that certain contents went along together with certain forms; there were certain genres of epos, of lyric, of drama, even of speech and of Socratic dialogue. To write a certain thing was to write in a certain way. And this was the impersonal continuity, preserving culture through generations and centuries. By the year 320, there would have been many dozens of works in circulation, in thousands of copies, all written in a closely related form. This form will define the future of the field. Mathematics emerges – and stays put – as a set of implicit rules for how mathematics *should be written*.

So far in this chapter, I have offered abbreviations or paraphrases of Greek mathematical arguments. Indeed, no mathematics from this era survives in its original, unmediated form. And yet, this much is clear: Euclid and Autolycus, writing a few decades later, already assume a very fixed genre; even earlier, many of the references by Aristotle to mathematics are explicit enough to get a sense of a form of writing that is almost identical to that of Euclid. It thus makes sense to read Euclid as an indication of what mathematics could have been like, already, in the mid-fourth century. Let me quote, now, a simple proposition in its original form. Because it is relatively simple, it is unrepresentative (many propositions will have a more complex structure). I chose one that, although simple, already displays some important structures. This is *Elements* 1.35, which I cite through Heath's translation.

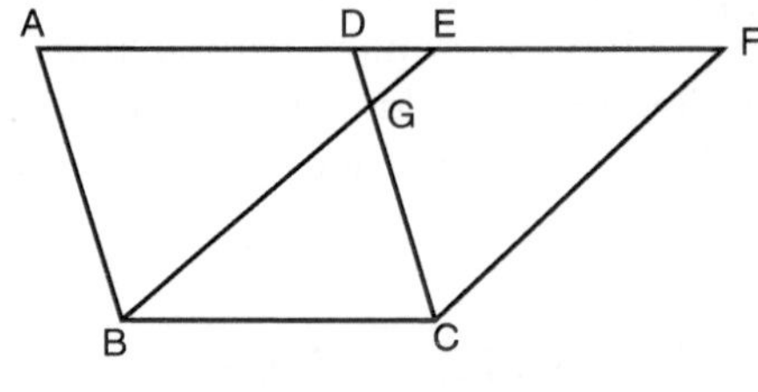


Figure 28

> Parallelograms that are on the same base and in the same parallels are equal to one another.
>
> Let ABCD, EBCF be parallelograms on the same base BC and in the same parallels AF, BC.
>
> I say that ABCD is equal to the parallelogram EBCF.
>
> For, because ABCD is a parallelogram, AD is equal to BC. For the same reason, EF is equal to BC, so AD is also equal to EF.
>
> And DE is common; therefore, the whole AE is equal to the whole DF.
>
> But AB is also equal to DC; therefore, the two sides EA, AB are equal to the two sides FD, DC, respectively, and the angle FDC is equal to the angle EAB, the exterior to the interior; therefore, the base EB is equal to the base FC, and the triangle EAB will be equal to the triangle FDC.

Let DGE be subtracted from each.

Therefore, the trapezium ABGD that remains is equal to the trapezium EGCF that remains.

Let the triangle GBC be added to each.

Therefore, the whole parallelogram ABCD is equal to the whole parallelogram EBCF.

Therefore, parallelograms that are on the same base and in the same parallels are equal to one another, which it was required to prove.

This is very *distinctive* writing. Not only is it unlike Homer or Euripides, but it is also, evidently, very much unlike Plato. When a Greek wrote in this precise genre, then, even if his audience might have consisted of philosophers, they had no doubt that what they read was, specifically, mathematical.

The most distinctive thing is, obviously, the combination of diagrams and letters. This is pervasive in all extant Greek mathematical writing and is one of the clearest things reflected in Aristotle's references to mathematics. Mathematical texts come as brief, distinct passages, each making an individual claim. Each is followed by a drawing, which is a simple network of lines, some straight, some curved. The drawing is interspersed with letters of the Greek alphabet, typically (with some exceptions) set next to some of the points of intersection in the drawn network. (In the diagram before us, each point where two lines meet gets its own letter. In some other diagrams, some of the intersections are missed; rarely, points are at positions other than intersections, and in appropriate contexts – referring to more abstract, quantitative objects – letters might be set not next to points but next to lines or inside areas.) These letters act as the written equivalent of a hand gesture. To write down "BC" is equivalent to pointing your hand at the particular line segment bounded by the points of intersection marked in the diagram by the letters *B* and *C*. The diagram thus has two faces. On the one hand, it makes the text concrete, and its reading becomes a reenactment, in the reader's mind, of the author's direct presence, pointing to a piece of drawing. On the other hand, it is consciously literate, not merely a written form but, indeed, one that explicitly refers to *letters* as its tool of choice.

There is only one layer of the individual mathematical argument that does *not* use Greek letters: the very external one, the peel of the onion – in Euclid's case, this is generally the first and last sentence, which form a precise repetition. In this case, this is "Parallelograms that are on the same base and in the same parallels are equal to one another." This is the point of the proposition, a general claim, true not only for this configuration but generally: *whenever* parallelograms have property X, they are equal. How is this generalization obtained? One constructs a particular case and assumes merely the property X – you are not allowed to take anything else for granted about these parallelograms. If, with this alone, you manage to prove *in this particular case* that those parallelograms are indeed equal, then you have, implicitly, proved this generally because, after all, you could repeat the very same argument whenever property X holds: nothing in your argument relied on anything other than the basic assumption, and so you could surely repeat it.

Thus, you are entitled to a general statement, not referring to a particular diagram. Such general statements become a fixed phrase one can recall and use elsewhere – a "formulaic expression." For an example of what I mean, I cite a brief passage from the previous argument: "and the angle FDC is equal to the angle EAB, the exterior to the interior." There is the echo here of a fixed phrase, "the exterior to the interior," which was coined a few propositions back. I cite now the general statement of an earlier proposition, underlining the repeated words (I.29):

> A straight line falling on parallel straight lines makes the alternate angles equal to one another, <u>the exterior</u> angle equal <u>to the interior</u> and opposite angle, and the interior angles on the same side equal to two right angles.

It is evident that Euclid, in proposition 35, seeks to repeat the verbal structure available from proposition 1.29. After all, in proposition 35, he is interested in comparing the triangles EAB, FDC, and he just stated, in this order (from left to right – the more common order in Greek references to diagrams), that the pair of sides EA, AB is equal to the pair FD, DC. It would be natural to have the angles stated in the same order, from left to right, but this is now reversed; the right angle, FDC, is said to be equal to the left angle, EAB. Why? So that we can assert that the exterior is equal to the interior, and not vice versa. And why? Because this is what we have extracted out of the argument of the preceding proposition 1.29: a certain result, in *a certain verbal formulation*. There is, in fact, a good reason for this: no one in antiquity would have referred to it as "1.29." I have been using this kind of shortcut throughout this chapter so far; it is a very convenient way of making references. It is also a modern way, perhaps made more natural by the format of the book (which invites readers to turn pages back and forth until one finds the looked-for position). Until the fourth century CE, most ancient books were in roll form, designed more naturally for a single, sequential reading. How would one recall texts, then? As one always did elsewhere: by recalling them. Greek literate culture would always remain close to the practices of performance, poetic or rhetorical, and to know anything well would be to know it as remembered speech. A Greek mathematician would be marked, among other things, by having this stored in his or her memory: a hundred or so fixed phrases, with the main results one always resorted to, again and again – quite literally, at the tip of one's tongue.

There would be more stored like this. A key fact about this form of writing – once again, one of its distinctive features – is its reliance on a fixed set of formulaic expressions. This is a subtle point – because the formulaic character of Greek mathematics, itself, is subtle. This is not like the explicit symbolism used in modern mathematics. And yet, when we see utterances such as the following:

> Let DGE be subtracted from each.
>
> Therefore, the trapezium ABGD that remains is equal to the trapezium EGCF that remains.
>
> Let the triangle GBC be added to each.
>
> Therefore, the whole parallelogram ABCD is equal to the whole parallelogram EBCF.

we should note the repetitive structures, such as "let X be subtracted/added from/to each," "therefore . . . therefore," and the fixed terms, such as "the X that remains" or "the whole X." Such expressions are not transparent in and of themselves, and although I imagine readers can piece together what they mean, it must be clear that the Greek mathematical reader did not need to do so. To him or her, the expressions were perfectly transparent because *they were always used* in the same context, with the same meaning. We can see a number of repetitions even within the confines of this small proposition, but in a book such as Euclid's *Elements*, expressions are repeated hundreds of times. In this proposition, for instance, the key claim concerning the particular diagram is preceded by the words "I say that"; the final restatement of the general claim is completed with the added words "which it was required to prove." Each of those expressions occurs exactly once in this proposition, but also, each of those expressions is repeated (with small variation) in nearly every proposition of the work as a whole. In general, there are but a few hundred phrases repeated again and again – out of which the entire fabric of Greek mathematics is produced. Once you are familiar with this stock of phrases, a Greek mathematical text fully makes sense. Try to recall your first days in middle school, struggling to make sense of what "$4x$" meant (remember how hard it was to figure out that $4x + 1$ does not equal $5x$?). This is how difficult the language of Greek mathematics now strikes you. Now, of course, as a grown-up, an expression such as $4x + 1$ reads itself: it is an obvious way of representing its mathematical idea. Understand, then, that to the mind of a professional Greek mathematician – which simply means one who immersed himself or herself in the writings of the mathematical genre – the expressions of Greek mathematical language are as transparent and as easy to read as "$4x + 1$." This, a Greek would feel – this system of formulaic expressions – is precisely how those mathematical ideas are naturally represented.

How do we use algebraic symbolism? We use it to make verifiable computations. We know that from

$$A + B = C,$$

we may deduce

$$A = C - B,$$

because – once we've trained to use this method – we can inspect the formula and see the equivalence, left and right, of the equation, B removed from each. This is a kind of visual-typographic verification, based on familiarity with a certain visual-typographic set of rules.

The ancient Greeks worked differently: they verified claims by reference to two separate sets of rules. They had a system of verification based on the diagram, as well as a system of verification based on formulaic expressions. Let us go back to the same phrases (adding in the immediate context):

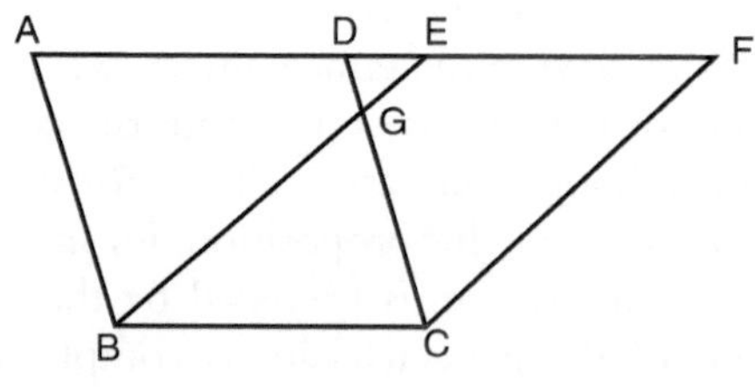


Figure 29

The triangle EAB will be equal to the triangle FDC.

Let DGE be subtracted from each.

Therefore, the trapezium ABGD that remains is equal to the trapezium EGCF that remains.

> Let the triangle GBC be added to each.
>
> Therefore, the whole parallelogram ABCD is equal to the whole parallelogram EBCF.

How do we verify the equality of the trapezia? By referring to the diagram, where we can see that the two equal triangles are indeed each composed of the trapezium together with the subtracted triangle. (This is a particular configuration that, typically in Greek mathematics, is simply seen in the diagram; generality is only slightly harmed, in that in principle, one can repeat variations of the same argument.)

How do we verify the equality of the parallelograms? By referring to the diagram, where we can see that the two equal trapezia, once the same triangle is added to each, indeed become the required parallelograms.

How do we know which is equal to which? By filling in the correct slots in the formulaic expressions ". . . which remains," ". . . common."

Perhaps the most significant feature of this proposition is its division into steps. One could, after all, have removed one common triangle and added in another simultaneously, moving at once from triangles to parallelograms. This would have been just as valid. But it would have been less transparent. This, after all, is what Descartes learned from his mathematics (ultimately, that is, from the Greeks!): divide problems so that the individual steps become, each, clear and distinct. Each time, we do a single, simple visual operation or operate on a single, simple, formulaic expression. We take the complexity of a mathematical train of thought and divide it into components until each individual component is at exactly the right size for our visual and verbal intuitions – trained, of course, by our familiarity with the mathematical idiom. And at this point, each proposition becomes not merely convincing but, indeed, compelling. It is impossible to understand and not to assent. And this is why Greek mathematical arguments are characterized not just by generality but also by necessity. This combination of generality and necessity makes mathematics an outlier in the entire universe of argumentation. Nothing else can be made so persuasive; all the more reason, within a culture that values the power of argument, to keep such arguments separate from others, maintained in their own genre. And all the more reason, indeed, for philosophers, such as Plato, to take notice.

This, then, is not a very difficult historical question. Quite naturally, once the genre was in place, it would not be changed. Indeed, it would largely survive into the many successor civilizations that inherited their mathematics from the Greeks, all the way down to early modern Europe. Which is important, in and of itself: the emphasis on persuasion – ultimately, the legacy of the democratic Greek polis – would survive through many later monarchic civilizations. It remains what most mathematicians emphasize to this day.

But it becomes even more interesting to try to pinpoint the creation of the genre. After all, this was one of the most lasting events in the history of science as a whole! We can be certain that the genre was in place by the end of the generation studied here. I am not sure that it was in place in the fifth century. In particular, looking at the extensive fragment we have from Hippocrates of Chios, I am not sure that he used Greek letters as diagrammatic labels (there are indications in the text that suggest that he could have referred to unlabeled diagrams). It is at least possible that the full form of Greek mathematics emerged later than Hippocrates.

On the other hand, once someone has written a substantial, influential piece of mathematics, with its own genre, it is hard to see why the immediate readers would wish to change this. For instance, one of the interesting features of Greek mathematical writings is that in number theory – as well as music – one relies on labeled diagrams, just as one does in geometry. The same is true, indeed, in the study of irrationals and of general proportion. This is strange and remarkable because, in such fields, the line-and-point diagrams of Greek mathematics are nearly useless. Had there been important authors in the early tradition, using labeled diagrams for geometry but avoiding them for arithmetic or music theory, I would imagine that such a bifurcation could have stuck. But that's not what we see: instead, the reliance on labeled diagrams is universal.

I thus assume that the early influential authors already used the labeled diagram for both arithmetic and music. In other words, I think this ought to be true for Archytas, especially because in his case (unlike that of Hippocrates), it is simply inconceivable that he could have presented his argument for the duplication of the cube without the use of lettered diagrams. An Archytas without the labeled diagram in geometry is impossible. But an Archytas with the labeled diagram in geometry but not in arithmetic would be significant enough, as an influence on later authors, for this practice to remain bifurcated. It is not bifurcated; hence, Archytas must have labeled his arithmetical diagrams as well. Now, everything in this chapter so far coheres around a picture with Archytas as the central, most influential node in the network of this generation of mathematicians. They all read Archytas.

In the first decades of the fourth century, Plato, in Athens, would record the memory of Socrates, broadcasting it throughout the Greek-speaking world, returning, again and again, to the very same form of writing. The genre was not unique to Plato – others commemorated Socrates as well, through recalling his words in conversation. Plato's legacy was the one that endured and came to define the genre, setting the terms, in many ways, for philosophy as a whole and for the more specific genre of philosophical dialogue. Many more would try to write in the same form – if none could quite rival the original author.

I imagine Archytas, through the same decades, doing something a little similar, returning, several times, across different disciplines, to the same form of writing: in geometry, in number theory, in music, probably in optics, perhaps several times in each. Time and again, he would keep to the same verbal texture, always writing discrete pieces of text referring to their separate diagrams, labeled by Greek letters. He was not the only one doing so, and in many ways, he simply continued writing rather like Hippocrates of Chios and the few others of Hippocrates's pioneering generation (although the form used by Archytas was even more specific, even more regimented). Those who read Archytas, however, knew that everyone else interested in the same field was reading the same works by Archytas, and so, to write in a style other than that of Archytas would be a marked departure. And for what? Clearly, the arguments were effective. And so, the genre began to pursue its own conservative logic. This is my proposed reconstruction.

What the Greek mathematical genre assumes is the existence of a specialized audience, one willing to immerse itself in writing of a special format, concerned with special, esoteric subject matters. And so we go back to sociology, to networks, to questions of scale. The Greek mathematical genre was formed when a certain group cohered around interests, shared in different ways by its members. Sufficiently many people, acquainted with each other, late in the fifth century or early in the fourth century, became interested in number, shape, the stars, vision – perhaps, above all, music. My own guess is that the realization that harmony is, in fact, number could have catalyzed interest in this cluster of topics. Inspired by this – Pythagoreanism? – a dialogue ensued – perhaps between Archytas and Plato? – giving rise to a new form of writing. And this, in a nutshell, was the first generation of Greek mathematics.

## Coda: The Final Contribution of the First Generation

Perhaps influenced by original ideas due to Eudoxus himself, Menaechmus – active, most likely, somewhat after the middle of the century – came up with a transformation of the problem of doubling the cube.

We recall that the doubling of the cube – or more generally, the scaling of a solid – was already transformed, by Hippocrates of Chios, into the problem of finding two lines between two given lines so that the four lines are in continuous proportion:

A : B :: B : C :: C : D.

Now, one of the techniques pursued in this period – one that, as usual, gained the attention of philosophers no less than of mathematicians – was to approach problems through *analysis*. In this context, *analysis* means that we solve a problem by assuming it solved – and then thinking through the implications of such a solution.

Let us assume, then, that we have already obtained a series of lines such that A:B::B:C::C:D. Think of what this implies. First of all, we may ignore the middle terms and just focus on A:B::C:D. This is interesting! Proportions can be transformed into shapes. Specifically, A:B::C:D implies that the two rectangles – the one contained by A, D and the one contained by B, C – are equal to each other.

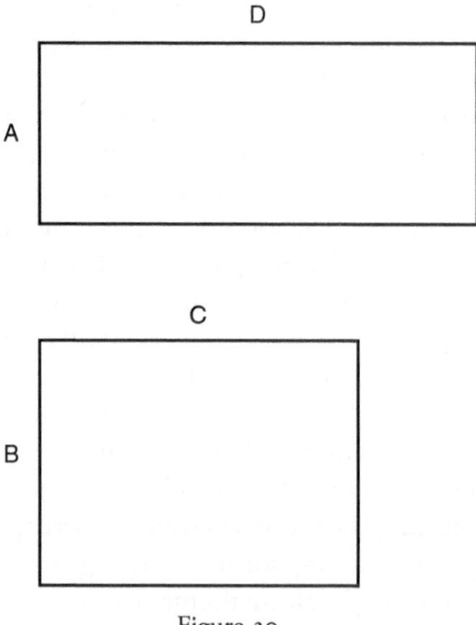


Figure 30

(A expands into B by the same ratio that D shrinks into C, with the two canceling each other so that the two rectangles are equal.)

This is interesting because the lines A, D – the extremes of this proportion – are the "given" or known elements. This means that we also know the magnitude of the rectangle contained by B, C. We do not know what B and C are, but we have a strong constraint: they must combine as a rectangle to have a fixed area.

By the same logic, we can also conclude that the rectangle contained by A, C is equal to the square on B. Once again, A is a given line: and so, what we learn is that B, C, have another interesting internal structure. The rectangle contained by C and a given line (namely, A) is equal to the square on B. To repeat: we do not know what B, C are, but they must conform to a relation where, as one of the lines – namely, C – is doubled, so is the *square* on the other one, B. B scales, we find, as the square on C.

Our two conditions are geometrical – having to do with rectangles and squares – and they are best seen as a geometrical configuration in a diagram. I reproduce it, once again, from Eutocius's catalogue of solutions to the problem of doubling the cube; this appears to have been Menaechmus's solution. (In Figure 31, the given lines are A, E, and the two lines found between them are B, G.[16])

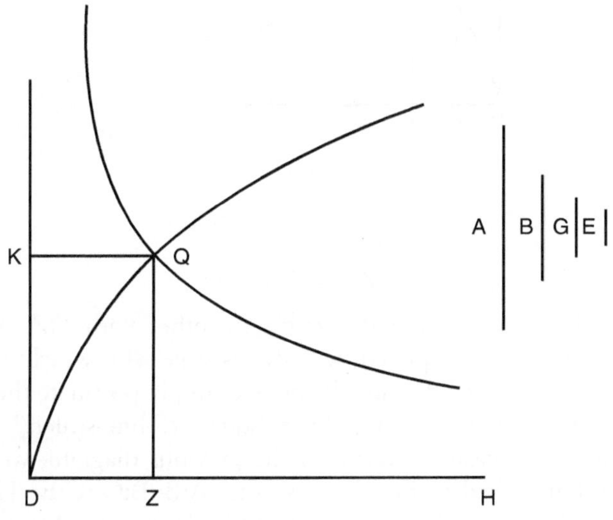


Figure 31

The rectangle KQZD is of a fixed area; specifically, it is assumed to be equal to the rectangle contained by the two lines (shown on the side) A, E.

This rectangle may be imagined as a transformable rectangle by moving the point Q along the curve so that the two sides of the rectangle always remain parallel to KD, DZ. (This curve obviously never touches the two lines, KD, DZ, because the rectangle's area is fixed and because, were the curve to touch KD or DZ, the rectangle's area would then become 0.) For the time being, we simply postulate the existence of such a curve, which we may call the "equal-rectangle-preserver."

[16] This solution is translated into English in R. Netz, *The Works of Archimedes: Translation and Commentary*, vol. 1 (Cambridge: Cambridge University Press, 2004), pp. 286–290.

Further, the square on the line QZ is assumed equal to a rectangle contained by the line DZ and the fixed line A. It is assumed that as we move along the other curve shown in Figure 32 – the one drawn through the points Q, D – we find new positions of Q′, Z′, such that the new Q′Z′, squared, is always equal to the rectangle contained by the new Z′D and the fixed line A. In our symbolism:

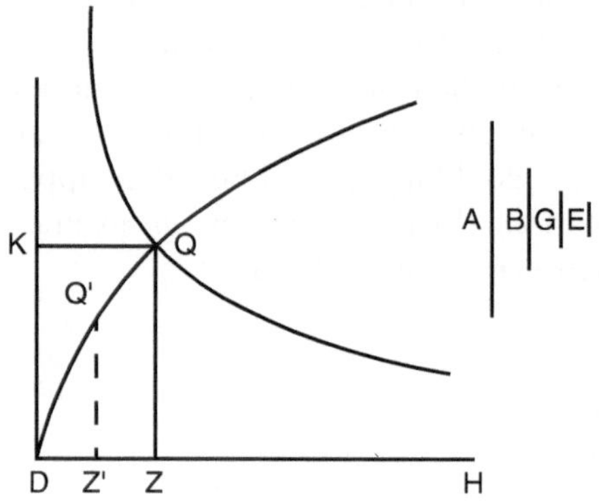


Figure 32

$$QZ^2 = ZD \times A$$
$$Q'Z'^2 = Z'D \times A.$$

QZ may be varied, along this curve, together with DZ. Whichever variation of QZ we get, squared, will always scale with which variation of DZ we get, the line. For the time being, we simply postulate the existence of such a curve, which we may call the "square-to-line-scaler."

The idea now is clear. In terms of the previous diagram, we are given A and E and are asked to find B, G so that A:B::B:G::G:E. This is how we proceed, then. We draw two orthogonal lines. We draw the equal-rectangles-preserver between them so that the rectangle it preserves is equal to the one contained by the two given lines A, D. We draw the square-to-line-scaler through the intersection of the two orthogonal lines so that the squares it obtains will be equal to a rectangle whose fixed side is equal to A.

The point at which the equal-rectangle-preserver intersects the square-to-line-scaler solves the problem. We draw everything, find the intersection Q, and complete the rectangle; obviously, the two mean proportionals between A, E must be the lines KD, DZ.

Now, there is a lot that we still miss for this to count as a satisfactory solution! It would be nice to find that the curves in question could be geometrically constructed (that is, not just *postulated*). It would also be nice to prove that the curves are indeed determined by the terms provided.

Somehow or other, Menaechmus did. You can *get* those curves, specifically, by cutting a cone.

Cut a cone one way, and you have a hyperbola. This is also an equal-rectangles-preserver; it can be determined by the magnitude of the rectangle it contains between two given asymptotes.

Cut a cone another way, and you have a parabola. This is also a line-to-square-scaler, and it can be determined by the combination of its vertex and the length of the fixed parameter of its rectangle.

And so, one can indeed solve the doubling of the cube by the intersection of a parabola and a hyperbola.

All those names are modern. We will return to all of this in the following chapter – and try to suggest, then, how, in fact, Menaechmus could have discovered that the conic sections display the required properties. Indeed, I left the discovery of the conic sections to the end of this chapter for two reasons. First, because it did, as a matter of fact, come right near the end of this period. (Eudoxus was the youngest of the major authors of the fourth century; Menaechmus, his even younger follower.) But second, because this would be the core of mathematics for the following generation. Mathematics pivoted from the generation of Archytas to the generation of Archimedes, and it pivoted, specifically, around the conic sections. And so, I ask for your patience. The story of the discovery of the conic sections would become clearer, as we bring in the evidence of their transformation, in the hands of later mathematicians.

But this should be noted, already, in the conclusion of this chapter – as we bid farewell to the first generation of Greek mathematics. What conic sections do, in general, is display an array of proportions. The interest in proportions, for an author such as Archytas, could well have sprung from metaphysical sources. This, certainly, is what mattered to Plato, reflecting on the mathematics emerging throughout his lifetime. But now, at around, say, 340 or 330, when both Archytas and Plato were dead, active mathematicians come across a new streak of potential geometrical discovery concerning proportion, highly technical and satisfying and of little obvious metaphysical import. This will fit right into the mathematical work of the next important group of mathematicians.

## Suggestions for Further Reading

Until recently, scholars of Greek mathematics were especially fascinated by this generation, where so little is known but so much is hinted. There is, in particular, a tradition of trying to use the hints in Euclid's *Elements* so as to reconstruct the process through which its contents were discovered. This tradition culminates in R. W. Knorr, *The Evolution of the Euclidean Elements* (Boston: Springer, 1975). Specifically, Knorr engaged with the discovery of irrationality, explicitly focusing on the indications in Plato's dialogue, the *Theaetetus*. Another wonderful book with similar ambitions is D. H. F.Fowler, *The Mathematics of Plato's Academy* (Oxford: Clarendon Press, 1987). There, the focus is not so much on irrationality as such as on the treatment of proportion, explicitly focusing on the meager indications we have for a pre-Eudoxean definition. More recent scholarship tends to focus on the better-attested mathematics of the generation of Archimedes, and I will have more books to recommend for further reading there. I did write, myself, on the genre of Greek mathematics, especially in R. Netz, *The Shaping of Deduction in Greek Mathematics* (Cambridge: Cambridge University Press, 1999).

Scholars today are indeed skeptical about the meaning of the few hints we have for early mathematics. For a good early critique of Knorr's use of such sources as Plato's dialogue the *Theaetetus*, see M. F. Burnyeat, "The Philosophical Sense of Theaetetus' Mathematics," *Isis* 69 (1978): 489–513. Burnyeat has contributed in other ways to our understanding of the interface of mathematics and philosophy. M. F. Burnyeat, "Archytas and Optics," *Science in Context* 18 (2005): 35–53, for example, provides not only the evidence for Archytas's contribution to optics but also a sense of what Plato could have liked – as well as disliked – about it. There is also a more general treatment of Plato and mathematics in M. F. Burnyeat, "Plato on Why Mathematics Is Good for the Soul," *Proceedings of the British Academy* 103 (2000):1–81; for this field, of foundational significance is G. Vlastos, "Elenchus and Mathematics: A Turning-Point in Plato's Philosophical Development," *American Journal of Philology* 109 (1988): 362–396.

This is mostly about Plato. For the other side of the encounter – the Pythagoreans of southern Italy – the key reference is two books by Carl Huffman: C. Huffman, *Philolaus of Croton: Pythagorean and Pre-Socratic* (Cambridge: Cambridge University Press, 1993), and C. Huffman. *Archytas of Tarentum: Pythagorean, Philosopher and Mathematician King* (Cambridge: Cambridge University Press, 2005). Greek theoretical reflection on music is a major field, largely alien to our own ways of thinking. The best entry point is A. Barker, *Greek Musical Writings, Vol. II: Harmonic and Acoustic Theory* (Cambridge: Cambridge University Press, 1990). Finally, there are few accessible treatments of Eudoxus's definition of proportion. This – together with the Greek treatment of irrationality – is best understood within the long history of the concept of number, for which an excellent introduction is L. Corry, *A Brief History of Numbers* (Oxford: Oxford University Press, 2015).

CHAPTER 3

# *The Generation of Archimedes*

## Plan of the Chapter

This chapter is a kind of rondo, with Archimedes as its main theme. In outline, with five subchapters:

- I Prelude to Archimedes
- II Enter Archimedes
- III Response to Archimedes: Apollonius
- IV Archimedes's Physics
- V Response to Archimedes: The Generation

And now, to the detailed plan of the chapter.

### *I Prelude to Archimedes*

I suspect that there was a hiatus, of sorts, between the generation of Archytas and that of Archimedes. We start with "A Brief Glance at the In-Between," the years between the two generations, and take a closer look at one indispensable figure in "Euclid, the In-Between Mathematician." Then, to prepare for the survey of the generation of Archimedes, I propose a historical account: how the science of this era could have become, to a certain extent, autonomous. This argument is based on the changing geography of Greek culture, and the section is titled "Athens and Alexandria: The Making of a Dual Mediterranean."

### *II Enter Archimedes*

With this, our main business begins. "Archimedes the Historical Person" is a section presenting the biographical information on this author (which, for once, is significant). "The Works of Archimedes" surveys the corpus (about twenty-five attested, about ten extant!). We need to choose our focus. The chapter first looks closely at Archimedes's pure geometry and begins to consider the response to Archimedes, then returns to Archimedes's more physical works and the response to them, and finishes with an assessment of the generation as a whole. For the pure geometry, we begin with a particular moment: "The Challenge of Archimedes" – the set of tasks proposed to Conon and gradually revealed to Dositheus. I focus in particular on one example: the area of the spiral. I provide an overview of Archimedes's proof of this measurement in the section "Closer to the Spiral." The following section, "Eudoxus, Inspiration to Archimedes," discusses the origins of this proof in one of the major achievements of Eudoxus (which I put aside in the preceding chapter). We may then understand Archimedes as taking his cue from a specific moment of the previous generation – but making it much more central to his own approach and perhaps subtly transforming it in the process.

### *III Response to Archimedes: Apollonius*

We understand something of Archimedes's pure geometry – and may consider the response to it. The most important mathematician to respond to Archimedes's achievement was Apollonius, and so the following section is "Archimedes, Inspiration to Apollonius." One work alone is extant in Greek – but what a work! We turn to the scope of Apollonius's conics in the section "Enter the *Conics*," which once again goes back in time. We return to Menaechmus's discovery of conic sections – and see how they were reinvented by Apollonius. A particularly intriguing possibility is covered in "Archimedes: Conversing with Apollonius?" – that is, can it be that Archimedes himself shows awareness of some of Apollonius's work? Can we get a sense of the working of the generation *as a generation*?

### *IV Archimedes's Physics*

To follow this, then, brings us back to "Archimedes: Conversing with the Physical World?" An entire series of works by Archimedes involves what we think of as "physical" and specifically the tool of (what Archimedes called) the "center of the weight." Is this tool physical – or geometrical? How was mathematical physics invented? This is a very long and detailed section, surveying the most sophisticated mathematics we will see in this book. It is divided into four smaller subsections, each on the main treatises dealing with the center of the weight: I discuss more briefly *On Balancing Planes* and *Quadrature of the Parabola* and then survey in a little more detail *Method* and *On Floating Bodies*. This is followed by "A Syracusan Coda: And More, on the Physical Sciences" – going back to Archimedes's death, his myth. He certainly used no burning mirrors on the Romans – but he may have been a major figure in the growth of ancient optics, perhaps his greatest lost contribution.

### *V Response to Archimedes: The Generation*

Against the background of Archimedes's achievement, surveyed nearly in full, we may finally consider its impact. The two last sections are "The Generation of Archimedes" – with a more detailed sense of the responses of the many authors in this generation, beyond Archimedes and Apollonius – and "The Significance of Ludic Proof," a note on the historical meaning of this generation.

## I. Prelude to Archimedes

### *A Brief Glance at the In-Between*

Even in the moment of efflorescence around such figures as Archytas, Theaetetus, and Eudoxus, the concrete foothold of mathematics was limited. There were no more than several dozens of mathematical authors active during, say, 360–340 BCE. And it seems that there were rather fewer who were active after that peak.

Here are the mathematicians whose activity was likely later than Eudoxus's but earlier than Archimedes's (by which I mean that they were likely born later than 380 but earlier than 280):

Autolycus
Aristotherus?
Euclid
Hipponicus
Aristyllus
Timocharis
Aristarchus
Dionysius
Pheidias
Conon

Perhaps not that much should be made of the precise numbers. I have nine to ten names in this century of mathematicians ("nine to ten" because

Aristotherus is a particularly flimsy reference[1]). The same list, for the preceding fifty years, had twenty-three. So, a ratio of 1:5, but perhaps this could be attributable to our sources. Eudemus, Aristotle's pupil, wrote a history of geometry, and so the mathematicians he commemorated had a better chance of being remembered later on. There was no similar history of later generations.

So, how do we know about those few authors? It varies. In the case of Autolycus and Euclid, their works are extant. (Those of Euclid range widely; Autolycus's extant works involve basic astronomy.) There is a tradition that Euclid worked for the first Ptolemy, and indeed, it seems as if later mathematicians take his works for granted – hence his dating to this era. We date Autolycus because a biography of a philosopher, Arcesilaus, mentions that Autolycus was among his teachers. The same philosopher also had another teacher in mathematics – which is all we hear of the otherwise-anonymous Hipponicus. Aristyllus and Timocharis are cited by later astronomers for their observations: it appears that they surveyed the fixed stars, and it is a standard assumption that they did so in the service of the early Ptolemaic kings. Aristarchus, once again, is extant, and he was even famous in antiquity. (He proposed, among other things, a heliocentric model! His extant work is a brilliant astronomical exercise, calculating the distances and sizes of the sun and moon – on all of this, more in Chapter 5.) Dionysius, too, is only known (and dated) thanks to a few observations, in this case of a planet (Mercury). We are sure that Pheidias is earlier than Archimedes; Archimedes mentions him as his "father." He is cited for a measurement – alongside that of Aristarchus – of the size of the sun. Conon, too, is mentioned by Archimedes (we will see much more of that in the section titled "The Challenge of Archimedes"). Conon is most famous, otherwise, from Callimachus's mention of him in a poem: something of a court astronomer, perhaps, Conon was the one who identified a group of stars as the reappearance, in heaven, of a lock from Queen Berenice's hair! (This subtle, ironic move combines poetry and astronomy in a manner very typical of the age.)

[1] We have a reference ascribing to a certain Aristotherus a critique of a certain cosmological feature of an astronomical theory, as well as some references to Aristotherus as a teacher of Aratus (on whom, see p. 187). The biographical tradition on Arstus is especially suspect, and one wonders if it may not be the source of the reference to cosmological critique, in which case, all we can learn is that a certain Aristotherus (a philosopher?) was said to be associated with Aratus. The evidence is summed up in G. L. Irby-Massie and P. Keyser (eds.), *Encyclopedia of Ancient Natural Sciences* (London: Routledge, 2008).

What matters, then, is not the fact that relatively few names are mentioned but that, even with such small numbers, a pattern emerges. All the authors for whom any interest can be established are interested in astronomy. (Even Euclid is no exception: his corpus does include basic astronomical works, although we cannot be certain of their authenticity. Even Aristotherus is dated to this era because of his name being mentioned – by late and unreliable sources – in connection with the astronomy of Autolycus and the astronomical poet Aratus.) For a surprising number of authors, it is possible to detect the traces (no more) of the Ptolemaic monarchy: Timocharis, Aristyllus, and Conon were perhaps something akin to court astronomers? Euclid, perhaps, commissioned by the kings?

We note, perhaps, falling numbers, an interest in astronomy, and courtly connections. Perhaps, all might be related. If, indeed, philosophers lost interest in mathematics – at the end of a generation where the very point of doing mathematics used to be to talk to such philosophers – then it is natural that mathematicians might get discouraged. The new interlocutors are laypersons, and it is perhaps not surprising that of all the forms of ancient mathematics, the one that always captured most people's imagination was the mathematical study of the sky (of which, more in Chapter 5). This field will, in time, give rise to the most successful court science of antiquity – that of astrology – but already in the Ptolemaic courts, pre-astrological research into the sky was sufficiently fascinating to justify the patronage of the courts. It was as such, then, that mathematics survived through the lean years. It would flourish again when a genius – Archimedes – broke upon the scene. Then – and perhaps because of the excitement he himself generated – there would suddenly be more mathematicians, perhaps enough for this new generation to find interlocutors within the mathematical profession itself. This is when mathematics, more than ever, becomes almost an autonomous, self-sufficient pursuit.

## *Euclid, the In-Between Mathematician*

Transitional, perhaps somewhat less creative, and yet, this is also the generation of the one Greek mathematician you've likely heard of: Euclid. Does he deserve his fame?

Of course he does. In the preceding chapter, I did – I hope, with caution – what many scholars in the past have done again and again: ransacked Euclid's *Elements* for evidence of the previous generation. Clearly, this work is composed of materials produced by past mathematicians. But it is a supremely well-crafted composition, and it is no less clear that this work is much more than the cut and paste of past works. Nothing in the *Elements* is simply due to Archytas, Theaetetus, or Eudoxus; the *Elements* are due to Euclid – a source of caution, then, in using the *Elements* for the recovery of the earlier history of Greek mathematics.

The first thing to note is that this is a very *big* work. It is divided into thirteen "books" – with "book" meaning something like "the typical length of a single papyrus roll." In practice, the sizes vary. Most books are fairly short – Book II has 14 propositions, but Book X, in particular – the mammoth classification of incommensurables, perhaps based on that of Theaetetus – has 115 propositions! Each has its own mathematical scope:

| | |
|---|---|
| Book I | Triangles and parallelograms |
| Book II | Quantitative relations between lines and the rectangles they define |
| Book III | Circles |
| Book IV | Regular polygons inscribed in circles |
| Book V | Proportion theory |
| Book VI | Application of proportion to geometry |
| Books VII–IX | Study of numbers (Proportions involving numbers, prime numbers, even and odd numbers – this sequence is not as neatly divided into books as the rest of the *Elements*.) |
| Book X | Incommensurables |
| Book XI | Elementary solid geometry |
| Book XII | Cones and cylinders |
| Book XIII | Regular solids |

The architectural achievement is palpable, literally building up from scratch: rectilinear figures built up in Books I and II, circles in Book III, combined in Book IV. So far, without proportion theory, which is introduced in Book V. Proportion and geometry are combined in Book VI (which provides some of the most important tools in Greek geometry). At this point, a pause is reached, and the other main forms of Greek mathematics are pursued: first numbers, then incommensurables. The final section recapitulates plane geometry, with solids: Book XI closely follows Book I; Book XII, with its treatment of curvilinear solids, recalls Book III concerning the circle. And as Book IV studied the inscription of regular polygons inside the circle, so Book XIII studies the inscription of regular solids inside the sphere – applying, along the way, a few results from Book X.

Many of the books are provided with evident capstones, adding to the sense of structure. Book I builds to the so-called Pythagorean theorem. Book IX ends twice: First in IX.20, which proves, in our terms, that there are infinitely many prime numbers. Then a new beginning is made with odd and even numbers, leading to the second capstone – the construction of perfect numbers. (Those are numbers equal to the sum of their factors, such as $6 = 1 + 2 + 3$.) Book XIII, a sort of capstone to the whole, ends with the proof that the constructed five regular solids (pyramid, cube, octahedron, dodecahedron, icosahedron – you might know them as the dice used in Dungeons and Dragons) are the only ones possible.

Euclid does not use explicit references from one proposition to another. As noted on page 96, the ancient Greeks relied not on textual markers such as "See Book I, proposition 47" but instead on the verbal echoes of formulaic expressions. Indeed, one of the key functions of Euclid's *Elements* must have been to put, in one place, all the important formulaic expressions for such cross-references. But with or without such references, Euclid's *Elements* is structured by the relation of logical dependence. Results are built up in an iterative way. Nothing is left for later in the book (it never happens, for instance, that a requirement for a claim made in Book I is delayed until Book V). The way in which Books I–IV add up to Book V to make Book VI, which then, together with Books X and XI, feeds into Book XIII – this is a very large-scale architecture, unique among extant Greek mathematics. It stands to reason that most of the Greek mathematics we have lost would have been, if anything, more modest in scale and ambition; against the background of the Greek mathematics available at his time, Euclid must have stood up quite prominently. Alexandria was famous for its lighthouse of unprecedented scale. So was the *Elements*.

The architecture is paramount, and it drives the mathematics. Especially significant is the choice to postpone proportion theory to Book V. One can only speculate that Euclid felt that starting with abstract proportion theory would be wrong – mathematics, after all, was primarily about concrete geometrical objects, and Eudoxus's proportion theory was too abstract. If proportion theory is delayed to Book V, and if one obeys the strict principle that results are allowed to depend, logically, only on previously established results, this creates genuine constraints on the manner in which the most basic geometry of Books I–IV must be developed. Consider, for instance, the most central result of all – *Elements* I.47 or, once again, Pythagoras's theorem.

Similarity between figures is, we noted, at the heart of Greek geometry: it brings together quality (similar shapes) and quantity (proportion between sides of those similar shapes). Specifically, using similarity for the finding of a single mean proportion (p. 64) suggests a very straightforward way of proving Pythagoras's theorem.

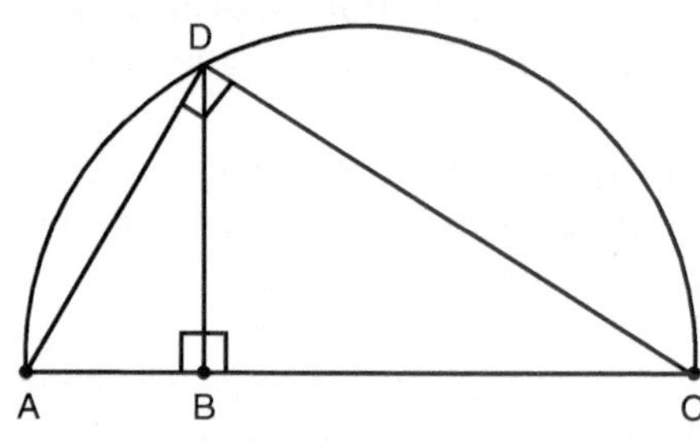


Figure 33

To recall, we work as follows: to find the mean proportional between the two lines AB, BC, we lay them side by side and set up a circle with the bisection of AC as the center. The perpendicular up from B hits the circle at BD, and then BD is a mean proportional between AB, BC because the two triangles ABD and DBC are similar to each other as well as to the whole triangle ADC.

More than this: the triangle ADC is a right-angled triangle. As we consider the similarities in this configuration, we may note several proportions:

> AC : AD :: AD : AB (hypotenuse to smaller side in similar right-angled triangles)
> AC : CD :: CD : BC (hypotenuse to bigger side in similar right-angled triangles)

At this point, we do the thing Greek mathematicians do all the time and convert a proportion into a concrete geometric relation (we noted this most recently in the context of Menaechmus's discovery of the conic sections). If we have

AC : AD :: AD : AB,

this also means that the rectangle contained by AC, AB is equal to the square on AD. And from

AC : CD :: CD : BC,

we can also deduce that the rectangle contained by AC, BC is equal to the square on CD.

Draw the two equalities on the diagram, and the conclusion follows: the square on the hypotenuse AC (divided into two rectangles, the one contained by AC, AB and the one contained by AC, BC) is equal to the sum of the squares on AD, CD because it is equal, respectively, to a rectangle directly underneath it.

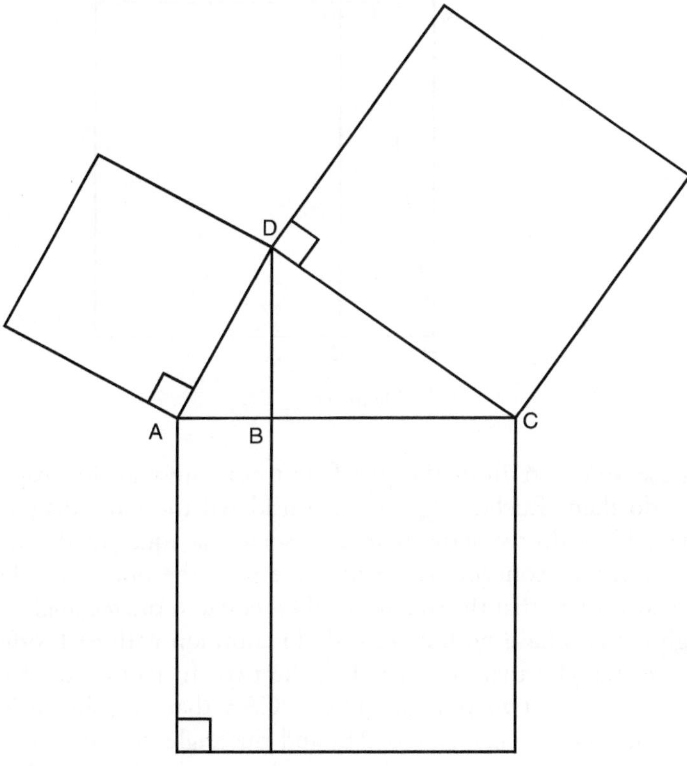


Figure 34

I assume this is how the Greeks most naturally would prove Pythagoras's theorem. I believe so not only because this is a reasonable story, based on what we know about early Greek mathematics, but also because as we read Euclid's actual proof, we can see this similarity-and-proportion argument visible, as it were, peering at us from behind Euclid's words.

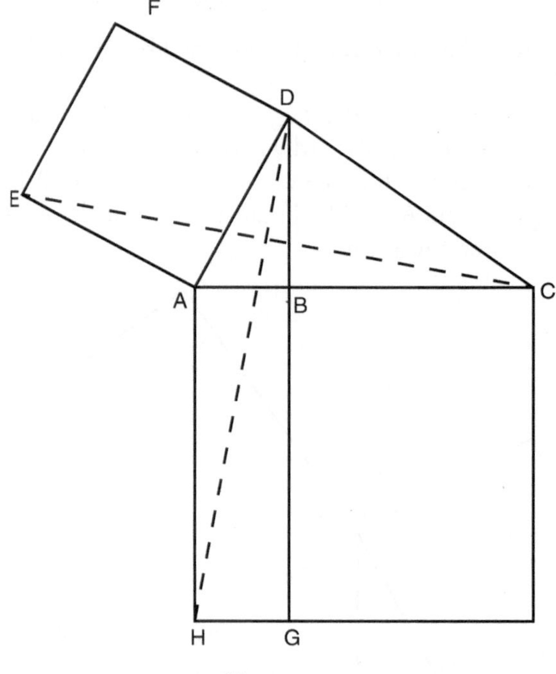


Figure 35

Suppose you start from the proof, but you must avoid proportions. What to do then? Euclid might have considered the same diagram with fresh eyes. How do we show that each square is equal to the rectangle beneath it? Let us concentrate on just one pair, the one to the left. We need to show, then, that the square AEFD is equal to the rectangle ABGH. Although the two have nothing directly in common with each other, it is easy to see triangles that are related to the two. In particular, if we join HD, EC, we have two triangles HAD, CAE that are obviously equal to each other (HA = CA, AD = AE, and the angle in between – HAD, CAE – is the same, simply rotated counterclockwise). All we need to show, then, is that triangles between two parallel lines (such as HAD, between the two parallel lines HA, GD, or CAE, between the two parallel lines AE, CF) are always half of the parallelogram on the same base between the same parallels (such as the parallelogram ABGH or the parallelogram AEFD). This, then, is the argument through which Euclid proves "Pythagoras's theorem": in his proposition I.47, the capstone to Book 1. Notice, however, that he needs, as noted, an important assumption:

triangles are half the parallelogram within the same parallels as them. This, Euclid indeed achieves earlier, in proposition 1.41.

But how to show the general relation between triangles and parallelograms? The key result required is proposition 1.34, which essentially states that the diameter of a parallelogram divides it into two congruent triangles. This is a straightforward enough result, indeed, if we can show that in a parallelogram such as ABCD, the angle ABC is equal to the angle BCD.

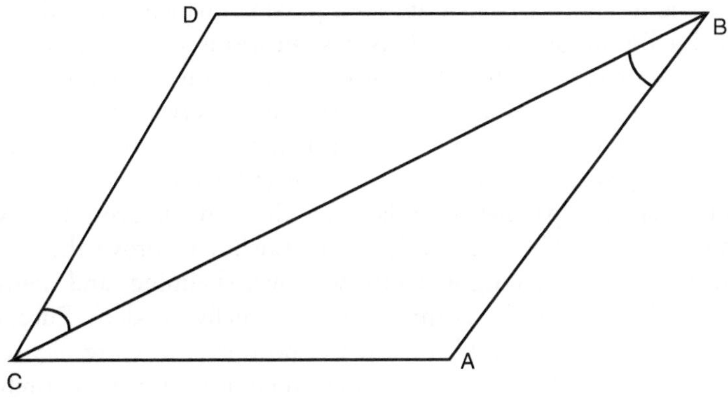


Figure 36

This, Euclid could not directly prove. Therefore, true to his architectural commitment – making sure everything is proved, based on previous results – he made the deliberate choice to introduce a special assumption, right in the introductory section of his first book. This was not a proved theorem but instead, simply a postulate – an assumption, stated as a requested fact:

> If a straight line falling on two straight lines make the interior angles on the same side less than two right angles, the two straight lines, if produced indefinitely, meet on that side on which are the angles less than the two right angles.

This is known as *Euclid's parallels postulate* or *Euclid's postulate 5*.

Based on this assumption, it is straightforward to obtain Euclid's proposition 1.29, which shows, among other things, that "a straight line falling on parallel straight lines makes the alternate angles equal to one another," and through this, to prove 1.34, hence 1.41, hence 1.47 or Pythagoras's theorem.

Euclid's parallels postulate defines what we now call *Euclidean geometry* (in *non*-Euclidean geometries, the two straight lines may fail to meet even if the angle is less than two right angles; or in other versions, they may meet, regardless of the angles). I believe Euclid came to this postulate – and came to position it so prominently – because of a practical consequence of his architecture. He needed to show Pythagoras's theorem, but he had to avoid similarity and proportion – the most natural tool that brings quality and quantity together. He thus had to work directly with the quantitative relations between triangles, based on congruences and thus based on the equalities between angles.

This reminds us that even Pythagoras's theorem depends, ultimately, on the parallels postulate. In non-Euclidean geometries, the relation between the squares on the sides of a right-angled triangle varies (if, indeed, multiple right-angled triangles are even possible, which is not the case in all non-Euclidean geometries). This has surprisingly many consequences and means, effectively, that our normal measurement techniques fail in non-Euclidean geometries. This is a very deep observation; I am not sure Euclid would have made it so prominent had he not committed himself to his particular grueling exercise in large-scale architecture. He did, with big consequences. The parallels postulate was there, made prominent within Euclid's structure. The urge to "correct" Euclid, to prove the postulate instead of merely assuming its truth, was overwhelming, and many later mathematicians – Greek, Islamic, and eventually modern European – would pursue this path until, finally, the postulate, in its resilience, gave birth to the non-Euclidean geometries, together with an entirely new conception of space and of what geometry is about.

But we are pushing ahead – and doing so anachronistically. For Euclid himself, the issue was not the nature of geometry. It was a particular task of arrangement: get lots of mathematics together, in a particular architecture.

Just what did this task mean, historically, for Euclid himself?

For this question, I will take my first hint from two contemporary mathematicians: Aristyllus and Timocharis. These astronomers are almost entirely unknown, save for a few reports of observations cited by later astronomers: this or that star, observed in this or that position. Such observations, put in the hands of historians of astronomy, can be made eloquent. A few involve the moon and so can be dated – we thus know that they were active from the very beginning of the third century (and so, just as the Ptolemaic kings become established). Most intriguing is the possibility that Timocharis's observations are all south of the zenith in Alexandria, whereas Aristyllus's are all north.[2] The implication is of an explicit division of labor in a systematic effort of mapping. This is usually taken to be a survey of the skies – perhaps with the aim of producing a globe? – commissioned by the kings. Perhaps it is relevant to mention,

[2] For more on such observations, see Y. Maeyama, "Ancient Stellar Observations: Timocharis, Aristyllus, Hipparchus, Ptolemy – the Dates and Accuracies," *Centaurus* 27, no. 3 (1984): 280–310.

then, that physicians active in Alexandria at the time, such as Herophilus and Erasistratus, were naming and surveying the entire body, based on dissections of bodies that were likely provided by the kings. Even closer to home, it is clear that the Ptolemaic kings engaged in massive book-collection efforts (there were many big libraries in Alexandria, but the royal library must have been particularly big). The greatest author of early Alexandria was a poet-scholar, Callimachus, and his greatest claim to fame was the *Pinakes*, a survey – in 120 books, putting Euclid to shame – of past literature. Was it, in some sense, a catalogue of the royal library? Once again, precise proof eludes us, but what we see, with greater clarity, is a pattern. Aristyllus and Timocharis, looking at the sky; Herophilus and Erasistratus, looking at the human body; Callimachus, looking at literature. This is the *look from the outside* – from the new vantage point of Alexandria. There is a project of producing exhaustive surveys, perhaps as royal commissions. It seems likely to me that Euclid was tasked with a similar assignment.

The metaphor of the catalogue is distinct from that of foundations. If indeed, Euclid's project should be put alongside those of Aristyllus and Timocharis, Herophilus and Erasistratus, Callimachus – then it should be seen as a catalogue that is only incidentally a study into the foundations of Greek mathematics. And this, indeed, becomes clearer as we bring in two further sources of evidence. First, the actual uses of Euclid's work. If his *Elements* were a catalogue, what were they a catalogue of? Second, Euclid's larger project: for after all, even his extant corpus extends further than the *Elements* alone.

What is the *Elements* a catalogue of? The first to realize the significance of this question – and to provide a persuasive answer – was Ken Saito. Let us look at a pair of results in the *Elements*, II.5 and II.6. *Elements* II.5 asserts (paraphrasing) that if C bisects AB, and D is somewhere on CB:

$$\text{(rectangle ADHK)} + \text{(square LHGE)} = \text{(square CBFE)}.$$

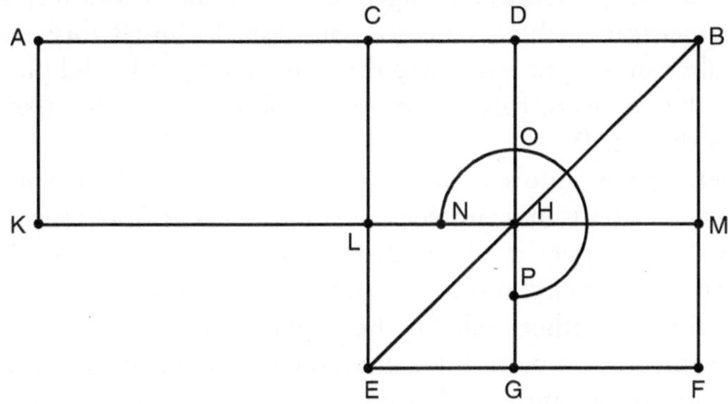


Figure 37

And 11.6 asserts (paraphrasing) that if C bisects AB, and D is somewhere on the continuation of CB:

(rectangle ABHK) + (square LHGE) = (square CDFE).

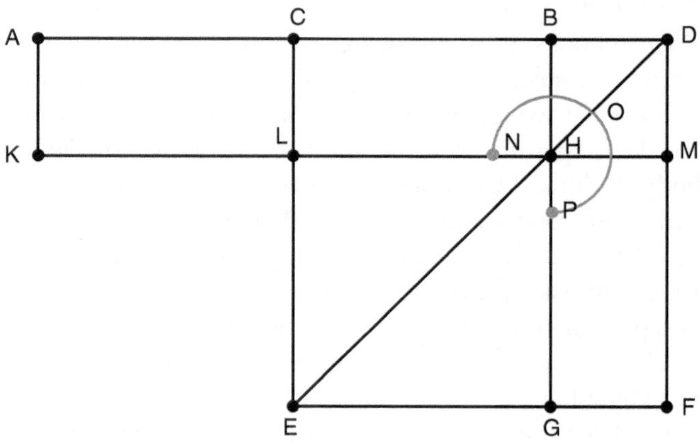


Figure 38

It is obvious that the two propositions are nearly identical. Why have both? Scholarship, prior to Saito, fixed on two observations. First, the claims here are very similar to algebraic identities (11.5 is rather like $(a + b)(a - b) = a^2 - b^2$; 11.6 is rather like $(a + b)^2 = a^2 + 2ab + b^2$). Second, the treatment of such algebra-like results in terms of operations on rectangles is reminiscent of Babylonian mathematics. This, then, belongs right in the tradition of Neugebauer, discussed on pages 20–21: Greek mathematics was originally influenced by Babylonian geometrical algebra!

Saito, following Unguru, no longer believed in the historical existence of such "geometrical algebra." And so, Saito asked himself this: not where those results came from, but where they were going. If Euclid put the two results in the *Elements*, this is because, for some reason, he *needed* both. What was this need?

And here comes Saito's simple observation – so simple that it must be the correct solution to our problem. In works such as Apollonius's *Conics*, one often has to apply results such as *Elements* 11.5 and 11.6. And they come up in distinct circumstances: some circumstances call for the application of the one, others call for the application of the other. A line is bisected; sometimes, a point is taken on the line itself, and sometimes, it is taken on the continuation of that line. In either case, it is often useful to know the consequences in terms of the equalities of rectangles. Hence, there are occasions when II.5 comes in handy, others, when II.6 does: not because they say such different things but because they are *applied in different circumstances.*

What the *Elements* is, Saito argues, is a catalogue of circumstances waiting to happen. It is the mathematician's toolbox. I imagine Euclid in Alexandria, at work on the *Elements*, perhaps not unlike Callimachus in Alexandria, at work on the *Pinakes*: surrounded by books. Euclid was going through the many dozens – or was it two hundred or more? – of mathematical books produced throughout the preceding century or longer. He picked, along the way, the tools required for this mathematics, then built the tools, together, within a cathedral.

This is clearly not the only thing Euclid did. A catalogue implies a certain reverence toward the past. (The same, indeed, is true of Callimachus.) It also seems that some effort was put toward preserving the past, even when the expectation of future applications was slim. Perhaps Book V is there (standing on its own, not just as a prelude to Book VI) to commemorate Eudoxus; Book X, for the sake of Theaetetus's memory. If so, perhaps, after all, the *Elements* should be used, within measure, as a historical resource? Perhaps the parts of Euclid most likely to reflect Euclid's past are those that were least applied in Euclid's future.

But this historical interest is localized and qualified. That Euclid's project was primarily that of gathering the tools becomes clear from a consideration of some of his other extant works. The *Data* is usually taken to be authentic, and in this case, the sense of tool collection is evident. There is relatively little architecture, and instead, we find the accumulation of fairly simple results, none too interesting individually, although many are useful. What is this usefulness? In general, a very basic technique of Greek mathematical study and proof is that of mathematical analysis. (We have caught a glimpse of this technique already in Menaechmus's solution to the problem of finding two mean proportionals.) In mathematical analysis, you seek to prove a result, but even before you do so, you engage in a thought experiment: Supposing the result obtained, what else would follow? The *Data* provides a set of tools for answering such questions: If this is given, what is given as well? So, for instance, proposition 42: if the ratio between the sides in a triangle is *given*, the shape of the triangle is *given* as well. A central result, underlying so many applications in Greek mathematics! Fundamentally, then: Euclid's *Data* is something that one applies. It is an extra toolbox, one more set of applications waiting to happen.

There are a few more works ascribed to Euclid, and it is hard to tell how valid the ascriptions are. Perhaps it became a habit, among medieval copyists, perhaps even in antiquity itself, to assume that very elementary works are due to Euclid. And still: we do have ascribed to him elementary treatises in mathematical astronomy, in mathematical music, in optics, in catoptrics (the study of reflections); he certainly also wrote an elementary treatise (now lost) on conic sections. A few works are attested and were apparently somewhat more ambitious (he had to gain his fame as a mathematician, somehow, so as to become the compiler in chief): *Divisions*, *Porisms*, *Surface Loci*, even a collection of fallacies. (What a pity that the last has been lost!)

What is typical to all such works is that they appear to be motivated, once again, by a certain impetus to classify and to exhaust. *Divisions*, for instance (which survives in a much-transformed Arabic translation), provides a series of solutions to tasks of the following character: given a rectilinear figure, draw a line that divides the figure in a required ratio. (We will note later on how Archimedes solves a much harder problem of this kind, involving the division of the sphere.) Those are simple solutions, perhaps already achieved, at least in part, by past mathematicians. Euclid adds them up, arranges them, and goes through them one by one.

Autolycus, perhaps contemporary with Euclid, is extant, and his own work is also such a compilation, in the elementary techniques of geometrical astronomy (more on this in Chapter 5). Aristarchus's extant work, however – once again, to be discussed in Chapter 5 – is a brilliant piece of geometrical ingenuity. Perhaps we are unlucky in terms of the works that survive from the generation in between. Perhaps there were, originally, more Aristarchuses, fewer Euclids and Autolycuses. I doubt this. The evidence we do have suggests that, quite possibly, at least as far as pure mathematics is concerned, the generation, itself, conceived of itself in such terms: the heir – and not much more? – to the brilliant achievement of the generation of Archytas, brokering such past achievements for new patrons.

Soon, mathematics will emerge to do much more. In particular, mathematics seems to cast itself in a new, autonomous mold. This means, above all, that – in contrast to the generation of Archytas – mathematics will become independent from philosophy. How to account for this? I argue that this goes back to a general divergence in Hellenistic culture, in turn dependent on a long-term trend in Greek culture – the canonization of Athens – and the variety of responses to this canonization, in the city and outside it. Mathematics will come to be a very non-Athens thing; philosophy will come to be a very Athens thing. This is my claim: and so, we need to delve, once again, into the trajectory of the history of the Greek city-state – going right back to the fifth century.

### *Athens and Alexandria: The Making of a Dual Mediterranean*

Cleisthenes gave Athens a new democratic constitution in 508 BCE. From then on, to get things done, you had to stand up in front of your co-citizens and *persuade* them. Soon after, it was Athens that led the Greeks against Persia – and many cities now looked to Athens for inspiration or, whether they liked it or not, came under its tutelage. Previously, perhaps, what happened in Athens stayed in Athens. But now, everyone noticed. Tragedies, comedies, speeches by the likes of Pericles, conversations by the likes of Socrates – what happened in Athens was moments of people speaking and performing in public. And what happened in Athens was instantly famous everywhere. Athens became the stage of Hellas.

Literary canons have a certain staying power. Suppose you are educated by reading Euripides (the most famous tragedian), Plato (the most successful author of Socratic conversations), Demosthenes (a somewhat later Athenian orator), Menander (a somewhat later still Athenian comic author). This was much of the education of literate Greeks, from as soon as the new literary canon, centered on Athens, was formed. And so, whether they knew Athens and its democracy directly or not, they continued to value the excitement produced by performance, to treasure certain public friction. After all, it all went back to engaging competitions: Who is going to win the first prize for tragedy? Will we choose peace or war – the two options espoused by competing orators? Is Socrates, finally, to be defeated in debate? Generations after generations of Greek children were reared on such verbal competitions, and so it was just taken for granted, for such readers, that this was what culture was mostly about. It was about words in violent encounters, flung out by strong individuals who mark themselves against each other in contest, rooted in the experience of the city in continuous political debate.

Meanwhile, political realities were changing. As new powers emerged in the periphery of the Greek world – above all, the kingdom of Macedon – Athens could not form, as it once did, a strong enough alliance against them. By the mid-fourth century, Macedon dominated Greece. Athens tried twice to resist and was defeated twice in battle, in 338 and 322. Following its final defeat, even its constitution was finally broken. Some form of democratic public ceremony would continue to be performed – in Athens and in many other Greek cities. But from the late fourth century onward, monarchy, not democracy, became the central form of Greek political life.

Those new kings ruled, indeed, over vast territories. Alexander of Macedon used his new power base in Greece to drive an expedition into the empire of Persia itself, and by the time the expedition was over – in 326 – Macedonians ruled over the entirety of the landmass between the Indus and the Mediterranean, all the way out into the Balkans, as well as Egypt. Following Alexander's death, a war broke out between his generals – warlords, chasing the spoils of the empire. Several new kingdoms ensued, and following the initial chaos of civil war, the new order of the Hellenistic era was surprisingly stable. Of course, the Near East always had kingdoms. What was new about the Hellenistic kingdoms was that they were all now ruled, for the first time, by monarchs focused on the *Mediterranean*. Egypt, in particular, was transformed. The Nile valley was the most fertile and internally connected region in the ancient world (no other river could compare as a means of transportation). It was a self-sufficient world unto itself, its wealth sucked up by governing monarchs setting up the antiquities we still admire. The new Macedonian rulers were known as the Ptolemaics (all male kings were called Ptolemy; the one exception was the last queen, Cleopatra, whose death in 30 BCE is often taken to be the symbolic end of the Hellenistic era). They sought to suck Egypt's wealth into a Mediterranean world. Previous Egyptian capitals were set on the Nile, but the Ptolemaic capital was on Egypt's Mediterranean coast, in the new foundation city of Alexandria. This was the most significant boomtown of antiquity. The cornerstones were laid in 331 BCE; fifty years later, it was already the biggest city in the Mediterranean, far overshadowing Athens itself.

The key thing is that the Athenian canon could not, now, be replaced. Everyone grew up on a literary canon that reinforced that culture was this: charged, violent, public debate. But there was little opportunity for this kind of behavior in a kingly court. The habits of culture – the historical pathway – can sometimes become entrenched, and then, when a gap opens up between the past and the present, one must overwrite the other. The compelling power of the Athenian canon was such that, even in the face of monarchic realities, it kept shaping the basic expectations of what culture was about. And so, because kingly courts could not create rivals to past public debates, it was the past public debates that were destined to remain. The Greeks, from this point onward, would stick with the same canon. As centuries would pass, the fixed position of that canon became even more entrenched. The Romans could not dislodge it and indeed found themselves imitating the same Athenian canon; it would be bequeathed further on and would ultimately provide the foundation for modern European culture. The original choice made in the fourth century – to canonize Athens – would never be undone.

But notice the consequence: the Greeks now had a cultural memory that was in sharp contrast to present realities. Hence, ultimately, the Hellenistic bifurcation.

Suppose you are away from Athens but heir to a culture dominated by the memory of Athenian performances. What is one's place, precisely? The city of Alexandria, in particular, was a city of expats. In its early generations, almost everyone was an immigrant. Daniel Selden wrote a famous article on this civilization of Alexandria, titled "Alibis" (looking at the major poet of the new city, Callimachus[3]). One was *away*. In this cultural position, it is natural to become reflective, self-aware of one's position relative to past culture. As one became reflective and aware of genre, it became common to engage in the deliberate *hybridization* of genre. Perhaps the most remarkable tendency, from our perspective, is the hybridization of pure poetry with more technical prose. One of the most successful works produced in Alexandria was Aratus's *Phaenomena*, a work that, among other things, took Eudoxus's astronomical map and put it into verse! This, indeed, already comes close to mathematics itself. The relation is reciprocal. Eratosthenes, an Alexandrian poet and mathematician, wrote a poem on the duplication of the cube; Archimedes produced a complex problem in numerical calculation as a puzzle in verse (this is known as the *Cattle Problem*). In more general terms, when one's culture emphasizes a reflective, self-aware play with genre, it is perhaps natural that one would emphasize such values as irony and surprise. Indeed, the poetry of the Hellenistic era came back into fashion in the late twentieth century. You see, the Hellenists were postmodernists. (Perhaps, literally so: they followed, after all, on the heels of Athens' breathtaking modernity.) Can mathematics be ironic and surprising? I would argue that it can and that such, in fact, was the mathematics of the generation of Archimedes.

What matters is not only what mathematics was like but also what it ceased to be. We therefore need to turn to Athens of the Hellenistic era. This, too, had to contend with a meaningful past. In Alexandria, the point was that one was no longer in Athens; in Athens, one was all too Athenian. Stripped of its political power but enshrined as a cultural model, Athens could not but fall back upon its cultural memory. It was partly a theater

[3] D. L. Selden, "Alibis," *Classical Antiquity* 17 (1998): 289–412.

district, partly a college town, and in both ways, it ended up, above all, conservative. First in tragedy and then also in comedy, one no longer favored new productions. What the audience wanted was revivals of the classics. And so it was, again and again: Euripides and Menander. And the same happened with philosophy, too. Following the topsy-turvy of philosophical innovation throughout the fourth century, each generation bringing forth multiple new doctrines, the city of Athens came to establish four philosophical schools, now finally as fully fledged institutions. The school would have a master, and many active philosophers surrounding him; many students, from across the Mediterranean, now came to Athens, to the extent, indeed, that philosophy almost disappeared elsewhere. Why settle for less than the place of Socrates's own memory? One could go to the Academy (based on Plato's memory), to the Lyceum (based on the fame of Aristotle), or to two somewhat more recent schools, the Epicurean Garden and the Stoa. The four schools differed greatly in the details of their philosophical doctrine but resembled each other in their growing attention to minutiae internal to the school's own philosophy. It matters, after all, whom you talk to (this is the sociological point we emphasized in the preceding chapter). The members of the Athenian philosophical schools – much like the members of contemporary academic departments – ended up talking, above all, to each other. For once, here was a fixed institution, with its own clear identity. If you were a Stoic, what mattered was the other Stoics, the members of your own immediate group, with whom you cohabited and together with whom you had spent so many hours learning the works of the schools' founders. Naturally, you would spend almost all of your life debating the detail of the founders' work.

What you did not do so much, then, was converse with others. Plato spoke with Archytas, shaping the encounter of an entire generation. But throughout the Hellenistic era, philosophers spent much less time on such conversations with outsiders, and they ceased, in particular, to be much interested in science. The heirs to Plato came to emphasize the dialectics of his dialogues, and so they turned ever more to methodology, argument – indeed, to skepticism (we shall return to notice this particular chain of changing Platonism in Chapter 6: Plato will eventually once again be significant for the history of mathematics, as, in fact, he still is today). The heirs to Aristotle focused almost entirely on ethics and on moral persuasion. The Epicurean Garden was founded upon Epicurus's shrill hatred of the sciences; from this, it didn't budge. The Stoics were attached to the naive science of their own founder, Zeno of Citium, and were only obliquely engaged with any new developments. And so, with the philosophers no longer in contact with anyone outside of Athens, it is perhaps not very surprising that those outside of Athens drifted away, in turn, from philosophy. One could perhaps choose to become a philosopher, and then one would go to Athens and dedicate oneself to one of the schools. But if your passion was science, you would neither go to Athens nor care very much about what the Athenians had to say. Alexandria, with its erudite, subtle poetry, was closer. The science of the Hellenistic era was subtle, surprising, poetic – and unphilosophical. What a contrast to the generation of Archytas!

## II. Enter Archimedes

### *Archimedes the Historical Person*

Now, for a change in pace. A lot of mathematics will be accomplished over the next few decades, and for once, a lot is known.

A few precise dates, even. The most precise is Archimedes's death – 212 BCE – at the fall of his city, Syracuse. Recall that Archytas's Tarentum was the leading Greek city of its time in the south of the Italian Peninsula. Archimedes's Syracuse was the leading Greek city of Sicily. Both areas – southern Italy and Sicily – were mostly Greek-speaking areas; both cities were Doric cities in the Greek west, perhaps meaningful.

Among our extant sources for the death of Archimedes is Polybius, a sober historian who lived through the second century while the memory of the events was still fresh. (We will revisit Polybius in the following chapters; he is an important early example for the transformation of Greek culture under Roman influence.) Thanks to him, we can be confident that Archimedes made an important contribution to the defense of Syracuse, specifically by his careful siting of catapults.

The age of Archimedes at death is often reported by modern historians as seventy-five.[4] This figure is based on a statement by Tzetzes, a Byzantine poet from the twelfth century. It is patent that Tzetzes knows no more than we do, and the precise age is certainly a poetic embellishment that needs to be discarded. Polybius, however, does refer to Archimedes as being an old man during the siege, which suggests he ought to have been at least about sixty years old. On the other hand, he was also clearly very active as a military engineer – which suggests he was, perhaps, not *much*

[4] The best synthesis of the biographical evidence concerning Archimedes remains E. J. Dijksterhuis, *Archimedes* (Princeton, NJ: Princeton University Press, 1938 [1987]), Chapter 1.

older than sixty. He was most likely born around 280, perhaps a bit later. We can add some more to that. For reasons we will consider in the later discussion, the bulk of Archimedes's work can be dated to the years following the death of Conon. As mentioned earlier, we know about one historical incident in the life of Conon – an astronomer, he named a region of the sky the "Lock of Berenice," referring to a Ptolemaic queen. The circumstances of this naming are known and can be dated to 245 BCE. The balance of likelihood is that this naming was not quite Conon's last gasp of life, and so he died probably closer to 240 or even later. The likelihood, then, is that the bulk of Archimedes's work was produced throughout the years 235–215. (Remember that he spent the last two years of his life in active war.) It is typical of Greek authors to emerge fully only in somewhat older age, when they have accumulated the status of respected elders (although a few mathematicians must have emerged somewhat younger: so Theaetetus, as well as Eudoxus). There is nothing surprising, then, in the notion that Archimedes's work was produced, or at least published, primarily in his fifth and sixth decades of life. He evidently was a patient author; we will see some direct evidence for this.

We glimpse something extraordinary: the course of a mathematician's life. Even with Archytas – an important political leader! – we did not have such precision. But there is much more: a series of anecdotes that reveals, at the very least, how the ancients *perceived* Archimedes. As I did with Thales and Pythagoras, my role here is primarily that of a buzzkill. The stories probably are fabricated. They should be read, however, for the light that such fabrication may cast on the reality from which they have been carved. Let me go through some of the stories that circulated fairly early on:[5]

1. The most famous anecdote (and the one that Galileo used to launch his own career): Archimedes once pondered the case of the goldsmith, suspected of mixing silver with the gold he was assigned for his job (pocketing the extra gold). Archimedes found his clever solution – based on the properties of solids immersed in liquids – while in the public bath. In his excitement, he ran out naked, crying, "I have found it! I have found it!"
2. The king once asked Archimedes to help launch the biggest ship in the world, in competition with those of Alexandria. Archimedes

[5] Each of these stories is told in many variations – as such stories are. A sensitive historical and literary reading is found in M. Jaeger, *Archimedes and the Roman Imagination* (Ann Arbor: University of Michigan Press, 2008). (This, however, only concentrates on the reception in Latin.)

devised a mechanism to obtain this end, and as he was pulling on it and launching the ship, he remarked, "Give me a place to stand and I shall move the earth."

3. As the Romans were besieging his city, Archimedes devised machines to protect the city; these were so effective that the very sight of a rope would drive the Romans away. The Roman general remarked: "Let us stop fighting this geometrical Briareus ... who plays with our sambuca" (a pun on an ancient war machine whose name also suggested a musical instrument).
4. As the city fell, Archimedes told the Roman soldier who had come to take him, "Move away from my diagrams," whereupon the soldier killed him in rage.

In general, when we read the biographies of ancient authors, it becomes clear that these are not even intended to be fully historical and that instead, the biographers concoct anecdotes, based on the contents of the authors' works. This is clearly the case here. The story of the crown is a clear echo of Archimedes's study of solids immersed in liquids, *On Floating Bodies*. (As Galileo pointed out, the anecdote drastically flattens the actual mathematical achievement of *On Floating Bodies*.) The story of the launching of the ship is a clear echo of Archimedes's studies concerning balances, among which we now know, extant, *On Balancing Planes*. (The principle of the balance, obtained in *On Balancing Planes*, is here transformed into the principle of the *lever*, although, once again, this is at some remove from the actual instrument envisaged in the launching of the ship, which must have been the *pulley*.) Both are works in mathematical physics, which the stories tend to make even more engaged with the concrete world; this is what such stories do. And yet, there is also a sense of Archimedes as the author of abstract, abstruse mathematics, at a remove from the world: the Archimedes who does not pay attention to the soldier, concentrating on his diagrams. All four stories involve the figure of the wise man, face to face with raw political power. This is simply the stuff of anecdote and, indeed, the stuff of the Greek cultural imagination (everything was about words, in violent, *political* encounter). But the image of Archimedes, somehow in touch with political power, probably had something to do with historical reality. One of his extant works, the *Sand-Reckoner*, is addressed to King Gelon of Syracuse, and after all, Archimedes did have a prominent position in the defense of his city. He was probably, at the very least, close to the court. Certainly a member of the elite, he also enjoyed one of the important advantages of this status – namely, contacts. Recall Plato and

Archytas: members of the Greek elite were always not just creatures of their own city but also members of relations of friendship that extended beyond political borders. Archimedes dedicated one extant work to King Gelon, but he also dedicated another to Eratosthenes, a librarian in Alexandria; we have an entire extant correspondence with Dositheus (more on this in the later discussion), and he refers again and again to his previous correspondence with Conon. It is likely that all three were related to the Ptolemaic court. A patriotic Syracusan, Archimedes was definitely part of the international culture based on Alexandria.

The key fact is that so much can be gleaned from the extant writings. With the generation of Archytas, we have only fragments. Much more is available from the authors in between the generations – we have extant works from Euclid, Autolycus, and Aristarchus, but these are all entirely impersonal and (in the cases of Euclid and Autolycus) largely derivative. But a large corpus survives from Archimedes, and with him, we pass to a different kind of writing. Indeed, the main body of the treatises is written in the same genre enshrined at least since Archytas – the same diagrams, the same formulaic language. But preceding many of the treatises, we have extant letters introducing the mathematical achievement: for once, the mathematician speaking in his own voice. This will be typical for much of the mathematics of this generation, and such letters are significant, already, in their social implication. The generation of Archytas may have been launched by the encounter between Plato and Archytas. The generation of Archimedes was launched by Archimedes addressing Conon, a *mathematician* colleague. And out of a letter, from one mathematician to another, an entire tradition of the writing of mathematics emerged: the mathematicians' mathematics.

That letter to Conon, and its aftermath, will provide our entry into Archimedes's mathematics. But before we get to that letter, we need to introduce the scope of the achievement – and of the evidence.

## *The Works of Archimedes*

The survey must start from a series of works sent to Dositheus. This is because all are introduced by detailed letters that help us fix a chronological sequence:

1. *Quadrature of the Parabola* (abbreviated as QP)
2. *Sphere and Cylinder*, Book 1 (SC 1)

3. *Sphere and Cylinder*, Book II (SC II) (The titles are misleading: the two books on SC share a subject matter, and SC II assumes certain results proved in SC I, but the two are essentially independent treatises.)
4. *Spiral Lines* (SL)
5. *Conoids and Spheroids* (CS)

All five books engage with the measurement of curvilinear objects: the segment contained between a straight line and a parabola (1), the surface and volume of a sphere (2 and 3), the area contained between and a spiral and its line of origin – and the length of the spiral (4), volumes of conoids of revolution (5). We shall return to discuss some of these in depth, especially *Spiral Lines*, which I will take as something of the paradigm of Archimedean measurements.

The introductions to the works in the series refer, from its beginning, to Conon's death. A few other works by Archimedes appear to postdate this sequence. If we assume that it comes fairly early following Conon's death and that the works were spaced apart by, say, at least two years (Archimedes emphasizes the passage of time), we can envisage this sequence through the years 235–225 or so.

Another extant work is addressed to Eratosthenes:

6. *The Method* (*Meth.*)

The standard English translation of this title is somewhat misleading (in the original, it means, rather, "the approach," "the line of attack"). The main thrust of the work is to review previously published work, mostly related to the series to Dositheus, and provide alternative proofs, based on a certain technique to which we will turn later on, when we discuss the treatise in detail. Archimedes explicitly states that he relied on such alternative proofs for the very finding of his previously published results, so he probably *knew* the contents of the work early on, but the publication of the work must have come late in his career, probably closer to 215. To a comparably late date, we can probably assign also:

7. *On Floating Bodies* (FB I and II)

This is Archimedes's hydrostatics (and much more), a work divided into two volumes (which, however, are, in this case, merely the two segments of a single work). The work, in its current state, is not addressed to any individual. It does assume significant mathematical results, obtained through works by Archimedes that are no longer extant. It must come from Archimedes's final years (once again, we will revisit this in detail later on).

Other extant works are harder to date. These include the following:

8. *On Balancing Planes* (PE I and II)

Archimedes's statics (and much more) is once again a single work, divided into two volumes – and in fact, it is, in many ways, parallel in structure to *On Floating Bodies*. In general, it seems likely that FB was designed to be an echo of PE. (We note how Archimedes plots his career, with works building up in sequences.) This would make PE earlier than FB – which, however, does not teach us much. In all likelihood, it is later than QP (because in PE II, Archimedes relies on the measurement of the area of the parabolic segment). It is hard to say anything about the dates of the remaining extant works:

9. *Measurement of Circle* (DC)
10. *Sand-Reckoner* (*Aren.*)
11. *Cattle Problem* (*Bov.*)

The first is a brief study of, well, the measurement of the circle (with Archimedes's good approximation of $\pi$). It is very brief and perhaps mutilated, and at any rate, it contains no introduction. It is somewhat simple for Archimedes, and so past scholars tended to assume it was early – indeed, why not start with the most famous problem of them all, the squaring of the circle? (I will return to say a little more about the overall strategy of this treatise on pages 333–335.) The second is the one addressed to King Gelon – a medley of sorts, involving both the theory of calculation and astronomy. Gelon was born about 270 and died in 216, so this, once again, tells us nothing about the date. The third is a poem containing an arithmetical problem, surviving through a collection of Greek poetry; it is probably genuine, but it is hard to situate it any further.

What we do note is that even just counting the extant works, Archimedes has written a lot. We also note two major themes: the measurement of curvilinear figures and the physical world. We note the planned structuring of works across an entire career. And we also note that there is a lot that is harder to classify!

This becomes clearer as we bring in other works that Archimedes certainly produced but that are now known only in fragmentary form or through intermediate testimony:

### 12. *Stomachion* (*Stom.*)

This is the most extant of all the nonextant works by Archimedes. A brief testimony survives in Arabic, but – much more significant – the first page survives within the Archimedes Palimpsest (on which, of course, more later in this chapter and in Chapter 7). This fragment was fully read only very recently, and it now seems possible that the treatise involves the counting of the number of ways in which a certain geometrical task may be fulfilled (a combinatorial study, then, although mostly a simple calculation, based on geometrical observations). This is somewhat reminiscent of:

### 13. *Semi-Regular Solids*

Pappus reports in detail the result obtained by Archimedes, according to which there are exactly thirteen semiregular solids. What are these? Well, the five regular solids (the topic of *Elements* XIII) are such where all the faces are the same regular polygon; all the faces meet in exactly the same vertex combination. A cube is a simple example (six squares, with each vertex having three squares orthogonal to each other). In semiregular solids, the demand that all faces be the same regular polygons is relaxed, and one has, instead, *various* regular polygons as faces. It is still required, however, that they meet, always, in exactly the same vertex combination. The rhombicuboctahedron – with each vertex as the meeting of three squares and one triangle – is probably the easiest to picture (see the following image), but many of the forms are much harder, and one wonders if Archimedes pictured them to himself at all. (How about a combination of thirty squares, twenty hexagons, and twelve decagons?) Pappus provides the specifications for the thirteen solids. Those specifications are correct, and those are all the nontrivial semiregular solids (otherwise, any prism based on any regular polygon is a semiregular solid in a somewhat trivial sense; it is not known how Archimedes ruled them out). One concludes that Archimedes must have obtained a proof that these are all the "real" semiregular solids there are. This proof Pappus does not report, and the earliest to rediscover this (very difficult) proof was Kepler, in the seventeenth century (and even he would have found it much harder to determine the proof, had he not had Pappus's report with its correct set of all such solids).[6] So, once again: a complicated – indeed, amazing – geometrical calculation.

[6] The story of the rediscovery of the Archimedean solids is told by J. V. Field, "Rediscovering the Archimedean Polyhedra: Piero della Francesca, Luca Pacioli, Leonardo da Vinci, Albrecht Dürer, Daniele Barbaro, and Johannes Kepler," *Archive for History of Exact Sciences* 50 (1997): 241–289.

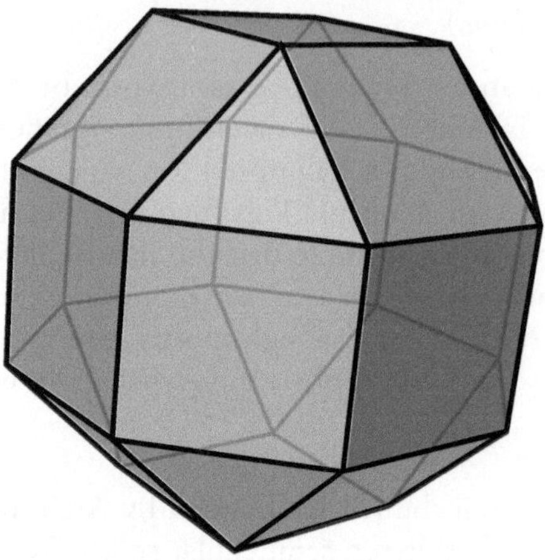

Figure 39

Another lost work is reported by Pappus, and in this case, it can be dated, in rough terms, to late in Archimedes's career:

14. *Alternative Spiral Lines*

Pappus provides, in this case, the outline of the main proof (and we will indeed consider this in detail later on). To recall, in the extant *Method*, Archimedes revisited previous results and described how they could have been obtained alternatively, through a less rigorous technique. This technique discussed in the *Method* applied for many of Archimedes's past results but *not* to *Alternative Spiral Lines*. Instead, Archimedes dedicated, apparently, a separate (small?) treatise to showing just that: how a less rigorous technique – different from that of the *Method* – was used by him to derive a result concerning spirals. Because this work evidently looks back at a (late) piece of the correspondence with Dositheus, it must be fairly late itself, as well. The opposite is true of the following:

15. *On Balances*

This produced results akin to those of *On Balancing Planes* but derived in a more directly mechanical way (considering not the abstract proportions of the law of the balance, but concrete balances). It appears that it was this treatise that could have been a tool used in *Quadrature of the Parabola*, the first of the works sent to Dositheus; if so, it could be a distinctly early work, which makes its more "mechanical" nature intriguing (note, however, that our sense of the work is very flimsy: all dependent on an interpretation of a few passages in *Quadrature of the Parabola*).

Three more works are also dependent on internal references, which, however, provide few clues:

16. *To Zeuxippus* (a treatise on calculation, referred to in *Sand-Reckoner*)

Unfortunately, we have no idea when the *Sand-Reckoner* was written! It does appear that Archimedes wrote a more formal statement of his original numerical system, however.

17. *On the Centers of Weights of Solids* (a treatise where one found the centers of the weight of solids, referred to in *Meth.*)

This tells us little chronologically because the *Method* is certainly very late. It is interesting (but not surprising) that Archimedes extended the study of the centers of weights of planes – the subject of *On Balancing Planes* – to the study of the centers of weights of solids. (It seems likely that this would be published later than *On Balancing Planes*, then.)

There is also the following:

18. An appendix to SC II, solving a particular problem implied by the treatise

This, finally, need not be considered entirely lost: Eutocius, in a late commentary to Archimedes, claims that he has found this treatise, and he quotes at least a portion of it verbatim, perhaps with his own additions. I shall return to this episode on pages 428–430. If this was at all published by Archimedes, the publication must have postdated that of *Sphere and Cylinder* II.

The following are even harder to place:

19. *Arbelos*

Pappus also reports that Archimedes has introduced, and found surprising results concerning, a special object, derived from certain combinations of circles (comparable, in a way, to Hippocrates's lunules), called the "Arbelos." (Archimedes's own naming? The word refers to a kind of knife. In general, note the fancy titles found throughout Archimedes's corpus – we catch, already, a glimpse of a certain playful attitude.)

20. *Heptagon*

Arabic sources report that Archimedes produced a construction of the regular heptagon. Although the attribution is late, Arab sources do seem to know a good deal about Archimedes. The construction is complicated and ingenious, and it is hard to see why, had it been by someone other than

Archimedes himself, it was not properly claimed. (A few more works are attested similarly via the Arabic tradition, of which the likeliest to have an actual Archimedean source are *On Tangent Circles*, *On Lemmas*, and *On Assumptions*. None is certain, but then again, had it not been for the Palimpsest fragment, we would likely have dismissed the Arabic testimony on the *Stomachion*, as well.)

Further, Greek authors sometimes refer to results by Archimedes that are not to be found in the extant corpus. They could be mistaken in each given instance, but the evidence does add up to suggest the loss of several additional contributions to pure geometry, perhaps as follows:

21. *On the Measure of a Circle* (Or are those references to an original, more expansive edition of what we now have as DC? But then again, we do note Archimedes's habit of returning to the same subject from different angles: Why not multiple works on the measurement of the circle, then?)
22. *On Plinths and Cylinders*
23. *On Irregular Surfaces and Solids* (Or is it the same as the reference by Pappus to semiregular solids?)

There are also references to work going beyond pure geometry, expanding Archimedes's scope even further: perhaps the following:

24. *Mechanics* (This would be one that engages not just with the theoretical concept of the center of the weight but also with its actual mechanical significance. Or is this the same as *On Balances*?)
25. *Catoptrics* (This, we are told, was an especially large work, and apparently, it was foundational to Greek thinking about the theory of reflection and refraction; I return to this on pages 204–206).
26. *Sphere-Making* (This, most likely, described the making of a mechanical planetarium, almost certainly comparable, in some sense, to the Antikythera mechanism, to which we will return in Chapter 5. It seems that Archimedes produced such a planetarium himself, and indeed, the idea of such a device became associated with his name. It is possible that ancient reports that Archimedes measured the length of the year belong to a treatise such as this, but it is also possible that he produced such measurements in yet other lost treatises: *Aren.*, too, includes its own astronomical measurements.)

There is some fuzziness in the numbers, surely with some double-counting already, so I will stop here. Even with such double-counting, twenty-six is surely an undercount of the original number of works (it is clear, from what we have seen, that the distribution of Archimedes's corpus carries a long tail of works attested only once or twice; there must have been many more, now attested *zero* times). Most of Archimedes's career, I argue, has to be fitted within twenty years or so. Throughout this period, he would have sent out, perhaps, two treatises a year – perhaps more? – sometimes very substantial, always breathtakingly brilliant. Let us go back to pursue this correspondence, starting with Archimedes's own, most explicit account. We start with the introduction to *Spiral Lines* and Archimedes's report on the letter to Conon.

## *The Challenge of Archimedes*

Here is Archimedes himself, in the letter to Dositheus introducing *Spiral Lines*:[7]

> Archimedes to Dositheus, Greetings.
>
> Of those theorems dispatched to Conon, about which you keep sending me letters asking that I write down the proofs – many you have, written down in the books conveyed by Heracleides, while some I send you, having written them down as well in this book.
>
> [By "Theorems dispatched to Conon," Archimedes means a set of statements that he has produced, as claims without proof, in a letter sent to Conon quite a few years back. These included, for instance, the bold claim that "the area of a spiral is one-third the circle" (more on this later on), stating that Archimedes possesses a proof of that bold claim, withholding the proof – and demanding others to show their mettle by trying to produce their own proofs. When Archimedes says that Dositheus has many of those already, this is because *Spiral Lines* is a late work: it was sent to Dositheus after *Quadrature of the Parabola*; *Sphere and Cylinder* Books I and II were sent already. Several of these works contribute to Archimedes's proofs of the theorems originally sent to Conon as challenges.]
>
> Nor should you wonder why I took such a long time publishing their proofs: this came about because I wanted to allow those who busy themselves with mathematics to take up studying those theorems first. For how many of the theorems in geometry appear not to go along the right lines at first, to the one who eventually perfects them? Conon passed away without taking sufficient time for their study; otherwise he would have made them all clear, discovering them as well as many others, while advancing geometry a great deal: for we know that his mathematical understanding was extraordinary, his diligence unsurpassed.

[7] In the following sections, Archimedes's *Spiral Lines* serves as entry point to Archimedes's scientific personality. The work is translated in R. Netz, *The Works of Archimedes: Translation and Commentary*, vol. II (Cambridge: Cambridge University Press, 2017). The passage cited here is on pages 17–19.

With many years now having passed since Conon's death, we are not aware of even a single problem being set in motion, not by a single person.

[The respect given to Conon seems real enough. It is also evident that the entire series of letters to Dositheus – starting from *Quadrature of the Parabola* – must have begun well after Conon's death, hence not earlier than about 235 BCE. What is most important is the overall competitive thrust. Archimedes names, and respects, the dead Conon, but toward his living contemporaries, his attitude is of fierce competition – and hardly a respectful one. Which is, indeed, how the letter to Conon came about: as a *competitive* challenge. That apparently no one took up! It was, after all, the in-between generation . . .]

I also wish to set out each of them, one by one. For it happens that [there are] a certain two of the [theorems] in it [= the letter to Conon], not distinguished apart but added at the end, so that those who claim to find all of them, but publish none of their [= the theorems'] proofs, would be refuted by promising to find solutions to impossible theorems. So I now find it appropriate to make clear which are those problems, and which of them are those whose proofs you have (which were sent to you already), and which I convey in this book.

[This is perhaps the strangest twist. The competitive nature of the letter to Conon was real and not very amicably intentioned, either. Archimedes, of course, made a bold claim: "I have provided proof to all those results." But at the same time, he asked much more of his potential respondents: to play the game, they had to provide the actual proofs explicitly. But of course, nothing prevented them from simply asserting – repeating Archimedes's own bold assertion – that they, too, have found those results, perhaps even independently. To prevent this – Archimedes now reveals *in retrospect* – he planted two poisoned results. Two of the claims were wrong, so anyone claiming to have proved them all *would be revealed as a fraud*. Furthermore, as a kind of insurance against anyone revealing the falsity of the poisoned claims and implying that Archimedes made an error, Archimedes started out by positioning the false claims in the wrong order (to provide some credibility to his counterclaim, in such case, that he knew all along that the claims were false and planted them deliberately).]

I go through this letter in detail because it is our most significant evidence for the sociology of Archimedes's mathematics: a competitive network. But of course, this means nothing until we get from the sociology to the mathematics. Just what are the claims made in this letter? And how are the two, sociology and mathematics, connected? How does the nature of Archimedes's communication help to explain what Archimedes found worthy of communication?

There is a clear line passing through the entirety of the correspondence with Dositheus – beginning already with *Quadrature of the Parabola.*

(Not on the list of promised results: the very first item in this correspondence, sent out while Archimedes was still waiting for responses to his challenge, was an "extra," so to speak, whetting the readers' appetite in preparation for the big reveals.)

In *Quadrature of the Parabola*, Archimedes's main result is that a segment of a parabola (the area contained between a parabolic line and a line cutting through it) is 4/3 the triangle it contains (whose base is the line cutting the parabola and whose vertex is the vertex of the diameter bisecting that base line). This result is obtained twice: once in a surprising approach that involves weighing slices of the parabolic segments on conceptual balances and then in another that is purely geometrical. (I will briefly return to this work – with its physical overtones – on pages 186–187.)

In the next item in the correspondence, *Sphere and Cylinder*, Book I – which creates the tools for dealing with a part of the challenge sent to Conon but does not yet solve any of its problems – Archimedes's key results are the measurements of the surface and volume of a sphere. Its surface is four times the great circle in the sphere; its volume is four times the cone whose base is the great circle; and its height, the radius of the sphere. This is then expanded to results concerning the areas and volumes of a *segment* of a sphere.

In *Sphere and Cylinder* Book II, Archimedes finally provided solutions to some problems set out in the challenge. Some are straightforward once SC I is available (for instance, to find a sphere equal to a given cone), but some are still very hard: for instance, to divide a sphere into two segments so that their ratio is a given one. (Recall that Euclid, in his *Divisions*, was solving this type of problem for a much simpler case: *plane*, *rectilinear* figures.) Especially difficult arguments are those refuting (implicitly) the wrong claims of the letter to Conon. For instance, one of the false claims was (in modern paraphrase) that if a sphere is divided into two segments, with volumes $V_1$, $V_2$ and surfaces $S_1$, $S_2$, then $V_1:V_2::S_1^2:S_2^2$. Archimedes shows that, in fact,

$$S_1^{1.5} : S_2^{1.5} < V_1 : V_2 < S_1^2 : S_2^2.$$

(Yes, Archimedes has a concept equivalent to "the exponent 1.5.")

*Spiral Lines* – from whose introduction we started – came next. It contains two key results. First, the area contained between a single rotation of a spiral line, as well as the line at the origin of the spiral, is one-third the circle circumscribing that spiral. (To recall: later yet in his career, Archimedes would reveal an alternative, heuristic way in which he had obtained this very result. All of this we will review in detail later on.)

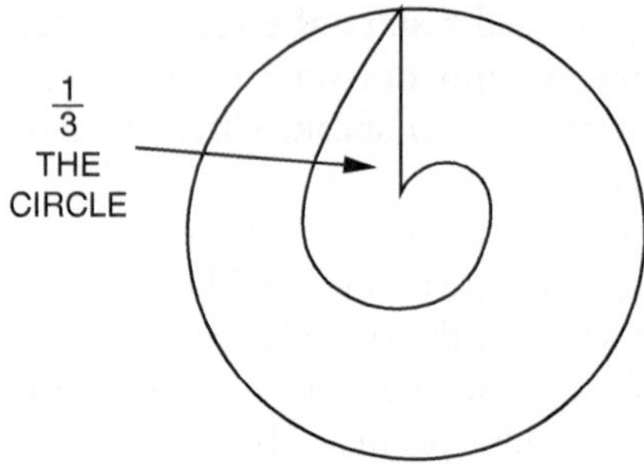


Figure 40

Second, in a configuration such as that in Figure 41, if QZ is a tangent to the spiral at the point Q and AZ is perpendicular to the line AQ at the origin of the spiral, then AZ is a line equal to the circumference of the circle that circumscribes the spiral line.

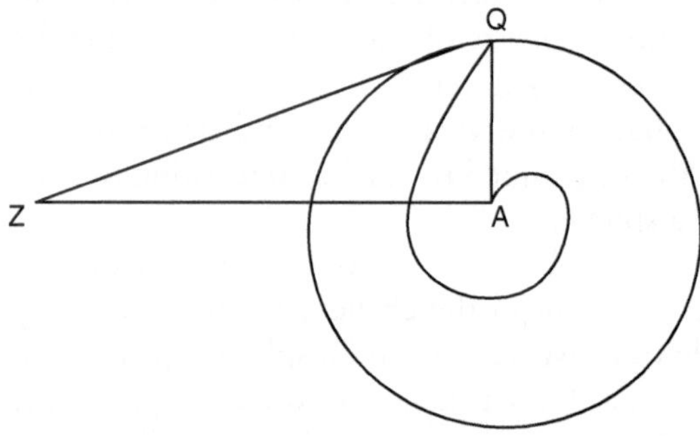


Figure 41

Both results are expanded to several more complex (but essentially equivalent) cases where instead of considering a single rotation of the spiral, we consider several rotations of the spiral or even positions obtained in between full rotations.

Finally – *Conoids and Spheroids*. The title refers to various solids obtained from the rotations of conic sections. Take a parabola – or a hyperbola – and cut it with a line passing orthogonally to the axis. Now rotate it around that axis, and you get a solid. In the case of a parabola, this will be of a more pointed shape; in the case of the hyperbola, this may be more bowl-like. Either way, you get a curved shape with a flat base – so, a little like a cone. Hence the name, *conoid*. Or you can take an ellipse and rotate it around one of its axes; the result, a symmetrical egg shape, will be curved all around and so a little like a sphere, hence the name *spheroid.* Archimedes is typically interested in segments of spheroids, which are therefore not all that different, visually, from conoids. Archimedes shows that a conoid is 1.5 times (in volume) the cone it contains (namely, the one having the same base and the same vertex), and he shows that a segment of a spheroid is twice (in volume) the cone it contains (namely, having the same base and the same vertex). All the results are shown first for the case discussed earlier – cuts perpendicular to the axes – and are then expanded to the case of an oblique cut.

In each treatise (but also through the correspondence as a whole), there is an emphasis on doing more than one thing. A proof for the area of the parabola – and yet another. The surface of the sphere – but also its volume – but also both, for sectors. The area of the spiral (but not only that – let's also throw in the line equal to the circumference of the circle), but also the same for cases with more, or less, than a single rotation. Conoids, but also spheroids – but also the same, for oblique segments. "You won't believe what I can do." Followed by several iterations of, "And yet, I can do even more."

The series ranges widely but is also narrowly defined: measure curvilinear objects in terms of simpler objects. The objects may be more or less far-fetched, but they all seem, at first glance, to defy measurement. Specifically, they all seem rather like the circle. And yet, Archimedes accomplishes their measurement – and then some.

And so, even at this overview, at some distance from the mathematical detail, a clear pattern emerges: engaging with maximally puzzling, apparently intractable tasks – with the supremely surprising, and therefore supremely satisfying, result that the tasks can be fulfilled. This is, indeed, competitive mathematics, seeking surprise above all. The mathematics – in line with the sociology.

## *Closer to the Spiral*

When I say that results are remarkable or surprising, this means nothing without a sense of the actual mathematics. It would be impossible to survey all of Archimedes's achievements, but we must have at least one closer look. I choose a relatively simple but spectacular achievement: the measurement of the area of the spiral (the simpler of the two key achievements of the treatise *Spiral Lines*).

Imagine an analog watch with a minute hand. Let it start at the top (the hour 12) and rotate for a full circle – for an entire minute. Meanwhile, imagine that there's a dot – you can imagine it as a minute ant – that starts out at the center of the watch, right at the bottom of the minute hand. Imagine that it crawls, at uniform speed, along the minute hand, from the bottom to the top, just as the hand keeps rotating. At the beginning of the minute, the ant is precisely at the center of the circle; at the end, the ant is precisely at its top. And through the minute, the ant had drawn a spiral line.

We can further imagine the area contained between: this spiral and the beginning (or end) position of the minute hand. This area, Archimedes proves, is exactly one-third of the circle.

Archimedes follows the basic proof – for the case of a simple rotation – with proofs for the more confusing situations arising with several rotations or the position of a spiral in mid-rotation. Although more confusing, the proofs of those cases are direct extensions of the same case, and so, to simplify my account, I will concentrate on just the basic proof for a single rotation. To clarify, then: what follows is a brief account of the simpler case, of the simpler proof, of Archimedes's *Spiral Lines*.

The result in question is obtained in proposition 24 of *Spiral Lines*: the curved area marked by the spiral and the beginning line is one-third of the circle. How is this achieved?

From the perspective of a reader trying to follow Archimedes's proof, it matters that there are two strands of proof – not only the one concerning the area but also the one (as noted on p. 140) concerning the finding of a line equal to the circumference of a circle. The two strands of proof are, in fact, independent – neither argument depends on claims required by the other (other, that is, than the basic definition of the spiral). That the treatise will touch on both results is made clear in the introduction. The information, however, that they are logically independent is crucially withheld from the reader; nor is it ever made clear, while reading the text, which train of thought is pursued at any given moment.

The reader is first provided with two propositions (1 and 2) in the geometry of points in motion on a line. (These have only a very general bearing on the results at hand, which is typical: an Archimedean treatise often begins with near-trivial, nearly irrelevant results, so to speak, lulling the reader.) This is followed by two propositions (3 and 4) concerning the finding of lines greater or smaller than given lines or circumferences, and then by very specific propositions (5–9) concerning the finding of lines, positioned between circles and other lines, so that they fulfill certain proportions. These are meaningful propositions already, but their function is not yet clear. Even less clear is the function of the following propositions 10 and 11, which are unlike anything else in Greek mathematics. To give a sense of the complexity, I will need to cite the general statement of proposition 10:

> If however many lines are set in order, exceeding each other by an equal [difference], and the excess is equal to the smallest [line], and other lines are set equal to them [= the lines exceeding each other] in number while, in magnitude, each is equal to the greatest [line] [= among the lines exceeding each other], the squares on the [lines] equal to the greatest [line] [= among the lines exceeding each other], adding on both: the square on the greatest [line], and the [rectangle] contained by both: the smallest [line] [= among the lines exceeding each other], and the [line] equal to all the [lines] exceeding each other by an equal [difference] – shall be three times all the squares that are on the [lines] exceeding each other by an equal [difference].

This is nearly impenetrable, and the proof for this claim is as opaque, if not more so. One has to read this proposition with infinite patience merely to understand what it means, let alone to follow its logic. Tucked at the end is a corollary; I shall return to explain its meaning immediately.

Following on this strange interlude of propositions 10 and 11, the text of *Spiral Lines* proceeds to the business of proving the result concerning the line drawn to the tangent. This is a difficult, subtle sequence of proofs, culminating in the series of propositions 18 (for the case of the single rotation), 19, and 20 (for more complicated cases).

Following that, in propositions 21–23, a new idea altogether is developed. This is best explained in the case of the single, complete rotation (proposition 21). I will follow this argument in paraphrase, avoiding Archimedes's rigorous (and less accessible) approach and supplying, instead, a more intuitive account.

Suppose you have the spiral in its rotation, inscribed within a circle. You can draw lines from the center of the circle, with fixed angles between them, which then divide the circle into equal sectors ("pizza slices"). Each also takes a slice out of the spiral area.

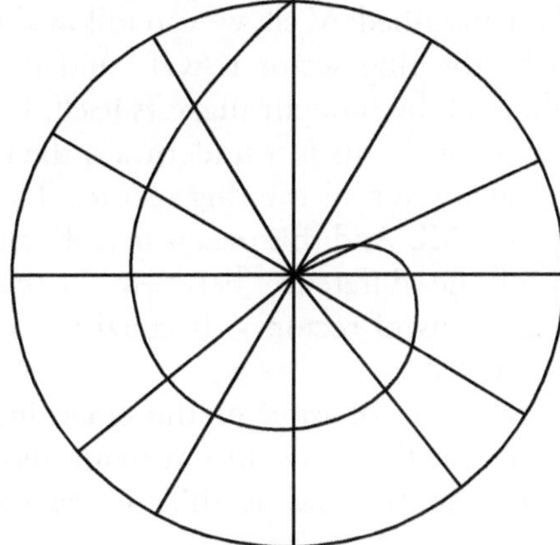

Figure 42

Now add the following: every time a line cuts through the spiral, draw there a two-arc circumference, such as OLM, around the point L; PNR, around the point N; and so forth. The outcome is that alongside the big sectors of the circle, we also have a sequence of smaller sectors surrounding the spiral line.

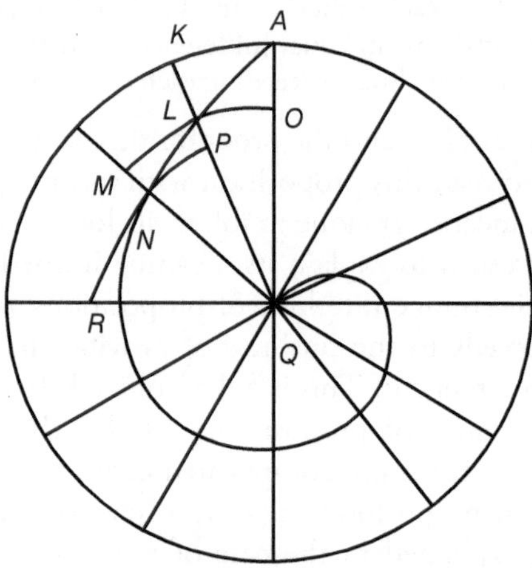


Figure 43

We may consider in particular two combinations of such sectors. We may consider the combination that exactly circumscribes the spiral area: adding up the sectors QAK and QLM and QNR, and so forth. And we may consider the combination that is exactly inscribed inside the spiral area: adding up the sectors QOL, QPN, and so forth.

What is the difference between the two combinations of sectors, the circumscribing and the inscribed? Well, we can follow this sector by sector. The difference will be the ring sector KAOL, and then the ring sector MLPN, and so forth. And the thought suggests itself: I can rotate the ring sector MLPN so that it snugly fits just underneath the ring sector KAOL. And I can do so repeatedly for all the ring sectors. Then they will all fit inside the single sector QAK. And this means that the sum of all such ring sectors – the sum of the difference between the circumscribing and circumscribed combinations of sectors – is equal to QAK or, simply, to a single sector of the circle.

Because one can take any division of the circle into however many equal sectors (one may cut the pizza into as many slices as one wishes), each individual sector can become as thin as we wish. And because the difference between the circumscribing set and the inscribed set is equal to a single sector, this also means that we can make the difference between the circumscribing set and the inscribed set as small as we wish. Thus, we have been given, right now, an algorithm that allows us to find sector combinations, circumscribing and inscribed within the spiral area, such that their differences are *smaller than any given magnitude*. (All of this, of course, rings somewhat familiar to anyone who has worked with limits and the calculus, which is, ultimately, because limits and the calculus are ways of making sense of Archimedes; more on this in Chapter 7.)

This, then, is achieved in proposition 21 (with more complicated variations for situations other than a complete, single rotation in propositions 22 and 23).

With this, Archimedes can finally spring into action, and proposition 24 proves the main result. Here is how this goes (once again, somewhat paraphrasing to make the argument more accessible). Our goal, to recall, is to prove that the spiral area is exactly one-third of the circumscribing circle. Now, says Archimedes, let us assume this is *not* the case. Well, then, the spiral area will either be greater or smaller.

Let us assume that it is smaller. (I provide an overview of only this horn of the dilemma; Archimedes deals with both exhaustively, but the one line of argument is, in fact, not hard to transfer into the other.) Well, because the spiral area is smaller than one-third of the circle, it is smaller than that by *a certain given magnitude*. We have an algorithm for that! As you recall, proposition 21 shows how to create circumscribing and inscribed areas around the spiral line so that their difference is smaller than any given magnitude. So, this is what we shall now do: draw the circumscribing and the inscribed figures so that their difference is smaller than the amount by which the spiral area is smaller than one-third of the circle:

(circumscribing figure) – (inscribed figure) < (1/3 the circle) – (spiral area).

(This kind of typographic setting out of the argument is necessary for modern readers; Greek readers, trained on the formulaic language of Greek mathematics, picked up the argument without such symbols.)

If the difference between the circumscribing and the inscribed figures is smaller than the given magnitude, then, of course, the same is true for the even smaller difference between the circumscribing and the spiral area itself:

(circumscribing figure) – (spiral area) < (1/3 the circle) – (spiral area).

The obvious conclusion is that the circumscribing figure, composed of sectors, is smaller than one-third of the circle. This much is settled:

$$\text{(circumscribing figure)} < (1/3 \text{ the circle}).$$

At this point, Archimedes quickly recalls the language of proposition 10 (still opaque to the reader!) and concludes that this is impossible. This is both amazing and frustrating, and so we will now make a long detour to understand what, on earth, Archimedes is even doing. I emphasize: the following explanation is all missing from Archimedes himself and has to be supplied by the reader!

And so, let us think a little more, to begin with, about the circumscribing figure. It is composed of sectors, all with equal angles but with different lengths. The lengths, in fact, form an arithmetical progression – they became longer by a fixed amount in each step. We may recall one of the very first observations we made, with Hippocrates of Chios's lunules: plane figures scale in a fixed manner, according to the square of their sides. If I double the side of the square, the square itself will be four times greater; by a related logic, if I double the side of the sector but keep its angle fixed, its area will be four times greater and, in general, a series of areas of sectors cut from the same circle, all with the same angle, will behave as the series of squares on the sides of those sectors. This means that the area of the series of the sectors, circumscribing the spiral area, behaves like the sum of squares on an arithmetical progression. If the area of the smallest sector in this series is 1, the area of the one following will be 4, then 9, then 16, and so forth..

But wait! Recall the impenetrable proposition 10. It starts out with, "If however many lines are set in order, exceeding each other by an equal [difference], and the excess is equal to the smallest [line]" – stop right there! That's an arithmetical progression; this is the same as lines with length 1, 2, 3, 4, and so forth. Proposition 10 then ends with a certain complicated combination: "three times all the squares that are on the [lines] exceeding each other by an equal [difference]" – that is, three times the series of squares on lines 1, 2, 3, 4, and so forth or three times the series 1, 4, 9, 16, and so forth.

In fact, the corollary tucked into proposition 10 provides the result we require. This is that the sum of the squares on all the lines equal to the greatest line is less than one-third of the sum of the squares on the lines exceeding each other.

Let us unpack this in our own algebraic way. Let us say there are *n* lines.

They form an arithmetical progression from 1 to $n$. The greatest, then, is $n$. And there are $n$ of them. The square on this greatest line is obviously $n^2$, and so adding together the $n$ of those, we get $n^3$.

The sum of the squares on the lines are like the following:

$$\left(1+4+9+16+\ldots+n^2\right).$$

The claim of proposition 10, transformed into our own algebraical symbolism, is

$$n^3 < 3 \times \left(1+4+9+16+\ldots+n^2\right).$$

Now, let us recall once again the precise arrangement with the circle and its sectors. In proposition 24, the circle is divided into many equal sectors; let us say $n$ sectors. There is also a sequence of sectors, gradually growing, circumscribing the spiral.

The sectors out of which the circle is made are all equal to each other and also equal to the greatest of the sectors circumscribing the spiral. (They are all equal to QAK.) There are $n$ of those.

The areas of the sectors circumscribing the spiral are, as pointed out earlier, like the squares on an arithmetical progression: they are like $1^2$, $2^2$, $3^2$, $4^2$, ... $n^2$, and their sum is obviously like $1^2 + 2^2 + 3^2 + \ldots n^2$. This is the area of the circumscribing figure.

Meanwhile, QAK is the biggest of those sectors, and it behaves, in this context, like $n^2$. (Always taking the area of the smallest sector as our unit.) The circle is made of $n$ of those, and so the area of the circle behaves, in this case, surprisingly, according to a cube value $n \times n^2$ or $n^3$.

To repeat: if the sum of the areas of the sectors circumscribing the spiral area is $1^2 + 2^2 + 3^2 + 4^2 + \ldots + n^2$, then the area of the circle is $n^3$.

What remains is obvious, although still remarkable. It has been proved in proposition 10 in general terms – in opaque, impenetrable, and unmotivated terms – that

$$n^3 < 3 \times \left(1^2+2^2+3^2+4^2+\ldots+n^2\right),$$

and from this, it follows that three times the circumscribed figure is *greater* than the circle. This, indeed, directly contradicts the hypothesis, and so Archimedes is justified in concluding that the spiral area is *not* smaller than one-third of the circle.

To repeat: most of this detour is an explanation added by me and is missing from Archimedes's text. This is not because Archimedes is less than rigorous: he clearly wants us to be flustered. Indeed, he very rigorously goes on to produce the analogous (but nonidentical) argument for the assumption that the spiral area is greater than one-third of the circle (one now will need to look at the sectors inscribed in the spiral area and apply proposition 10 somewhat differently). And once both sides of the argument are completed – once both options, "smaller" and "greater," are eliminated – only one option remains: the spiral area must be precisely one-third of the circle.

This is the argument of *Spiral Lines*, proposition 24. It is clearly harder than most of the mathematics we have seen so far; for instance, all of the foregoing relies on proposition 10, which I have black-boxed. The actual proof of proposition 10 is devilishly hard, and I did not clarify at all how Archimedes came to suspect that it might be true (more on this to come). So far, even complex arguments, such as Archytas's duplication of the cube, could be made accessible at one go, so to speak. This is no longer the case – providing the sense that this is the mathematicians' mathematics, the kind of mathematics that demands professional commitment.

But not only that: the argument is not merely hard; it is also subtle and surprising.

It is subtle in that it makes the most, in an inventive way, of small details. There is an economy to the argument: everything flows out of the key clever thought – considering the circle as a series of sectors with equal angles. This gives rise to the series of sectors circumscribing and inscribed around the spiral line, which, marvelously, contributes twice to the main argument. First, the sectors are arranged in consecutive, equal increments, and those increments of increase can be added up and be made equal to a single full circle sector. Hence, the two series – the circumscribed and the inscribed – differ from each other exactly by a *single* full sector, from which it follows, finally, that the difference between the circumscribed and the inscribed can be made as small as we wish. This is the argument of proposition 21, and it provides the possibility of a proof by contradiction, by "squeezing" an area in between the two areas that are assumed (impossibly) to be unequal to each other. But more than this: because the sectors are arranged in consecutive, equal increments, we can use an entirely distinct line of attack and apply proposition 10, considering the properties of (in modern terms) the squares on an arithmetical progression. The circle, reimagined as a configuration of sectors, is a brilliant conceptualization – with a double payoff.

But more than this: the conceptualization is not incidental to our object of study; in fact, it gets right to its essence. The entire point about spiral lines, we get to realize, is a certain geometrical proportion: equal angles correspond with equal linear increments.

Said like this, a modern reader may begin to have a glimmer of an understanding of the logical order in which *Spiral Lines* is produced. But such a statement is, in fact, anachronistic in a sense. Of course, Archimedes does get to the essence of spirals, and he teases out their properties in a brilliant, subtle way. But he does not aim to help the reader understand how all that is done. Instead, Archimedes aims at *surprise*. The key point is that as proposition 10 is introduced, Archimedes makes all efforts to *disguise* its potential application. It is sandwiched between unrelated results, and it is presented in opaque language, its key takeaway relegated to a corollary. The key observation – that the sectors in a circle behave as the series of squares on an arithmetical progression – is not asserted in advance. Instead, the application of proposition 10 is postponed and revealed only at the very last minute when, introduced in the middle of proposition 24, it finally makes sense of the argument. (To recall, in my own presentation, I made your life too easy: Archimedes never *explains* how proposition 10 applies.) Everything is designed for the sake of this denouement where, finally, the narrative of the treatise would make sense in a surprising turn. Ugly, misshapen proposition 10 is really about sectors in spirals: the duckling was a swan all along!

The idea that this treatise has a *narrative* is not a metaphor. One comes out of treatises such as *Spiral Lines* with the sense that these are carefully crafted works. Which indeed fits together with Archimedes – the author of the letter to Conon – devising a plan and then gradually, patiently executing it in the series of treatises sent to Dositheus. Across his career as a whole, and then in the detail of each individual argument, we see the same, again and again: an extraordinarily precise author, aiming throughout at subtlety and surprise.

### *Eudoxus, Inspiration to Archimedes*

We are historians, and we look for explanations. And so, the context matters. Recall our vignette of cultural life in the third century BCE, away from Athens, centered around new cities such as Alexandria. It was a time, and place, of ironic reflection. The literature of the Hellenistic era was "postmodernist," I said. Subtlety and surprise: those would come naturally. Callimachus sent out pointed epigrams, teasing rival poets; he was famous for his clever and subtle juxtaposition of historical lore in his poem, the *Aetia*, a mosaic of various pieces of erudition, wandering across the Greek Mediterranean, seemingly haphazardly, the one constant being the ironic point of view of the scholar-poet. The last of these vignettes told the sublimely ironic tale of Conon finding a lock of hair in the sky.[8] Why would Archimedes not be inspired by such poetry – which clearly inspired his entire generation?

This is certainly a part of the answer. But we need to do better than this. We need to find Archimedes's *mathematical* inspiration. In this regard, perhaps we should trust Archimedes himself. It is not surprising, I would say, that this inspiration does not point at the more immediate past – that of what I call the in-between generation. Archimedes was gracious to Conon's memory, but he does not refer to his mathematics at all. Aristarchus gets a mention from Archimedes for a particular suggestion (see Chapter 5); Euclid, apparently, is never mentioned. The in-between did not matter, to Archimedes, all that much. On the other hand, twice in the extant introductions, Archimedes conveys his sense of accomplishment by comparing his own results to those of *Eudoxus*. Indeed, both times, he refers precisely to the same result. The introduction to SC I was probably written early in Archimedes's published career:[9]

> I would not hesitate to compare them [= the results concerning sphere and cylinder] to the properties investigated by any other geometer, indeed to those which are considered to be by far the best among Eudoxus' investigations, concerning solids: that every pyramid is a third part of a prism having the same base as the pyramid and an equal height, and that every cone is a third part of the cylinder having the base the same as the cylinder and an equal height.

The introduction to the *Method* was written perhaps twenty years later:[10]

> For it is more feasible, having already in one's possession, through the method, of a knowledge of some sort of the matters under investigation, to provide the proof – rather than investigating it, knowing nothing. For this reason, the discoveries of those theorems, of which Eudoxus was the first to publish the proofs – the proof of both the cone and the pyramid, that it is a third part: the cone of the cylinder, the pyramid of the prism (having the same base and an equal height) – one should assign not a small share to Democritus, being the first to state the claim – without a proof – about the said figure.

The context is different for the two quotations. The first is straightforward: Archimedes emphasizes the value of Eudoxus's proof. The second is much more complicated: Archimedes wishes to explain why it is useful to make even

[8] For this fascinating literary vignette, see K. Gutzwiller, "Callimachus' Lock of Berenice: Fantasy, Romance, and Propaganda," *American Journal of Philology* 113, no. 3 (1992): 359–385.

[9] R. Netz, *The Works of Archimedes: Translation and Commentary,* Vol. 1 (Cambridge: Cambridge University Press, 2004), p. 32.

[10] This is my translation of the text found in R. Netz, W. Noel, N. Tchernetska, and N. Wilson, *The Archimedes Palimpsest* (Cambridge: Cambridge University Press), p. 71, ARCH15v col. 2 ll. 15–32.

nonrigorous claims (as he does in the *Method*, to which we will turn in the following discussion), by bringing in Democritus as the first to offer the claim to which Eudoxus provided the proof. (As noted in Chapter 1, I believe Democritus merely asserted that cones are essentially ledged pyramids whose ledges are so small as to become invisible.) The very fact that the two quotations are so differently motivated is reassuring, as is the fact that the two introductions have different addresses. All of this suggests that we have found something real about Archimedes's thought process. Eudoxus's result concerning the cone and the cylinder was, for him, a paradigmatic piece of mathematics.

And we are fortunate enough to possess it! Well, in the mediated form of Euclid's rendering into the *Elements*. Book XII of the *Elements* is entirely dedicated to just this combination: finding the measurement of a pyramid, in terms of a prism, which is then extended to the measurement of a cone, in terms of a cylinder. Undoubtedly, this conserves, in some way, Eudoxus's achievement. On, then, to this argument: the key, perhaps, to Archimedes's career?

The part of the argument involving the pyramid and the prism is surprisingly simple. You recall that Euclid's way of arguing for Pythagoras's theorem crucially relied on the bisection of the parallelogram by the diagonal: pass a line between two opposite angles of the parallelogram, and you have two congruent triangles. This is especially obvious in the cases of the rectangle and the square, where this bisection property must have been part of elementary education from Babylonian times on. Remarkably, we may get a similar result – although one that is far less obvious – with solids, as well. Take, for simplicity's sake, a cube. A little effort of imagination shows that it can be dissected into three components, each a slanted pyramid with the same square base (a face of the cube) and the same height (a side of the cube). Perhaps the Greeks knew this already in Democritus's time (conceivably, Democritus could have been the one to notice this), or perhaps Eudoxus was the first to notice this (and Democritus merely asserted, as a philosophical, atomistic observation, that the cone and the pyramid are indistinguishable). Either way, this is a challenging observation but one that calls for no mathematical background at all. One then needs to extend it in all sorts of ways – to show that all pyramids with the same base and height are equal to each other, regardless of their shape (hence, it doesn't matter that the pyramid is slanted); to show that, once this result is true of the cube, it can be extended to all other prisms; and finally, to show that, if true for pyramids whose base is a square, this will be true for all pyramids, regardless of which polygon they have for a base. All of this is child's play for the mathematicians of the generation of Archytas and, in fact, conceivably could have been proved rigorously even before. This is all simple and is not what inspired Archimedes. The crux was the result concerning the cone and the cylinder.

Here is how this goes, in Euclid's telling of the *Elements*, Book XII.

First, I note that the argument relies on an elaborate proof that circles are to each other as the squares on their diameters (XII.2). It is likely enough that this kind of result was always taken for granted (it is implicit in Hippocrates of Chios's treatment of lunules), and so the essence of Eudoxus's intervention, here, was to patch a logical gap – a matter of *foundations*. One is reminded, then, of Eudoxus's intervention in proportion theory. The proof of such a foundational result is difficult, and Eudoxus's breakthrough was to show how this can be achieved by a double proof of contradiction.

Assume the circles in question are in greater, or lesser, ratio; one can then derive a contradiction by considering polygons inscribed within the circles, whose ratio to the square on the diameter is independently established. Now, to prove a proportion by considering the cases of a greater, or smaller, ratio is directly reminiscent of Eudoxus's definition of proportion, and it is conceivable that geometrical cases of this kind informed Eudoxus's general proportion theory.

All of this is very abstract (indeed, there is a similar sense of abstraction in Euclid's text already: this is a study in logical foundations, after all!). It all becomes clearer and more concrete as we reach the core result of *Elements* XII: proposition 10. We need to show that a cone is one-third of the cylinder with the same base and an equal height.

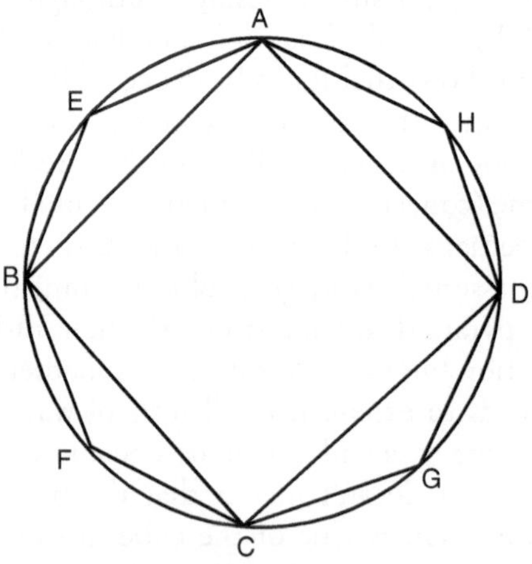


Figure 44

We look at a circle ABCD, which serves as the base of a (merely implied, in Euclid's diagram) cone and cylinder with the same height.

And once again: we assume that the cylinder is not triple the cone. In this case, it is either greater or smaller. Let us take the case that the cylinder is *greater* than triple the cone. Euclid proceeds to develop a long argument showing that it is possible to inscribe a polygon in the circle ABCD, for instance, AEBFCGDH, so that the prism set up on that polygon is smaller than the cylinder on the circle ABCD – smaller by an amount less than that by which the cylinder is greater than triple the cone. This might be confusing for a modern reader without symbolism, so just to clarify the point, let us restate this in an anachronistic format. We assume that

$$\text{cylinder} > 3 \times \text{cone}.$$

Or,

$$\text{cylinder} - X = 3 \times \text{cone} \quad (\text{where } X \text{ is some magnitude}).$$

We establish that it is possible to find a prism AEBFCGDH such that

$$\text{cylinder} - \text{prism AEBFCGDH} < X, \text{or}$$
$$\text{cylinder} - X < \text{prism AEBFCGDH}.$$

Hence,

$$\text{prism AEBFCGDH} > 3 \times \text{cone}.$$

At this point, however, we bring in the observation that

prism AEBFCGDH $= 3 \times$ pyramid AEBFCGDH (that is, the pyramid on the same base and of the same height as the prism). Hence,

$3 \times$ pyramid on polygonal base AEBFCGDH $> 3 \times$ cone on circular base ABCD; that is, pyramid on polygonal base $>$ cone on circular base.

And this is obviously false because *the cone contains the pyramid and therefore cannot be smaller.* Hence, we can rule out the possibility that the cylinder is greater than triple the cone.

A similar argument is then extended for the opposite case, and as the two alternatives are exhausted, the remaining option – the cylinder, equal to triple the cone – must be correct.

This, then, is precisely the proof mechanism adopted by Archimedes in his measurement of the spiral area. If indeed we may glimpse, via Euclid's formulation, something of Eudoxus's original approach, then we note that it might have been somewhat differently motivated from Archimedes's. The emphasis, in Eudoxus, seems to be foundational. This is all about plugging a gap in the way in which proportions are treated in the case of curvilinear objects. The point of the exercise is rigor. There is little effort to surprise: the result is first obtained for the pyramid, and then the telling of the main result, for the cylinder, is straightforward. The bulk of the proof itself is on the interim step (that one can inscribe the prism fulfilling the required conditions), which is integrated into the main proposition (Archimedes, in *Spiral Lines*, separates this lemma so that the key result proceeds much more smoothly). Finally, Eudoxus falls back on a very simple analogy, which is directly highlighted – the cone, as related to a prism (but with a circle, rather than a polygon, for a base). Archimedes falls back on an extremely opaque analogy, made further opaque by the positioning of the results throughout his *Spiral Lines* (the series of sectors around a spiral area, as the squares on an arithmetical progression). For this reason, Archimedes's result is surprising, whereas that of Eudoxus is merely impressive.

What the two share is the result: 1:3. This is, remarkably, not a coincidence – and may serve as yet another confirmation of Eudoxus as Archimedes's inspiration. And so, starting from Eudoxus, we may now understand, finally, how Archimedes could have discovered his result.

As noted previously (item 14 in the list on p. 134), later in his career, Archimedes produced a treatise with an alternative proof of the key result concerning the spiral area. This alternative-proof treatise does not survive, but it is reported in some detail by Pappus.[11]

Here, then, is the nonrigorous proof with which Archimedes became convinced that the spiral area is one-third of the circle. All we need to do is consider the following diagram:

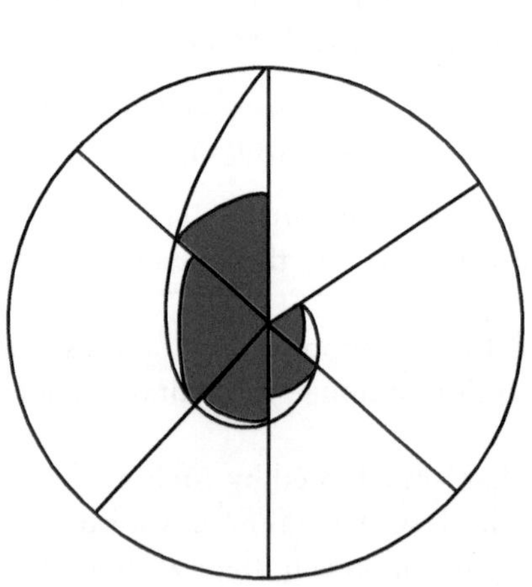

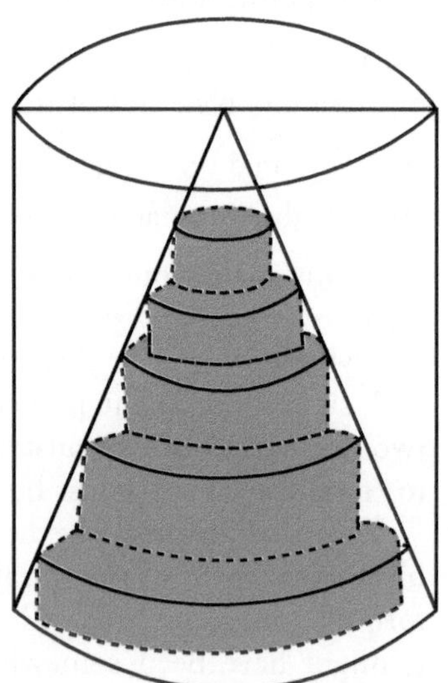

Figure 45

[11] An English translation is given in R. Netz, *The Works of Archimedes: Translation and Commentary*, vol. II (Cambridge: Cambridge University Press, 2017), pp. 181–186.

On the left, we have sectors inscribed inside a spiral line, as we are familiar with already from *Spiral Lines*, all within a circle.

On the right, we have a cone, inscribing a set of thin cylinders, all within one big cylinder.

The observation is that the two arrangements are entirely isomorphic. (The same observation can then be extended to sectors and cylinder, circumscribing the spiral/cone.)

Arrange the sector pairs (circumscribing and inscribed) from biggest to smallest.

Similarly, arrange the thin cylinder pairs (circumscribing and inscribed) from biggest to smallest.

It becomes clear, for instance, that each circumscribing thin cylinder is to its paired inscribed thin cylinder, as is the associated circumscribing sector to its paired inscribed sector. (In both cases, they stand in the ratio of the square on line segments that are in the same ratio.)

Similarly, the ratios between consecutive thin cylinders (whether circumscribing or inscribed) – arranged in sequence from biggest to smallest – are the same as those between consecutive sectors (whether circumscribing or inscribed), arranged in the same sequence. All of this is mind-boggling but ultimately exceedingly simple: you wouldn't think that, but just *look*! The set of magnitudes composed of cylinders and the set of magnitudes composed of sectors carry exactly the same internal proportions.

And once this isomorphism is seen, it becomes clear that as the series of thin cylinders (whether circumscribing or inscribed) is to the big cylinder encompassing them all, so is the series of sectors (whether circumscribing or inscribed) to the circle as a whole. And because we know already, thanks to Eudoxus, that as the thin cylinders multiply and become closer in shape to the cone, they come closer to one-third of the cylinder, it is clear that the same must be true for the spiral area. So the spiral area must, ultimately, be simply one-third of the circle.

This "must, ultimately" argument is already nonrigorous, and for this reason, Archimedes held off and never published this argument – until, that is, he already had published the rigorous proof of *Spiral Lines*. And we have to rely on Pappus's report of a claim by Archimedes, a sly and misleading author. But the claim makes sense, and I think we should trust it. For now, indeed, we may finally understand how Archimedes had obtained his rigorous proof.

Consider the sequence of thin cylinders: if they are indeed arranged as a series of concentric circles, it is clear that they stand, consecutively, in ratios defined by the square on an arithmetical progression. The smallest cylinder will have a base radius of 1, the next will have a base radius of 2, and so forth, and they are all under the same height so that their ratios are as the squares on their radii.

Adding up all those cylinders, then, is rather like adding up

$$1^2 + 2^2 + 3^2 + 4^2 + \ldots + n^2.$$

Further, the greatest of them all is like $n^2$, and $n$ of these constitute the big cylinder, which is therefore like $n^3$.

And because we've learned from Eudoxus that the cylinder is triple the cone, we have also found that, "ultimately,"

$$3(1^2 + 2^2 + 3^2 + \ldots + n^2) \sim n^3.$$

Three times the series of squares on an arithmetical progression to $n$ is somehow the same as $n$ times the square on the greatest term of that progression. This result is instantiated in the case of the cone – and therefore it must be true for any isomorphic series and therefore for *the series as such*. Thus, even though Archimedes has nothing like our algebra, he knows that the general result must be true and therefore can set out to prove it, hard as it is without algebra – and this becomes proposition 10 of *Spiral Lines*. At this point, all Archimedes had to do was present this result, tantalizingly and inexplicably, on its own – and use it to derive the maximal surprise and subtlety of proposition 24 of *Spiral Lines*.

All the while, inspired by Eudoxus. We may perhaps follow the line of inspiration. Eudoxus's motivation might have been very different from that of Archimedes. Surprise, I suggest, may not have been the goal for him. The goal was rigor. However, rigor in the measurement of curvilinear shapes necessarily involves subtlety. Ultimately, this is because the curvilinear measured involves the summation of infinitely many rectilinear terms, and the rigorous study of the infinite most naturally proceeds indirectly. Hence, to prove rigorously results for curvilinear figures, one moves, first, so to speak, sideways – to the consideration of associated rectilinear or simpler figures, and then one moves, so to speak, upward – to the meta-claim of the proof by contradiction, showing the result by ruling out the sequence of alternatives. Sideways and upward and only then back to our original goal: the subtle, indirect approach required for measuring curvilinear objects.

Looking back at the mathematical past, this particular achievement caught Archimedes's eye: the combination of a powerful result proved subtly. It is possible that all Eudoxus cared about was foundations – he simply wanted to plug a hole in the application of proportions to geometry. I suggest that Eudoxus stumbled upon subtlety. Archimedes glimpsed that subtlety and – motivated by a culture that prized subtlety above all else – made it the cornerstone of his own science. All it took was a certain narrative rearrangement, hiding the grounds of one's argument, springing them upon the reader. But more than this – a lot more could be achieved, Archimedes realized: the same proof technique could yield the measurement of many other curvilinear figures, all obtained through subtle, surprising routes. With this realized, Archimedes turned to craft his letter to Conon.

## III. Response to Archimedes: Apollonius

### *Archimedes, Inspiration to Apollonius*

> "With many years now having passed since Conon's death, we are not aware of even a single problem being set in motion, not by a single person."

These are Archimedes's words in the introduction to *Spiral Lines*. The line is reasonable enough, perhaps. According to my account, Archimedes started out at a low point for Greek mathematics. There were perhaps only a few mathematicians, most of whom thought of themselves primarily as astronomers. Perhaps there is little wonder that only a few even cared about Archimedes's esoteric challenge. The one most likely to propose his own solutions would have been, perhaps, Conon himself. But he soon died, and those who remained were perhaps not up to the challenge.

And then we need to imagine a mathematical milieu – a mere few years later – where several brilliant works by Archimedes circulate. Archimedes, in this telling, found a way of repurposing the achievement of the first generation. The subtlety required for Eudoxus's study in foundations could be remade into works of maximal narrative surprise – striking a chord with contemporary Alexandrian culture. And so, perhaps, the challenge of Archimedes revived the field, and soon there was not a handful but dozens of authors directly engaging with the kind of mathematics championed by Archimedes.

This, then, is my narrative. It is constrained by our sources and might be distorted by them. And yet, the implication of the evidence is that a group of mathematicians did indeed work primarily in response to Archimedes. This was *his* generation. For most of those authors, only a few results are now known – and those in mediated form. From only one author – Apollonius – there is a substantial survival. However, there was much more that Apollonius wrote and is now lost, and the likelihood, indeed, is that he was as creative as Archimedes himself. The little we do know suggests that Apollonius set himself precisely this task: to rival Archimedes.[12]

Consider, for instance, Apollonius's *Okutokion*, or "Fast-Delivery" (the metaphor refers to childbirth). The very tantalizing title is reminiscent of some works by Archimedes, and the key result is clearly meant to over-Archimedize. In his *Measurement of the Circle*, Archimedes found close bounds on the value of $\pi$; Apollonius, in his *Fast-Delivery*, has found much closer bounds. What were those bounds? We do not know – and have merely the statement, from a late commentary to Archimedes's *Measurement of the Circle*, that Apollonius's bounds were closer. Apollonius's memory – surviving, barely, in the shadow of Archimedes. Or consider Archimedes's *Sand-Reckoner*. This playful exercise starts out from the poetic trope that some things are as "innumerable as the sand." In a mock-serious way, Archimedes tries to make precise mathematical sense of this phrase. He defines this to mean "the number of the grains of sand is such that no number can be named to exceed it." He then develops a calculatory method of number-naming, allowing him to name very big numbers, and pursues an astronomical calculation whereby the universe is assumed to be very big, showing he can name, then, a number bigger than the number of the grains of sand that fill the entire universe (I briefly return to this in Chapter 5). Such is the mathematics of this Alexandrian generation: playful, engaging with poetry, trying out calculation for its own sake. (Also, as we recall, courtly: this work by Archimedes is addressed to the *king* of Syracuse!)

This, once again, served as a foil to Apollonius. Pappus – the source for so much of our knowledge of Archimedes's lost works – wrote a commentary on an otherwise-lost work by Apollonius, of which the last few propositions survive (as a fragment of the extant Pappus's *Collection*, Book II). There, Apollonius develops his own calculatory method, very clearly rivaling that of Archimedes (both involve, effectively, calculations with exponents), and then Apollonius shows how this method can be used to calculate ... a line of poetry! Specifically, what Apollonius does is to take a hexameter poem as a series of numbers (as will be explained in the

12 As noted in the Suggestions for Further Reading, the best study of Apollonius is M. N. Fried and S. Unguru, *Apollonius of Perga's Conica: Text, Context, Subtext* (Leiden: Brill, 2001). This, however, is really focused on Apollonius's major work, the *Conics*, and we simply do not have a study surveying the totality of Apollonius's corpus. For this, the best resources remain Heath 1921 II.175–II.196, supplemented by the article in the *Dictionary of Scientific Biography* by G. J. Toomer.

following chapter, the Greeks often operated with a number symbolism whereby letters represented numbers). To calculate a line of poetry, then, was, for Apollonius, to multiply its letters-taken-as-numbers by each other. Because a line of poetry has several dozen such numbers, their multiplication easily gives rise to spectacularly big numbers. Once again, our sources are what they are, and we cannot *prove* the relation, yet it seems clear enough that Apollonius, here, directly tried to rival Archimedes's *Sand-Reckoner*.

But what, then, of Apollonius's study of the *Cochlias* – entirely lost save for a brief notice by Proclus? It appears that Apollonius studied (and also invented and named?) this curved line, a three-dimensional spiral drawn on the surface of a cylinder: such a treatise, I suggest, would inevitably be read by Apollonius's contemporaries – familiar with Archimedes's own *Spiral Lines* – as an attempt to rival the master. Our evidence is always inferential, but it adds up, and so we may bring in yet another connection, more obscure but also, ultimately, more important. As we shall return to consider in Chapter 5, Archimedes produced a planetarium that, quite likely, involved not merely a method of calculating the positions of the planets but also a geometrical model. Whatever Archimedes may have offered in terms of such a mechanism, Apollonius made a contribution of lasting implications. As we will note in Chapter 5, Apollonius suggested, or commented upon, two possible elaborations of circular motions (the combination of the motions of two circles, superimposed on each other, or a single circle, rotating around a point other than its proper center) – and then demonstrated the equivalence of the two. This pair of techniques will remain to define astronomy for almost two millennia more – although nothing remains of Apollonius's work, save for a few words in Ptolemy's *Almagest*. But this is already suggestive: indeed, it is likely enough that Apollonius was reflecting, directly, on the operation of mechanisms. An astronomical mechanism, proposed by Archimedes – answered by a key theoretical observation concerning such mechanisms, offered by Apollonius? A speculation, made likely by the context. We shall return to all of this in Chapter 5.

Apollonius, doing so much – with so little now extant. We know of many other works by Apollonius. He was far from a mere echo of Archimedes. Indeed, specifically, it seems that many of his works were directed not only at Archimedes but also at *Euclid*. (In this, indeed, Apollonius differs from Archimedes: perhaps the passage of time made Euclid – now entrenched as a central resource – loom larger. We shall return to the place of Euclid in mathematical education in the following chapter.) Many of the works now attributed to Apollonius have an air similar to that of the Euclidean work *Divisions*, mentioned on page 122. These are systematic surveys based on the minute variation of a single problem. In his *Collection*, Pappus summarizes the contents of no less than six such works by Apollonius: *Cutting Off of a Ratio*, *Cutting Off of an Area*, *Determinate Section*, *Inclinations*, *Tangencies*, *Plane Loci*. The first of these works is even extant, in Arabic translation. It is not necessarily an inspiring work (Archimedes would never write such a laborious text). But it is also, at places, quite hard-going and subtle. The sense of mathematical craftsmanship is evident, even throughout the ponderous elaboration of case after case. We must have lost rather more mathematics – possibly, more entertaining mathematics – in other responses to Euclid, which are now merely reported. A commentary by Pappus to Euclid's *Elements* Book X is extant in Arabic translation, and a brief comment there reveals that Apollonius did engage with this Euclidean material (almost the only one to do so, in antiquity), obviously in a competitive way: Apollonius's own work was titled *On Unordered Irrationals*. Elements X concludes with the observation that its classification is not exhaustive; it seems that Apollonius could have taken this as his starting point, providing a subtler and nevertheless not-yet-exhaustive, alternative classification (we note a tendency in Apollonius: subtlety above all, produced by sheer precision and elaboration). Another work took as its starting point Elements *Book* XIII, the study of the five regular solids. We know about this because Hypsicles – a later mathematician (Hellenistic? of the Roman era?) responded to Apollonius himself in a work that is still extant. (It survives in the manuscript tradition, appended to Euclid's *Elements* and titled, simply and misleadingly, "Book XIV.") Hypsicles tells us that Apollonius, in his original study, had a dodecahedron and an icosahedron inscribed in the same sphere (this is the arrangement assumed by *Elements* XIII, which is all about the inscription of solids in the sphere). From this, Apollonius derived the surprising result that the ratio of the surfaces of these two solids is the same as the ratios of the solids themselves (what we would call their volume). This interest in the ratio between surfaces and volumes in solids is already directly reminiscent of Archimedes's letter to Conon. But we also recall Archimedes expanding on Book XIII, in his own way – through his study of semiregular solids; was Apollonius responding to both Euclid and Archimedes, here? Perhaps. This sense of a double response, to Euclid and to Archimedes, is clearest in Apollonius's greatest work – and the one best attested today: the *Conics*. Indeed, we have mentioned the conic sections frequently in previous pages. It is time for them to retake center stage.

## *Enter the* Conics

We last left the problem of finding two mean proportionals – or of "duplicating the cube" – at the hands of Menaechmus (p. 104). To recall, Menaechmus looked for line segments B, G that, with A and E given, satisfy

A : B :: B : G :: G : E.

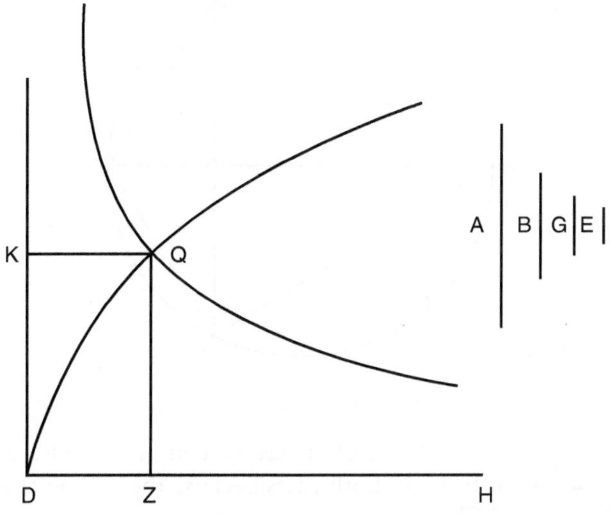


Figure 46

His breakthrough was to consider the other conditions satisfied by such line segments and realize that these would give rise to two types of curves (which, therefore, can be invoked to solve the problem at hand). One curve is such where the square on a determined line (such as QZ in Figure 46) is always equal to the rectangle contained by another associated line (such as DZ) as well as by some fixed line (such as A). Another is a curve that ensures that all rectangles drawn against it in a particular way (such as KQZD in Figure 46) always have the same fixed area.

It is possible that at first, Menaechmus simply left it at that and was satisfied to show that curves such as these *would* solve the problem, whether or not they in fact "existed" (whatever existence means, in this context). But the thing is, the curves are very real.

Having found the description of the beast, let us now go, so to speak, and hunt for it in the wild.

Let us concentrate on the example of the curve, where a certain square is equal to a certain rectangle. This looks promising because this kind of arrangement arises very naturally and is, in fact, a recurrent theme in our considerations of the problem of finding proportionals. Consider, once again, the diameter of the circle and the property of similar triangles arising when two chords in a circle cut each other. Specifically, one of the chords we take to be the diameter, and the other we take to be perpendicular to it:

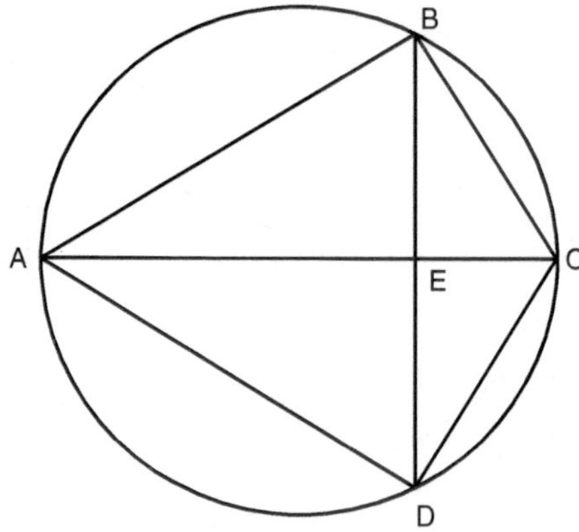


Figure 47

Because the triangles ABE, DCE are similar (the angles at A, D are equal, as are those at B, C, in both cases because they lie on the same arc), we have BE:AE::EC:ED, or

$$\text{rectangle (BE, ED)} = \text{rectangle (AE, EC)}.$$

To recall, in Figure 47, we have BD perpendicular to the diameter AC, and so BE = ED, so that we also have the following:

$$\text{square (BE)} = \text{rectangle (AE, EC)}.$$

And here we have precisely a determined square, equal to a determined rectangle (see also pp. 64–65). Unfortunately, this is not *yet* the beast we are looking for. As we move the position of the line BE to the left or to the right (that is, draw the perpendicular from different points along the line AC), we change *both* lines: AE, as well as EC. In the curve we look for, only one of the sides of the rectangle should change, and the other should remain fixed. Our goal should be to preserve this type of equality – but so that one of the sides of the rectangle is projected into some fixed value.

Projecting a ratio so that one of its sides becomes fixed . . . Obviously, we think of triangles on which we can construct similarity. The simplest thing would be to raise an isosceles right-angled triangle on top of our circle. Let us do this, as the right-angled triangle AFC; to fix EC (for instance) in a single value, we would look for some line emerging when we draw a parallel such as GE (parallel to FC). To complete our construction, let us also draw GH, parallel to AC.

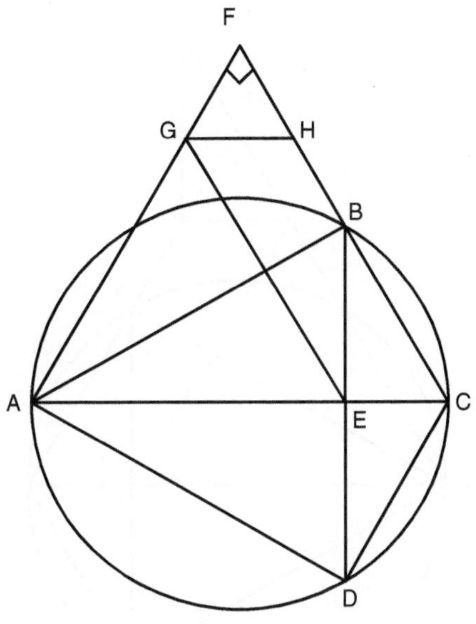


Figure 48

Now we have a few more proportions and equalities to work with. GHCE ended up a parallelogram; this, together with ACF, GHF being isosceles, establishes several relations:

EC = GH;

AG = GE; but also,

triangle AGE ~ triangle GHF; and so

AE : GE :: EC : GF.

We recall that we are interested in fixing one of the sides of the rectangle AE, EC. Well, it turns out that if we transform

rectangle (AE, EC)into the

rectangle (GE, GF),

we do the same transformation twice: from AE to GE and from EC to GF, both in the same ratio. Which is, in fact, the very simple ratio of the diagonal to the side in the square (what we would think of as $1/\sqrt{2}$). Applying the same ratio twice is simply halving.

It turns out that the rectangle contained by AE, EC is twice the rectangle contained by GE, GF. The line GF is fixed, and so if I simply double it (extending the line AF to Q where QF = GF), I get

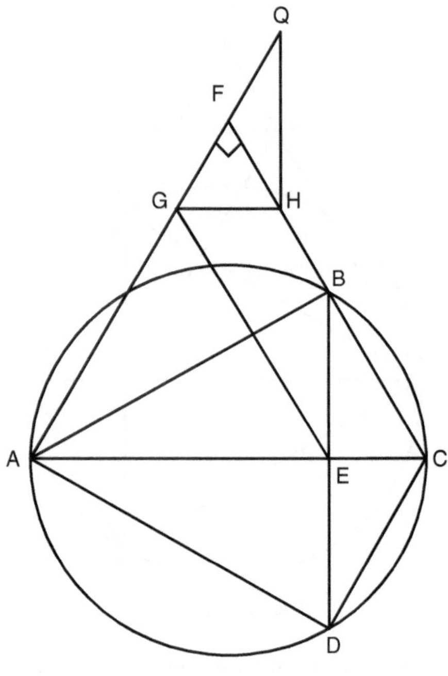


Figure 49

rectangle (AE, EC) = rectangle (GE, GQ).

We have now found a very useful rectangle, *which is always equal to the square on BE.* It is true that it is no longer clear what is meant, at this point, by "moving the line BE while preserving the relation," but a moment's reflection shows how this can be done. We may imagine the right-angled triangle extended into space, above the original circle (this is very Archytean), and then choose different circles, picking up different planes intersecting with that triangle (always parallel to the same base). At each plane, they intersect with the same plane defined by the points G, B, D. And as they intersect in this manner, the same relation arises: the square on the line such as BE (for instance, B′E′) is equal to the rectangle contained by a line such as GE (for instance, GE′) and the fixed line GQ. The proportion defining the arrangement is as follows:

square (BE) = rectangle (GE, GQ) (BE, GE covariable, GQ fixed).

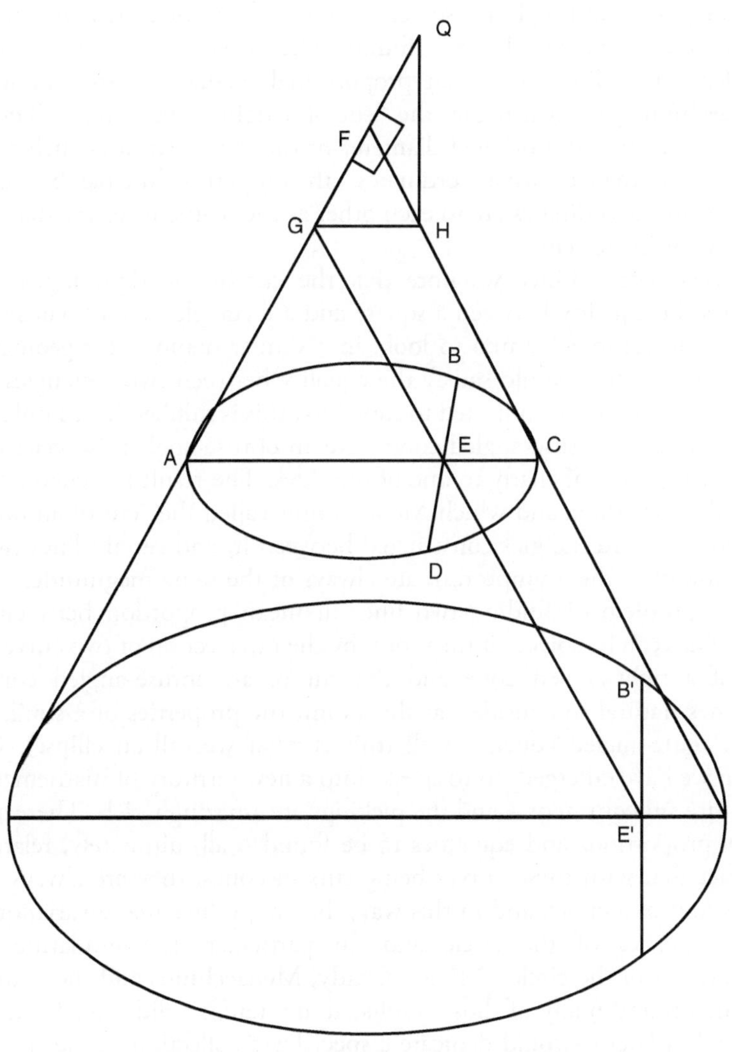


Figure 50

From a modern perspective, we have the arrangement where a square on one variable is equal to the line on another, multiplied by a fixed parameter:

$$y^2 = px,$$

which is what we call a parabola. From Menaechmus's perspective, this is the cut passing through an isosceles, right-angled cone, perpendicular to one of the sides. The beast – found. The curve required to solve the problem of finding two mean proportionals actually *exists*. Finding it, Menaechmus gave it a name: the "cut of a right-angled cone." The line GEE′ is the axis and original diameter of the parabola; lines such as BE, B′E′ are said to be drawn as "ordinates"; the property of the parabola is that squares on the ordinates are to each other as the segments of the diameters that the ordinates cut.

It gets better. Once we note that the cut of the right-angled cone satisfies the equality between a square and a rectangle, one of whose sides is fixed, it becomes natural to look, in a similar manner, for geometrical arrangements that would satisfy the equality between two rectangles. It is only slightly more complicated to show that this is fulfilled by a similar cut, this time to an obtuse-angled cone: take an obtuse-angled, isosceles cone, and cut it perpendicularly to one of the sides. The result is a curve, which we call a hyperbola and which Menaechmus called the "cut of an obtuse-angled cone." Rectangles constructed between it, and certain lines (which we think of as the asymptotes), are always of the same magnitude.

The problem of finding two lines in mean proportion between two given lines can be solved, it turns out, by the intersection of two curves: the cut of a right-angled cone and the cut of an obtuse-angled cone. It becomes natural to consider, at this point, the properties of a similar cut of an acute-angled cone, as well (this is what we call an ellipse). Once again: we have emerged, so to speak, into a new territory of mathematics – an entire subcontinent – and the pickings are amazingly rich. There are so many proportions and equalities to be found – all, ultimately, related to the fact that with these curves being cuts in cones, they are always some projection of a circle, and in this way, they reproduce many variations on the properties of the circle and, in particular, the similarities and proportions of the circle. Perhaps already, Menaechmus and those around him had found many of those results; at any rate, Euclid – likely, active a generation later – would dedicate a specialized collection of *Elements* just to this field alone. This is no longer extant – for reasons that will become apparent.

Archimedes was heir to this tradition. An important fraction of his work is based on the properties of conic sections. He refers to them invariably by their early name. I have referred earlier to such works as *Quadrature of the Parabola*, but Archimedes himself named this treatise *Quadrature of the Section of the Right-Angled Cone*. Conoids and spheroids, for Archimedes, are produced from the rotations of the sections of right-angled, obtuse-angled, and acute-angled cones. This is all a matter of names and, as such, is perhaps innocuous: after all, I often modernize some of the symbolism so as to make it more accessible. So, for instance, I try to provide my diagrams labeled with Latin, not Greek, letters, and I hope you understand that this is my tacit transformation.

But in the case of the nomenclature of the conic sections, this is not so innocuous, and I must ask you now to reimagine everything written about Archimedes, so far, substituting back in the ancient terminology for conic sections. Because the terminological transformation, in this case, is not merely a patina of modernization. It is, instead, an active mathematical contribution, introduced in antiquity itself. Here is the rare case of a mathematician asserting himself by intervening within the established mathematical language (otherwise, as we noted, usually very conservative – it was a genre, after all). And so, we circle back to Apollonius's own contribution: his *Conics*.

We have noted Apollonius responding to both Archimedes and Euclid. Something similar may be suggested for Apollonius's *Conics*: it aimed to replace – no less – Euclid's conics in the *Elements*; at the same time as Apollonius was writing his *Conics*, the name of Archimedes was established as the most significant author of spectacular results concerning conic sections. To provide one's own spectacular results concerning conic sections would be the most natural way to assert oneself, then, vis-à-vis the master.

We can judge some of this because the *Conics* is mostly extant. It is a massive work, consisting of eight (mostly, big) books. As Apollonius clarifies in his introductions, this work is bipartite. The first four books seek to replace Euclid's conics in the *Elements*. Even there, however, it is clear that Apollonius sought not merely to provide foundations but also to achieve new, remarkable results. In fact, Apollonius is very critical of Euclid as a mathematician. (Centuries later, Pappus will take offense: one should not behave like this toward the teacher of one's teachers! This, indeed, will mark a very different Alexandria, more teacherly and more reverential toward its past; see the section "Teachers, Commentaries, Books" in Chapter 6.) The following four books are original studies. In my terms, the first four books are Apollonius's response to Euclid; the following four, his response to Archimedes. The first four books were re-edited, with a commentary by Eutocius, in the sixth century CE (on which, more in Chapter 6), and so they survived in the original. A translation into Arabic of the entire *Conics* (save the final book, lost probably already in Late Antiquity) is still extant and was published in Latin for the first time in 1710. We thus have direct knowledge of Apollonius's most spectacular work – Books v–vii (among which, especially important is Book v).[13] The seventeenth-century makers of the scientific revolution knew of Apollonius only through the first four books. But this was enough, in their minds, to establish him as the equal of Archimedes.

Recall that Apollonius renamed and redefined the conic sections. But even this redefinition was much more than a mere terminological intervention. Menaechmus and Euclid, as you recall, defined the three conic sections with reference to a single type of cone and a single procedure of cutting it – an isosceles cone, cut perpendicularly to one of the sides. There is only one variation allowed – the angle at the top of the cone – and the three sections arise from that variation. Apollonius, however, takes up *any* cone, cutting it in *any* way. He is serious about taking the cone in the most general, difficult sense, and for him, a cone is not simply a right-angled triangle rotated around one of its sides. Instead – and producing a much greater variety of shapes – it is the surface composed of all the lines passing through (i) a given point (the vertex) and (ii) an arbitrarily chosen circle in space (that does not pass through the vertex). He then shows that all the resulting cuts must produce one of the three conic sections known before. This takes, in and of itself, considerable ingenuity in three-dimensional geometry (one recalls, in this regard, Apollonius's study in the spiral, drawn on the surface of the cylinder). What determines a conic section now is not the cone from which it was produced but purely the relations of magnitude it gives rise to. In the curve previously known as the "section of the right-angled cone," the equality is that of a certain square with a certain rectangle. The square applies *precisely* along the rectangle, so this forms a perfect "application" or, in Greek, *parabolē*. In what was called the "section of the obtuse-angled cone," the application exceeds by a certain amount (excess is *hyperbolē*); in what was called the "section of the acute-angled one," it is deficient by a certain amount (deficiency is *elleipsis*). Hence, the modern terms (only lightly Latinized) *parabola*, *hyperbola*, and *ellipse*.

[13] Toomer translated Books v–vii from Arabic into English, and so these are now the most accessible to an English reader: G. J. Toomer, *Apollonius: Conics Books v to vii* (New York: Springer-Verlag, 1990).

It will also be noticed that, as conceived in Apollonius's way, a conical surface has two components on each side of the vertex. It is now seen that a hyperbola cuts the conic surface on both sides of the vertex, producing a strange beast, the "opposite sections," a single, *discontinuous* geometrical entity.

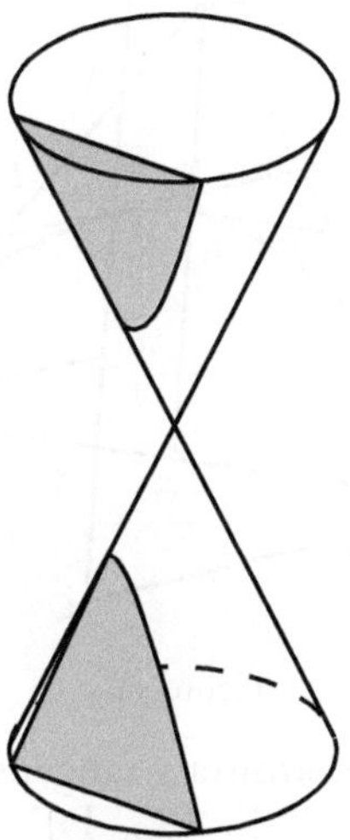

Figure 51

All of this is explored in detail – and all of this occupies no more than the first few propositions of the first book. By proposition 14, the three-dimensional geometry is fully surveyed, and the four sections (parabola, hyperbola, ellipse, and opposite sections) are already in place. From that point on, the discussion is purely planar. By proposition 21, Apollonius proves the key proportions defined by the sections.

He is not modest in his introduction: he makes clear that the definitions of Book I are original; he also proudly asserts that many of the detailed results, already in Books III and IV, are new. Conic sections are such that, even at the so-called elementary level, the results are already striking. These, especially the results in Book III, are "elementary" only in the sense that, besides their intrinsic beauty, they are also widely applicable: because, just as Menaechmus originally noted – solving the problem of two mean proportionals with the aid of conic sections – it is always useful to have meaningful equalities and proportions associated with determined curves. The more equalities and proportions, the better. And so many can be derived! I will give just one example from Book III. The choice is arbitrary, and this in itself is typical of the reading of this book: an arcade of continuous surprises – fireworks, exploding almost in no particular order. To III.41, then:[14]

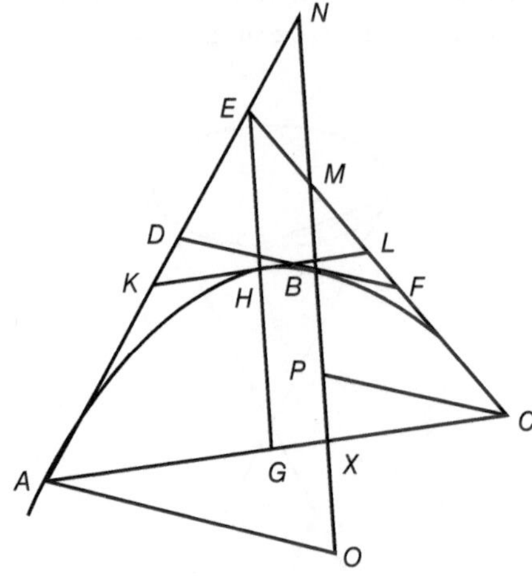


Figure 52

On a parabola, draw three arbitrary tangents so that each intersects with the other two, for example, AE, EC, DF. The points of contact and intersection determine, on each line, two line segments: AD, DE on the line AE; EF, FC on the line EC; DB, BF on the line DF.

Apollonius applies the truly elementary results of the *Conics* – those whose main purpose is to derive the more surprising results, such as those of III.41. First, use Apollonius's *Conics* II.29: if we bisect AC at point G, the resulting line GE is the diameter of the parabola. In terms of the construction of the parabola from page 165, it is like the line GE in Figure 49.

This requires a detour – and is the moment to mention a crucial surprise concerning parabolas. We constructed the parabola, in Figure 49, with a particular axis GE, one that runs through the middle of the cone. It is a "diameter" in the sense that there is an infinite array of parallel lines such as BD in Figure 49, all of them bisected by the diameter. In a parabola, we consider any such line to be drawn "ordinate" to the diameter. In the aforementioned case, BE is an ordinate on the line GE, and in this case, it

[14] The books of the *Conics* surviving in Greek can be read in English translation in M. N. Fried and R. C. Taliaferro, *Apollonius of Perga/Conics I–IV* (Santa Fe, NM: Green Lion Press, 2013). This brings Fried's recent and scholarly translation of Book IV together with the much older translation of Books I–III by Taliaferro, which was originally produced to answer the pedagogic needs of St. John's College. The version of Books I–III reprinted in 2013 has been revised from Taliaferro's original, but it remains not very good, although not quite as bad as to justify the enormous labor a new translation would entail.

happens to be perpendicular to it; the important feature of the parabola, of course, is that the squares on the ordinates scale as the segments by which they cut off the diameter. Now, here is the surprising thing: it turns out that we can take any line parallel to the original diameter, and this, too, will be a different diameter; its end point, a new vertex. The only difference is that the ordinate lines to such derived diameters are no longer perpendicular to the diameters and are instead defined by the property that they are *parallel to the line tangent to the parabola at the vertex of this particular diameter.* (The original diameter, it turns out, has the special property that it is perpendicular to the tangent at its vertex.) It thus follows that if we take any arbitrary line in a parabola, bisect it, and draw a line through the bisection, parallel to the original diameter, we find a new diameter. This closes our detour, and now back to III.41: G bisects AC and GE is drawn (as is shown separately) parallel to the diameter; hence, GE is a diameter.

Next, find the vertex H where the diameter cuts the parabola and then draw KL, tangent at H. We then have (I.32) KH = HL and also (I.35) AK = KE and LC = LE. But wait – there's more! Let us draw the parallel to EG via point B, MBP (CP drawn parallel to FB). Not only is this a diameter, but we can also establish (I.35) MB = BP. Now we have a great many equalities and parallelism to work with, and we can immediately set forth our starting point:

MC : CF :: EC : CL.

(In each ratio, in fact, the first term is double the second.)

An incredibly elaborate and opaque set of transformations yields from that:

CF : FE :: XC : AX.

Apollonius follows similar, although distinct, lines of transformations and substitutions on two more ratios of line segments, until each is shown to be in proportion with the same XC:AC; hence, all are proportionate to each other:

DE : DA :: CF : FE :: BF : BD.

The three tangents all cut each other, producing the same ratio!

A surprising result, yet again, and also a useful tool. If you wish to find the tangents to a parabola, you know they need to satisfy this property. Conversely, and perhaps more significant, if you know that a particular problem requires the division of three lines in the same ratio (and remember all the many treatises dedicated to the ramifications of dividing a given object by given ratios!), you know that you can take the tangents to a parabola as your starting point. These are *elements* – you do things with them – but so much more than just elements, the typical Apollonian combination.

Apollonius is as brilliant, as surprising as Archimedes. In the remarkable parts of the elementary books – in many of the results of Books II and III – the sense of surprise comes from the repeated application of the mechanism of proportion so that at the end of the long sequence of iterations, one hardly even sees the reason why the end result is indeed true. This is the typical way in which Apollonius asserted himself as a master: the bombardment of subtlety.

And yet Apollonius does produce a great deal more, often involving a mathematics of a different, more qualitative character. For instance, Apollonius has an entire series of results on the conditions under which straight lines cut, or fail to cut, the conic sections (I.17–I.31). A similar example: much of Book IV concerns the conditions of intersection between conic sections. After all, these curves are more complex than just the straight line and the circle, and so there is a variety of them, so there is a considerable combinatorial variety to the number of times that a given conic section may touch or intersect with another. Much of Book VI is about the *similarity* between conic sections. In all of this, we find attention to more or less "topological," nonquantitative properties of the conic sections. One pays attention to shape or to the presence or absence of points of intersection.

This brings us to the most impressive of Apollonius's extant works. This is Book V, that is, a non-"elementary" part of the treatise. Its subject matter is *greatest and smallest lines*. What does that even mean? Well, as usual, we need to start from the circle. A very simple set of results (proved by Euclid as *Elements* III.7 and III.8) shows the following:

a. From a given point – not the center – inside a circle, the greatest line to the circumference is the one passing through the center, the smallest is the complement to the same segment (in Figure 53, AF is greatest, and FD is smallest).

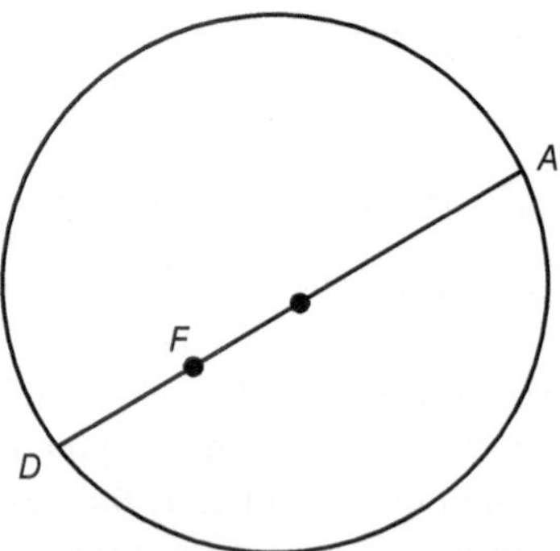


Figure 53

b. From a given point outside the circle, the smallest line to the circumference is the one passing through the center; the tangent is the greatest.

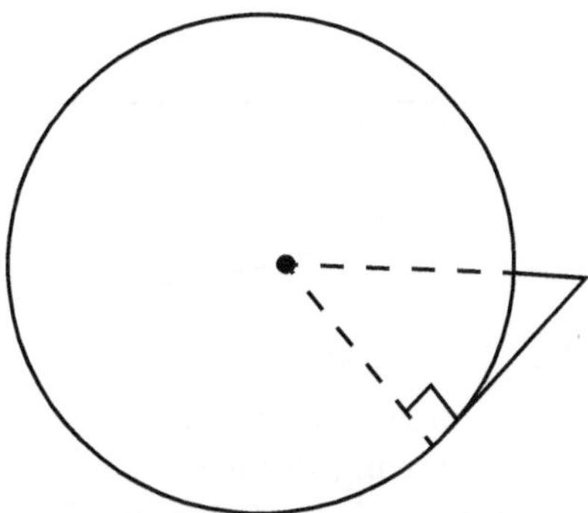

Figure 54

In the case of the circle, all of this is simple and intuitive, and as usual, as we ask the simple question – and what if one has a conic section, instead of a circle? – intuition nearly breaks down. Apollonius studies this question thoroughly and precisely through an especially massive treatment – seventy-seven propositions, often of considerable difficulty.

The very concept of "smallest line" or "minimum," extended to conic sections, is difficult and subtle. To begin with, the idea is simple enough: if we take a point on the axis of a parabola, very close to the vertex, it is fairly evident that the smallest of the lines drawn from that point to the curve is the line segment of the axis, the one drawn directly to the vertex (this is analogous to Euclid's *Elements* III.7). Conversely, if we take a point on the axis very far from the vertex, it becomes clear that the smallest of the lines drawn from that point to the curve no longer is the line segment of the axis; one can find shorter lines to the curve. In V.8, Apollonius finds the position at which the point, so to speak, "escapes the gravity of the vertex": where the parabola stops being like a circle. The position is at the distance of *one-half of the parameter* of the parabola from the vertex. This value – one-half of the parameter – will return several times again in this chapter. (To clarify: by "parameter," we mean the line segment D, which satisfies, for any point B chosen on the diameter of the parabola, $D \times BC = AB^2$.)

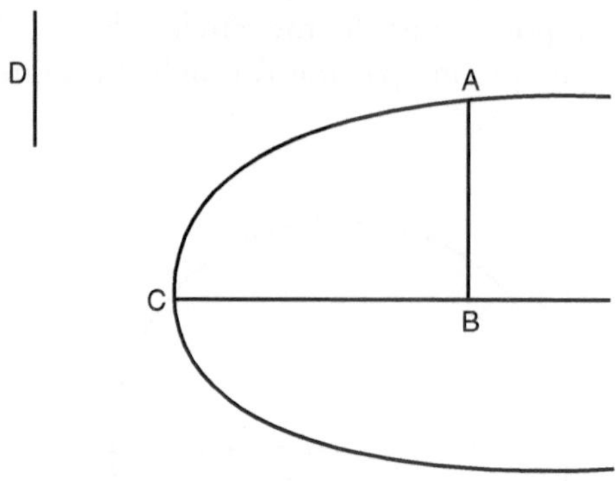


Figure 55

So far, so clear – although already difficult. It becomes even more difficult to consider the following problem: For a given point P on the curve, are there points on the axis, such as some X, so that the line XP is the smallest of the lines drawn from X to P?

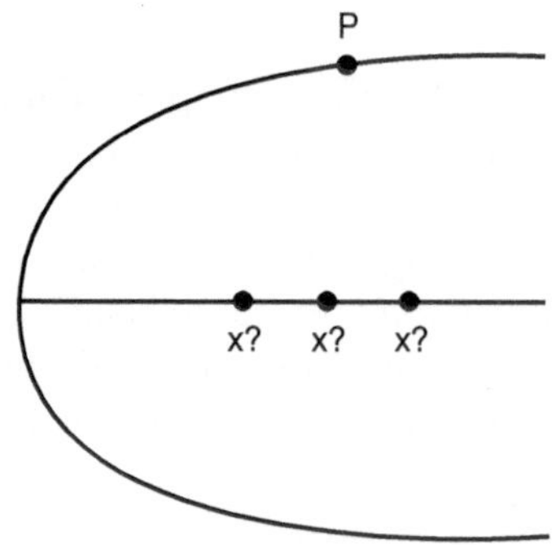


Figure 56

The meaning of "smallest line" suddenly becomes counterintuitive. We take point P on the curve as a given and scan the axis. For each potential point X on the axis, we ask the same question: Is the line PX, drawn from X to P, the smallest among all other lines that could have been drawn from X to the curve? Note carefully: a "smallest line," defined relative to point P on the curve, means not the smallest of the lines from P to the axis (this, quite obviously and uninterestingly, is the perpendicular from P to the axis). It is a line, such as XP, so that, *relative to all other lines drawn from that X under consideration*, it is the smallest. Because there are many *X*s to consider, it becomes obvious that, in this sense, there could be more than a single "smallest line" at a given point P on the curve – but also that there could be none. Now our study has developed a clear "qualitative" sense and has also become much subtler. To take just one example (and always staying with just the simplest case, that of the parabola): we have noted that the "escape from the vertex" happens at the distance of *one-half of the parameter*.

This value, indeed, has a more general import. It turns out that in general, a line is a minimum from a point P on the curve to the axis if and only if the segment of the axis cut off between that line and the perpendicular dropped from the same point P is equal to the same *one-half of the parameter.*

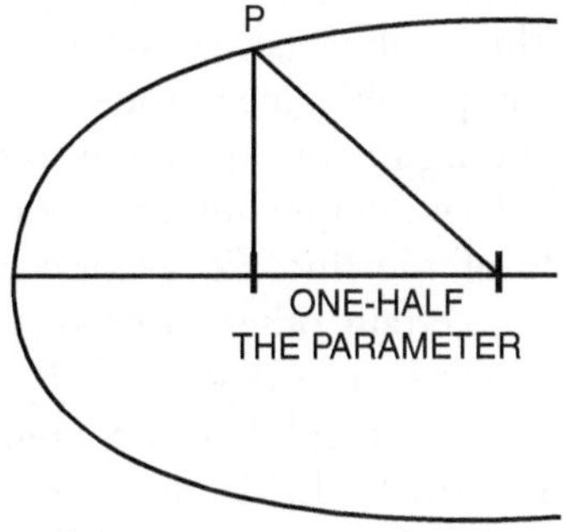


Figure 57

This is a precise value, and we can use it to determine the points through which the smallest lines can be drawn. For instance, you might wish to ask: Given a point E not on the curve of the parabola, but somewhere else on the plane on the parabola, can we find the smallest lines that pass through that point? (To be fair, you probably would not wish to ask that. But once Apollonius asks this, the question becomes interesting! I shall return to this issue of *motivation* in later discussion.)

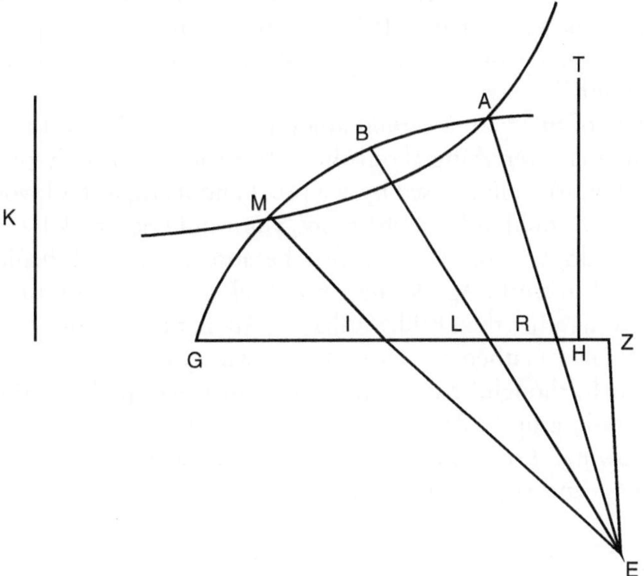


Figure 58

This is achieved in the extremely elaborate proposition V.51. Take point E and draw the perpendicular from it down to the axis to fall on point Z. Obviously, we are only interested in positions where GZ – the distance of Z from the vertex – is more than one-half of the parameter (otherwise, the smallest line is a segment of the axis and certainly does not pass through E). With this established, we assume, through the course of the proposition, different possible magnitudes for the line EZ. (In an interesting economy of the argument, Apollonius produces different proofs for the different claims, always employing the same diagram – even while the assumed magnitude of the line EZ keeps varying.) A complicated calculation gives rise to a certain other magnitude, determined by one-half of the parameter, which is here designated by the line K (the calculation of the magnitude of this length K is too complicated to be summed up here). The key result is this: if EZ is greater than K, no lines can pass from E to make a smallest line (no line such as LB can be a smallest line). If EZ is exactly equal to K, then there is exactly one smallest line passing through point E (indeed, the line LB). And if EZ is smaller than K, then there are two smallest lines, namely, MI, AR; their positions are found by drawing a hyperbola whose asymptotes are GH, HT (the finding of the line HT is a separate problem, which I set aside here). The hyperbola intersects the parabola at the points M, A that generate the smallest lines. And once again, I remind you: when we say that MI is a smallest line, this means that from point I, the smallest line to the parabola is MI; when we say that AR is a smallest line, this means that from point R, the smallest line to the parabola is AR. And from any other point on the axis – other than I or R – the smallest line to the curve does not pass through point E.

I relay all of this with a certain amount of guilt. I demand you to have faith in me that everything about this argument is rigorously proved, and then I ask you to be impressed by the proof and its rigor. I ask you to trust me that, if you could only see the proof, you would be struck by its beauty. And yet, I do not show it to you because it is so difficult. A fine matchmaker! In truth, Apollonius's proof takes time – impressive subtlety and precision, rather than brilliant flashes. Apollonius was not Archimedes. He was his own mathematician with his own style.

This much, though, is clear. If mathematics was, perhaps, dormant in the years leading up to Archimedes – it certainly came back to life in the years following. One could write a mathematics as hard as that of Apollonius – and yet count on an audience!

### *Archimedes: Conversing with Apollonius?*

Two different ancient sources say that Apollonius was active under Ptolemaic kings: the one says that he was active under Ptolemy Euergetes (247–222), the other that he was active under Ptolemy Philopator (222–205). This "active under" locution was probably often taken in antiquity to mean "forty years old," so what we learn is that some ancient sources thought Apollonius was born somewhere between 287 and 245. There are many more complications to this problem of dating, and the scholarly position on this question has moved back and forth (see the Suggestions for Further Reading). But we can sum up the situation as follows. Apollonius was almost certainly younger than Archimedes, but perhaps not by all that much. The impression holds: surely Apollonius, at least among other things, did respond to Archimedes.

When Apollonius started making a name for himself, was Archimedes still alive? The evidence is less decisive, and yet, if Apollonius was born at around, say, 260–250, he would be in his forties, more or less, by the time Archimedes died – perhaps already an established scholar but at any rate, almost certainly one who has already published his first works and made some of his contacts. I envisage Archimedes conversing with Conon at around 245, when he was perhaps thirty years old or a little more. Archimedes might have noticed Apollonius, then, the way Conon noticed Archimedes.

We usually think of *On Floating Bodies* as among Archimedes's most mature works. It relies on the lost *Centers of the Weight of Solids*, which was probably later than *On Balancing Planes*, which probably relies on *Quadrature of the Parabola*. It is certainly a major achievement: the second book, in particular, is arguably the most remarkable piece of all extant Greek mathematics. The object under study is, once again, the conoid: we take – to use the modern term – a parabola and rotate it around its axis to produce a more-or-less bullet-shaped object. Then – which brings this into the realm of "floating bodies" – we assume that it is made of some homogeneous matter lighter than a liquid and proceed to conceive of it immersed in the liquid. We know that it will float (it was assumed to be lighter than the liquid). But will it *balance*? This is a difficult question, as we will return to discuss in greater detail later on. One needs to identify two segments of the paraboloid – one immersed in the liquid, the other floating above it – and find the centers of the weight of the two segments; stability depends on the qualitative relation between these two points.

All of this varies, depending on the precise shape and relative weight of the paraboloid. A complicated study, giving rise to many potential combinations.

Almost all of the discussion in this treatise is done via a plane section taken of the configuration of solids (this is typical of many other studies in Greek mathematics, where one starts with a three-dimensional space but then simplifies by considering just a single "representative" plane; recall the treatment of the cone and the cylinder in *Elements* XII, for instance, where we only looked at the base). The study is therefore, in practice, of a parabola or, literally, in the terms employed by Archimedes, it is a study of a section of a right-angled cone. More than this: Archimedes treats many iterations of geometrical combinations and so draws the diagram again and again and assigns the labeling again and again. Generally in Greek mathematics, such labeling is done "randomly," the deck of cards reshuffled between one proposition and the following so that what was a B in one proposition can become an A in the following, usually depending on the order in which elements are introduced in the construction. Finally, parabolic segments are usually labeled, naturally enough, with three letters.

It is therefore intriguing that Archimedes's labeling in this treatise is anomalous. A certain element of the labeling is never varied – never follows an alphabetical logic, always has an extra, unmotivated letter. This occurs as the main parabolic segment is introduced, in the following nomenclature:

> Section of the right-angled cone, APOL.

The section of the right-angled cone is always labeled APOL. All things considered, I think it is likely (although, of course, unprovable) that Archimedes, in this late work – his last? – referred, obliquely, to Apollonius's renaming of the conic sections. Now, to refer to that which Apollonius recently called a "parabola" as a "section of the right-angled cone, APOL" is, I propose, *funny*. A career, begun with cunning, false propositions, sent in a challenge, concluding with a cunning pun, dismissing an upstart's terminological innovation.

Archimedes, responding to Apollonius? The story is almost too good (which is why it should be treated with some skepticism).[15] But this thing stands out – whether or not a response to Apollonius: the work that we usually take to be Archimedes's most impressive, in some sense the culmination of his life project, was not, in the end, a study in the

[15] I discuss all of this in R. Netz, "Nothing to Do with Apollonius? Concerning the Style and Chronology of Late-Archimedean Mathematics," in *Philologus* 161 (2017): 47–76.

measurement of volumes. The series of works sent to Dositheus does not really sum up Archimedes's achievement. That culminated, instead, with a study in the mathematization of the physical world.

We recall the generation of Archytas – where much of the inspiration for mathematics itself came from the recent idea of mathematizing music. This is key to the way in which Plato, and philosophers like him, engaged with mathematics – and the way in which mathematicians of that generation engaged back. And so far, we have considered the conversation of the generation of Archimedes as much more internal to mathematics. How does the mathematization of the physical world fit into this conversation?

## IV. Archimedes's Physics

### *Archimedes: Conversing with the Physical World?*

Our interpretation of the generation of Archytas and its engagement with philosophy has relied on mere fragments (and on the more substantial evidence from the *philosophical* side). With Archimedes, we have the mathematics itself. Four "physical" works by Archimedes are extant, and although they are not strung together in a single correspondence (as are those sent to Dositheus), they follow a clearly defined conceptual path. Those are the four works engaged with the center of the weight: *On Balancing Planes*, *Quadrature of the Parabola*, *Method*, and *On Floating Bodies.*[16] (Chronologically, I suspect that *On Balancing Planes* followed *Quadrature of the Parabola*; however, Archimedes relied, in the latter work, on a different version of the claims of *On Balancing Planes*, and, conceptually, therefore, we should consider *On Balancing Planes* first.) This sequence is crucial to the rise of modern science, and we should survey it in somewhat greater detail.

### *On Balancing Planes*

*On Balancing Planes*, or PE I–II (a single work, divided into two books), begins with a proof of the famous result tied to the phrase "give me a place to stand and I shall move the earth": the law of the balance. To paraphrase:

[16] Volume III of my translation of Archimedes includes these four treatises, and I hope that by the time you read this book, it is already published by Cambridge University Press. Until then, the only English version of these works is found in (really, a paraphrase) T. L. Heath, *The Works of Archimedes* (Cambridge: Cambridge University Press, 1897/1912).

> Magnitudes balance around a fulcrum when their distances from the fulcrum are reciprocally proportional to their weights.

What does all of that even mean? Archimedes is not very explicit. In the extant form of the treatise, it carries no introduction other than the statement of a handful of postulates. We effectively have to reconstruct Archimedes's concepts from the way in which they are applied. (The contrast to Eudoxus's interest in conceptualization as such, Theaetetus's interest in classification as such, is striking.)

"Weight," in practice, has, in the works of Archimedes, no meaning independent from geometry and is not really a physical term. Usually, we assume that all objects are made of the same homogeneous matter, so that *weight* simply means "magnitude." The weight of a polygon is its magnitude or (we would say) its area; the weight of a solid is what we would call its volume; even lines possess weight, which is what we call their length.

Now, what about "distance"? The implicit definition is that it is the distance or the length of a certain line segment. This is the line segment drawn from the center of the weight of the given magnitude to the fulcrum. However, the term *center of the weight* is never clarified. The implicit definition is that if two magnitudes balance around a point, then that point is the center of the weight of the magnitude composed of the two. So much for conceptual clarity! "Center of the weight" depends on the operation of balance, which conceptually depends on "distance," which in turn requires, for its definition, the notion of "center of the weight." This, in practice, is not so much a matter of circular reasoning but instead a matter of conceptual bootstrapping. To get started, we need some presuppositions concerning particular centers of weight – and once we get them, we can generate all the rest. But what centers of the weight *are* always remains opaque. This opacity is significant: Archimedes is able to maintain a certain ambiguity. In some sense, the "center of the weight" is a purely geometrical point, serving as a ratio-summing mechanism that transforms many ratios between many geometrical objects into a single ratio (this will become clearer as we see examples in practice). In another sense, the "center of the weight" is a physical term whose existence is assured by the physical intuition that objects do, in fact, balance around some point, and it can then be used to derive further consequences as we assume that the motion of rigid objects is determined by that of their center of the weight.

In fact, we can prove quite a lot about centers of the weight without knowing at all what they are. Here is how this works, then, in Archimedes's proof of the law of the balance, *On Balancing Planes* proposition 4 (the number of the proposition is erroneously given as "6" in modern editions).

First of all, we set out the assumptions. Produce two simple objects – say, squares – and have them arranged around the fulcrum X. Make A the center of the weight of the left square and B the center of the right one, so now we can measure the distances AX, BX. (Note, once again, that for this exercise, we do not need to know the meaning of the phrase "center of the weight.")

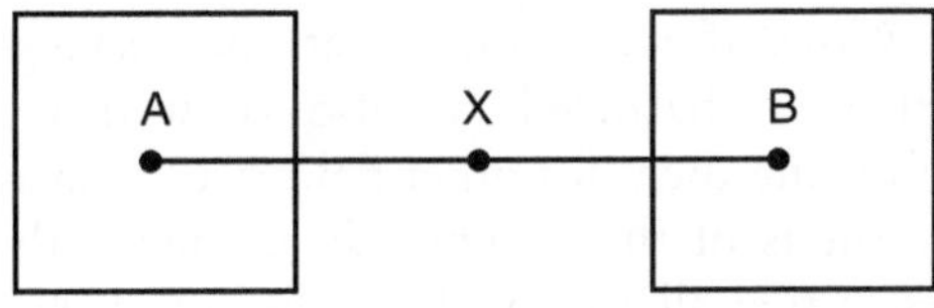


Figure 59

A brief set of postulates right at the beginning of the treatise establishes, directly or indirectly, that if magnitudes balance at equal distances, the magnitudes themselves must be equal, and vice versa. If the two squares are equal and they balance around the fulcrum, we must have AX = BX. X is therefore the center of the weight of the composite magnitude AB. (It is crucial for this operation that we recognize the existence of noncontiguous magnitudes, composites of distinct continuous segments.) It is further postulated that "if magnitudes balance at certain distances, the [magnitudes] equal to them will also balance at the same distance." What this means, apparently, is that if we substitute the squares for some other equal magnitudes – but preserving the same centers of the weight A and B – the center of the weight of the system as a whole would remain the same. So, for instance, cut both squares in half horizontally along the middle, push the lower resulting rectangles down by a notch, and push the upper resulting rectangles up by the same amount. We have not yet proved this, but assume that, in fact, following such a manipulation, the center of the weight of each of A and B should, in fact, remain the same as it was before. The consequence would be that the distances AB, BX would not budge either; hence, the system should remain in balance around the same point X.

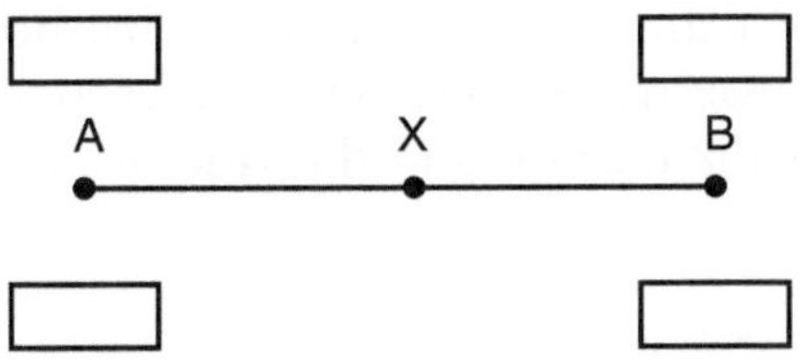


Figure 60

This is enough for the main derivation. Assume that the two magnitudes in consideration have the same shape, a rectangle. (From the postulate, it follows that shapes do not matter as long as the center of the weight is preserved, so the choice of the rectangle is merely an illustration chosen by me and is not, in fact, specified by Archimedes.) Once again, let their center of the weight be at A and B, and let the center of the weight of the system as a whole be at X. Let us drop, however, the assumption that they are equal: for instance, let A have the plane area of 2 units; B, the plane area of 7 units. Now, if we assume that the rectangles A, B are so positioned that they are balanced, we also know that they will remain balanced if we substitute them for other figures, as long as we preserve the magnitudes and centers of the weight. Let us divide the rectangles into thinner rectangles so that all are equal, on the A as well as on the B side. A will thus be divided into two parts; B, into seven parts. Let us position those resulting nine parts as nine smaller rectangles, each balanced around the original positions: so we have A in the middle between two small rectangles, and B is right at the position of the central small rectangle, surrounded by three on either side.

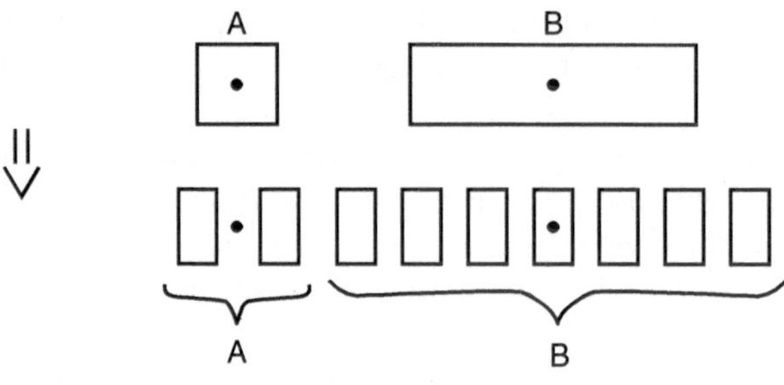


Figure 61

Because all small rectangles are equal to each other, it is evident (and indeed follows from the postulates) that their center of the weight, taken as a whole, would be the center of the arrangement of all nine, which is the middle of the central rectangle.

This is one unit of distance away from B and 3.5 units of distance away from A. And the same must have held even before the big rectangles got dissected because, by definition, the system was balanced before we did the dissection, and the dissection preserved the centers of the weight! The ratio of distances 1:3.5 is the reciprocal to the ratio of the weights 7:2. This is my example of the argument that Archimedes produces, of course, in geometrical generality (and also extending this, in a Eudoxean flourish of rigor, to the case where the two magnitudes are incommensurable). But even without reproducing Archimedes's general proof, it is clear that the same will hold no matter which numbers are taken. We thus learn that magnitudes balance, around a fulcrum, when their distances from the fulcrum are reciprocally proportional to their weights. This is it. Almost no mathematical tools at all – indeed, not even a clear articulation of what is meant by "distance from the fulcrum" (*certainly no physical experiment*), and a fundamental physical claim is established.

Impressive, right? But clearly not impressive enough for Archimedes, and it is evident that, as far as Archimedes was concerned, this result, early in the treatise, served merely as a tool. The treatise proceeds to find centers of the weight of individual figures. Here is where the bootstrapping really happens.

Specifically, Archimedes postulates that wherever the center of the weight of a polygon is, the center of the weight of a similar polygon must be similarly situated. This is all rigorously defined, but for our purposes, we may take a very simple consequence: if two triangles are congruent, their centers of the weight must fall on the same position relative to the vertices of the triangles. Once again: we do not know what centers of the weight are and where to find them – but we can postulate such obvious assumptions and, with this, bootstrap the results we actually need.

And so, proposition 8 (misleadingly labeled "10" in modern editions). A parallelogram is dissected by its diagonal into two congruent triangles. (This, indeed, is how Euclid got his own proportion-less derivation of Pythagoras's theorem: an elementary result if ever there was one.)

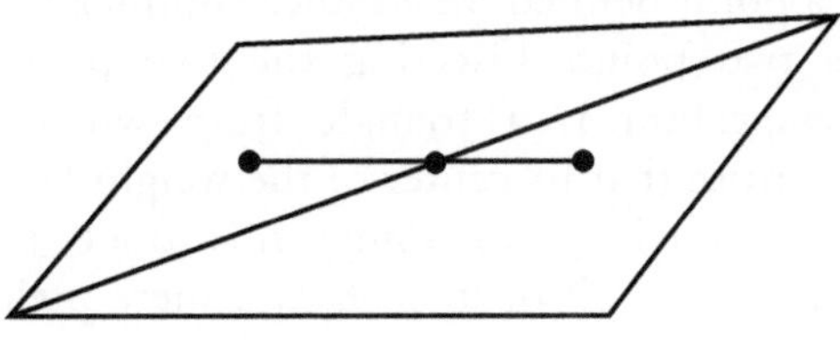

Figure 62

And here's the simple and yet surprising observation: if we take any two points in the parallelogram so that each is identically situated within its own triangle, then the bisection of the line joining those two points must fall on the diagonal of the parallelogram. Now, one of those pairs of points (we don't know which!) must be that of the two centers of the weight of the two triangles. Because the two triangles are equal (they are congruent, after all), their combined center of weight must be at the bisection of the line joining their individual centers of their weight. Thus, the center of the weight of the composite of the two triangles – that is, the center of the weight of the parallelogram – must be on the diagonal. It is immediately apparent that because a parallelogram has two diagonals cutting each other, their point of intersection is, in fact, the center of the weight of the parallelogram as a whole. This is takeoff. We've got an entity in the world for which we've been able to find the center of the weight – the parallelogram – which we did based on the triangle *but* without knowing the position of the center of the weight of the triangle (which we have no way of finding, as yet) and, strangest of all, without having any clear sense at all of what a center of weight *is*!

Once again, impressive – but trivial for Archimedes, who proceeds immediately to find the center of the weight of a triangle: the circle of bootstrapping is completed and indeed reversed as the triangle, now, is decomposed, in turn, into parallelogrammical components. I skip this argument, but the result, it turns out, is analogous to that of the parallelogram. Just as the center of the weight of the parallelogram is located on the parallelogram's diagonal, so the center of the weight of the triangle is located on the median. And once again, this establishes the position of the center of the weight: all the medians of the triangle meet at a single point. Specifically, this is the one dividing each median line in the ratio 2:1. This, then, must be the center of the weight of the triangle. Even more difficult is the finding of the center of the weight of the trapezium – a point defined, in a rather complex manner, along the line joining the two points bisecting the two parallel sides of the trapezium. (A trapezium is a triangle from which a triangle was removed; little surprise that its center of the weight falls on the median shared by those two triangles.) Thus ends Book I – breathtaking, seemingly impossible, and all of it, in fact, a mere prelude to the main action, that of Book II.

Book II begins with a statement illustrating how the principle of proof used for showing the law of the balance can be extended for such cases where we do not know how to dissect the original shape into a number of equal, smaller shapes: specifically, how to do so with a parabolic segment.

Archimedes then develops a remarkable argument for finding the center of the weight of the parabolic segment. It turns out that this is on the diameter, divided in the ratio 1.5:1. That is: in a parabolic segment ABGD, the center of the weight Q satisfies BQ:QD::1.5:1. The median of the triangle is divided in the ratio 2:1; the diameter of the parabolic segment is divided in the ratio 1.5:1.

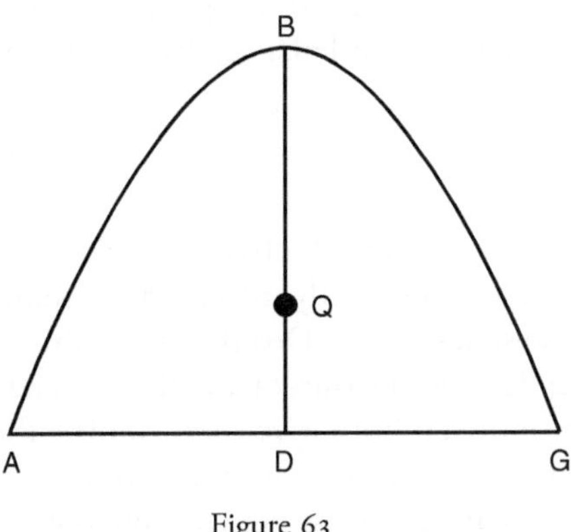


Figure 63

But even this is not enough for Archimedes! The treatise ends, instead, with a very complex and subtle argument (this argument works as we come to expect by now: Archimedes first develops a seemingly unrelated lemma and then, in a surprising move, finds how to apply it), finding the center of the weight of a parabolic "slice," that is, the subsegment of a parabolic segment enclosed between two parallel lines. This center of the weight is found on the line joining the bisections of the two parallel lines, at a point defined in a complex way that is not identical but very similar to that for finding the center of the weight of the trapezium. The parabolic segment is not a triangle, but it is akin to the triangle; the parabolic slice, it turns out, is akin to the trapezium.

As we begin to read *On Balancing Planes*, with our modern training, we are convinced that this is, effectively, intended as a contribution to what we now think of as physics. And indeed, if we stop reading almost at the beginning – once we have achieved the proof of the law of the balance – we may conclude that this is its intended goal. (In fact, the few modern readers who pick up this treatise rarely go beyond that point.) But the treatise does proceed much further, and its great bulk is dedicated to questions whose entire motivation is geometrical, indeed, most closely related to Archimedes's, otherwise attested, interest in pure geometry: a display of how a geometrical technique – which is intuitively applicable to simple, rectilinear figures – may be extended to much more complex, curvilinear ones. In the pure geometrical treatises – especially the series addressed to Dositheus – this is done with measurement; in *On Balancing Planes*, this is done with the finding of a center of the weight. The operation, in this case, is assigned (implicitly) with an extra layer of physical meaning – that of the balance. But it is, in practice, a purely geometrical operation.

*Quadrature of the Parabola*

The geometric impetus is indeed obvious with the *Quadrature of the Parabola*: this, after all, is explicitly about measurement (in fact, this is the work opening the series sent to Dositheus). In this treatise, Archimedes provides two routes for the measurement of the parabolic segment. The second and perhaps more authoritative route applies the standard geometric toolbox, dividing the segment into components and finding their measurements (and then, of course, relying on the Eudoxean technique of following the consequences – in either direction – of the segment *not* being four-thirds of the triangle).

The first route is "physical." The parabolic segment is imagined as hung down and put on a balance. But then again, the operation – once this "balance" manipulation is assumed – is strictly geometrical. The segment as a whole is enclosed within a triangle and is also dissected into strips, each enclosed in small trapezia. Everything then follows from the fact that we know the position of the center of the weight of the triangle, enclosing the segment, as a whole. We can therefore surmise where it would balance with other, specified weights. Results concerning the parabola can then be used to extend this to the system of trapezia enclosing the parabolic segments. This finally translates – once again using standard Eudoxean techniques – to measurement of the area of the parabolic segment as a whole. "Weight" really means, in all of this, geometrical area; "balance" really means, in all of this, a proportion involving areas and line segments; the whole exercise is, explicitly, about geometrical measurement. Center of the weight, to be sure: but is it *physics*?

*Method*

Perhaps all of this becomes clearer as we consider the direct version of the same argument provided in Archimedes's *Method*. Indeed, we can say a little more about Archimedes's motivations in this case, based on the introduction to this work. We should linger longer on the *Method*.

The conceptual richness of this treatise is perhaps related to the identity of the addressee: the *Method* is addressed to Eratosthenes, the leading poet-scholar of Alexandria of his day. Note the contrast: Dositheus, the recipient of the main series of geometrical works, seems to have been a minor author; King Gelon, recipient of the *Sand-Reckoner*, had no scholarly aspirations. But Eratosthenes was a serious author – who made, as we will note later on, his own serious mathematical contributions. It is appropriate that this would be an especially difficult and enigmatic work.

Archimedes broaches, as noted earlier, a claim about mathematical practice: that it is useful to know in advance that a statement is true – regardless of how one comes up with this knowledge – so as to be able to find a rigorous proof. Archimedes heaps praise on Eratosthenes and then suggests that he, Eratosthenes, is especially capable of judging (and applying?) a certain technique, "proceeding along which it shall be possible to possess starting-points to enable one to see, through those of mechanics, some of the theorems in mathematics." This, then, is presented as a less-than-rigorous technique – specifically, less than rigorous in that it applies mechanics to geometry.

I mentioned that Archimedes heaps praise on Eratosthenes; the reason I did so is because such praise strongly suggests, in my view, the possibility of irony. We should be on guard. With this in mind, let us proceed with the key examples.

The very first proposition, as mentioned earlier, is the measurement of the parabolic segment. Draw a parabolic segment and, inside it, the triangle having the same base and the same vertex, both ABGD (BD diameter). Draw AZ, parallel to BD, and GZ, tangent to the parabola at G. Extend the line GB so that it cuts AZ at K, and then extend it further so that QK = KG. Now draw an arbitrary parallel to AZ, such as XONM. (It quickly follows that GK is a median in the triangle AZG and that the triangle ABG is one-fourth the magnitude of the triangle AZG.)

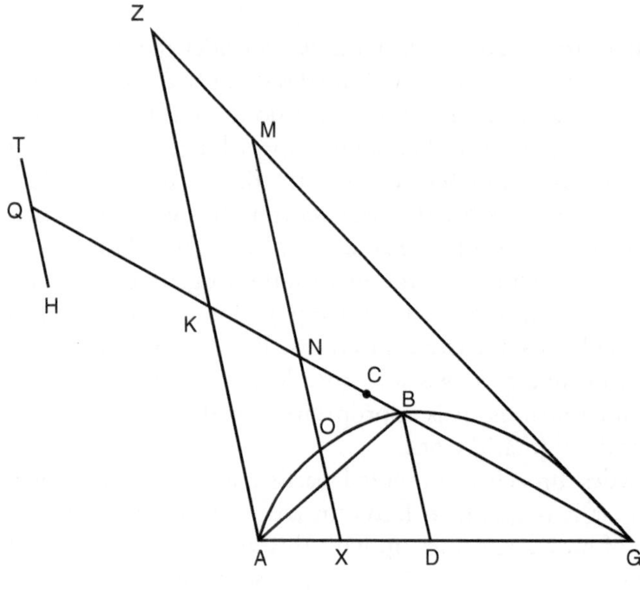


Figure 64

Archimedes now takes for granted a rather complex, although elementary, result in conic theory (which he has proved in QP): in this kind of arrangement with a parallel to the diameter, intercepted by the tangent, we have

MX : OX :: AG : AX.

The entire line (arbitrarily chosen, parallel to the diameter) intercepted between the base and the tangent is to its section, intercepted between the base and the parabola itself, as the entire base is to the segment of the base intercepted by the arbitrarily chosen line.

With this, it only takes a couple of transformations to derive

MX : OX :: GK : KN or (by construction)
MX : OX :: QK : KN.

We may now take OX and produce a copy of it (a line with the same length) and position it so that its middle will lie on point Q (this is line TH in the diagram). We now have

MX : TH :: QK : KN.

And so, we find that K is the point around which, as a fulcrum, the two lines MX and TH balance.

The argument proceeds directly, if very adventurously, from here. The result we have just found could be repeated for any arbitrarily chosen line.

Hence, it will be true for all of them. (The power of this "hence" is left ambiguous.) Take all the lines in the triangle, such as MX, *in the position in which they are now*. Then take all the lines in the parabolic segment, such as OX, *transported so that their center of the weight is on point Q*. The triangle would balance the parabola around the fulcrum, K. Now, we know that the center of the weight of the triangle is at point C (where CK is one-third of GK), and we have constructed, in this thought experiment, point Q as the center of the weight of the transported parabolic segment – with QK = GK. The center of the weight of the triangle AZG is distant from the fulcrum by the distance CK, but the center of the weight of the parabola segment is distant from the fulcrum by the distance GK. Hence, the parabolic segment is, in area, one-third the triangle AZG, or equivalently, the parabolic segment is 4/3 the enclosed triangle ABGD.

Eratosthenes must have been impressed, if puzzled: Just what makes the argument – however brilliant – so problematic? The words of the introduction clearly imply that the issue has to do with the application of mechanics, but here, as noted previously, irony should be suspected – not to mention the possibility of downright deception (Archimedes, back to his old tricks!). After all, the very same argument – involving, however, not lines but, instead, thin trapezia – was employed in the already-published *Quadrature of the Parabola*. That was indeed a suspect, nonrigorous argument, but note that Archimedes suggests that the technique is revealed, here, for the very first time: mechanics cannot be, then, the essence of the technique. Indeed, we can do better than just comparing with *Quadrature of the Parabola* – where the attribution of a "physical" ground for suspicion appears valid. We can compare within the *Method* itself – to see another theorem that does not apply a "physical" consideration and yet is equally suspect.

We need to get a fuller sense of the structure of the *Method*. It begins with general "methodological" statements – and proceeds to provide various examples for the application of a certain technique. The measurement of the parabola (area) is followed by the measurement of the sphere, the spheroid, and the conoid (volumes), followed by the finding of various centers of the weight. All of those were proved already, more rigorously, elsewhere.

There is more. The method starts out by promising not only the revelation of a new proof technique but also of two new measurements, this time of absolutely surprising objects. One is the shape produced by cutting a cylinder obliquely through a diameter of a base circle; the result is a lipstick-tip-shaped curvilinear object. The other is the result of having two equal cylinders intersect perpendicularly: this is an even stranger object, a box bounded by eight symmetrically curved surfaces. Archimedes says in the introduction that he will first display the new technique through several of its applications, and then he will measure the two new objects, first by providing nonrigorous proofs based on the new technique and then through a valid proof. Following the various results concerning already-established measurements and already-known centers of the weight, then, Archimedes proceeds to the new objects.

Archimedes's treatment of the box is now lost, but we have not two but three separate proofs for the "tip-of-the-lipstick" cylindrical cut. The first of these is similar in character to the proof we have seen in the *Method*'s treatment of the parabolic segment, involving the centers of the weight of individual slices. The third is indeed a valid geometrical argument, based on circumscribing the shape between thin prisms and applying the type of Eudoxean technique we saw with *Spiral Lines*.

The second is neither: clearly intended as less than a rigorous proof and yet involving no mechanics. Appropriately, this is the last of the examples of the new technique, produced through the *Method*: the hint, as it were, tucked in, once again, right at the end.

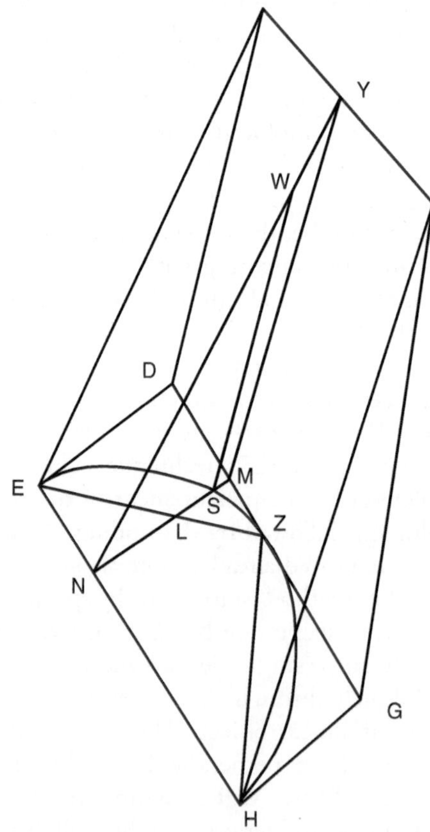


Figure 65

The object that interests us is a cylindrical cut; let us think of the cut as enclosed within a triangular prism. The base of the cylindrical cut is a semicircle; the base of the triangular prism is a rectangle. Let us label the rectangle EDGH and the semicircle EZH. Archimedes throws in an extra construction, and of course, this is a parabola! We draw a parabola through the three points – the number required to determine a parabola H, Z, E (the parabola is represented in Figure 64 by a triangle). We now draw an arbitrary plane parallel to the line GH and perpendicular to the rectangle EDGH, such as NMY. At the base, this is a line NM, cutting the circle at point S and cutting the parabola at point L.

At this point, we have two triangles, one associated with the triangular prism, namely, NMY, and one associated with the cylindrical cut, namely, NSW. We also have three interesting lines at the base: one associated with the rectangle, namely, NM; one associated with the circle, namely, NS; and one associated with the parabolic segment, namely, NL.

The crux of the proof is a very surprising result in (elementary) conic theory. It turns out that in the arrangement of the base with the parabola passing through those three points of the circle, we always have the following proportion:

$$NM : NS :: NS : NL$$

The line associated with the circle is the mean proportional between the two other lines. Remarkable enough (the circle – in some sense, the mean between a parabola and a line!) – but note the consequences. We immediately see, through the operations on proportions that Hippocrates of Chios already knew well, that we have also established that

$$(NM : NS)^2 :: (NM : NL).$$

That is: the ratio of the line associated with the rectangle to the line associated with the circle, *applied twice* or *squared* – is the same as the ratio of the line associated with the rectangle to the line associated with the parabola.

Now, a ratio of lines, squared, is also simply the ratio of associated similar plane figures. The triangle on top of the line of the rectangle (NMY) to the triangle on top of the line of the circle (NSW) is like the lines to each other, squared:

$$NMY : NSW :: NM : NL.$$

We have now reached a ratio involving, on the one side, triangles, on the other side, lines.

Archimedes, at this point, does something rather incredible. It is so surprising that no one, in fact, suspected it before it was finally read, twenty years ago, by the application of digital technology to the Archimedes Palimpsest. I will return to the Archimedes Palimpsest (all too briefly) in the final chapter, and so, for the time being, no more on how we read this and instead just a quick note on the mathematics we find.

What Archimedes does, now, is to gesture, briefly, at a calculation with infinity. He points out that if we consider all the planes drawn parallel to GH, then we may note that "the lines in the [rectangle] … are equal in multitude to the triangles in the prism; and those triangles, in the [cylindrical cut], are equal in multitude to the triangles in the prism; and the lines [in the parabola] … are equal in multitude to the lines [in the rectangle]." This is asserted, not proved, but clearly, Archimedes's thought – eerily reminiscent of modern set theory – must involve the one-to-one relationship between planes and planes, planes and lines, and lines with lines: asserting, in this way, an equality between *infinite multitudes*. What is even the point of all that? This observation on the equalities between multitudes is only part of a wider set, all of which, taken together, allow Archimedes – but only to the extent that he trusts a certain operation, as it is extended to deal with infinities – to sum up a set of proportions. In this case, Archimedes sets up the infinitely many different proportions, one for each of the arbitrary chosen planes and associated lines, and derives from them one single proportion, involving the sums of all the planes and lines taken together. Thus, Archimedes concludes this operation by asserting that the triangular prism as a whole (consisting of all the triangles in the prism) to the cylindrical cut as a whole (consisting of all the triangles in the cylindrical cut) is as the rectangle (consisting of all the rectangle lines) to the parabolic segments (consisting of all the parabola lines).[17]

Let us pause to make a quick comparison. Archimedes, starting with proposition 1, always did sum up, in this treatise, infinitely many proportions. He did this by transforming each proportion into a system of magnitudes balancing around a fixed point. Then, instead of invoking some kind of rule with which the summation of proportions may be achieved, he simply assumed that if one adds up many objects, even infinitely many, so that each balances at a fixed point, so will all of them, taken together. In this

[17] For the full technical detail on this episode, see R. Netz, K. Saito, and T. Tchernetska, "A New Reading of *Method* Proposition 14: Preliminary Evidence from the Archimedes Palimpsest (Part 1)," *SCIAMVS* 2 (2002): 9–29; R. Netz, K. Saito, and T. Tchernetska, "A New Reading of *Method* Proposition 14: Preliminary Evidence from the Archimedes Palimpsest (Part 2)," *SCIAMVS* 2 (2003): 109–126.

final proposition, this is no longer relevant (the obtained proportion is not transformed into a statement of physical balance). Hence, one operates, instead, via a general argument concerning proportions.

The result itself is clear: prism to cylindrical cut, as rectangle to parabolic segment – but we know the ratio of the rectangle to the parabolic segment! Archimedes has found this much earlier, in the *Quadrature of the Parabola* – and again, just now, in the first proposition of the *Method* itself. So many circles are now closing. The rectangle to the parabolic segment is 6:4 (after all, the parabolic segment is 4:3 the enclosed *triangle*), or if we wish, the entire prism, surrounding the entire cylinder – four times this triangular prism – is 24:4 or six times the cylindrical cut. The lipstick has been measured. So much is clear, and so much else remains difficult – and tantalizing. Just what was Eratosthenes supposed to make out of this?

But then again: we should bear in mind, above all, the possibility of *irony*. Likely enough, being tantalizing and difficult was not some failure on Archimedes's part – as if Archimedes tried hard to convey to Eratosthenes just what was valid, and what wasn't, and to demarcate the borders as clearly as possible, and then failed in doing so. No: everything suggests that Archimedes wanted the question of what makes his technique problematic to be, itself, problematic. For the same reason that he engaged with problematic techniques in the first place – for the same reason that he made it maximally difficult to realize why the result of *Spiral Lines* was true, by making the route there as surprising and opaque as possible – for the same reason that he sent out a challenge with two false claims thrown in. Archimedes sought valid proofs, of course. But his goal was not to achieve incontrovertibility. His goal was the precise edge of contestability: the point at which one's result would be shielded by the power of mathematical proof – and yet would invite disbelief and even controversy. And we can see the various layers of history – sociology – leading up to this moment. The role of debate and controversy in defining the very nature of intellectual engagement, from the very beginning of Greek science and philosophy, and then, the value put on irony, subtlety, and surprise in the Hellenistic culture centered on Alexandria.

### *On Floating Bodies*

The value put on irony . . . which is why I believe that "section of the right-angled cone, APOL" really ought to be an intentional swipe at Apollonius. Perhaps – this we cannot tell. (Even if it was intentional, it was also intended to be *hidden*, after all!) The treatise itself is there: *On Floating Bodies*, the culmination of Archimedes's project of mathematizing the physical world.

The first thing to note is that this treatise forms a close parallel to *On Balancing Planes*. There, as you recall, Archimedes first proves, in a first book, a key general result in "physics" (law of the balance), then applies it to relatively simple results (center of the weight in triangle and trapezium); the second book restates, somewhat differently, the general result and then applies it in a much more complex way, by bringing in the parabola (the center of the weight of the parabolic segment – and the parabolic slice).

Similarly, in *On Floating Bodies*, the first book derives a key general result in "physics" (law of buoyancy), then applies it to a relatively simple result (conditions of stability of a spherical segment). The second book restates the key result in a different way and then applies it in a much more complex way by bringing in the parabola (the conditions of stability of a segment of a paraboloid). The parallels run deeper – both treatises, in their current form, do not have an introduction other than a very brief statement of postulates. (*On Floating Bodies*, incredibly, manages with only a single postulate at the start – and then another, snuck in following the proof of the law of buoyancy.) Quite obviously, both are divided, somewhat arbitrarily, into two books. And of course, both deal, in a purely mathematical way, with a question about the physical world. I find it incredible that such a parallelism was not intentional.

And so, everything in *On Floating Bodies* is like that previous achievement – but *more*. As noted: a *single* postulate will do. "Let the liquid be assumed having such a nature so that the less pressed of its parts – being set equally [with each other] and being contiguous [with each other] – is pushed out by the more pressed, and that, further, each of its parts is pressed by the liquid above it which is along a perpendicular." Just as with *On Balancing Planes*, this cryptic statement gains its meaning only through its applications. Figure 66 is from Book 1, proposition 5.

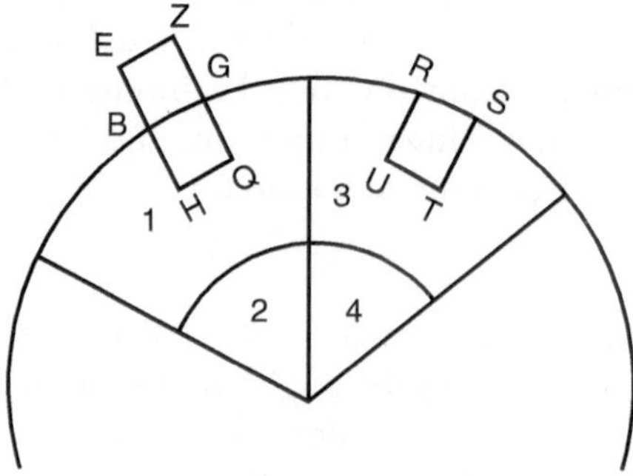


Figure 66

We consider a planar section to a liquid sphere – a circle, then. (Note, at this point, that Archimedes takes for granted – as did all educated Greeks – that the earth is spherical. The spherical liquid is an approximation, then, of the sea.) We conceptually find a sector within this circle, divide it in turn into two equal sectors, and draw a smaller arc inside, so that now we have four regions, which I will label, in this figure, 1, 2, 3, 4. The key observation (in terms somewhat modernized from Archimedes's) is that region 1 presses down on region 2 while region 3 presses down on region 4. The forces pressing down are simply the weights of the bodies of liquid in regions 1, 2. (You notice that it is taken for granted that liquids press down toward the center of the earth. Further, liquids – and later on, solids immersed in them – are tacitly taken to be homogenous; this is, of course, the same as the assumptions used in PE.)

The postulate works in practice through the claim that if a liquid is stable, the forces acting down on adjacent and equal sectors are equal to each other. (This is a little bit like the balance of equal magnitudes equidistant from the fulcrum, postulated for PE.) That is, if we assume stability, it must be the case that the forces pressing down on regions 3 and 4 by regions 1 and 2 are equal. (From this, indeed, we can quickly deduce – as Archimedes does at the start of the treatise – that the liquids form a sphere.) Now, what if – as in the figure of proposition 5 – a solid, lighter than the liquid, is immersed in region 1, such as EZGQHB? A moment's reflection shows that if indeed region 1 is to remain with the same weight as region 2, it must be the case that the solid will float at the top of the liquid, partly immersed, in such a way so that its overall weight (EZGQHB) is equal to the weight of the liquid it has displaced. The liquid it has displaced is, of course, its immersed portion (BGQH; equal to RSTU in region 3).

We have found that vessels floating on the sea do so in such a way that they displace a liquid whose weight is equal to the vessel's own weight. I state this as a physical result to emphasize the sheer unlikelihood of our achievement. Pure geometrical thought – and no more than a single postulate – and so much can be proved about the world! A magic trick, already. In some ways, this is much more impressive than the achievement of the law of the balance. The generalization of the experience of the balance (equal weights, balancing at equal distances) to the general case was certainly known, empirically, prior to Archimedes. His achievement was "merely" the finding of a pure mathematical proof. But nothing suggests that anyone prior to Archimedes was aware of the law of buoyancy. This, after all, was the original "Eureka" moment!

And yet – and just as with *On Balancing Planes* – this is barely the beginning. Proposition 5 itself is further transformed – but relying on the same type of reasoning throughout – to a form that allows us to calculate the force with which an object, immersed in the liquid, is pushed up or, equivalently, the weight it loses by being immersed. Following that, Book I concludes with a relatively simple case of the determination of the stability of a particular solid immersed in a liquid, namely, the segment of the sphere. (This is the analogue of the way in which, in PE I, Archimedes finds the centers of the weight of the triangle and the trapezium.) I will go through this (proposition 8) because it is much easier to explain than the main development of Book II and because it is clear that this was Archimedes's intention, as well: I.8 is a useful preparation for what follows, not in mathematical terms (Archimedes does not refer back to it) but in cognitive terms. Without it, Book II would have been totally illegible.

In the arrangement of the diagram (which I slightly simplify and modernize), ABGD is the spherical liquid, with L its center, EZHQ – a light solid spherical segment (greater than a hemisphere) partially immersed in the liquid (BNGZ being the immersed portion). ZQ is the axis of the segment, and K is the center of the original sphere (of which EZHQ is a segment).

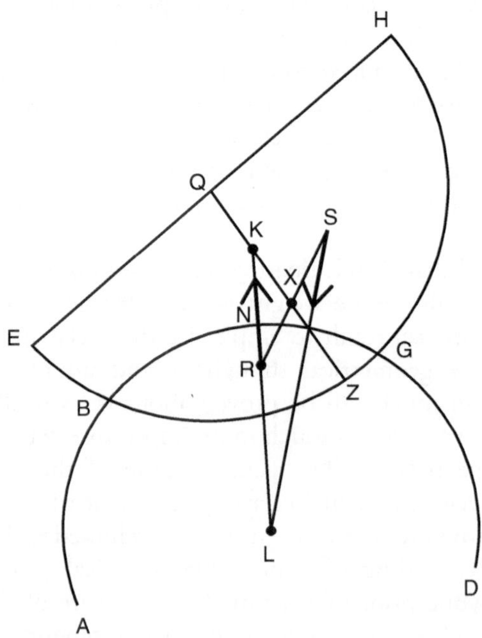


Figure 67

An added postulate – snuck in right before this proposition – is that a light solid, immersed in a liquid, is moved up along a perpendicular (this phrase is left undefined) passing through its center of the weight. It is also taken for granted (perhaps understood to be in no need of a postulate because it is already assumed in mechanical discussions) that a solid, outside of a liquid, is moved down along a perpendicular passing through its center of the weight. We thus have two portions: BNGZ and EBNGHQ. The first is immersed and moved upward; the second is nonimmersed and moved downward.

Archimedes states his intention to prove that the segment comes to a stand upright so that its axis is along a perpendicular (the phrase here means – as we see from its application – *verging toward the center of the earth*). To do so, he asks us to assume that the segment is not upright, that is, that its axis does not verge toward the center of the earth. This is explicitly – and meaningfully – presented as a proof by contradiction. Thus, the line joining the center of the earth and the center of the spherical segment is assumed not to pass through the axis of the spherical segment.

A moment's reflection shows that if two spheres intersect, their conjoint volume (a kind of lens shape) is symmetrical around the line joining the centers of the two spheres. Thus, the center of the weight of this region must be on that line, joining LK; we assume it to be R.

The center of the weight of the spherical segment as a whole must lie on the axis QZ, on the side of K that is closer to Z (originally, the sphere had its center at K; we obtained a spherical segment by removing part of the sphere, farther away from Z, so after the removal, the center must have moved toward Z). Let it be X. The center of the weight of the *nonimmersed portion of the segment of the sphere* (EBNGHQ) must lie at a point such as S, which is on the continuation of the line RX on the other side of the axis from R. (This is because R and S balance out as point X; it is the fulcrum of the two segments. Notice that Archimedes applies *On Balancing Planes* here.)

This is all we need. The immersed portion is moved up along the line LR. The nonimmersed portion is moved down along the line SL. Because the lines do not coincide, the solid is not stable, and in fact, it will rotate. Hence, the solid will be stable, Archimedes concludes, when it stands upright. This final transition is rather fast, and Archimedes elides the need to show the existence of a path leading from instability to stability. (He does so, perhaps, because he thinks of this problem strictly within the terms of a proof by contradiction where once the alternative has been ruled out, the affirmative result is thereby immediately obtained; this is, after all, the basic technique Archimedes had learned from Eudoxus and applied so often since!)

What this proposition does is clarify our approach to problems of stability. A light solid is stable in a liquid when the centers of the weight of its portions (immersed and nonimmersed) are aligned. The meaning of "aligned," in this case, is that the line joining those two points passes through the center of the liquid sphere; this will be slightly (and implicitly) revised in Book II.

The case of the segment of the sphere already involves some interesting geometry and is perhaps surprising. That is: it is not obvious that however slanted the segment is at the start, it will always end up standing upright. (Note, incidentally, one constraint: we are not allowed to slant the segment such that its base is immersed in the liquid; Archimedes's tools, in fact, would fail then because the immersed portion, in such a case, is no longer a symmetrical lens, and so its center of the weight cannot be determined.) But we can already expect that this would be trivial compared to what comes next.

Book II is an extended study in the conditions of stability – derived from the positions of the centers of the weight – in the segments of paraboloids of revolution. Let us take a look at FB II.2 (the first – and least difficult – of the cases).

We consider, once again, a plane section. APOL is a parabolic segment ("section of a right-angled cone," of course!). IS is the surface of the liquid, so IPOS is immersed and AISL is nonimmersed. Note that Archimedes moves to treating the surface of the liquid as flat and to treating "perpendicular" not as "verging to the center of the earth" but as "perpendicular to the surface of the liquid." (Almost incredibly: this happens without a word of explanation!)

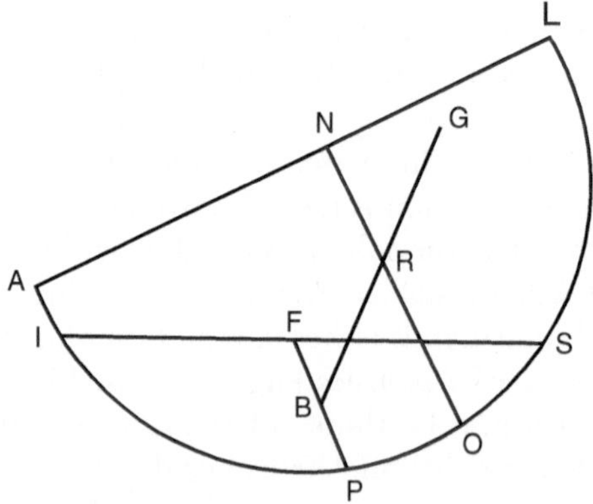


Figure 68

We need the centers of the weight of paraboloids of revolution, which Archimedes had found in a previous (lost) work. Unsurprisingly, they lie on the diameter of the segment, and remarkably, they are found at a simple cut: one-third the diameter. NO is the diameter of the original parabolic segment, and it is easy to show that if we bisect IS at F and draw FP parallel to NO, FP is then the diameter of the parabolic segment IPOS. We thus divide NO at R, so that OR:RN::2:1, and FP at B, so that PB:BF::2:1. R is the center of the weight of the entire paraboloid; B, of its immersed portion; point G can be found as the center of the nonimmersed portion. Clearly, the segment will be stable when the line BG is perpendicular to the surface of the liquid. In practice, we may as well consider just the line segment BR. Is it, or is it not, perpendicular to the line IS?

This, it turns out, depends on two questions.

1. What is the shape of the original parabolic segment APOL: Is it elongated, like a bullet? Or flat, like a bowl? This is a degree of freedom, as we are allowed to cut the parabola however far from or close to the vertex as we please, and such a cutting produces a variety of shapes. "However far or close" has to be measured by some unit inherent to the parabola, and Archimedes measures this, naturally, in terms of the ratio of the main diameter NO to the fixed parameter of the parabola. (This is already reminiscent of Apollonius's Book v, where the position of the smallest lines was dependent on the distance from the vertex, expressed in terms of multiples of the parameter; more on this to come.)
2. What is the immersed fraction of APOL? This is a degree of freedom that, remarkably, can be given a precise geometrical as well as physical meaning. In the first proposition of Book ii, Archimedes restates the law of buoyancy to show that the immersed fraction is determined by the ratio, in weight (in the sense of what we now call *specific gravity*), between the liquid and the solid. If the solid is one-fifth the specific gravity of the liquid, one-fifth of it will be immersed; if it is 99 percent of the specific gravity of the liquid, 99 percent of it will be immersed. This is a ratio that we can simply assume as we wish. To make this even more interesting, Archimedes had proved that the volumes of segments of paraboloids of revolution are to each other as the square of the diameter. The ratio of specific gravities between the solid and the liquid is therefore the same as the ratio, in square, between the line segments FP, NO.

We thus have an arrangement such as in Figure 68, where we may vary two values:

1. NO, in terms of the parameter (producing different shapes for the figure APOL)
2. $FP^2:NO^2$ (producing different fractions of the immersed portion)

The question then becomes, What happens to the relative positions of the points R, B, as those two values are gradually varied? This is the single question studied throughout FB II.

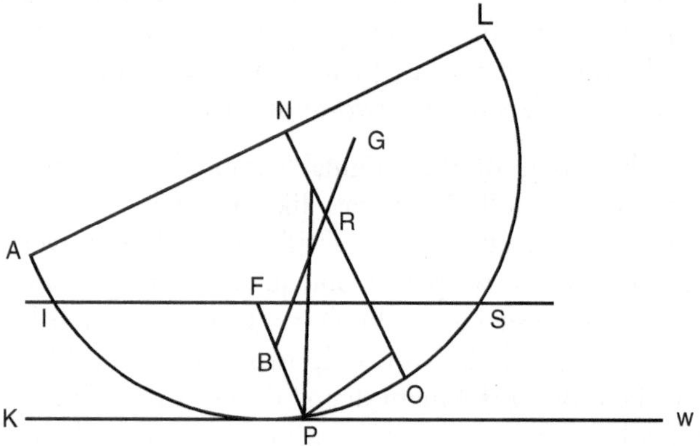


Figure 69

The way Archimedes deals with this question here, in FB II.2, is by drawing a tangent at point P, such as KW. Because it is a tangent at the vertex of the diameter PF, it will be parallel (by elementary conic theory) to the line IS. This provides us with an opening. Draw the perpendicular from P to the tangent/surface of the liquid, and if this perpendicular passes through both points B, R, the figure is obviously stable. If it passes between the points B and R, with B on its left and R on its right (as in Figure 69), we find that the perpendiculars drawn from B and R, respectively, do not coincide. The nonimmersed portion will move down more on the side of L; the immersed portion will move up more on the side of A; and overall, the segment will rotate in a clockwise direction and thus right itself. If the position of B and R is replaced, the direction of the rotation will be counterclockwise, and the figure will not right itself but instead capsize (I return, later on, to this point to explain the sense of "capsizing" here).

We are now looking at the relative positions of points B, P, R. In the simplest case, studied in FB II.2, the axis of the main segment NO is taken to be small compared to the parameter, by a precise value: the diameter NO is smaller than 1.5 times the parameter. With such a bowl-shaped parabolic segment, it can be shown that regardless of the ratio $NO^2:FP^2$, the segment behaves rather like the spherical segment, and indeed, when the segment is tilted, the perpendicular drawn up from P must always fall between B and R so that the segment then rotates and, presumably, stabilizes when the points R, B, and P all fall on the same axis common to both segments, APOL and IPOS.

Why is that? Because NO is less than 1.5 times the parameter, its portion RO is obviously smaller than the parameter. It can be shown, however, that for any arbitrary line drawn parallel to the diameter, such as FP, if we draw two lines from P, one perpendicular to the diameter NO and one perpendicular to the tangent KPW, they will intercept a segment, on the diameter NO, exactly equal to the parameter. (This is precisely analogous to the reasoning used by Apollonius in Book v of his *Conics.*) Thus, the line drawn from P, perpendicular to KPW, must cut the diameter NO above the point R (or otherwise, the intercept will be smaller than OR and so smaller than the parameter).

All of this is trivial compared to what comes next. Indeed, I am not quite sure how to continue our guided tour because, from this point onward, I cannot think of a way of making the arguments at all accessible. I focus, then, on the results alone.

Archimedes goes, in the remainder of the treatise, through the combinatorics of length and weight, adding in the extra consideration that the segment may start out with its base entirely outside the liquid (the starting point we have considered so far) or entirely inside the liquid (which brings its own, slightly different complications). This is elaborated across nine propositions, the last of which breaks down into many subcases. It turns out that the moment we allow the axis to be long, the ratio $NO^2:FP^2$ does matter. Specifically, the final proposition 10 shows that with a very long, bullet-shaped parabolic segment, a certain paradoxical result ensues. If the segment is almost as heavy as the liquid, almost all of APOL will be covered by IPOS, and so the segment will still be stable. That is, if you tilt it (while keeping the base entirely outside or entirely inside), it will right itself.

Make it somewhat lighter, and more and more of it will extrude outside of the liquid, and so it will be less robustly based in the liquid. At a given weight, the segment will obtain its stability – excitingly! – no longer upright, but when it is tilted at a certain angle. This already is very counterintuitive because, at first glance, it would appear that a symmetrical object such as the paraboloid must be stable upright – or not at all!

Indeed, this stability is rare and obtained only at a precise weight. Make the solid even lighter, and it will capsize.

But, and here is the greater paradox, if you make it *even* lighter – a feather now, only slightly touching on the liquid – it will, counterintuitively, regain its tendency to stabilize. Make it light enough, and once again, it will stabilize at an angle. Make it super-light, and it will stabilize perfectly along the axis.

The tale of the Roman soldier who approached Archimedes while he was hard at work and was asked to "leave my diagrams alone," then killed Archimedes in rage, is there to broadcast a particular image. However involved with the practical world, Archimedes was also at a certain remove from it. I find it hard to read *On Floating Bodies* any differently. It is, in a sense, a supreme application of mathematics so as to derive results that – through that magic of mathematics that philosophers, rather than historians, need to explain – are, in fact, true of the physical world. But it is clearly motivated, ultimately, by something much less practical, not just in the sense that it is not about physical reality but that even within the terms of mathematics, the point is not to furnish widely applicable results but rather to furnish subtlety and paradox, as such. No ancient ships were designed, in antiquity, based on Archimedean mechanics, and as a matter of fact, it is clear that the ultimate motivation, for Archimedes, lay within conic theory itself.

I have mentioned Apollonius's *Conics* Book v, and it is indeed, in some ways, comparable to Archimedes's *On Floating Bodies*. It is about finding meaningful points on conic curves, often based on considerations having to do with the point at which a line equal to the parameter is cut off from the axis. (In the most difficult proposition 10, Archimedes, too, operates by producing an extra, auxiliary conic section, in this case, a parabola.)

But here lies also a significant difference. The question posed by Apollonius is right at the edge of geometrical significance. We are looking at, first, the smallest line from a given point on the axis to the curve (so far, so clear). But then we move to another object altogether, the property that a given point on the curve has, serving as the end point for the smallest among the lines drawn from some point on the axis. And then even at a further remove – the property of a given point on the plane, of having some smallest line to a given curve (in the sense defined previously) passing through it. This is already hardly motivated and can only work because we are fascinated, indeed, by the very subtlety of the exercise.

Archimedes, in this treatise, is even further removed from geometrical significance. This is all about the relative positions of almost-arbitrarily chosen points on a parabolic segment and its subsegment; one point is A, and another is L; yet another point is R, located at two-thirds of the diameter NO; B is the final point, located at two-thirds of the diameter of the subsegment. We are provided with this general arrangement and asked to find, for various values of segment and subsegment, the relative positions of the points from left to right. This, in a nutshell, is the problem. It is a genuine problem in pure conic theory, a very hard one – and a surprising one, too. But who would even care?

This is what the physical interpretation does for Archimedes: it provides the problem with meaning. Of course, I do not intend to say that Archimedes got obsessed with the relative positions of A, B, R, L and then looked for a possible meaning for this geometrical question. Rather, he noticed in general that this *thing*, the "center of the weight," applied to the problem of stability in a liquid, can endow geometrical configurations with meaning and thus expand the field of what is geometrically meaningful. Indeed, my suggestion is that this is why Archimedes got interested in the center of the weight, itself, to begin with: because it made certain relations rich with significance. It was always Archimedes – turning back toward his diagrams. Because, you see, it was the makers of diagrams – other mathematicians – with whom Archimedes was in conversation: Conon, Dositheus, even Eratosthenes; Apollonius, perhaps, his final, most important interlocutor.

### *A Syracusan Coda: And More, on the Physical Sciences*

In the end, the Roman soldier came. Archimedes got entangled in Mediterranean history, indeed, in its most spectacular showdown. In 218, Hannibal crossed the Alps; in 217, at Cannae, he completely destroyed the Roman army. Rome had almost nothing, other than its walls, with which to defend itself. Its allies in Italy were defecting; soon enough, it was clear: Hannibal's Carthage might rule over the western Mediterranean. Averting immediate catastrophe, the Romans began to raise a new army, and Hannibal was preparing for the final push. What he needed, above all, was fresh troops from Carthage, and he needed them fast: they had to cross via Sicily. To prevent this, the Romans wanted Syracuse on their side. Roman resources were thin, but this hole in the defenses, above all, had to be plugged at all cost. When Syracuse demurred, Marcellus laid siege to the city. We need to imagine the Roman flotilla approaching the city – a few last ships leaving the harbor of Syracuse, bound for the eastern Mediterranean. I imagine one of those holding in its bow a tiny piece of cargo – two rolls, addressed to Alexandrian readers, containing Archimedes's *On Floating Bodies*. How eagerly this cargo was awaited by those in the know!

And meanwhile, Archimedes stayed back. Syracuse was holding. Everyone in the Mediterranean, now, had catapults (as we shall return to note in the following chapter). But those of Syracuse were especially well sited by Archimedes, their ranges calculated precisely in advance so that the besiegers would find no safe place. (This is asserted in clear terms by the historian best positioned to know, Polybius, on whom more in the following chapter; I see no reason to doubt any detail in this account.) And then, a rapid series of events. (Apollonius in Alexandria, meanwhile – allow me to dream – was plotting his comeback, his own subtle response to Archimedes's APOL!) A traitor was found – they always did find one in those long sieges. The gates were opened in the dead of night; Syracuse fell; Hannibal's army was never replenished. The future would be Roman. Archimedes died. He was old, to be sure, with perhaps many dozens of books to his name. Later still, the future would be Archimedean. The last two rolls coming out of Alexandria were more enduring than Roman power. (More on this in the epilogue to this book.)

But enough with this romantic note and now to a sober observation: one thing that did not happen in Syracuse was any mirror burning any ship. To be clear, such a story would circulate, starting in Late Antiquity. It is patently impractical with anything like ancient technology. But as with "Give me a place to stand and I shall move the earth," as with the story of Archimedes jumping out of the bath, this story, too, was probably invented out of a reality, rooted in Archimedes's own writing.

To be clear, I do not think that Archimedes studied the geometry of burning mirrors, for reasons that I will explain later on. But I do think he was a pioneer in optics and was remembered for that – and that, years later, the study of burning mirrors would come to be a dominant part of the study of optics. And so, years later – in Late Antiquity – it became natural to encapsulate, in narrative form, the idea that an author was a master of optics by inventing a tale where he puts mirrors to good burning use.

What can we say of Archimedes's optics? Of ancient mathematical optics in general? It now has its moment in my book – a history of Greek mathematics, where I try to do justice to the variety of the exact sciences as a whole. And yet I shall add no more than a few brief notes (this one; another one, later in this chapter, on burning mirrors; and yet another one, very briefly, in Chapter 5, as I introduce Ptolemy's character as a scientist). Ancient optics, quite simply, is unlucky in its survival. It is likely that its three main protagonists, in antiquity, were Archytas, Archimedes, and Ptolemy. Of the first, we have but the vaguest indications, which suggest, however, that he might have invented the discipline. Of Archimedes's optics, what survives is no more than a couple of highly mediated and late testimonies. Ptolemy survives – as Books II–IV and some of Book V of an originally five-book treatise, all of this extant only in a Latin translation from an Arabic translation of the Greek original (!).[18]

A couple of works – on optics and on the field of catoptrics, which is the general theory of (nonburning) mirrors – are extant under the name of Euclid. Whether or not they are indeed by him (or even should be dated to his general era), they likely represent the elementary, and yet sophisticated, results already obtained by Archytas or by other authors of his generation. One quickly finds there the basic properties of perspective, and the key property for catoptrics is that the line of vision is reflected by a mirror at equal angles. This may, or may not, have been extended, even early on, to the study of non-flat mirrors; we do not know.

To repeat, Archimedes did *not* study burning mirrors (which are one of the more natural applications of the study of non-flat mirrors). And it does seem likely that his main contribution lay elsewhere. The two positive, tiny bits of information we have on Archimedes's optics both involve Archimedes engaging with the phenomenon of *refraction*. Both come from late ancient commentaries: Theon of Alexandria – seemingly a well-informed author – cites Archimedes on the manner in which the size of the stars could be affected by the refraction, through the passage of light through (what we call) the atmosphere. Olympiodorus, a commentator to Aristotle – perhaps less reliable but also hardly one to be able to invent this kind of information – cites Archimedes for the observation that a ring tossed into an empty vessel appears, once water has been poured, to become displaced. There is a complication with that (the very same claim is asserted – without any further elaboration – in the extant text of the Euclidean catoptrics). The simplest account, in my view, is not that Olympiodorus was confused between Euclid and Archimedes but that someone took a well-known Archimedean observation and inserted it as an addition to the Euclidean catoptrics (because, in fact, this observation is *not* otherwise motivated by the text of Euclid). The study of refraction came to be seen, in antiquity, as part of the general study of mirrors, and

[18] A. M. Smith, *Ptolemy's Theory of Visual Perception: An English Translation of the Optics* (Philadelphia: American Philosophical Society, 1996).

the reason is clear enough. When one studies mirrors, one usually studies them for their optical illusions: the distortions produced by concave and convex mirrors. And the tiny bit preserved by Olympiodorus, not coincidentally, involves just that – an illusion, an optical trick.[19]

To be clear, I doubt that the tossed ring was the point of Archimedes's lost treatise on optics. Later readers tend to remember the most elementary bits of an Archimedean treatise. (I would seriously hazard the guess that, if anything, Archimedes went on to study refractions produced by oddly shaped bodies of water, or the image, through refraction, of oddly shaped bodies: How about a paraboloid?) But it is nice to part with Archimedes on this note of parlor magic. This is what this is all about: mystery, surprise.

And yet, there is one more Archimedean magic trick we have forgotten to perform! Once again, I will be brief and merely hint at this one right now. But let us remember Archimedes's contribution to astronomy, which consisted, above all, in the making of a wondrous device. We will, of course, look further at the Antikythera mechanism – when we return to discuss this as evidence for Greek astronomy. This much can be gleaned, however, from Cicero's report (ultimately based on Archimedes's writing?) of the machine. It put in place a subtle, carefully calculated combination of hidden gears – so that, by pulling on a single handle, the mechanism derived the entire motion of the sky and all the planets. It is hard to reconstruct much more than this concerning the place of Archimedes's device in the history of ideas – or of technology. But what we know is enough to put it, tentatively, alongside our sense, so far, of what Archimedes was like, as an author. We imagine him displaying the machine to his friends, perhaps to his patrons, but also conveying it, more indirectly, as a written work. Conveying, throughout, the surprise of a magic trick.

Throughout of all of this, we see, for sure, a communication pattern different from that of Archytas conversing with Plato. The intended audience is not philosophical. This is not conceptual analysis, abstraction for the sake of the quest for hidden ontological truths, logical classification. I imagine Archimedes seeking a group of like-minded readers, immersed in the same culture of subtlety and surprise. I provided a number of examples, but before we move on to consider, finally, the totality of this audience – Archimedes's generation – a quick word on one final, negative

[19] A somewhat more expanded survey of ancient optics is offered in R. Netz and M. Squire, "The Limits of Ancient Optics," in M. Squire (ed.), *Sight and the Ancient Senses* (London: Routledge, 2015), pp. 68–84.

bit of evidence. We have seen engagement with the physical world – or perhaps some kind of "mechanics" – and for sure (although now less well attested), we have seen astronomy and optics. What we do not find in Archimedes – and almost without exception, in his entire generation – is any hint of music. This – the hallmark of the application of mathematics – seems almost to have vanished in the Hellenistic era, *alone of all the mathematical fields*. Is it possible that the association between mathematical music and a particular philosophical attitude became so pronounced that mathematicians, from now on, who wished to avoid the implication of philosophical allegiance, would prefer to *avoid* that discipline?

## V. Response to Archimedes: The Generation

### *The Generation of Archimedes*

This has mostly been – appropriately – a chapter about Archimedes. I have shown enough, I hope, to provide a sense of the scale of his achievement. And in considering the figure of Apollonius (whom I have *not* adequately represented), we have seen something of the generational response. For indeed, if Archimedes was so important, this is because so many significant scientists read and responded to him over the centuries. This began right then.

The following is a list of the authors whom it is reasonable to place in the generation of Archimedes (that is, let us say, born later than Archimedes but before about 225 BCE, and so active until the middle of the second century). Our sources differ from those for the generation of Archytas: we do not have anything like Proclus's historical summary, apparently organized chronologically. Therefore, I will not try to arrange these authors by date. But our sources are better: for many of these authors, we have actual reports, sometimes fragments, of their works. A few are even extant.

For each author, then, I note the state of our knowledge concerning the mathematics: a mere attestation of mathematical activity (attested); fragmentary survival of some works, at least in a report that provides a sense of the contents of the mathematical achievement (fragmentary); or extant work(s). I also note the likelihood that the authors in question are, in fact, from this generation: whether we have direct ways of dating them (dated) or we merely assume, based on the character of writing, that it best fits this period (extrapolation). The list is arranged alphabetically.

| | | |
|---|---|---|
| Apollonius | Extant | Dated |
| Apollonius (son) | Attested | Dated |
| Attalus | Attested | Dated |
| Basilides | Attested | Extrapolation |
| Charmandrus | Attested | Extrapolation |
| Crates | Attested | Extrapolation |
| Demetrius | Fragmentary | Extrapolation |
| Diocles | Extant | Extrapolation |
| Dionysodorus | Fragmentary | Extrapolation |
| Dositheus | Attested | Dated |
| Eratosthenes | Extant | Dated |
| Erycinus | Attested | Extrapolation |
| Eudemus | Attested | Dated |
| Heraclides | Attested | Dated |
| Heraclitus | Attested | Extrapolation |
| Hippias | Attested | Extrapolation |
| Hypsicles (father) | Attested | Extrapolation |
| Hypsicles | Extant | Extrapolation |
| Naucrates | Attested | Dated |
| Nicomedes | Fragmentary | Dated |
| Nicoteles | Attested | Dated |
| Perseus | Attested | Extrapolation |
| Philo of Byzantium | Extant | Dated |
| Philo of Tyana | Attested | Extrapolation |
| Philonides | Attested | Dated |
| Protarchus | Attested | Extrapolation |
| Puthion | Attested | Extrapolation |
| Scopinas | Attested | Extrapolation |
| Theodosius | Extant | Extrapolation |
| Thrasidaeus | Attested | Dated |
| Xenagoras | Attested | Dated |
| Zenodorus | Extant | Extrapolation |
| Zeuxippus | Attested | Dated |

There are (adding in Archimedes himself) thirty-four names, of which almost exactly half (sixteen) are securely dated. Surely, at least some of those whom I extrapolate to have been working in this era were, in fact, active at some other time (if so, most likely in the imperial era; I shall return to discuss this briefly in Chapter 5 as I comment, in the context of the history of astronomy, on the character of mathematics under the Roman Empire). But the number is significant: compare it with what we found for the preceding century (nine or ten?) and the generation of Archytas (twenty-three). Once again, we have no advantage in terms of historiography. There is no extant work setting out to record the mathematicians of this era. The only thing that makes the memory of this era survive is the sheer survival of its own works – which is very remarkable. Aside from the two main authors, Archimedes and Apollonius, works may now be extant in full, from this generation, by up to six other authors: eight extant authors is already something like a third of the entire set of mathematical authors surviving from antiquity. (A very large group in itself: of the roughly three hundred extant ancient Greek authors, almost 10 percent are mathematical! I shall return to discuss this in the last chapter.)

So, to begin with, it is likely that there were indeed many mathematicians. The ratio of attestation – how many of the extant authors are actually now known – cannot be better than that for the generation of Archytas, and it is therefore likely that there were at least a hundred mathematicians in this period. Two – Archimedes and Apollonius – must have been the most prolific, with perhaps fifty to a hundred works just between the two of them; most must have written just a handful of works or perhaps only a single one, but the conclusion is that there were hundreds of mathematical treatises produced in this generation. About twenty are now extant, even though almost all of the extant works from this generation are difficult pieces of mathematics, definitely not part of the educational curriculum. They are extant not because they were in educational use but because they were valued. Already, the ancients knew: something important was taking place then.

And indeed, the sense of a qualitative identity is what provides us with the sense of a generation and justifies our ascription of otherwise hard-to-date authors to this era. The core is, indeed, the response to Archimedes. In some cases, it can be very direct. In *Sphere and Cylinder* II.4, for instance, Archimedes sets out a problem akin to that of the duplication of the cube, although considerably harder than it. To recall: *Sphere and Cylinder* II came about halfway through the series of treatises sent to Dositheus, discharging the original challenge to Conon. My own interpretation of the evidence is that within this work, Archimedes set out an extra challenge: see if you can solve that cubic problem! There was no one to answer the original problem set out in the letter to Conon, but now mathematics was revived, and there were, indeed, at least two solutions offered, one by Dionysodorus and the other by Diocles. We also have extant Archimedes's own solution. (All three are collected and cited by Eutocius in the sixth century CE; I return to this episode in Chapter 6.) The problem is subtle and calls, specifically, for ingenuity with conic sections. This is all typical to the challenge posed by Archimedes – and to the responses to it. Everyone, now, was brushing up on their parabolas. There is a wider interest in other inventive curves.[20] Nicomedes describes a curve defined by a complex geometrical condition that is best constructed by a geometrical machine. (Is the machine purely hypothetical, or was it built?) This is the conchoid, now known through a detailed account preserved, once again, by Eutocius. Diocles proposed a different curve, which was perhaps called the cissoid. A conchoid ("shell-like"), a cissoid ("ivy-like"): the rather flamboyant nomenclature is in itself Archimedean. But the key, of course, is the Archimedean tenor of the mathematics itself. A certain Perseus studied the curve produced by the intersection of a plane and a torus (a rotated circle) – this is already reminiscent of Apollonius, studying the cylindrical spiral. Dionysodorus measured the volume of this torus by finding that the ratio of this volume to a certain cylinder is as that of a circle to a parallelogram. (This is directly related to the approach we noted in the *Method*: Were Archimedes's infinitary methods applied by other mathematicians?) A certain Hippias invented a curve now known as the quadratrix, distinguished by a certain proportion defining its construction as a point is moved along a line. (In this case, then, a very obvious variation on the idea of the spiral. Just like the spiral, the quadratrix – as its name suggests – can be used for the squaring of the circle.) Zenodorus is now extant, in very mediated form, in a study of isoperimetric figures (figures that have different shapes but the same perimeter or surface). The precise approach taken is a direct continuation of Archimedes's studies on the surface of the sphere and is related, in fact, to the original challenge sent to Conon. It all goes back to there!

Perseus celebrated his curves with an epigram:[21]

> Finding three lines on five sections
> Perseus, honoring these [the gods?], propitiated the demons

He must have set up an inscription with an image of the curves, a demon-pleasing diagram! Eratosthenes did something similar with his *Mesolabion*. This small piece is extant (again, cited by Eutocius). Eratosthenes describes a machine that solves, directly – no need for extra curves – the problem of finding mean proportionals. He criticizes past solutions (Nicomedes, in turn, would criticize this solution by

[20] I refer, in the Suggestions for Further Reading, to W. R. Knorr, *The Ancient Tradition of Geometric Problems* (Boston: Dover, 1986). The kind of authors briefly surveyed here are discussed there, in greater detail, in Chapter 6.

[21] G. R. Morrow, *Proclus/A Commentary on the First Book of Euclid's Elements* (Princeton, NJ: Princeton University Press, 1970/1992), p. 91 (p. 112 in Proclus's critical edition).

Eratosthenes) and concludes with a celebratory epigram that was inscribed on stone, accompanying a model of the machine. This is indeed all reminiscent of Archimedes's own mathematics-in-a-poem, the *Cattle Problem*, a puzzle – yet another challenge! – asking its readers to solve a problem of calculation that is presented twice, in a simpler and in a more difficult version; the simpler version is hard, and the difficult version, as far as we can tell, would have been impossible for an ancient reader. (It was first computed only with the aid of modern computers. Yet another poisoned bait?) We already mentioned the poetic premise of the *Sand-Reckoner*, to which, as noted earlier, Apollonius himself responded with a calculatory treatise that culminated with the calculation of the numerical value of a hexameter line! This pattern of an interface between mathematics and poetry is perhaps natural in a culture revolving around the literary center of Alexandria, with its distinctive poetic tradition. The goal is to obtain surprise, through subtlety, in the manner of the poets. We could go on, but such few strokes already establish the sense of the generation. There were many mathematicians, all of a sudden – always under the spell of the culture on which they grew, that of Hellenistic Alexandria, with its emphasis on irony and subtlety. This is what I have characterized, in the title of a previous book, as the mathematics of "ludic proof."

This is all a very rapid overview and therefore does not do justice to this very active generation. As usual, I try to amplify such rapid overviews with at least one example – and surely, we must have the burning mirrors!

And so, to Diocles. As is typical of the work of the minor figures of Archimedes's generation, it survives in a mediated form. Not only is it in Arabic translation only, but the treatise, as it is stands, is a medley of three distinct problems, each treated fairly briefly and all, taken together, amounting roughly to the scale of a single Archimedean treatise. Likely, the compilation is late (although, conceivably, it could have been produced by Diocles himself; our evidence is too meager to rule out the possibility that authors in this generation did produce such hybrid works).

I will now concentrate just on the problem of burning mirrors and on one of its proofs alone – the one that shows the burning-mirror property of the paraboloid. This gives us, first of all, a clue for Archimedes's optics: this treatise strongly implies that Archimedes did *not* study burning mirrors. Diocles strikes us as a very able geometer, and so I think we should trust his statement that he is the inventor of the construction he proposes. Furthermore, the paraboloid of revolution was so central to Archimedes's geometry that I find it incredible that Archimedes would have studied burning mirrors without producing, specifically, the construction proposed by Diocles. (Indeed, it is precisely for this reason that Diocles's treatment reads like the fulfilling of a task, worthy of Archimedes himself: the clear sense of the generational response.) But if so, we find that Archimedes's optics did not study burning mirrors, and so – as suggested by our scattered reports – it probably focused on the geometry of refraction. Diocles, then, must have been a deeply original mathematician – but also, as we will see, one deeply indebted to Archimedes.

The study of burning mirrors involves not refraction but simple reflection. We take it for granted, to recall, that rays of light and lines of vision are reflected by mirrors so that the angles of incidence are equal.

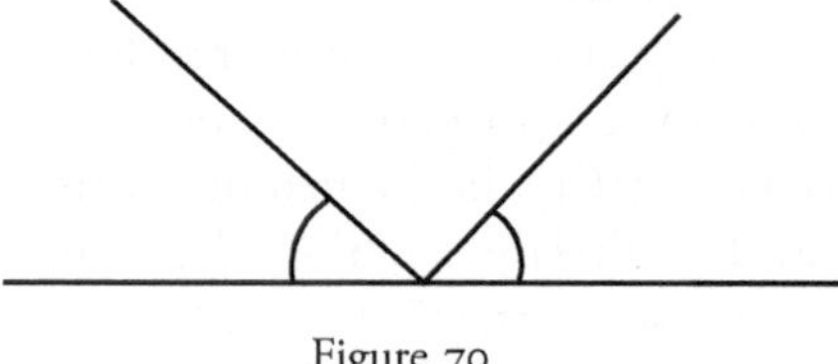

Figure 70

It is an ingenious but straightforward thought to consider the reflection of light and of lines of vision on *curved* mirror surfaces. The angle here will be measured with the *tangent* at the point of incidence.

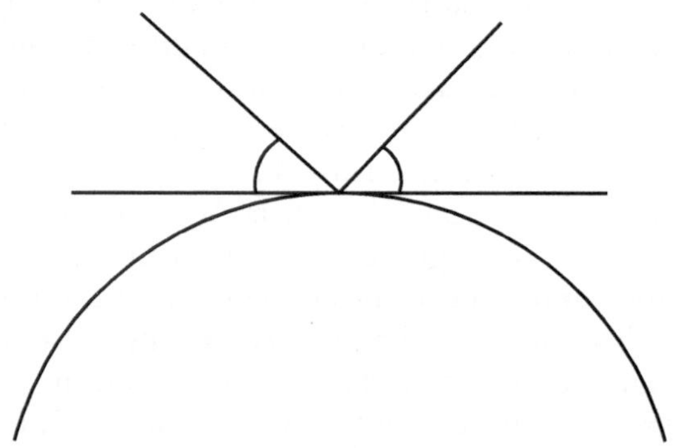

Figure 71

With this ready, let us now dive into Diocles's argument.[22] We consider a paraboloid mirror surface, and as usual, we argue based on just

[22] Diocles can be read in English translation in G. J. Toomer, *Diocles on Burning Mirrors* (New York: Springer, 1976).

a single plane section so that the diagram shows a parabola. The mirror is turned so that its concave side faces the sun, specifically in such a way that the rays of the sun are all taken to be parallel (a simplification that is, of course, justified and is analogous to the simplification in FB II, where the surface of the liquid is taken to be a flat plane).

In Figure 72, KBM is the parabola, and BZ is the axis of the paraboloid, hence, the axis and diameter of the parabola. We take an arbitrary point on the curve, namely, Q.

At this point, we add in an extra construction. First, draw the tangent at Q, as the line XA, meeting the axis and main diameter of the parabola at point A. Second, draw QZ perpendicular to the tangent, meeting the axis at Z. Finally, draw QG from Q, perpendicular to the diameter. The next bit is a little more unexpected, but this, too – and this is crucial – is by now traditional. We further add point E so that BE is equal to the parameter (we recall the significance of this position from both Apollonius's *Conics* V as well as Archimedes's *On Floating Bodies* II). The argument is now ready to proceed (the diagram drawn here is schematic only).

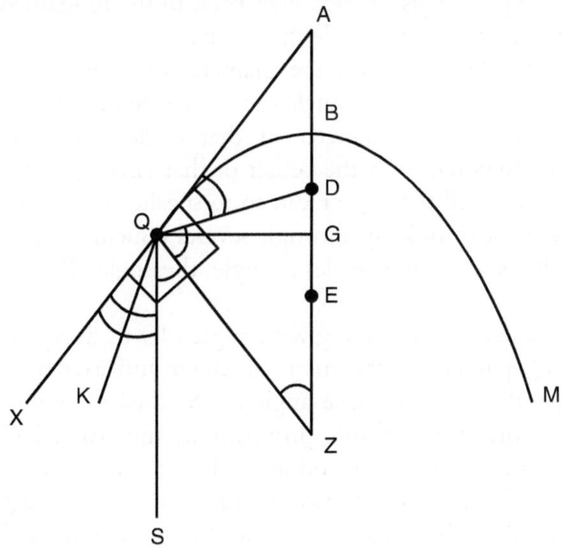


Figure 72

The key observation of the argument is the one we are familiar with: with the construction at hand – QZ perpendicular to the tangent, QG perpendicular to the diameter – the intercept on the diameter between the two perpendiculars, namely, GZ, is equal to the parameter. In this case, this means that GZ is equal to BE.

From more elementary considerations of conic theory, we can also get the equality AB = BG. But from GZ = BE. we can also directly conclude that EZ = BG, and so EZ = AB. At this point, we bisect the line BE at D, and it is obvious that AD = DZ. D, we find, bisects the line AZ. It is defined as the bisection of the line BE, which is itself a given (because it is defined by the parameter, itself given with the parabola).

We thus find that wherever we choose to draw the tangent to the parabola at a point such as Q, we always find a different intercept of the diameter between the tangent and the line perpendicular to the tangent – such as the line AZ. This line keeps changing, depending on the position of Q. However, even as Q and AZ keep changing, the *bisection* of all the lines, such as AZ, will always be at the fixed point D.

Now let us go back to the diagram. We know that AD = DZ, and by construction, the angle ZQA is a right angle. Diocles asserts that in such a construction, QD = AD = DZ. This is merely asserted, and we are asked to provide the proof on our own. (Archimedes would often leave such key steps unproved: instead of dotting the "i"s, one leaves out such details as subtle, surprising challenges.) This, in fact, goes back to the fundamental geometry of the semicircle and the right angle that we have seen ever since Hippocrates of Chios himself. The angle on the diameter of a circle is a right angle, and conversely, a right-angled triangle can always be inscribed inside a circle, its hypotenuse being the diameter of that circle. Obviously, then, the bisection of the hypotenuse is the center of that circle, and for that reason, QD = AD = AZ. (All three are radii, so to speak, of an implied circle.)

We can now move quickly through several conclusions. Because those are angles at the base of an isosceles triangle, the angle DQZ is equal to the angle DZQ.

But wait, because we're playing with angles, let us draw a parallel! QS is drawn from Q, parallel to the main diameter and axis BZ. Because QS and AZ are parallel, we also have angle ZQS equal to angle DZQ. (Note that the key protagonist of this proposition, the parallel QS, has been sneakily – Archimedes-like – introduced.) Hence, obviously, angle ZQS is equal to angle DQZ, as well. However, the angles XQZ, ZQA as a whole are both right angles (by construction), and so if we remove from each the equal angles ZQS, DQZ, the remaining angles XQS, DQA must also be equal to each other.

Point D, to recall, is fixed, regardless of the choice of point Q. This means that we have just found that for any point chosen on the curve, the line passing through that arbitrary point, parallel to the main diameter, will create a pair of angles equal to each other. One of those equal angles will be between the parallel line and the tangent. The other equal angle will be between the same tangent and the line drawn from that arbitrary point to the fixed point D.

But obviously, each arbitrary parallel line SQ, hitting the curve at an arbitrary point such as Q, can be taken to represent a ray of the sun: we started out by positioning our mirror so that it faces the sun in that direction! Meanwhile, the equality of the angles determines the reflection of such rays from the surface of the mirror. Hence, a paraboloid mirror turned toward the sun will direct all the rays to a single point D, at which, it is now stated, the mirror will act as a *burning mirror*.

I went through this in some detail, among other reasons, because it provides us a rare opportunity to follow the precise mathematical detail of the achievement of the generation. A typical Archimedean result (let alone a proof by Apollonius) is simply harder to follow. This is more straightforward. Although relatively simple, this is also palpably Archimedean. The same emphasis on subtle surprise – down to the intentional delay in the construction of the parallel line, so that, throughout the argument, we do not yet see the relevance of any of it for the optics of rays of the sun. And not only do we see here the mere concern with the mathematization of physics: indeed, the mathematization follows precisely the same route as in *On Floating Bodies*, using precisely the same properties associated with the parabola: the intercept between tangent and ordinate. In fact, the two diagrams – of FB II.2 and of Diocles – are essentially identical. The same geometrical configuration – with a radically different meaning. In Archimedes, the secondary parallel diameter is used to find the position of the center of the weight of the portion of the paraboloid immersed in the liquid. In Diocles, the secondary parallel diameter is the ray of the sun. In Archimedes, the tangent to the parabola defines a line parallel to the surface of the liquid. In Diocles, the tangent determines the optical property of reflection. In both Archimedes and Diocles, one starts from the magnitude of the intercept to derive results concerning that tangent, relying on the property that the lines defining the intercept have been drawn as perpendiculars. In Archimedes, this perpendicularity combines with specific metrical assumptions concerning the parabolic segment so as to determine the position at which the angle of that perpendicular will contain, or fail to contain, the center of the weight. In Diocles, this perpendicularity combines with a particular result concerning the equality AB = BG to find a fixed equality of angles (this is the chief original contribution by Diocles, which, we now notice, is really a – brilliant – variation on an Archimedean theme).

And so, in the treatments we just saw from Archimedes and from Diocles, the geometrical figure and its key relations are the same, whereas the physical meaning is different. But this, in fact, is the key similarity between the two proofs. Both gain their meaning through the assignment of a particular physical meaning. (And in this, they both differ from Apollonius's Book v, otherwise a rather similar work.) And that physical meaning is essentially irrelevant to the argument of the proof, other than in providing it with a meaning – so that essentially the same mathematics can be repurposed, with ease, for distinct physical "applications."

To be clear, I do not suggest that Archimedes and Diocles had ready a set of pure geometrical investigations and were shopping for potential physical interpretations. ("Eureka!" Archimedes exclaimed. "Two-thirds the diameter in a parabolic segment can stand for the center of the weight of the paraboloid of revolution and so defines its behavior in a liquid! I can finally publish my pure geometrical result on the relative positions of points two-thirds the way of the diameters of various parabolas!") This, obviously, is absurd. Archimedes was thinking in parallel about the properties of solids immersed in a liquid and the properties of paraboloids of revolution, and then he must have realized that the fact that (assuming a flat surface for the liquid) the immersed portion of a paraboloid of revolution is itself a paraboloid of revolution opened interesting avenues for bringing the two together and developing a relevant geometry of the parabola. With this example in place, it was easy for Diocles to notice another opening, in considering the same tangent, now not as parallel to the surface of the liquid but as the line with which to measure angles of incidence and reflection.

And so, perhaps, one of the ways in which Archimedes inspired a generation. Not merely in that there was an example of a great mathematician, so one wanted to emulate him. More than that: the example of Archimedes suggested specific ways in which one could create surprising and subtle mathematics. The significance of Archimedes, then, is that he found ways in which subtlety and surprise could play out mathematically. These can be classified mostly under two headings: the indirect approach for the study of especially difficult (so, in practice, curvilinear) measurements and the application of mathematics to the study of the physical world.

## *The Significance of Ludic Proof*

Let us first of all note the historical significance of the two main avenues opened up by Archimedes. Perhaps the case of the mathematization of physics is obvious: clearly, this will lead (through authors such as Galileo,

Kepler, and Newton) to no less than the making of modern science. But exactly the same authors, we should now note, relied also, crucially, on the measurement of curvilinear figures. This, too, is a key theme of the history of science, for a good reason. The thing that makes the measurement of curvilinear figures difficult is that they involve, effectively, the summing up of infinitely many components. This type of summing up is, then, equivalent to modern calculus. The technique pioneered by Eudoxus – but then perfected by Archimedes – is roughly equivalent (because it is its direct ancestor) to the delta/epsilon in modern calculus. Besides providing this essential idea, Archimedes and his generation provided many examples of the application of this technique, such that the makers of the modern calculus would eventually generalize and deduce from more basic principles. I will return to note this in the final chapter, but for the time being, we may conclude with this: Archimedes opened the door to calculus and to mathematical physics, hence to modern science. It was just a crack; much more had to be done. But the door, firmly shut before, was now open.

But our goal as historians is not to be impressed by Archimedes. Our goal is to understand him. And here is why it is important for us to situate Archimedes within his generation, within the historical moment of Hellenistic civilization. A literary culture of irony and surprise; science and philosophy, drifting away from each other. This is important to note because otherwise, we could easily misunderstand the route that led Archimedes to his mathematics.

It is often suggested – as a rather uncontroversial statement, even – that "science" came to be in the eighteenth century or even the nineteenth century. Indeed, there isn't a Greek word neatly translating the modern concept "science," and more to the point, something does change throughout the eighteenth century itself. Before that, throughout the seventeenth century and well into the work of Newton himself, many authors of science think of themselves as philosophers, and they engage with wide metaphysical concerns. It is only in the eighteenth century that something like the modern institutions of modern science begin to form and, indeed, only early in the nineteenth century that the career pattern of the "scientist" becomes established.[23] In early modern Europe, all the way down to Newton – or even beyond – what we now call "science" was pursued in the context of a wider pursuit, best understood – as indeed it was understood

[23] For a standard statement on this sociological development, see E. Mendelsohn, "The Emergence of Science as a Profession in Nineteenth-Century Europe," in K. B. Hill (ed.), *The Management of Scientists* (Boston: Beacon, 1964), pp. 3–48.

by its very practitioners – as "natural philosophy." This is all clear and correct. The mistake is that, from our own modern vantage, we tend to take this relatively recent transition and extrapolate it all the way back into the past. *It was always natural philosophy*. This is a common fallacy: that the part of history we know less about is more or less linear, more or less the same as the part we know better. And so we assume that because it used to be that what we now call science was once mixed up with philosophy, it must have been even more so, further back in the past: surely, they were all philosophers, back then!

Clearly, they were not. Linear extrapolations are often wrong. From Newton onward, science becomes ever more autonomous from philosophy. But prior to Newton, the path was more complicated and zigzagging. In Archimedes's time, in particular, science was quite autonomous, and in his way, I would say that Archimedes was a scientist.

But "his way" is not ours. Archimedes was a scientist – as opposed to a natural philosopher – not because he pursued the distinctive career path of a scientist, defined by the institutions of science. In this institutional sense, indeed, science did not exist. There were no royal societies, no journals of scholars, no university departments of science, no science chairs to fill. The identity of Archimedes was provided, first, by his choice of a literary genre – the one pioneered by Hippocrates of Chios and then by Archytas, of proofs provided with the aid of diagrams and a technical language. Further, at this particular juncture of time, this literary genre was written with a particular audience in mind – that of fellow mathematicians and, to a much lesser extent, the elite audience as a whole (the philosophers in Athens, least of all: they did not read the literature written away from Athens), written, always, against the background of wider literary currents, emphasizing subtlety and surprise.

And it is those motivations, this sociology – and *not* the ideas of an Aristotle of a Plato – that account for the specific scientific route taken by Archimedes. He picked up a particular technique, first offered by Eudoxus, because its subtlety (required, ultimately, by its subject matter) made a certain kind of surprise especially satisfying. Hence the infinitary methods. And he saw the possibilities of applying geometry to a seemingly unrelated field – the study of centers of the weight in solids, balancing outside and inside liquids – because there was a particular payoff of subtlety and surprise to be obtained by the bringing together of apparently irreconcilable, maximally distinct fields of study. This was rather like Callimachus's poetry! Hence the mathematization of physics.

This is best seen by considering the autonomy of the mathematical research, as shown by the topics selected for study. The objects discussed by Archimedes – and by the following generation as a whole – are hardly motivated by concrete applications. There seems to be a special value placed, if anything, precisely on somewhat outré objects: a conchoid, a spiral; above all, again and again, the conic sections. Other cultures simply did not consider such objects as part of their mathematics: Why should they? One studies, normally, things such as rectangles and triangles, squares and circles, perhaps cubes and spheres. This is natural when mathematicians are at the service of a society. You refer to those objects that are already known as part of more or less established practical usage. The study of, let us say, the paraboloid of revolution can arise only when mathematicians can make up their own objects to study and essentially no longer even worry about an outside audience. Of course, one does not just invent objects for study on a whim. The typical route we see – and one that is found time and again in the later history of mathematics – is that the thing that was first studied as a tool for other purposes gradually becomes the subject of research for its own sake. Study the finding of proportions, and therefore invent the tool of conic sections – until, at some point, you are no longer studying proportions as such, but rather, you simply study the conic sections themselves. Why is this possible? Because your audience shares, with you, the interest in the tools: because it is an audience of fellow mathematicians.

And so, this is what the generation of Archimedes did – to the point that, even when studying physics, the preferred object of study was, precisely, conic sections: the center of the weight of a parabolic slice; the stability of a segment of a paraboloid; a paraboloid mirror.

Something crucial took place in the making of science in the generation of Archimedes: the making of a well-defined field, pursued apart from philosophy, based on its own concerns and techniques. The manner in which this happened favored, precisely, an engagement with a ludic realm, of toy objects of no obvious practical significance, pursued purely for the aesthetic value of their study. And it is, finally, this very purity of aesthetic concern that gave rise to a particular brand of abstraction, typical to what would ultimately become the mathematization of the world as studied by science. This mathematization was based not on Plato's conceptual preference for the abstract over the material. It was based, instead, on the *autonomy of mathematical research.*

## Suggestions for Further Reading

In this chapter, I had to be selective: I concentrated on the major author – Archimedes – and even with him, I did not survey the contents of his works in full. This is the point where readers should turn to T. L. Heath, *A History of Greek Mathematics* (Oxford: Clarendon Press, 1921). The final part of volume I surveys Euclid exhaustively; Archimedes, Apollonius, and the minor figures of this generation occupy the bulk of volume II.

We are indeed lucky to have several outstanding surveys now available. A very rich interpretation of the architecture of Euclid's *Elements* is provided by I. Mueller, *Philosophy of Mathematics and Deductive Structure in Euclid's Elements* (Cambridge, MA: MIT Press, 1981). The best account of the mathematical achievement of the era is W. R. Knorr, *The Ancient Tradition of Geometric Problems* (Boston: Dover, 1986). For Archimedes, the simply titled E. J. Dijksterhuis, *Archimedes* (Princeton, NJ: Princeton University Press, 1938/1987) is incomparable. For Apollonius, the best starting point is M. N. Fried, and S. Unguru, *Apollonius of Perga's* Conica*: Text, Context, Subtext* (Leiden: Brill, 2001).

My own interpretation of Hellenistic mathematics is offered in R. Netz, *Ludic Proof: Greek Mathematics and the Alexandrian Aesthetics* (Cambridge: Cambridge University Press, 2009). The historical context for the autonomy of science from philosophy is a main theme of the second part of R. Netz. *Scale, Space and Canon in Ancient Literary Culture* (Cambridge: Cambridge University Press, 2020).

CHAPTER 4

# *Mathematics in the World*

## Plan of the Chapter

This chapter takes a step back and considers mathematics from a different angle: the mathematics closer to day-to-day practice. At the same time, the chapter brings forward the historical narrative by beginning to consider the science of the Roman era. The same duality will characterize the following (longer) chapter on astronomy, and both for a reason: in the centuries following the two breakout generations of Greek mathematics, most work in the exact sciences was concentrated on more "applied" mathematics and in particular on astronomy. It is therefore fitting that we begin with the author (not a mathematician but a historian) most emblematic of the turn from Greek to Roman: "Polybius, Introducing Applied Mathematics."

We then move into a deeper background to the applications of mathematics: What did "everyone" know? The first section, "Number," describes the (considerable) numeracy of the Greek civilization. The following section, "Education," touches on the little we know (mostly based on papyrus finds) about the dissemination of such knowledge.

With basic mathematical knowledge fairly widespread, we could expect it to be applied. This is briefly surveyed, and qualified, in a number of fields. The first is "Geography," which is a surprisingly rich area for the display of mathematical knowledge. The following section, "The Engineering of an Empire?," is headed by a question mark: it is not clear that antiquity had a profession similar to our "engineer" – but it is also clear that the place in which to look is the Roman Empire. A related field is covered in "Tactics" – that is, the theoretical study of military operations – because whatever engineering existed in antiquity was primarily military. Indeed, as the following section, "Catapults," emphasizes, the ancients connected mathematical knowledge specifically with a particular problem in mechanics: the scaling of war engines – and the general scaling of solids, that is, the problem, so familiar to us now, of finding four lines in continuous proportion!

This brings us into the general field of "mechanics." We look at it from two perspectives. The first section, "Machines of Marvel," considers a feature of several ancient mechanical treatises: the contemplation of toys, ludic devices meant to amuse or to impress. The following section, "Mechanical Theory?," considers what remains of ancient mechanics, after we have removed from it both military and ludic applications: Is there anything such as a science of mechanics? With this in place, the final section, "Mathematics, Real and Symbolic," concludes by emphasizing the qualified sense in which mathematics, in antiquity, was at all "applied."

## Polybius, Introducing Applied Mathematics

Much of the foregoing chapter was about Archimedes. After all, he survives not merely as the name on the title of abstract works but as a figure of historical memory. This was established thanks to Polybius.

Polybius had the finest gallery seat to Roman history. A Greek general, fighting for the freedom of his native city Megalopolis, he ended up exiled to Rome in 167 BCE (a mere forty-five years after Archimedes's death). Once again: we should remember the sociology of ancient friendship. Ancient aristocrats tended to let bygones be bygones, putting the nexus between elite members above the enmities between states. For many years, exiled Polybius would be a protégé of the Scipios – Rome's most illustrious family, leaders of its victory over the Carthaginians and of its ascent to world domination. Polybius was keen to learn from them how Rome took over the world in the space of two generations.[1]

It is clear what it was about the Romans that impressed Polybius. They won, after all. But they were impressed by him, too. For generations now, Romans were aware of the great sophistication of Greek civilization. Polybius filled a need: he was a Greek approaching the Romans from a position of exalted sophistication and yet acknowledging their superiority. And so, Romans paid attention to him. After all, everyone recognized that there was Greek knowledge out there to be had – among which, practical knowledge about politics and war. Romans were the true masters in politics and in generalship, but Greeks had something crucial to impart about the theory of politics and about many questions having to do with tactics.

[1] Polybius's relation to the Romans is, to us, strange, and it is not as easily understood as my brief outline suggests. For a (more jaundiced) perspective, see A. Erskine, "Polybius among the Romans: Life in the Cyclops' Cave," in C. Smith and L. M. Yarrow (eds.), *Imperialism, Cultural Politics and Polybius* (Oxford: Oxford University Press, 2012), pp. 17–32.

Which is what Polybius set out to do, by writing a *useful* history: one that gave the information required for those wishing to go to war and to administer a state. This was a massive history – forty books, now extant only in (very substantial) fragments and epitome.

At the end of Book VIII, Polybius reached a decisive moment in his narrative: Syracuse fell; Archimedes died. And early at the beginning of Book IX – was this proximity intentional? – Polybius went on an interlude that, in effect, sums up a part of the motivation for his project: generals, he explained, ought to be *knowledgeable* (IX.14)[2]:

> Skill ... can perhaps be acquired by a general just through military experience ... but what depends on scientific principles requires a theoretical knowledge more especially of astronomy and geometry, which, while no very deep study of them is required for this purpose at least, are exceedingly important and capable of rendering the greatest services in projects such as we are speaking of.

For the following sections, Polybius goes on to identify the mathematical knowledge required by a general. The scenario involved is mostly that of an attack upon a walled city – the key theme of Polybius's world. Why does the general need astronomy? Above all, to find the precise length of days and nights so as to calculate the timing of marches that are designed to catch a city by surprise. Why does the general need geometry? This is explained, above all, through the example of Philip V of Macedon, who attempted to scale, by surprise, the walls of the city of Melitaea in Thessaly. You see – Polybius is almost spitting with contempt, telling this – the ladders he brought with him were *too short*. As a matter of fact, Polybius explains, there is a technique for measuring the required length of a ladder, based on the empirically established rule that the ladder should be placed on the ground so that the distance between its foot and the wall is half the wall's height. (Position it farther away, Polybius explains, and it is liable to break under the soldiers' weight; position it closer, and it becomes unstable.) And then, of course, we are supposed to apply – Polybius leaves this implicit – Pythagoras's theorem.

And so, Archimedes, a Greek in the city of Syracuse, devising fantastically effective catapults that almost fend of the Romans, and then another Greek, Polybius, picking up the story a little later. A captive in the city of Rome, he weaves together Greek history and Greek knowledge: it is all useful, you see.

[2] W. R. Paton, *Polybius: The Histories*, vol. IV (Cambridge, MA: Harvard University Press, 1925), p. 35.

There are a number of themes we can pull out from this vignette. First, it appears that – by the second century BCE, at least – many Greeks could see the exact sciences as part of their cultural patrimony. Second, it appears that technical knowledge was often valued, in antiquity, in connection with war. Third, and most important, we note a historical development. With the failure of Hannibal – exactly, that is, with the death of Archimedes – the Mediterranean came to be, for many centuries to come, under Roman control. The influence of the Romans was much more subtle than the random killing of a mathematical genius. Rome took over the Greeks precisely by its accommodation and embrace of the Greek elite. Eager to join a world defined by Roman patronage, Greek intellectuals refashioned themselves as masters of Greek wisdom, brokering their culture to a new audience. From Polybius onward, we see the emphasis on new, creative, eclectic mixtures produced out of the great variety of the Greek legacy. And for centuries to come, mathematics would be pursued as one element among other varieties of the Greek experience, transmitted to Rome.

An eclectic mixture, calling for an eclectic chapter: mathematics was useful in many ways. We start early, indeed, from even before mathematics itself: with number.

## Number

The generation of Archytas seems to have cared deeply about numbers. The integer proportions of musical harmony may have been the key inspiration. They seem to have inspired, in particular, an interaction between the few mathematicians – and the rather more numerous philosophers. When Archytas talked about number and music, nonmathematicians could follow. This, then, tells us something about ancient education. The audience must have known about both music and number. Indeed, throughout classical antiquity, the elite youth was brought up by learning the old songs. Educated Greeks could play the aulos – a double flute – or the lyra, a small handheld harp. "Tuning," "harmony" – these were not distant abstractions but a lived experience. The same, apparently, was true for number, as well.

What did Greeks learn when they learned their numeracy?[3] The first thing to realize is that Greek numbers are different from their modern counterparts. They were more complicated: the Greeks used two separate methods for manipulating number, both no more than resembling our

[3] I provide here a brief survey of ancient numeracy; a more thorough introduction is provided by S. Cuomo, "Exploring Ancient Greek and Roman Numeracy," *BSHM Bulletin: Journal of the British Society for the History of Mathematics* 27 (2012): 1–12.

own decimal and positional systems. The one we find in the papyri – and then in most of the literary record – is, as to be expected, the more literary. It is based directly on the alphabet and hews close to natural language. The letters from A to Θ are used for the numbers 1 to 9, those from I to ϟ are used for the numbers from 10 to 90, and those from P to ϡ are used for the numbers from 100 to 900. (The Greek alphabet as standardly used has twenty-four characters, and three extra ones, only rarely used outside of this context, are utilized in this system: digamma, F, for 6; koppa, ϟ, for 90; and sampi, ϡ, for 900.) Thus, what we pronounce as "thirty-two" was written, in this script, as

ΛB,

and what we pronounce as "two hundred and twenty-two" was written, in this script, as

ΣKB.

I emphasize "pronounce" because the Greek language has specialized words for "twohundred," "threehundred," and so forth (unlike English, in which "two hundred" is transparently, indeed, just the modifier "two" preceding the distinct number word "hundred"). Thus, a written form such as

ΣKB

was, to the Greeks, a fairly transparent abbreviation of spoken language ("twohundred, thirty, two"; I avoid the complication of word order). We recall Apollonius playfully considering a line of poetry as if it were a sequence of numerical values – and then multiplying them all. This, we note now, is not really so out of bounds, given this standard, literate way of keeping records of numbers. A line of poetry is made of letters, and to the Greeks, letters *were* numbers.

Note that this system is obviously limited to less than a thousand, but a simple apostrophe-like mark turns each number into a thousand times that number – an alpha with an apostrophe becomes a thousand instead of a unit, and so forth – so that the system can easily accommodate numbers up to a million, which is more than enough for any practical purpose. (Archimedes and Apollonius, of course, went all-in for the *im*practical purposes and so invented their own systems.)

Such is the literary system of numerical records. Another system altogether was that of the practice of calculation, and this survives not on papyri but on inscriptions. It will be, in fact, more familiar to readers today because it is very similar to what we think of as Roman numerals. In this epigraphic system, the simplest vertical stroke stands for 1; separate symbols, similarly used, stand for 5, 10, 50, and so on:

I – 1

Π – 5 (for *pente*, "five")

Δ – 10 (for *deka*, "ten")

$\Pi^{\Delta}$ – 50 (a special symbol combining Π and Δ)

H – 100 (for *hekaton*, "hundred")

The archaeological record shows that this system is no more and no less than a record of the *abacus*. By this, we should understand not the East Asian, fairly recent invention with beads moved along strings, but a much simpler method. Lines are scratched on a surface to mark vertical regions – the ones, the fives, the tens, the fifties, the hundreds, and so forth. Once the surfaces are ready, you need small tokens or pebbles (any small, movable piece would do), which are moved through the board in a kind of backgammon game. The rule is that any cluster of five in a "unit-type" row may be removed while a single pebble is placed instead in the row beyond; any cluster of two in a "five-type" row may be removed while a single pebble is placed instead in the row beyond. Suppose you add 33 to 22. You start out with the 22, as follows:

| I | Π | Δ | $\Pi^{\Delta}$ | H |
|---|---|---|---|---|
| XX | | XX | | |

And then you add the 33:

| I | Π | Δ | $\Pi^{\Delta}$ | H |
|---|---|---|---|---|
| XX | | XX | | |
| XXX | | XXX | | |

Now you can transpose:

| I | Π | Δ | $\Pi^{\Delta}$ | H |
|---|---|---|---|---|
| | X | | X | |

So, ΔΔII + ΔΔΔIII = $\Pi^{\Delta}\Pi$.

If we wish to double this, we just need to add a pebble next to each one on the board so that we get the following:

| I | Π | Δ | $\Pi^{\Delta}$ | H |
|---|---|---|---|---|
| | X | | X | |
| | X | | X | |

Which we may immediately transform into

| I | Π | Δ | $\Pi^{\Delta}$ | H |
|---|---|---|---|---|
| | | X | | X |

So, twice $\Pi^{\Delta}\Pi$ is HΔ.

This method of calculation is quite straightforward. Everything is based on the ease with which the human eye identifies small clusters. (This is known as "subitization" because it happens *subito*: see a cluster of five, and you do not need to pause to count how many members it has; you see the "five-hood" directly.) Apparently, this system was widely taught and employed. Indeed, we can say more: it was widely spread through the variety of Greek practice with numbers. Here is what I wish to stress. To us, it is almost the essence of number that it is a written trace. What is "3"? To us, quite close to the core of our conception is the notion that this is a couple of semicircles vertically juxtaposed. This is, of course, absurd: "3" is an abstract concept that is merely denoted by a particular symbol, but it is very hard for us to "unsee" the symbol – especially because this script is shared across most language boundaries, so we do not even think of the symbol "3" as in any way "English." It is simply the script of number, and it is how numbers are manipulated in practice: keyed into a computer or traced on a piece of paper.

Greeks had no such notion of the scripted number. The literate record – the letter Γ – was a mere convention for abbreviating the *spoken* word *treis* or "three" and was not primarily attached to number, whereas the epigraphic record, III, was merely a stand-in for the form through which numbers were universally manipulated. This manipulation, finally, did not involve traced symbols but instead involved the transposition of small tokens on a surface. And it was as such – as clusters of small tokens – that numbers primarily resided. Numbers, for the Greeks, were not written traces. They were, instead, the thing held between thumb and index finger and moved about.

And the thing is that the Greeks had plenty of these. In practical, daily life, the Greeks developed many contexts in which one did calculation – which meant moving about small pieces between the fingers. Above all, this happened in the economic sphere. As noted earlier: whereas ancient Near Eastern kingdoms relied on heavy bars of precious metals, the Greeks created the coin – a smaller denomination and much more widely circulating. The system through which such coins traded was based on measures and weights already established in the ancient Near East, ultimately of Babylonian origins. Thus, the number 60 was key, and in many ways, Greek coins were not decimal (or "metrical"). But they traded by simple integer equivalences: six obols would get you a drachma. An obol would buy you a drink; a drachma would buy you a small book (so Socrates says, explaining how easy it is to get ahold of the views of Anaxagoras: spend a drachma and you have them all written in a book!). Six obols for a drachma, six drinks for an Anaxagoras. Coins, and books, were widespread. Instead of the ancient Near East – with its narrow scribal elite, experts in calligraphy and complicated counting for the state – the Greeks had a fairly large literate population that was also familiar with economic calculations done on a daily basis, as one counted one's coins. This spread of literacy and numeracy is indeed obvious in the Athenian constitution, which is markedly based on various encounters with number. In democracy, one always counts. There were exactly five hundred members in the jury, randomly selected on each day of court (through a complicated process involving, of course, the manipulation of small counters). They listened to speeches and then voted – by submitting their small voting counters, which were then assembled and tallied. In many cases, it appears that Greek democracies relied merely on a show of hands or even acclaim (most votes were less tightly contested), but when votes really mattered, a count was made, and so counters were used. It may be that democracy actively expanded the reach of numeracy, and it was an ancient commonplace that Spartans – with their undemocratic institutions – knew less about numbers than did the Athenians. But even as democracy was eroding, coins remained, and the ancient Mediterranean always had a widespread, working familiarity with number.[4] I started this book with the ethnomathematical context: what we learn, through anthropological

[4] Indeed, Serafina Cuomo emphasizes the relationship of numeracy not just with democracy but also with imperialism: S. Cuomo, "Accounts, Numeracy and Democracy in Classical Athens," in M. Asper (ed.), *Writing Science: Medical and Mathematical Authorship in Ancient Greece* (Berlin: De Gruyter, 2013), pp. 255–278.

studies, of the widespread knowledge of mathematics among simple societies. It should be clear that by the classical era, the Greek-speaking Mediterranean was not a simple society, and some knowledge of mathematical concepts and practices was shared far wider than just a small group of experts. Polybius probably had a point: for many practical purposes, one did know, and use, what is – in cross-cultural comparison – quite sophisticated mathematics.

## Education

The ancient Mediterranean was not a simple society with mere knowledge of numbers and shapes. Sophisticated, theoretical mathematical knowledge was fairly widespread. But it was not like the empires of the ancient Near East, either. Lacking specialized bureaucracy, the Greeks did not develop any specialized scribal schools. Athens was not Babylon. The city was organized by a dispersed network of knowledge – many citizens coming together to pool their experience.

With or without democracy, tools such as the alphabet and the papyrus roll made literacy more accessible, and the centrality of public performance to Greek civilization meant that to be a full member of one's society, one had to share in its culture. So, many households sought to provide their children (mostly, their sons) with the education required to make them good – indeed, prominent – citizens. Our evidence on the contents of such education, in the early days, is meager. It seems that it mostly took place at home and that a lot of it – aside, of course, from the obvious emphasis on sports – was "literary" in character. This, once again, had the typical Greek preference for public performance. As we noted, in the early days, Greek education was, above all: learning *the songs*. (This is the reason, in fact, that so much early Greek poetry still survives: people made an effort to memorize it and, eventually, to put it into writing.) But perhaps already throughout the classical era, and certainly by the Hellenistic era, Greek education would become much more formally structured around writing. It came to be mostly about learning the texts of the past literary masters: how to read and perform Homer, Euripides, Menander, Demosthenes. We actually know this and no longer have to rely on sheer guesswork. This is because there are plenty of papyri and other written materials – altogether, about a thousand – used in the ancient classroom, dug up from the sands of Egypt. Almost all, unfortunately, are from the Roman era, but the few earlier ones are of the same character as the later ones. Education is a conservative enterprise: in 200 BCE or 200 CE, Greeks probably educated their children the same.

Among those plentiful papyri dedicated to literary education, we also have a much smaller group of papyri – about a hundred – dedicated to *mathematical* education. Many of those are dedicated to number words and symbolism and to the simplest operations, such as addition. Several dozen are more demanding, although even these are often quite elementary, for instance:[5]

> Concerning stones and things needed to build a house, you will measure the volume according to the rules of the geometer as follows: the stone has 5 feet everywhere. Make 5 × 5! It is 25. That is the area of the surface. Make this 5 times concerning the height. It is 125. The stone will have so many feet and is called a cube.

Or, at a higher level of complexity, let us look at a relatively early example: a piece of papyrus from the third century BCE. (This is inscribed not in Greek but in Demotic, the local language of Egypt; this hardly matters because the culture of mathematical education seems to have been much more widely shared, as will be obvious from the example itself):[6]

> A plot of land that [amounts to] 60 square cubits, that is rectangular, the diagonal [being] 13 cubits. Now how many cubits does it make to a side? You shall reckon 13, 13 times: result 169. You shall reckon 60, 2 times: result 120. You shall add it to 169: result 289. Cause that it reduce to its square root: result 17. You shall take the excess of 169 against 120: result 49. Cause that it reduce to its square root: result 7. Subtract it from 17: remainder 10. You shall take to it 1/2: result 5. It is the width. Subtract 5 from 17: remainder 12. It is the height. You shall say: "Now the plot of land is 12 cubits by 5 cubits."

And now, in many ways, we are back in Babylon. The very grammar is the same as that of the Mesopotamian clay tablets – a second-person imperative, the teacher provides instructions to the pupil as he goes along finding a numerical value based on given terms. The basic mathematical arrangement – a rectangle, with area equal to 60 units! – is directly Babylonian, as well. The key trick here, however, is somewhat closer to Pythagoras's theorem than is the case in extant Babylonian mathematics. We are given the area of the rectangle and its diagonal. Clearly, then, the square on the

[5] K. Vogel, *Kleinere Schriften zur Geschichte der Mathematik* (Stuttgart: F. Steiner Verlag Wiesbaden, 1932), p. 117. This is at the beginning of a papyrus that quickly moves into much more complicated measurements and volumes.

[6] This is taken from the verso of P. Cairo J.E. 89127-30, 89137-43. (Why verso? Because the set of problems from which I cite one example seems to have been produced informally on the back of a separate document, as a kind of scrap paper – this is not a book but rather something like the notebook used in a classroom.)

diagonal together with two times the rectangle finds the area of the big square in which the square on the diagonal is inscribed; its side is the root of that number. This is how we get

$$(13)^2 + 2 \times 60 = 289, \sqrt{289} = 17.$$

But then, the inner square inscribed within the square on the diagonal has as area the difference between the square and the diagonal and twice the rectangle, and as side, the root of that:

$$(13)^2 - 2 \times 60 = 49, \sqrt{49} = 7.$$

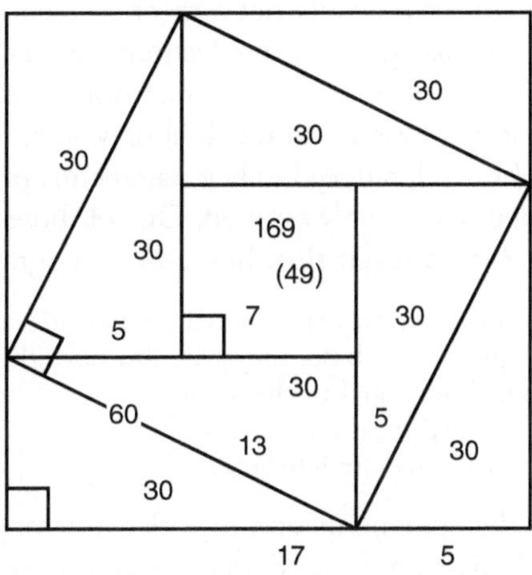


Figure 73

The difference between the side of the big square and the side of the small square is two times the width of the rectangle; hence, the width is 5, and the length has to be 12. (Astute pupils would probably have guessed all along that the answer must be in such simple integers; presumably, they had to show their work.) This is the third century BCE; by the millennium and a half, mathematical education did, in fact, incorporate Pythagoras's theorem. But the key theme is not change but continuity.

Not an impressive piece of mathematics, perhaps, in and of itself – but very impressive if we think of it as something that *many people knew*. Mathematical education had not changed much since Babylon, but it most definitely escaped the confines of the scribal school. Polybius, once again, provides a useful testimony. In the course of one of his arguments – to which I will return later on – Polybius refers to right-angled triangles and then applies what we call Pythagoras's theorem. Remarkably, he refers to this as the "schoolchild's measurement." And so yet another proof, to put alongside the papyri. By Polybius's youth, elementary education would be expected to include – at least for members of the elite – knowledge of basic geometry, perhaps comparable to results such as the first book of Euclid's *Elements*, perhaps capped by Pythagoras's theorem.

Or did Polybius learn Euclid directly? There are some indications that Euclid's *Elements* became – in some simplified form – part of this very mathematical education. Seven fragments related to Euclid's *Elements* were dug up from the sands of Egypt. Two are related to the last book, with its Platonic solids. Of the remaining five, four belong to Book I, and one belongs to Book II; this is probably not a sheer accident, and the likelihood is that if Euclid at all circulated, it was in selections, based mostly on the earlier parts. One of these fragments must have come from an actual book or text of Euclid, but the remaining four are all much abbreviated. In fact, what we see in them is the text of Euclid, reduced to key statements, probably meant to be memorized. The proofs are entirely omitted. One of those papyri contains the sequence of Book I.8–I.10 (even this, however, is fragmentary):[7]

> If two triangles have the two sides equal to two sides respectively and also have the base equal to the base, they will also have the angle contained by the equal straight lines equal to the angle.
> To bisect a given rectilineal angle.
> To bisect a given limited straight line.

These papyri have the statements accompanied by rough sketches, representing the original diagram from Euclid's text but without the labeling (which, of course, is only required for the proof). A more extensive survival, but still with only statements and rough figures, is seen in Plate 2 (mostly Euclid's Elements I.18–25).

These are probably the schoolmaster's tools, and we can readily imagine him teaching his Euclid much as he taught his Homer. Interpretation was not the point; the point was to learn the text. As one went along, presumably, the more sophisticated schoolmasters would actually show how to produce a construction, explain why a particular statement was true. But to the extent that Euclid became part of ordinary mathematical education, this must have been in some ways continuous with the old

[7] This papyrus is P. Berol. inv. 17469. In the Suggestions for Further Reading, I refer to N. Sidoli, "Mathematics Education," in W. M. Bloomer (ed.), *A Companion to Ancient Education* (Chichester: Wiley-Blackwell, 2015), pp. 387–400. This includes the best up-to-date discussion of mathematical papyri. A very engaging discussion is found in D. H. F. Fowler, *The Mathematics of Plato's Academy* (Oxford: Clarendon Press, 1987), Section 6.2.

traditions of Babylonian-like education, in other ways continuous with the Greek traditions of literate education. Euclid became Babylonian-like, in that one bothered less about reasons and instead just learned a rule. And Euclid became literate-like, in that the point became to know him *as a text*.

It should make sense, on first principles, that Euclid would be read in the same way that Homer was: namely, that the schoolmasters did so much more Homer. As pointed out, the order of magnitude of papyrus fragments related to literary education is about a thousand; that of papyrus fragments related to mathematical education is about a hundred. There is no indication that some Greeks went, so to speak, to a *Realgymnasium* focused on the exact sciences. Wherever you went in the ancient world, you would have been judged by your cultural refinement, your education or "paideia." No upscale parent would omit that, and the strong likelihood, then, is that the evidence we have for mathematical education comes from within the same setting as that of the literate education. About 90 percent of the curriculum was dedicated to the literary classics. The rest of the time included, then, numbers, sums, and figures, culminating with a few weeks of mastering a small fraction of the text of Euclid.

Is this a little? Is this a lot? It depends on what comparison we are making. Most Greeks did not have access even to that education; only about 10 percent of the entire population, perhaps, was literate. Most of those literate people did have some, at least, of this education, and they had a smattering not just of the street smarts of converting obols to drachmas but also some sense of a theoretical science out there, with which one can calculate and measure – a glimmer, indeed, of the sense that someone had *proved* it all. The elite, quite often, at the very least, knew *about* Euclid. When Polybius emphasized that one needed to know mathematics, his readers knew, roughly, what kind of thing one ought to know – even if they knew rather little of that. And it made sense to them that such knowledge could be supremely useful. A certain amount of mathematical education was spread across a fairly wide stratum of Greek society. They were not the master-scribe calculators of Mesopotamia, but ordinary Greek citizens could count their coins, tally up votes in the court, and set up camp in a military expedition.

## Geography

The "schoolchild's measurement." Let us, then, look at the relevant passage from Polybius. How did one *apply* one's youthful, elementary training?

This passage is taken not from Polybius's massive historical work but from a distinct work, a geography. Now, Polybius's geography is lost; we are lucky, in that it was cited by a later author, Strabo, a geographer writing during the reign of Augustus. Strabo often cites previous authorities, usually to criticize them. So did Polybius, before him. The passage we have from him, in fact, is a response to several past authors in geography from the fourth and third centuries – Pytheas, Dicaearchus, and Eratosthenes. We are by now familiar with all of that: our knowledge of the Greeks is based on thick webs of criticism. (We know so much about the Greek tradition not because the Greeks venerated their own tradition but because they continuously tore it apart.)

The critique of Dicaearchus is cited by Strabo in detail. In fact, in this particular case, it seems as if Strabo is citing Polybius word by word.

Dicaearchus's claim was that the length of the western Mediterranean, from the Peloponnese to the Pillars (our Gibraltar), was ten thousand stades. The size of the ancient stade is not certain, but perhaps it was about 180 meters; this is a considerable undermeasure, for which, see the discussion that follows.

Polybius criticizes Dicaearchus as follows:[8]

> The coast-line [of the northwest Mediterranean] is most like an obtuse angle, standing on both: the strait and the pillars; having Narbo [= Narbonne] as its vertex. So that a triangle is set up, having the line through the sea as a base, and, as sides, those making the said angle, of which the one from the strait to Narbo is more than 11,200 Stades, while the remaining is a little less than 8,000. And let the depth of the gulf at Narbo be 2,000 Stades, as a perpendicular from the vertex to the base of the obtuse-angled [triangle]. Now, it is clear . . . from a schoolchild's measurement, that . . . if we add [the 3,000 stadia] it would be more than twice what Dicaearchus said [i.e., more than twice 10,000 or more than 20,000].

"More than 20,000 . . ." – the real value of the distance from the Pillars to Ionia (although, of course, we are not sure of the exact positions Polybius had in mind for his measurement, or even of the length of the ancient stade) is perhaps about thirteen thousand stades, curiously close to the midpoint of the two calculations.

Polybius's point – surely attuned to a Roman and Roman-minded audience – was that past Greeks badly underestimated the size of the western Mediterranean. (You will notice that he errs on the other side from Dicaearchus and roughly in the same ratio.) Polybius's calculation is terrible. The perpendicular from Narbonne to the African coast is taken to be two thousand stades, a very bad undermeasure (more on this to follow). He then extrapolates the length of the sides of the triangle based,

[8] Strabo, *Geography* II.4.2.23–II.4.2.55.

presumably, on length of travel by sea, and he comes up with very bad overmeasures, perhaps because his estimate essentially assumes that the routes are more or less straight (in fact, the coasts are anything but: the eastern one is very concave; the western one, very convex). I am not even sure Polybius calculated using Pythagoras's theorem. The impression from the text as it is written is that Polybius realized the theorem would be applicable in principle but also that his real thinking was much more elementary. I suspect his train of thought went as follows: because the length of the base would have to be quite near the length of the two sides added together (because the perpendicular from Narbonne to Africa is taken to be so small), then it would be at the range of 19,200, and so, with an extra 3,000 thrown in from Sicily to the Peloponnese (everyone seems to agree on that figure), surely this would add to more than 20,000.

The one truly scandalous bit here is Polybius's estimate of the length of the perpendicular from Narbonne. Polybius had an axe to grind: to get his very wide western Mediterranean, it helped to make it narrow so that the length of travel along the shores would be translated more purely into west–east distances. But the thing is that here – in the north–south estimate of distances – Polybius really had no excuse. Greeks *knew* how to calculate this. We need to anticipate some of the materials of the following chapter: geography, you see, is intertwined with astronomy.

By looking at the sky, it is possible to determine absolute positions such as due south. It is thus in principle possible to determine, within reasonable boundaries, that one's travel is on a straight line on the south/north direction, that is, along a longitude or a meridian. And here is the thing: suppose you fix observers at two positions on the same meridian and ask them to look at the sky. Compare their observations. Both will see the same stars rising above the

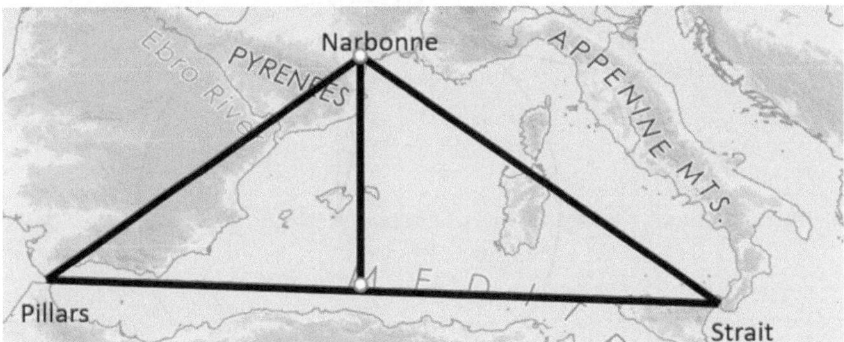


Figure 74 Polybius describes the coastline of the northwest Mediterranean as being like an obtuse angle, standing on the strait and the Pillars and having Narbonne as its vertex.

horizon at the same exact moment. Both will see the shadows at their shortest at the same exact moment. In short, we can synchronize their observations because – as we will return to explain in the next chapter – the sky simply is an outstanding time-keeping device. And so, looking at the sky from two different positions along the same meridian, one can find the angles at which the same object is viewed at the same exact moment. (I simplify by considering the case where we know the two observations on the same meridian; little precision is lost if this assumption is dropped.)

So let us get back to an author already mentioned several times in this book: Eratosthenes. He wrote a study of the finding of two mean proportional lines. He was the recipient of Archimedes's *Method*. And he also made the most famous geographical measurement in antiquity – perhaps you've heard of it. This, too, is recounted by Strabo.

Eratosthenes's measurement is based on the sun, observed at noon, in the middle of the summer, from the two positions of Alexandria (of course) and Aswan. (The two are not exactly on the same meridian, Aswan being a degree or two – depending on the exact location used by Eratosthenes – to the west of Alexandria. This does not interfere materially with this particular calculation.) Aswan lies nearly precisely on the Tropic of Cancer, which means that in the middle of the summer, the sun is directly above it. This gives rise to a spectacular and easily detectable effect. At noon in Aswan, in the middle of the summer, the sun shines all the way down to the bottom of a well. The angle of the shadow cast by the sun is zero.

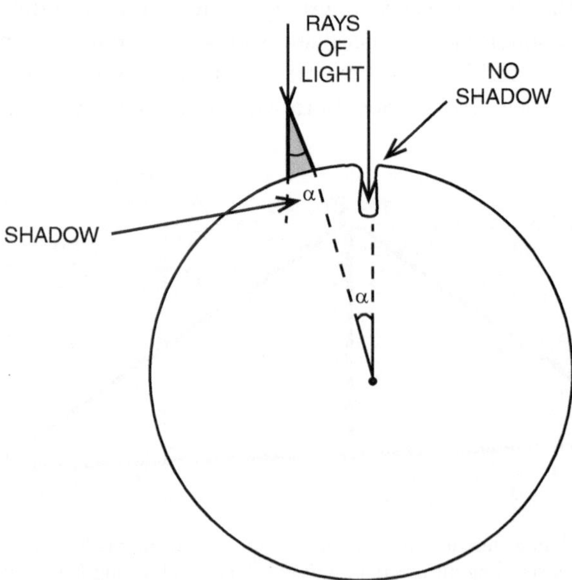


Figure 75

In Alexandria, at exactly the same time, the angle is non-zero. In fact, the angle of the shadow cast by the sun at noon in Alexandria in the middle of the summer was measured by Eratosthenes to be 1/50th of a circle or 7° 12'. Now, assume that the rays of the sun are all parallel to each other. (We recall that Diocles made the same assumption in his theoretical construction of a burning mirror. This is obviously justified in the small scale of a man-made mirror but can be justified for the earth as a whole only if we believe the sun is very distant – which the Greeks did!) It is immediately clear, by considering the diagram, that the angle of the shadow cast at Alexandria at noon in the middle of the summer is the same as the north–south arc between Alexandria and Aswan (where the shadow, at the same time, is zero). Now, Eratosthenes further stated that the distance on the north–south axis between Alexandria and Aswan was five thousand stades. (This is, in fact, very difficult to estimate, even for travel by land, and is clearly meant as no more than a rough estimate.) We thus find that the circumference of the earth is roughly 250,000 stades, which Eratosthenes transformed to 252,000 so as to get a round figure for the length of a single degree of latitude – exactly 700 stades. This is a very good approximation (surprisingly so, indeed, given the difficulty of measuring the precise distance from Alexandria to Aswan), likely about 10 percent or so above the correct value – depending on the precise length assumed for a single stade. Greek mathematical geography was that good!

This calls for a few observations.

First, although there are issues one may raise with Eratosthenes's calculation, it is sound in principle, and ancient authors essentially followed Eratosthenes, sometimes replicating his results (with variations) with different kinds of astronomical observations. Significantly, a later author – Posidonius – chose, for ease of calculation, an even simpler approximation, with the degree equal to five hundred rather than seven hundred stades. This implied a rather smaller circumference of the earth as a whole, which Christopher Columbus would follow – an error with famous consequences.[9]

Second, we now see that Polybius effectively assumed that Narbonne is about three degrees north of the coasts of Africa. This is risibly wrong (the distance is about seven degrees) and is, in principle, very easy to refute by simple astronomical observations, easily available to the ancients.

[9] The net result of Posidonius's value was to have Eurasia significantly "stretched" on later maps. See L. Russo, "Ptolemy's Longitudes and Eratosthenes' Measurement of the Earth's Circumference," *Mathematics and Mechanics of Complex Systems* 1 (2013): 67–79. For more on Posidonius and Eratosthenes, read C. M. Taisbak. "Posidonius Vindicated at All Costs? Modern Scholarship versus the Stoic Earth Measurer," *Centaurus* 18 (1974): 253–269.

Third, we should consider a comparandum. In the eighth century, I Hsing, a Buddhist monk in the service of the Chinese Tang dynasty, produced a series of measurements along the length of the Chinese empire. Those were essentially the same as those made by Eratosthenes. (In the Chinese case, there was no longer an assumption that the earth was necessarily spherical. Instead, the supposition was more directly that differences in the length of shadows at different positions correspond to distances, perhaps precisely because the sun is at a finite distance so that its rays are not, in fact, parallel.) The net result of this project was a fairly precise measurement of distances in central China (where a number of observations were made) and some glimpses of distances reaching as far as present-day Siberia, to the north, and present-day Vietnam, to the south.

What is most striking is that the Romans never did anything similar to the Tang dynasty. It would have been not just theoretically feasible but even straightforward to have a group of moderately trained observers fan out across the Mediterranean and record a few basic observations of the sun and the stars from various locations so that latitudes could be established for many locations. Further, as ancient astronomers pointed out, observations of synchronized events – best of all, observations of lunar eclipses – could be used to determine the difference between *longitudes*. (Suppose that the onset of a particular lunar eclipse is observed at the fourth hour of the night in Rome and at the fifth hour of the night in Alexandria: we have established, then, that they are "a time zone," or about fifteen degrees, apart!) Although this is mentioned by ancient astronomers as a theoretical possibility, no one did anything of the kind. Specifically for the calculation of longitudes: the two mentions of this technique come from (1) Hero of Alexandria, who reports a mere made-up observation proposed for the sake of theoretical calculation (see p. 254 n. 20), and (2) Ptolemy – who could find historical reference to the timing of a *single* lunar eclipse, seen from various positions.[10] Which one was that? The one immediately preceding the battle of Gaugamela, in 331 BCE, at which Alexander the Great vanquished Persia: an omen recorded by various historians. Ptolemy is, in fact, explicit that this was the *only* such record

[10] See the discussion in J. L. Berggren, and A. Jones, *Ptolemy's Geography: An Annotated Translation of the Theoretical Chapters* (Princeton, NJ: Princeton University Press, 2000). Berggren and Jones believe that a single error in this observation entirely transformed the shape of the Mediterranean in ancient maps. A serious program of observing lunar eclipses would have meaningful consequences!

available. The conclusion is clear: no one bothered to record lunar eclipses *for the sake of geographical measurements*, even though the theoretical significance of such measurements was well known.

In short, even the Roman Empire did not produce anything like the Tang court astronomer, carefully measuring the earth. The theory was all there, and the ancients could have mapped the Mediterranean with mathematical precision. That they failed to do so should tell us something significant about the limits of their ambition, the limits of the application of mathematical knowledge.

## The Engineering of an Empire?

This lacuna – the Roman Empire, foregoing the survey of its own lands even though the science for the survey was there for the taking – is made even more remarkable when we consider that land surveying was a constant of Roman life. There is, in fact, an entire array of ancient treatises in Latin, almost entirely from the imperial era or later (the "imperial era" refers to the time that Rome was ruled by emperors, so from Augustus's ascension in 31 BCE onward), known as the Corpus Agrimensorum ("Corpus of the Roman Land-Surveyors"; *agrimensores* means "land surveyors"). When new towns were set up – and the Roman Empire was often busy setting up new towns or "colonies" for its veterans – the Romans followed a precise code. Specialists would find the true north – so, a little astronomy – and lay out streets arranged by a grid with precise measures – so, a little geometry and calculation.

It used to be asserted that a certain Hippodamus, from Miletus, invented the gridded city in the fifth century. Historians still repeat this, but in fact, it is clear from archeological evidence that planned cities emerge across the Mediterranean much earlier. Already in the eighth century BCE, Greek sites such as Megara Hyblaea (a city in Sicily, founded by Megara) are organized by a regular orthogonal system. It seems possible that the Greek model came first – orthogonal planning is typical when new cities are set up, and this is something the seafaring, expanding Greeks did much more often than others – but Phoenician settlers followed a similar path, and one finds early orthogonal planning among the Etruscans, as well. (Those, indeed, are likely to have been the proximate inspiration for the Romans.) What lies behind this town planning is not any mathematical theory. People of the Mediterranean, arranging their cities along orthogonal lines, are rather like the fishermen of Mozambique, arranging their fish along a circle: reflecting a pre-theoretical facility with geometrical shapes. What may be distinctive about all those Mediterranean civilizations – Greek, Phoenician, Etruscan, and Roman – is the culture of the city-state. Eventually, the Mediterranean would be ruled by larger political units. But to begin with, even Rome was a city-state. And although monarchies would end up ruling everywhere, the early city-states were more self-ruling (whether in the form of democracies or of aristocracies). Because it was the city that governed itself, it was also natural that it would plan itself. This habit of town planning became entrenched and so was kept in the foundations of the imperial era.[11] Indeed, Romans were particularly traditionalist, especially in the matter of ritual, and to some extent, the practice of land surveyors should be understood on those terms, of religious practice (they did not merely survey a land; they did so according to a prescribed and sacred ritual). One needed to have specialists in this kind of ritual – just like the Romans had specialist priests, officiating over sacrifice, practicing divination, they had specialist land surveyors, conducting the precise procedure for the setup of a new city.

It is not surprising, then, that the Romans had a specialized profession of those land surveyors. What is interesting is that in the imperial era, several land surveyors reached for the status of authors, producing treatises that presented their skills. This is perhaps typical of a more general phenomenon we get to see a lot in the imperial era: the quest for status and patronage. Technical practitioners wish to become technical authors, in this way expanding the scope of scientific writing.

And if so – if the point is to assert status – it is less surprising that here and there, the writings of the land surveyors included a few traces of Greek mathematical learning. Balbus, for instance, early in the second century CE, wrote a treatise for the education of land surveyors. The title, "Expositio et Ratio Omnium Formarum" (The Account and Calculation of All Figures), is ambitious enough. But – at least in its current, fragmentary form – it is made mostly of definitions translated from Euclid's *Elements*. (So this is also yet another reflection of the way in which Euclid – in much-abbreviated form – has penetrated general education.) It should be emphasized that Balbus was the exception, not the rule, and that most writings in Roman land surveying only briefly touched on geometry and

[11] The best study of geometrical town planning in the Mediterranean is a PhD dissertation: S. D. Ehrlich, "Greco-Roman Urban Form in Its Global Context" (PhD dissertation, Stanford University, 2018).

instead emphasized law. After all, land surveying was about landed property – the most important form of wealth. Romans may have touched only superficially on geometry, but jurisprudence was a Roman science par excellence. The land surveyors should perhaps be understood as ancillary to the jurisprudence of landed property, a corner of Roman law where geometry, briefly, could matter.

Whether with their lawyers or with their land surveyors, there is no doubt that the Roman state, and the Roman peace, gave rise to a certain specialization and professionalization. The late Digest of Roman law sums up, in abbreviated form, many earlier precedents, rules, and lists: the Roman legion, we are told in the Digest, was said by one Tarruntenus Paternus, of the second century CE, to possess a panoply of masters of various professions. Included are surveyors, physicians, and architects, but also much humbler professions: swordsmiths, wagon makers, butchers, stoneworkers, and many more – I believe I count forty-three professions in total.[12] I have mentioned that Mediterranean city-states often engaged in the foundation of new cities, an act that involved careful reflection concerning the nature and form of the city – hence the orthogonal grid. More than this: an ancient military campaign, and in particular the Roman legion, was essentially a city in transit, forever set up and then transported again. The constitution of the legion involved a reflection on the practical needs of the city, and so it had something of a permanent engineering corps (although never conceptualized as such). There was a specific military role – the *praefectus fabrum*, "master of the works" – responsible for the construction needs of the legion (setting up the camp; making sure supplies are in place; getting all those surveyors, butchers, and stoneworkers into action). As was typically the case in Rome, this role was essentially political, an early stepping-stone in a career that might lead to military command in battle and to civic duties in the city of Rome. And so, the entire group of professionals is made of two groups: on the one hand, craftsmen, trained by apprenticeship; on the other hand, members of the Roman elite, learning the military profession on the job. Neither the master of the works nor any of his underlings were trained engineers. The point remains, though, that the constant practice of military life made sure that many Romans did become proficient in the technical details of management and construction.

[12] Justinian's Digest, l. 6.1.6.

This comes out clearly from Frontinus's *On Aqueducts*: an amazing survival of an extremely technical piece of writing. In 97 CE, Frontinus, by that time a leading figure of the Roman political elite (already a consul, as well as a past governor of a province), was appointed to one of the most sensitive positions in the empire: supervisor of the waterworks of the city of Rome. This he took seriously, producing an account of the entire aqueduct system. In places, this reads as an ambitious literary work, proceeding through the history and geographical description of the water system. In other places, it becomes almost entirely numerical and technical, indeed, fundamentally a piece of accounting. (Frontinus pays considerable attention to the way in which water flow can be measured, which he uses to detect the considerable amount of theft of water along the route of its transmission.) At almost exactly the same time, we have Balbus impressing upon his readers that he is not just a practical surveyor: he also knows Euclid. There is real technical knowledge in both land surveying and the building and measurement of aqueducts, and this technical knowledge certainly involves a certain understanding of, and experience with, shape and number. All of this owes something to a Greek legacy, even if in a polemical way. (Frontinus begins, in fact, with a proud comparison between Rome's waterworks and the "the idle Pyramids or the useless, though famous, works of the Greeks".) How much does this Roman, self-conscious practice owe to its Greek, scientific antecedents?

Polybius's entire point was that the practical man needed knowledge, which was why, essentially, the Romans needed the Greeks. The Romans won – but they did so, according to Polybius, because they were, in their own way, theoretical. The professionalization of the Roman army was, in Polybius's eyes, an example of theory in practice: science, in war. War, indeed, was what mattered most to Polybius and to his readers, Greek and Roman.

## Tactics

Which brings us, indeed, right back to geography. Polybius's general would use astronomy to know how long the night is at a given latitude at a given time of the year – so as to know how far he can march his soldiers through the cover of darkness. This is reasonable and correct – which makes it even more curious, then, that no effort was made to ground this kind of application in an accurate mapping of the

Mediterranean. Would it not be helpful to have available tables that listed, for a given location, the length of the day throughout the seasons? And yet, no Roman emperor gave any resources for accurate, mathematical mapping. Nor indeed does it seem that any general really carried with him an astronomer to calculate the length of the night. You see: people just *knew* – roughly, but well enough – how long the night was in different places and at different times of the year. It was just like the butchers and stoneworkers in the Roman legion: they had to be trained, but surely their training was a matter of apprenticeship, not of technical schools.

But if this is so, what is Polybius doing? Well, this is not so hard to tell. He is making a correct assertion – there is indeed astronomical understanding implicit in the knowledge of the length of the night – and by putting this to the fore, he establishes the point that animates his project as a whole. Underneath practice, there is knowledge; this knowledge has been accurately described by past Greek authors, and it is Polybius's job to synthesize it and present it to an audience ruled – and ruling – by Roman power. He is showing how power and knowledge are intertwined. But notice carefully: when we say "underneath practice, there is knowledge," we do not describe an educational system, analogous to the modern engineering school, where one studies theoretical knowledge and then learns how to apply it – knowledge first, practice later. The knowledge that underlies the practice is discovered by theoretical scholars who reflect upon the practice. It is practice first, knowledge later. To the practitioners themselves, this knowledge is mostly an invisible part of their daily routines.

Polybius's most extended excursus is a book-length study of the Roman constitution: it is dedicated, almost in its entirety, not to Rome's political regime but to its military practices, with very precise details, such as the following (concerning the setting up of the military camp, VI.28):

> They now measure a hundred feet from the front of all these tents, and starting from the line drawn at this distance parallel to the tents of the tribunes [= military commanders] they begin to encamp the legions, managing matters as follows. Bisecting the above line, they start from this spot and along a line drawn at right angles to the first, they encamp the cavalry of each legion facing each other and separated by a distance of fifty feet, the last-mentioned line being exactly half-way between them.

This is the Roman legion as a city in transit – and quite clearly employing the language of the Roman surveyors. In Polybius's words, though, this is also, quite obviously, intended to evoke Greek mathematical language.

Indeed, one finds a similar quantitative emphasis throughout Polybius's account of the Roman constitution. Rome was always an aristocracy, but perhaps under Greek influence, perhaps following on a similar logic of a constitution built for (originally) a small city-state, the Romans had many functions and roles defined in strict numerical terms: an assembly divided into precisely 193 voting blocks, of which the wealthier parts had 98 . . . As Polybius recounts some of these details, it is only a natural transition from the numbers of consuls and quaestors and military tribunes to the geometrical encampments of the latter. Such was the Greco-Roman Mediterranean: even if not highly theoretically mathematical, it was awash in number.

It was awash in war, above all. Polybius's history was a military history because this was what mattered, politically: the key question debated throughout ancient politics is one and the same: Who should we ally with going to war? Having written forty books on military history and then also a geography, Polybius was also the author of a theoretical book on military tactics. This is lost, but we have an entire array of Roman-era military treatises, all so close to each other as to imply a common source, probably by Posidonius.[13] This author was active in Rhodes in the early first century BCE. At this point, Greeks did not need to be held captive in Rome so as to care about Roman audiences: Roman presence was everywhere (it was now, mostly, direct Roman rule). Both Pompey and Cicero came to visit Posidonius at Rhodes. He wrote precisely on the Polybian themes – history, geography, war – with the significant addition of philosophy: Posidonius was the leading Stoic philosopher of his generation. Away from Athens, at places like Rhodes, encouraged by Roman patronage, philosophy and science came together again. A Stoic philosopher – and the author of a tactical manual?

A surprising combination, made somewhat less surprising when we read the version probably closest to that of Posidonius himself – written by his immediate pupil, Asclepiodotus.[14]

The most striking thing about such manuals is how divorced they are from real, historical war. Unlike Polybius (and most earlier Greek writings on war), they include no anecdotes, no tales of military success and failure. Instead, everything is abstract and theoretical.

[13] The ascription is uncertain; see P. A. Stadter, "The *Ars Tactica* of Arrian: Tradition and Originality," *Classical Philology* (1978): 117–128.

[14] An English translation is in J. Henderson, *Aeneas Tacticus, Asclepiodotus, Onasander* (Cambridge, MA: Harvard University Press, 1923).

The *Tactics* by Asclepiodotus is structured by an overarching conceptual taxonomy: the arrangements for war are landed or naval (we are to discuss just the landed), which are then either footed or mounted, each further subdivided until one reaches such units as the hoplite phalanx. Then, the author is engaged primarily with the numbers to be deployed within given units, especially concerned to make sure that the numbers will be divisible by two; hence, everything is governed by the powers of 2, and the ideal phalanx is declared to have $16{,}384 = 2^{14}$ soldiers. Chapter II is almost entirely given to the various divisions arising out of this number. In Chapter III, the parts of the phalanx are arranged by military valor, and we are told that the rightmost fourth of the line should be composed of the best soldiers, the leftmost and second leftmost should be composed of those who follow, and the worst should stand as the second rightmost because in this way, the right and the left are balanced ("the geometers say") – clearly a reference to a:b::c:d $\longleftrightarrow$ ad = bc. (This is the application of proportion to geometry we have seen again and again, in the finding of two mean proportionals and in the study of conic sections!) Later follows a very abstruse discussion (Chapter VII) of the geometry of cavalry formations, which are, of course, more complex than the purely rectangular phalanx. It is in this context that diagrams are added in great profusion, typically lettered (each letter, in this case, standing for an individual horseman). This now gains even the look of a mathematical text! Once again, the author touches briefly on terminology, and then the most substantial discussion in the treatise (Chapter X) involves motions – typically rotations, either of each individual separately or of formations as a whole. Once again, this discussion is carried through extensive reference to diagrams. This brings in a kind of geometry that is not typically discussed in explicit terms in Greek mathematics, and Asclepiodotus makes apposite, if trivial, observations: for instance, that following three forward quarter-rotations, it is more convenient to return to the original position via one further forward quarter-rotation than by three backward quarter-rotations (X.9). Chapter XI combines, in a sense, the discussions of formation and of motion to discuss formations in marching – once again, of course, considered via lettered diagrams.

What is the purpose of all that? In fact, we are puzzled by the purpose of such treatises as a whole. Asclepiodotus's (or Posidonius's) is surely not a practical manual. It could have some function for the reading of other military texts, perhaps of a historical character. It is indeed strangely behind the times, taking the Macedonian phalanx as its primary reference. Remarkably (and the sure sign this treatise does not go back specifically to

Polybius), Romans are not in evidence at all. As mentioned earlier, we have quite a few of those treatises extant from the Roman era: a thriving Greek genre, written in a world dominated by Roman military might. That it invokes a past of Greek military greatness suggests a specific attunement to a cultural moment. Here are Greek authors invoking a lost military Greek grandeur while acknowledging Rome's ascendancy. If we understand such military manuals in this context, then perhaps the mathematical components of the style might be seen as yet another element of specific Hellenism. Mathematics is an exotic genre, resonant of Greekness; it comes naturally in this evocation of the lost, exotic Macedonian phalanx. And so – the mathematization of war, not because mathematics is really useful but because it is evocative of the Greek legacy?

## Catapults

The Greek word *mēchanē* – the subject matter of "mechanics" – very often was used to mean, simply, a device used in war. In practice, nearly all engineers and authors in antiquity who engaged with mechanics either wrote about or made catapults. (This includes, indeed, even Archimedes himself!)

We are still familiar with the bow and arrow. In the simplest form, the archer has only his or her hand with which to pull the string. Now, devise an instrument that multiplies this hand power – and you have a war machine. One way to improve this would be to prevent the string from premature release: you pull the string and "lock" it in its pulled state. Thus, the string can store more potential energy than the hand, on its own, can manage. Another way to make it possible to apply more force to the pulling of the string (for instance, by allowing the full weight of the human body to apply pressure that transforms into pulling the string). Combine these two ideas, and you get the crossbow. There remains a limit to the amount of potential energy that may be stored simply by pulling a string – ultimately, a string might break. Such limits can be extended by the application of torsion: instead of pulling a string, twist it. The idea now would be to twist strings and block them from unwinding. Then, remove the block. The string will unwind with a violent motion. Find a way to harness this to a missile-throwing device – and you have a "ballista" or "catapult." (Note that the terms are also used with various other, more general or more specific, meanings. Also, I add in passing that there is yet a superficially similar instrument that stores the energy not of torsion but of gravitation: a device pulls up a weight and blocks it from falling; when the block is removed, the power of the fall produces energy that, harnessed into a motion, flings a projectile. This is known as a *trebuchet*, and surprisingly, it is not clear that it was even known in Greek antiquity. The trebuchet was widely used in medieval siegecraft from China to Europe, and it is the trebuchet, not the catapult, that gunpowder ultimately displaced.)

Even the crossbow can throw sharp projectiles to a distance of hundreds of yards; add in the torsion of the catapult, and you can throw not merely light (if deadly) sharp objects but even heavy stones. These, in turn, have the capacity to break down walls. The tendency of both the crossbow and the catapult, then, is to make large-scale hand-to-hand combat less significant. Long-distance projectiles make smaller, professional armies more effective. Remember that the classical Greek city relied on citizen-soldiers who, working together, could face equal citizen bodies or, if their luck ran out, defend their cities from within their walls. Wall-breaking devices make it harder for armies to hide behind the safety of walls in preparation for pitched battle. As the battlefield came to be dominated by mechanically hurled projectiles, more professional armies, relying on more expensive gadgets, gained the upper hand. The only way to prevent your enemies' catapults from tearing down your walls was to set up catapults on your own walls, to drive the enemy away from catapult range. You can't bring a knife to a catapult fight. A classic arms race ensued – and its consequence: war, dominated by mercenaries and machines.

Athens, in the fifth century, created an empire based on hoplites and sailors. Over the fourth century, the military balance moved to those who could master the most war instruments (the elephant – a living war machine – would join in the third century). This military revolution is perhaps one of the contributing factors to the rise of monarchies, displacing the early city-states. At the end, even the most ingenious engineer – even Archimedes – would be defeated by those who had more resources at hand and could simply accumulate more machines. Power told.

But how did ingenuity even matter? Polybius is clear about Archimedes: his main contribution was the successful *ranging* of his catapults, leaving no "blind spots" uncovered by missiles from within the city of Syracuse. This represents, at the very least, important evidence concerning what was expected (whether realistically or not) from the application of mathematical skill to the making of catapults. In fact, we can say much more because – remarkably – we have works in mechanics that explicitly discuss the theory of catapults. The earliest extant – by Biton and by Philo – are, even more remarkably, roughly from the age of Archimedes himself. There was, early on, a genre of writings about catapults: precisely because theory – book knowledge – was understood to be important. Presumably, you had the better contracts, as a military engineer, if you were known to be good at writing about your craft!

It is striking, indeed, just how *literary* Philo is, how much he immerses catapults within a wider cultural endeavor:[15]

> Many who have undertaken the building of engines of the same size, using the same construction, similar wood, and identical metal, without even changing its weight, have made some with long range and powerful impact and others which fall short of these. . . . Thus, the remark made by Polyclitus, the sculptor, is appropriate for what I am going to say. He maintained that *perfection was achieved gradually in the course of many calculations*. . . . Therefore, I maintain that we must pay strict attention, when adapting the design of successful engines to our private construction, especially when one wishes to do this after either increasing or diminishing the scale. . . .
>
> Engineers drew conclusions from former mistakes, looked exclusively for a standard factor with subsequent experiments as a guide, and introduced the basic principle of construction, namely the diameter of the circle that holds the spring. . . .
>
> The object of artillery-construction is to dispatch the missile at long range, to strike with powerful impact. . . . Reduce to units the weight of the stone for which the engine must be constructed. Make the diameter of the hole as many finger-breadths as there are units in the cube root of the number obtained. . . . It is possible, from one number, to determine the remaining diameters by geometrical construction by the doubling of the cube.

There are many things to unpack here. The key idea in the first of the paragraphs quoted is that even though there is an important theoretical knowledge – which is the point of Philo's book – one cannot just rely on book knowledge because the making of a catapult requires precise know-how in each of the many measurements. This much is clear, but even this "craftsmanship" point is made with reference to a special kind of cultural authority. Philo refers to Polyclitus, a sculptor author. This calls for a detour: the application of mathematics to art, we will see, is directly relevant for understanding how it was understood to apply to catapults.

[15] E. W. Marsden, *Greek and Roman Artillery: Technical Treatises* (Oxford: Oxford University Press, 1971), p. 107. This book by Marsden also remains the best general survey of ancient catapults.

Polyclitus was active in the fifth century – roughly, perhaps, a contemporary of Hippocrates of Chios. His wonderful statue the Spear-Carrier (Plate 3) (known via many later, Roman-era copies) was accompanied by a treatise, a kind of an artist-authored self-catalogue.[16] The treatise had a theoretical point to make. Titled *The Canon*, it provided numerical rules, in the form of proportions, for the crafting of perfect beauty. Such rules were then modified and contested throughout antiquity, sometimes given more numerical, sometimes more geometrical form: the famous drawing by Leonardo called the "Vitruvian Man" is a reflection of one of those sets of rules. There is an idea – that beauty is in number – that continues to fascinate throughout the centuries. Perhaps this has something to do specifically with the kind of inspiration that musical harmony had on authors such as Plato (as mentioned briefly on p. 90 when we touched on Silanion, Plato's portraitist and yet another author on proportion in art). Polyclitus's work is probably roughly contemporary with the original discovery of mathematical musical harmony. By the time we reach Vitruvius, there is already an enduring tradition, clearly inspired by music, that monumental architecture should be guided by number. (Vitruvius, an architect, wrote on the mathematical proportions of the human body; this is where Leonardo got his inspiration.) For centuries, the education of architects would be organized by a precise system of numerical proportions underlying, above all, the various types of columns. In all of them, one finds a unit of measurement – the "module" – and then one determines the various measures based on that. (The three types of columns are called Doric, Ionic, and Corinthian. In the Doric column, for instance, the module is half the base's diameter; the height, for example, is fourteen times that, although calculations become quickly much more complicated as we get to the details of the capitals.)

Architecture is already not all that far from engineering, and it turns out that something similar is proposed by Philo, and by other authors in mechanics, for the making not of statues or buildings but of catapults. The idea, once again, is the same: provide a module, and then provide, in numerical terms, the many module-based measures of the various functioning elements of the catapult. In practice, one takes as a module the

[16] We get a sense of the original statue from its many copies, but the original treatise left only a handful of traces (in fact, one of the most significant testimonies – tantalizing as it is – is the passage we have just read from Philo). There is a great room for speculation and interpretation; see, for example, G. M. Warren (ed.), *Polykleitus, the Doryphoros, and Tradition* (Madison: University of Wisconsin Press, 1995).

diameter of the hole through which the torsioned string passes – not unlike the diameter of the column as used in architecture. The rest of the machine is described in terms of this measure.

In this sense, the catapult is governed by a system of numerical ratios. Philo's emphasis, however, is not on the ratios between the parts *inside* a single catapult. His emphasis is on the proportion between *different* catapults. The central working assumption in all of this is that the range of a catapult is proportional to its weight, and so, to its volume, and the result is that transforming the catapult is essentially a special case of scaling the solid, or "duplicating the cube," or to be precise, finding two lines in continuous proportion between two given lines. If we want to transform the range of a catapult, we need to transform the module of the catapult – that innocent-looking diameter, therefore a linear measurement – by the cubic root of the ratio of the ranges.

The literature of ancient mechanics is dominated, to a surprising extent, by this point: the making of catapults is somehow intimately related to that particular problem in advanced mathematics. Not just any problem, indeed, but the most central one – the one famously first solved by Archytas, the one leading to the discovery of conic sections.

Eratosthenes was a many-faceted scholar: the addressee of Archimedes's *Method*, he was trained as a philosopher, poet, and geographer. He was not a military engineer. But he did address to King Ptolemy, as noted on pages 210–211, a certain mechanical solution to the problem of finding multiple mean proportionals (his solution is of greater generality), and it is thanks to him, indeed, that we learn much of the history of the problem. Having described the solution, he explains how it is important for finding measures and then elaborates:[17]

> The conception will be useful also for those who wish to enlarge catapults and stone-throwing machines; for it is required to augment all – the thicknesses and the magnitudes and the apertures and the boxes for the strings and the inserted strings – if the throwing-power is to be proportionally augmented, and this cannot be done without finding the means.

I think we can be confident that there is something wrong with this characterization. First of all, it seems clear that the making of catapults was much more widespread than the knowledge of technical solutions to the

[17] R. Netz, *The Works of Archimedes: Translation and Commentary*, vol. 1 (Cambridge: Cambridge University Press, 2004), p. 295.

problem of finding two mean proportionals. Second – and here I reveal my true colors – I am quite convinced that geometrical solutions to the problem of the finding of two mean proportionals would have been quite irrelevant. If a particular structure required scaling (a significant undertaking that one would take up only rarely), then this indeed required the finding of a cubic root. But here is the thing: if I did need to find the cubic root of, say, 3, I would not look for complicated geometrical constructions (how would I even be sure my parabolas and hyperbolas were correctly drawn?). It would be much simpler just to look, by trial and error, for numbers whose cube approximated the number 3. This takes some patience and skill in calculation, but even so, this is straightforward, and a good calculator would come up with very precise answers within minutes.

Calculating by trial and error is a matter of skill. It shows that one has gone through the elementary education of literacy and through the apprenticeship of a technical profession. This is how one became a maker of catapults, after all. Of course this involved many numbers! But as one turned to write about this calculation, one described not calculation, not craftsman's knowledge, but something else: the theoretical, book knowledge of famous Greek authors. Greek authors of mechanics frequently turn to the problem of finding four lines in mean proportion – in exactly the same way that the authors of tactics conjure a technical, mathematical terminology – as a claim to elite, literate sophistication. Is this the major way in which applied mathematics mattered?

## Machines of Marvel

Authors of mechanics wrote about their catapults: apparently, there was an audience. Machines were something to talk about. Here is Hedylus, a poet writing in Alexandria, early in the third century BCE:[18]

> Come, lovers of strong wine, and behold the *rhuton*
> In the temple of the venerable Arsinoe, dear to the West Wind:
> It represents the Egyptian dancer Besas, who trumpets a shrill
> Blast when the stream is opened up, allowing the wine to flow.
> . . . honor this clever invention of Ctesibius –
> Come, young men! – in this temple of Arsinoe.

The poem is quoted many centuries later, perhaps early in the third century CE, in a book by Athenaeus, the *Deipnosophistae* (The Dining

[18] Translation adapted; the context can be read in C. B.Gulick, *Athenaeus: The Deipnosophists* (Cambridge, MA: Harvard University Press, 1933), pp. 219–221.

Wise Men). This work belongs to the genre of "sympotic literature," strange to us but quite common in antiquity. Those are descriptions of banquets (Plato's famous *Symposium* comes to mind) where highly educated people exchange pleasantries and quote old literature, above all quoting the literature that is appropriate to a banquet (stories having to do with wine and food, for instance, are naturally favored). It appears that this kind of dinner party was always central to ancient culture, and one wanted to be able to shine in such events. Hence the genre: wise, erudite men, shining in conversation around wine. And here is a conversation piece, then. The icebreaker at the cocktail party: a clever mechanical contraption where, once a certain valve is released and wine begins to flow, a trumpet sounds.

In this case, we do not know how the device worked because all we have is a poem quoted in a symposium. But there are still extant from antiquity sources that describe devices of this sort in a more technical manner. These are two treatises on pneumatics: one by Philo of Byzantium, whom we have met already as an author on catapults, the other by Hero of Alexandria, whom we have already met for his observation that one could find longitudes, in principle, by observing eclipses. (Notably, our sources begin to circle back to a smaller group: not that many "applied" mathematical authors.) The title of such treatises, *Pneumatics*, refers to *pneuma* or "spirit," and the subject at hand is the motion of subtle, nonsolid bodies. Effectively, what is under discussion is devices where *liquids are put in motion.* The wine poured, and the trumpet blown. Hero's very first concrete example is straightforward (which I repeat in paraphrase): Immerse a vessel with holes at the bottom and at the top within wine. As it is filled with wine, close the top hole with your finger. Lift the vessel, and the wine will not be poured. (You can try this in your bathtub, perhaps with water instead of wine, but you've probably tried this already as a kid, with soda and straw. Hero essentially intuits the correct explanation, which is the difference between air pressure and vacuum.) And now a further trick: you can let the wine pour, simply by releasing your finger from the top hole! Now that's truly hilarious. It is hard to imagine that such tricks were not first tried at a symposium, and the entire literature reads as a series of tricks to be performed, or discussed, in sympotic conversation. Drachmann, who has studied this literature extensively, sums this up as "parlour-magic"[19] (although remember: this is mostly

[19] A. G. Drachmann, *Ktesibios, Philon and Heron: A Study in Ancient Pneumatics* (Copenhagen: Munksgaard, 1948).

wine magic, sympotic magic). Past scholars have been very dismissive of this literature, in particular because it constitutes such a giant miss. The ancient tradition of pneumatics used the effects of different pressures applied by liquids, sometimes even heated to produce gas. They proceeded right next to the key technological breakthrough of all history – the steam engine! – and completely failed to notice it. (We will, in fact, return to the steam engine in the epilogue to this book; it does have ancient roots, although mostly distinct from Hero's *Pneumatics.*) It seems that the ancients did not really care about steam, as such: with very few exceptions, the pneumatic devices are liquid marvels and, most often, wine marvels. Liquids are usually a key part of the visible effect of the device, and they are almost always presented as the driving forces. Even steam is understood in terms of water, brought to a boil. It is always the *liquid* that does the magic. So, Hero's famous "steam-engine toy" – the one that gets modern scholars mad. Let water boil inside a sphere with two tubes extending out of it. As the steam escapes, the sphere begins to turn. The visible marvel is of water, heated, rotating spheres. But even this is the exception. By far the more common type of marvel is very directly liquid. Again and again, we find devices that pour libations on an altar, or that dispense wines of different kinds, or that combine (as in Hedylus's epigram) the flow of wine with some other observable phenomenon. Why this insistence on liquid marvels? Because of the function of this entire literature: the sympotic conversation piece.

Why should we even bring such wine marvels into our own, sober account of the history of mathematics? Because the ancients did. The ancient treatises on pneumatics were clearly understood to be integral to the discipline of "mechanics," which was clearly understood to be integral to mathematics. Indeed, the very presentation in those pneumatic works is mathematical: the objects are described tersely, with a technical language suggestive of the mathematical genre; most artifacts are presented with schematic diagrams, labeled by letters. The description of a mechanical device reads, then, like a geometrical construction. In the symposium, one merely performed the trick or recalled it; turned into book knowledge, the devices of pneumatics had to be brought into a specific, mathematical form. The same is true for another treatise by Hero, the *Automata.* The title, referring to self-moving devices, is correct, but its scope should be limited immediately. What Hero discusses is the making of a puppetry device that, pulled by a string, gives rise to a complex series of actions (essentially, rather like our attractions in amusement parks; Ctesibius's blown trumpet at the temple is actually not far from that). What makes this treatise mathematical is the schematic drawing of the arrangement pulling the puppets, labeled with the letters of a geometrical diagram and described with the language that is often that of pure geometry: a wheel ΟΠ passes at the middle of the side ΓΔ, revolving around the axis ΦΧ . . . The device is meant to amuse, as a tableau of puppets, but if it is to be read seriously, it becomes a geometrical construction. Perhaps this, in and of itself, is part of the thrill: a magic trick, as it were, where geometry and puppets are interchangeable.

This is an attractive account, where the application of geometry to mechanical contrivances is more "ludic," more in line with the literary emphasis seen in the mathematics of the generation of Archimedes. When Diocles contemplates burning mirrors, he effectively engages with just this kind of equivalence between the theoretical geometry and concrete magic. But in Diocles's case, the theoretical geometry is significant in and of itself. The direction with texts such as Hero's *Automata* is different: a piece of real-world machinery (whether actually built or just conceived) gets presented in a mathematical format. It seems reasonable that for Hero's readers, mathematical presentation would be of a higher status, and if so, the move has a natural sociological explanation.

Of all ancient mechanical authors, Hero occupies a unique place: we ought to get to know him. Not that this is easy! There are essentially no biographical data (it used to be thought that we could tell his date, but really, all we can say is "likely from the Roman era"[20]). And indeed, his corpus – even though it does present serious questions of authenticity, concerning entire treatises and then within each treatise – seems to project a clear, consistent persona. This may teach us more about the identity of ancient mechanics.

First, this corpus contains works engaging with mechanical devices, sometimes more practical, sometimes more toylike. These include, besides the *Pneumatics* and the *Automata* already mentioned, also the *Catoptrics*, or study of mirrors: as is usually the case in ancient studies of mirrors, this once again belongs in the field of the study of marvels, in particular the fun-house effects of non-flat mirrors. More sober is the *Dioptra* (a study of a survey device – which provides the context, incidentally, for Hero's note concerning the finding of longitudes based on eclipses). Hero's treatise on

[20] As noted previously, Hero made an observation concerning the manner in which eclipses can be used for finding longitudes. For this exercise, Hero made up a set of theoretical values for a fictional eclipse. Neugebauer was pranked, so to speak, and used the values to deduce Hero's date (62 CE, by Neugebauer's calculation): clearly, unfounded. N. Sidoli, "Heron of Alexandria's Date," *Centaurus* 53 (2011): 55–61.

another device – the water clock – survives in fragmentary form. Not surprisingly, Hero wrote also at least two treatises on catapults (one of which is extant only in fragmentary form) – so, obviously, he provided a solution to the problem of finding two mean proportionals! – and also about machines in more general terms, in a treatise that might have been titled *Mechanics* (and is extant only in Arabic). In all of these works, we see Hero presenting devices, of more or less remarkable function, through geometrical principles.

Alongside those works, there is another line of more "pure" mathematics: a *Geometry*, a *Stereometry*, a *Metrica* (a study, specifically, of the measurement of areas and volumes), and also a *Definitions* – consisting mostly of citations from Euclid's *Elements*, with additions from later geometers. We recall how Euclid was stripped, in the classroom, to bare statements without proofs so that pupils were asked, probably, to learn a little bit of Euclid, beginning with the definitions, by heart. This seems to be the context for the *Definitions* and indeed for Hero's "pure" geometry as a whole. Throughout his geometrical treatises, Hero uses the same geometrical vocabulary and the same geometrical diagrams as we find with pure geometers, but his emphasis is on measurement, which he effects with concrete numbers. Those treatises then look very much like the evidence, provided by papyrus fragments, for mathematical education. This, then, also explains why we have evidence (through Arabic sources) that Hero wrote a commentary to Euclid's *Elements* (was he the first?). He surely must have *taught* Euclid, then? There is no direct connection between craftsmanship and teaching, but in both wings of his work, we see Hero starting out from a profession – the maker of machines, the schoolteacher – and then seeking, for this relatively modest profession, the higher status of a more elaborate, bookish science. Is this how we are to understand the mathematization of mechanics in general?

To some extent, this would be to downplay the real ingenuity of ancient mechanics, which, on some rare occasions, did involve theoretical sophistication. I do not believe that the making of a catapult really involved the solution, via techniques such as the use of conic sections, of cubic equations. But mechanical devices could demand the solution of other, genuine theoretical problems. Perhaps the most spectacular case is that of the Antikythera mechanism, to which I will turn in more detail in the following chapter. At its essence, the Antikythera mechanism is a system of gears whose turning imitates astronomical cycles. I set aside the ingenuity – in itself, nothing short of amazing – required for the production of the precise gears. I do not know how the makers of the mechanism produced a gear with fifty-three nearly precisely equally spaced triangular teeth. I am sure, however, that they did not solve the theoretical problem of constructing the 53-gon. Craftsmanship trumped theory. (Some evidence for this, indeed, is that one of the displays, which is clearly meant to have a spiral form, is produced as a series of independent circular arcs. The spiral effect is good enough, and geometrical precision was not the key point.) What is deeply theoretical about this mechanism, however, is the basic construction of its train of gears, which requires both astronomical as well as calculatory theory. One needs to have a model of arithmetic regularities in the sky – and then to understand that to turn those into gears, one needs to factorize such regularities in an economical manner, which then calls for arithmetic ingenuity, similar in spirit to the finding of musical ratios. In this case, craftsmanship and science go hand in hand. Not surprising, after all: the very purpose of the machine was to display a piece of scientific theory. In this sense, it is perhaps comparable to a much simpler astronomical device – the sundial. Even in its very simplest form, as a single needle casting shadows – together with marks on the ground, to interpret those shadows cast at different times – this instrument encodes both astronomical as well as geometrical theory: the changing position of the sun in the sky, along the day and the year. It is significant, then, that there are now hundreds of traces of ancient sundials. (Often, all that survives is an inscription, letting us know that a sundial was set up.[21]) Because this is, of course, merely a fragment of all ancient sundials, the conclusion is that in the High Empire – when most of those inscriptions can be dated – there must have been many thousands of such sundials dotting the urban landscape across the entire Mediterranean. We started by noting the ubiquity of some mathematical education: many members of the literate, urban class knew how to calculate and how to measure basic geometrical figures; perhaps they even remembered a few basic facts from Euclid up to and including Pythagoras's theorem. And they lived in a landscape where scientific, mechanical devices were not scarce: indeed, in the minimal form of the sundial, they were ubiquitous. But this perhaps relates most to the central place of astronomy in the ancient imagination, and we will therefore need to return to this in the next chapter. With all the applications to the real world, nothing ranks as far high as the applications to that most distant slice of the real world – the heavens.

[21] S. L. Gibbs, *Greek and Roman Sundials* (New Haven, CT: Yale University Press, 1976). This study catalogues the 256 Greco-Roman sundials published at the time; a significant number have been added since.

## Mechanical Theory?

The Antikythera mechanism is not a mere symposium conversation piece, and it continues to amaze. And it circles back to one of the main figures of this book. Had you shown it to an ancient Greek, they would likely think that it was made by Archimedes. In fact, as noted in the preceding chapter, there is a credible tradition, according to which Archimedes even wrote a book about this type of device, titled *Sphere-Making*. This report is from Pappus, whom we have mentioned several times already: an author of the fourth century CE – who quotes, in turn, Carpus of Antioch, who might *perhaps* be from the first century BCE.

What Pappus quotes is the statement, precisely, that Archimedes wrote *only one book on mechanics – Sphere-Making* (because, in general, he preferred geometry).[22] A couple of observations follow.

First, indeed, the likelihood is that this is true. If so, Archimedes contributed to and perhaps initiated the tradition of the Antikythera mechanism. We shall, of course, return to this in the following chapter.

Second, and perhaps more surprising, we find that the ancients did not consider Archimedes's seminal contributions to mathematical physics – the study of the statics of the balance, and the hydrostatics of solids immersed in liquid – as part of "mechanics."

There was an ancient science the Greeks called "mechanics," and this was the study of *mēchanē*, "machine," above all the war engine but also understood to include any other device that might pique a reader's curiosity. Then, there was geometry. Geometry was universally seen as applicable to mechanics, as a matter of genre – mechanics would be written, to a large extent, in the formulaic geometrical language and with the aid of the geometrical diagram – and as a matter of a specific, albeit very important, piece of theory. This was the problem of finding two mean proportionals, correctly said to be relevant to the problem of scaling solid artifacts (although the practical significance of this applicability was wildly exaggerated).

What was *not* seen as an application of geometry to mechanics was the study of the mathematical principles underlying physical force – exactly the thing that, from Galileo onward, came to define the discipline.

To see this, let us now sum up our overview of ancient mechanical writing with yet another table of authors. This table lists all the credibly attested authors in the ancient genre of mechanics (many of whom I have

[22] Pappus's *Collection*, VIII.3 1026 Hultsch.

mentioned already). For each, I report the century; whether or not they are extant; whether or not they touch, in their writing, upon military machines; and whether or not their writings include a theoretical element comparable to Archimedes's studies in mathematical physics. It often happens that all we know about a certain ancient mechanician is that they are credited with a famous device or construction. It may or may not be that this became famous through their writing; in this list, I only report such cases where there is some evidence suggesting an actual writing, thus concentrating on the small but clearly defined group of the ancient authors on mechanics.

| Name | Century | Extant | Military? | Theoretical? |
|---|---|---|---|---|
| Callistratus | 4 BCE | N | ? | ? |
| Diades | 4 BCE | N | Y | N |
| Polyidus | 4 BCE | N | Y | N |
| Ps.-Aristotle | 4/3 BCE | Y | N | Y |
| Ctesibius | 3 BCE | N | Y | N |
| Archimedes | 3 BCE | Y | N | Y |
| Daimachus | 3 BCE | N | Y | N |
| Philo | 3 BCE | Y | Y | N |
| Moschion | 3/2 BCE | N | Y | N |
| Biton | 2 BCE | Y | Y | N |
| Agesistratus | 1 BCE | N | Y | N |
| Athenaeus | 1 BCE | Y | Y | N |
| Democles | 1 BCE? | N | ? | ? |
| Diphilus | 1 BCE? | N | ? | ? |
| Vitruvius | 1 BCE | Y | Y | N |
| Hero | 1 CE? | Y | Y | N (or slightly Y) |
| Apollodorus of Damascus | 2 CE | Y | Y | N |
| Pappus | 4 CE | Y | Y | Y |
| Urbicius | 5 CE | Y | Y | N |
| Marcellus | Late Antiquity? | N | ? | ? |
| Quirinus | Late Antiquity? | N | ? | ? |

Once again, we have a list with about twenty figures. (The numbers could have been expanded somewhat by including figures known only for their famous devices; I also exclude a small group of physicians reported to have invented various medical devices. Had I expanded the list, its emphasis on actual mechanical construction would have been more obvious.) It is clear enough that the field is associated mostly with war engines, and indeed one can glimpse something of a chronological pattern based on this association:

First, a significant group in the fourth century around the time of the military revolution of the catapult. (This group is reflected even later by the figure of the mechanical author as a member of the Hellenistic court.) Later, we find another significant group in the moment of Polybian exchange, as Greek intellectuals present the achievement of their culture to the Romans, mostly around the first century BCE.

Theory is extremely rare and is, as it were, the opposite of mainstream mechanics: the authors who write on theory are those who do not write on catapults. They are, quite simply, idiosyncratic. But of course, this does not take away from their importance!

One idiosyncratic author is Archimedes, whose physical, "mechanical" science we have seen, in great detail, in the preceding chapter. Even though he did produce catapults – because he had to? – he did not write about them.

Let us now turn to Pappus, who is otherwise our main source.

Here is an author who dedicates most of his efforts to compilation and commentary – an impetus we will return to understand in Chapter 6. His *Collection*, in eight books, surveys a wide variety of sciences: Book VI is dedicated to astronomy; Book VIII, to mechanics. Throughout the *Collection*, Pappus produces mathematical introductions to the field at hand, usually through comments on important works from the past. The attitude is sophisticated, intended for a very well-prepared mathematical reader, so the emphasis is on difficult mathematical proofs. Book VIII is not structured through past treatises, but the spirit of an advanced mathematical introduction is preserved, and so Pappus structures this treatise according to the requirements for the application of mathematics to mechanics. (It is in this context, introducing his project, that Pappus cites Carpus's observation on Archimedes's mechanical work.)

Pappus claims there are three applications of theoretical mathematics to mechanics. As usual, we have the problem of finding two mean proportionals, but remarkably, Pappus adds two more problems. One is, given a toothed cogwheel with a given radius and number of teeth, to find the radius required for another cogwheel, given its desired number of teeth and the desired ratio of rotations. It is hard to imagine that the second problem did not arise from a treatise dedicated to a device such as the Antikythera mechanism, and it is tempting to speculate that it is ultimately due to Archimedes's study in *Sphere-Making*.

The third and final one is, given the force required to move a weight along a horizontal plane, to calculate the force required to lift it along an inclined plane. Now this is exciting: this is, in fact, as close as ancient science ever gets to what we now understand as the elementary application of physics to mechanics (and is, in fact, a standard problem of the scientific revolution). This is not entirely isolated: even though Hero's discussions are mostly tied to concrete machines, his treatise titled *Mechanics* does make several comments of a more theoretical character, and Hero, too, mentions this problem of the force required to move a weight on an inclined plane. It thus seems that both Hero and Pappus reflect a longer tradition.

This tradition, as far as we can see, is interestingly wrong. The ancient authors did not develop a correct formula for dealing with the inclined plane. Pappus's approach is to conceive of the object, set on the inclined plane, as if it is a composite made of two segments, one higher and one lower. The force required to keep the weight in place is that which, added to the naturally present forces acting on the two segments, makes them balanced. (The force required to move them, then, would obviously be the force *bigger* than the balancing force.) Pappus's task, then, becomes to consider the force that the higher segment of the body exerts on the lower one.

This is all evidently a crude adaptation of Archimedes's *On Floating Bodies*. There, one considers the solid immersed in the liquid as constituted of two components, the one fully within and the other fully outside the liquid, stability obtaining when the two precisely counteract each other. The extension of this approach to the problem of the object, stable on an inclined plane, is clearly deficient. (What is even the point of dividing the body into segments, now? The lower part is moved downward, just like the upper part.) But the main point is not just that this approach, presented by Pappus, is mistaken. The point is that had there been a robust exchange and critique surrounding such applications, preceding Pappus, such errors could have been filtered and corrected. It appears that there were none because so few even attempted this kind of research. It is not that there were zero authors on mechanical force, understood mathematically. Archimedes led the way (albeit in a purely geometrical context); Hero and Pappus reflect a tradition of trying to extend it. But it must have been a small, negligible sidestream of ancient writing in the exact sciences. There was no ancient Galileo to pick up from where Archimedes had left the study of force. The ancient tradition was a false start. (We shall return to this – antiquity, as the first moment when one could have extended Archimedes, but didn't – when considering Hipparchus in the following chapter.)

Writing about theoretical mechanics (as opposed to concrete machines) was simply not a recognized genre. One only wrote about it in the rare case that one explicitly strove – as Pappus, after all, did – to write about *everything*. This brings us to our final example of ancient theoretical mechanics, coming from yet another context where an author – or better put, a group of authors – sought universal scope.

The treatise in question is titled *Mechanics* and is ascribed, in our manuscripts, to Aristotle. It is likely that the attribution is not precise and that the philosopher Aristotle did not write this treatise. (Quite simply, this treatise does not carry the imprint of Aristotle's genius; this view is widely shared, and it is usually said to be by "pseudo-Aristotle."[23]) The language and the overall attitude of wide-ranging curiosity are indeed reminiscent of Aristotle, and the consensus is that this is the work of an author trained in the Aristotelian school, close in time to the master. An early text, then, perhaps roughly contemporary with Euclid's *Elements* – and surely preceding Archimedes.

The Aristotelian project was, indeed, universal. A follower of Plato, Aristotle shared his master's fascination with the sciences. Unlike Plato, Aristotle went on to expand this project systematically, striving to survey a wide array of phenomena and to provide them with rational accounts. The Aristotelian corpus is like a running commentary to the universe: it discusses logic, animals, speeds and motions, sense perception, comets, rainbows, ethics and rhetoric, sleep, the Athenian constitution, tragedy, chemical mixtures, predicate logic . . . Individual works very often devolve into nearly unstructured lists of observations, followed by explanations. (Why is it the case that . . . ? Maybe because . . . ?) The series of observations and explanations in Ps.-Aristotle's *Mechanics* all involve the working of various human artifacts (notably, including no military devices). They are mostly accounted for on the model of the lever – no surprise there. Much more surprising to us, the lever is understood not on the basis of anything such as Archimedes's geometrical proportions but, instead, in terms of the geometry of the *circle*. The central model of the Aristotelian mechanics is of a rotation around a fulcrum point, and most of the results reduce to the fact that longer circles cover greater distances. (This is supposed to explain, essentially, why the weight further away from the

[23] The question of the authorship of the treatise remains complicated. A recent and nuanced survey of the evidence is found in I. Bodnár, "The Pseudo-Aristotelian Mechanics," in M.-L. Desclos and W. W. Fortenbaugh (eds.), *Strato of Lampsacus: Text, Translation and Discussion* (New Brunswick, NJ: Routledge, 2011), pp. 443–454.

fulcrum exerts the greater force; such texts remind us that Archimedes was indebted to a tradition – and how much he improved upon it.)

Once again: the Aristotelian *Mechanics* is, on the whole, more wrong than right. And once again, it appears to be an isolated text. The latter is true, in fact, for many other works within the Aristotelian corpus. Aristotle produces a comparative study of the animal kingdom; a predicate logic; a theoretical study of poetics. In all such cases, it is difficult to find any authors, away from Aristotle's immediate disciples, who add anything new to these uniquely Aristotelian projects. This might have been especially true in the case of mechanics, where Archimedes's achievement quite evidently supplanted the effort by Ps.-Aristotle.

Ancient civilization was large, and even the things that happened very rarely would happen from time to time. The exceptions occurred, and they mattered: Ps. Aristotle's mechanics, Archimedes's works in mathematical physics, Pappus's Book VIII – all those works were not merely composed but were also copied and preserved. In time, they would provide an important impetus to the scientific revolution. This is a significant perspective but also an anachronistic one. In antiquity, the field of mechanics remained, on the whole, the study of *mēchanē*. Some of the engineers making catapults chose to present themselves as mathematicians – which they obtained "on the cheap," so to speak, simply by inserting a solution to the problem of finding two mean proportionals. Indeed, the association between mechanics and *mēchanē* was so strong that even the greatest achievement of the ancient mathematical account of forces – Archimedes's mathematical physics – was understood by the ancients to belong not to mechanics but to pure geometry instead. Which may well have been correct!

## Mathematics, Real and Symbolic

This was a many-branched, meandering chapter. Mathematics was embedded in ordinary life: numeracy was widespread, and many members of the elite were exposed to elementary mathematical education. Mathematics also features in many technical traditions: geography and land surveying; engineering, sculpture, and architecture; above all, the practices of war – the setting of a military camp; tactics; most practically, the making of catapults. But then again, mechanics also involved the making of various devices of marvel. I have merely glanced at the most significant area of the application of mathematics to the real world – that of astronomy, which we will take up in the following chapter. And I have omitted from this review two of the main areas of such application, simply because we have noted them already in preceding chapters. The idea that mathematics is applicable to music – because numerical ratios determine harmony – was crucial to the generation of Archytas and will always remain a chief reason for philosophers to care about mathematics. And one of the specialized fields of mathematics, particularly in the generation of Archimedes, was that of optics: the analysis of light and vision in mathematical terms. In both music and optics, what is at stake is not precisely an "applied science": it is not as if craftsmen built their stringed instruments, that artists fashioned their mirrors, based on theory. Theory was posterior to practice – it added an element of theoretical understanding to an area of practice, pursued independently of any such theory. But the same, after all, was the case with almost all the applications we have discussed in this chapter.

The breadth is significant. Mathematics was not marginal in antiquity, nor was it of low status. To the contrary: it reached across many fields where, quite evidently, it brought prestige to those who could claim its knowledge. And yet, the theme of this chapter is not just breadth but also superficiality. Yes, mathematics was widely applied, but it was applied, perhaps, in the sense in which lacquer is applied to furniture. This is true even in the most practical subdivision of ancient mechanics, namely, catapult making. On the one hand, there is the craftsmanship, and then, when this comes to be written down, knowledge is handed out with the thin added layer of advanced mathematics (in the form of the problem of finding two mean proportionals). It is exactly like music, where singers sing, flutists play, instrument makers ply their craft – and then, when it is all written down, the author would pay homage to number theory.

This seems like a tension in our survey. I do emphasize that mathematics was fairly widespread. This was not the ancient Near East, with a handful of professional scribes, masters of abstruse numeracy. Most everyone could count, and many could even remember Pythagoras's theorem. How to square this with the mostly superficial application of any more advanced mathematics?

But perhaps the two elements – breadth and superficiality, the wide access to numeracy and the symbolic use of advanced mathematics – go hand in hand. We recall how Philo emphasizes the need for precise professionalism:

> Many who have undertaken the building of engines of the same size, using the same construction, similar wood, and identical metal, without even changing its weight, have made some with long range and powerful impact and others which fall short of these. ... Thus, the remark made by Polyclitus, the sculptor, is appropriate for what I am going to say. He maintained that *perfection was achieved gradually in the course of many calculations.*

And surely Polyclitus – and Philo after him – were right. It took great skill to make an individual artifact, and one had to make many precise calculations. In this sense, then, one had to be a very good practical mathematician so as to become a sculptor or an engineer. But what was at stake here was to know about bronze, wood, nails, and ropes and to know, mathematically, about the basic properties of number and shape. One definitely needed mathematics, but one did not need the finding of two mean proportionals; one really needed to know, instead, how to calculate – one needed to be familiar with the basic properties of geometrical figures.

And this was exactly what the members of the educated Greek public were supposed to know from their elementary education. This is how the widespread reach of basic numeracy goes hand in hand with the superficial use of advanced mathematics. I suggested that the most practical way to find a cubic root would have been to calculate cube values by trial and error – certainly a more reliable technique than, say, finding the intersection of conic sections. Trial-and-error calculation calls for skill, and even today, without a calculator, many would fail! But this type of calculation would have been entirely underwhelming for an educated Greek reader.

The French sociologist Bourdieu emphasized the role that *distinction* plays in social life. We all live in the same material world; elites mark themselves – distinguish themselves – by a certain detached, aesthetic attitude to material reality. Perhaps one should not rush to generalize: Bourdieu studied modern and contemporary society, where the distinction between mass and elite is much more contentious and hence needs to be marked in various subtle ways. But the fundamental logic is clear and probably should be extended to our case. The problem of finding the two mean proportionals, applied to mechanics, is a case of Bourdieu-like *distinction*. Any good craftsman can calculate the required lengths for the pieces of a catapult. A mechanical author is marked as a true member of the elite – part of the polite conversation between authors – by being able to marshal pure geometry. And if so, it is quite natural that applied mathematics would be posterior to practice: its main function, in truth, is in the sociology of literary culture, not in the epistemology of mechanical knowledge.

When we say that advanced mathematics was applied, in antiquity, as a form of distinction, we do not take away from its value. The point is not that mathematics was "merely of symbolic value" – for there is nothing "mere" about symbols. Symbols provide meaning, and what we see is that mathematics, in antiquity, could endow domains of practice with significance. One did not just move through geographical space; one moved on the surface of a sphere, measurable with the tools of geometry, and this was important not because, thanks to this knowledge, the ancients had superior maps but because, thanks to this knowledge, space gained in meaning. This observation, indeed, is at its most obvious with astronomy. The stars are, above all, vehicles of *meaning*. Two thousand years later, Immanuel Kant's mind would still be "filled with admiration and awe, reflecting on the starry heavens above." To the ancients, the stars were divine; for many of them, they were also auspicious. Not merely divine, they were also the key tool for divination. Astronomy, and its sister science, astrology, were the most important applied sciences in antiquity because they were the most laden with symbolic meaning: precisely because so far away, in the figures of the sky, the concrete and the purely mathematical could most naturally be joined together.

## Suggestions for Further Reading

The chapter touches on many themes, several of which have benefited from recent scholarship. An essential general reading is S. Cuomo, *Technology and Culture in Greek and Roman Antiquity* (Cambridge: Cambridge University Press, 2007). Another recent book approaches mechanical writing as a genre: C. Roby, *Technical Ekphrasis in Greek and Roman Science and Literature: The Written Machine between Alexandria and Rome* (Cambridge: Cambridge University Press, 2016). Mechanics is approached from a philosophical perspective in S. Berryman, *The Mechanical Hypothesis in Ancient Greek Natural Philosophy* (Cambridge: Cambridge University Press, 2009).

Older but very reliable studies are J. G. Landels, *Engineering in the Ancient World* (Berkeley: University of California Press, 1978) and E. W. Marsden, *Greek and Roman Artillery: Historical Development* (Oxford: Oxford University Press, 1969). Similar in character (but somewhat more dated) is O. A. W. Dilke, *The Roman Land-Surveyors: An Introduction to the Agrimensores* (New York: David & Charles, 1971). There is now growing, excellent scholarship on ancient geography, and a good introduction to the field is D. Dueck, *Geography in Classical Antiquity* (Cambridge: Cambridge University Press, 2012).

There is not yet a monograph-length treatment of mathematical education in antiquity (the author of such a monograph must be not only a good social historian of mathematics but also a trained papyrologist, as the evidence of the papyri is yet to be tapped in full). An excellent brief survey is N. Sidoli, "Mathematics Education," in W. M. Bloomer (ed.), *A Companion to Ancient Education* (Oxford: Wiley-Blackwell, 2015), pp. 387–400.

CHAPTER 5

# *Mathematics of the Stars*

## Plan of the Chapter

A one-volume history of Greek mathematics has to be selective and compressed. All the more so, a one-chapter history of Greek astronomy. We follow an entire trajectory – from the beginnings to Ptolemy (second century CE). Chapter 3, "The Generation of Archimedes," was a rondo with Archimedes as its recurrent theme. Here, we progress through three subchapters for our three main cast members:

I. Rising: Greek Astronomy to Eudoxus
II. Conjunction: Greece, Babylon, and Hipparchus
III. Culmination: Ptolemy

### *I Rising: Greek Astronomy to Eudoxus*

The first section, "What Everyone Knew," begins from ethnoastronomy (the astronomical knowledge of simple societies). We dwell on this more than we did on ethnomathematics in Chapter 1. Throughout history, people knew about the sky *more* than we – most modern city-dwellers – now usually do. To explain the history of astronomy, then, one must also provide some of that once-elementary astronomical knowledge. The following section, "Bring in the Bureaucrats," visits the astronomy of early states across the globe, ending with Mesopotamia. We find a common set of themes – an interest in events seen on the horizon, as well as repeating cycles and, above all, the calendar. This will remain at the background of Greek astronomy itself, and Mesopotamian astronomy, specifically, is itself a major protagonist in our story. It was on a par with Greek astronomy, and eventually, the science of Hipparchus and of Ptolemy is best understood as a synthesis of the two.

"Prelude: The Threshold of Greek Astronomy" brings us to Greece. The earliest Greek astronomy was, perhaps, no more than whatever it was of Mesopotamian astral science that came to be, already, the shared knowledge of the Mediterranean. Against this backdrop, Greek astronomy proper began to develop. This, too – just as we saw for mathematics – may have taken the form of generational waves. In the section "The First Generation of Greek Astronomy: The Extent of Our Knowledge," we consider what may be safely asserted concerning Eudoxus's innovation of an astronomical model where the planets are carried by nested homocentric spheres. The following section, "The First Generation of Greek Astronomy: The Limits of Our Knowledge," is more speculative. I emphasize the possibility that Eudoxus could have relied on mechanical models (such visible displays will remain as a theme of Greek astronomy throughout).

### *II Conjunction: Greece, Babylon, and Hipparchus*

"The Astronomical In-Between?" notes how – following the generation inspired by Eudoxus himself – the study of homocentric spheres seems to disappear. A few works do survive from late in the fourth century and early in the third, and their interests are, in some cases, closer to those of other ancient societies (once again: horizon phenomena and calendrical cycles). We also find, however, traces of the more ludic science of the Hellenistic era. The following creative generation of Greek astronomy culminates with Hipparchus (second century BCE), but it is harder to say how it originated. This is the subject of the next section, "A New Astronomy: Whodunit?," which considers the possible contributions of Archimedes, Apollonius, and Hipparchus – and also brings in the unique witness of the Antikythera mechanism. The following three sections revolve around Hipparchus's astronomy – which opens up to much more besides. "Hipparchus and the Babylonians" discusses the broad outlines of Hipparchus's contribution: astronomical models that integrate precise observations. The section also picks up again the story of Mesopotamian astronomy (the origin of many of Hipparchus's observations, parameters, and techniques). Hipparchus's creativity is hard to define in any simple formula, even considering astronomy alone. One of his contributions must have been, more or less, the foundation of an entire mathematical field – trigonometry. This is, among other things, a reflection of the role of precise detail in Hipparchus's scientific project and is the subject of the section "'Industrious': a Detour Regarding Trigonometry." But this does not nearly sum up the intellectual project as a whole, and the following section, "'A Lover of Truth' and Hipparchus's Character as a Scientist," considers

Hipparchus's scientific personality and brings in the evidence not just for his astronomy but also for his contributions across mathematics as a whole.

### *III Culmination: Ptolemy*

Hipparchus was unique and, possibly, isolated. There seems to be much less creative activity in the generations following him, and we indeed note, once again, the sea change marked by the coming of Roman patronage. In the history of astronomy, this sea change shows itself in the spectacular rise of Greco-Roman astrology. "Before Ptolemy: Signs and Tables" starts, once again, from Mesopotamian astronomy, now considered as a science of astral omens. As we consider the rise of astrology, we also consider the practice of the astronomical table. The use of tables is crucial for our last major protagonist – Ptolemy. "Ptolemy: Toward the Canon" considers the major outlines of Ptolemy's science, "A View of the Monument" concentrates on an example (Book III of the *Almagest*). Ptolemy aimed at monumental finality, a science of astronomy where everything is computed, concretely, from first principles – but he also aimed to achieve the same in other sciences as well, the project as a whole integrated within a metaphysical program. Remarkably, he was successful, his project canonized soon after his death. Eventually, he would be one of the pillars of the scientific canon emerging in Late Antiquity – the subject of the following chapter.

## I. Rising: Greek Astronomy to Eudoxus

### *What Everyone Knew*

Why should simple societies not have an astronomy? We understand that a people will know a lot about their environment, the lay of the land, its animals and plants, its seasons. Star lore is a part of that knowledge woven into a society's fabric of life. This is often, literally, *myth*. The Gamilaraay people – in what is today New South Wales – knew the tale of the Mayi Mayi sisters, who, angry at the rape of one of their own, brought frosts to the land; repenting of the harm to the people, they went up to the sky, as stars, to bring back the warmth. Their appearance on the horizon foretells the coming of dingo pups and of the warm season. The patterns of appearance and reappearance are always associated with beliefs about the weather, hence about the most basic life cycle; so, a widespread proverb in Africa, south of the equator, reminds that "if the Digging Stars set in sunny weather, they rise in rain; if they set in rain, they rise in sunny weather." (Why "Digging Stars," then? Because they remind you when to dig – specifically, to hoe – for the planting of new crops.) Star lore can be maintained by tales but also, very often, by concrete artifacts: the Navajo still make string figures that represent constellations, for instance, the Dilyéhé, the constellation whose first reappearance marks the end of the planting season. Other artifacts try not to just to imitate constellations but instead are observational tools, used to align the viewer with the sky. Large earthworks, believed to date somewhere in the range from about 500 BCE to 500 CE, are found all over the Ohio Valley, with single structures, on a few occasions, measured in miles. Lacking any historical context – and with the archeological evidence itself mostly eroded beyond recovery – no certainty is possible, but the likelihood is that such structures served, at least among other things, as permanent objects relative to which one could site horizon phenomena, perhaps mostly those of the moon.[1] This is, of course, directly reminiscent of Stonehenge, which, once again, is but one of many megalithic structures found all over Eurasia, from a period preceding any written history. I traveled rapidly across the globe, and with Stonehenge, we're finally back on the same continent as the Greeks. And this is as good of an opportunity as any to dispel a common misunderstanding. Stonehenge – and its like – tends to impress us too much. Indeed, the most superficial familiarity with the evidence confirms that the layout of Stonehenge must have been aligned with astronomical observations. You take this fact in, and your mind explodes: wow, they knew *astronomy*! This must have been some kind of Atlantis – possessing some weirdly advanced science, since lost (were they visited by aliens?) The premise of such reasoning is that astronomy should be a type of science; hence, it should be rare. The concrete facts of Stonehenge suggest, to our mind, a lost Newton. In fact, astronomy – in the sense of paying close attention to the changing appearances of the main features of the sky (especially along the horizon) – is as common as it gets. It is remarkable that the people of Stonehenge lifted all those heavy slabs. But there is

[1] R. D. Haynes, "Astronomy and the Dreaming: The Astronomy of the Aboriginal Australians," in H. Selin (ed.), *Astronomy across Cultures: The History of Non-Western Astronomy* (Berlin, Germany: Springer, 2000), p. 77; K. V. Snedegar, "Stars and Seasons in Southern Africa," *Vistas in Astronomy* 39 (1995): 529–538; T. Griffin-Pierce, *Earth Is My Mother, Sky Is My Father: Space, Time and Astronomy in Navajo Sandpainting* (Albuquerque: University of New Mexico Press, 1992), pp. 165–166; R. Hively and R. Horn, "Geometry and Astronomy in Prehistoric Ohio," in A. Aveni (ed.), *Foundations of New World Cultural Astronomy* (Boulder: University Press of Colorado, 2008), pp. 39–60.

absolutely nothing remarkable about the fact that they cared deeply, and knew a lot, about what one sees on the horizon just before sunrise. This is what people always did, everywhere.

Let us consider, then, what one sees on the horizon just before sunrise. (Incidentally, the word *sunrise* implies a geocentric perspective; I will switch freely, in my exposition, between geocentrism and heliocentrism. I will also generally assume – this is Greek astronomy – a position in the Northern Hemisphere). I explain basic things because I feel a need for them, myself. For instance, I was told as a kid that the sun rises in the east, and therefore I used to believe that it always rises at the same spot, due east. In fact, as you look to the horizon from a fixed spot, day after day, you will notice the point where the sun rises, relative to fixed elements of the environment. It is "just right of that tree over there." (In a place such as Stonehenge, slabs of stone fix those references.) This keeps changing. In the dead of winter, it rises at a more southern spot. The next day, it will rise a little to the north ("Now it's behind that tree!"), and so on until the middle of the summer. The sun will then rise at a much more northern position. That rising position will start to migrate south again the very next morning, all the way into the southernmost position reached, once again, at the dead of winter. This north–south pendulum of the points of rising forms a stable pattern, repeated year after year. When we say that many civilizations know astronomy, we mean that they know such patterns.

The pendulum of the points of the risings of the sun is one set of events associated with sunrise. As we look to the horizon, waiting for the sun to rise, we can pay attention to the stars, as well. What stars are visible at the horizon, just before sunrise?

Once again, this is perhaps not obvious to city-dwellers: *the stars keep changing*. First, and more obviously, as the earth, in fact, rotates throughout the night around its axis, we see this from our perspective as a rotation of the stars. The Greeks ended up thinking of this as a big sphere – on which the fixed stars are set – turning around an axis. (By "fixed stars," we refer to stars outside the solar system, all so enormously far away from us that they do not seem to change their relative position at all even as we change our position on the earth or as the earth rotates around the sun.) The nightly rotation of the fixed stars around their axis is quite rapid, and even by looking at the sky for a fraction of an hour, we notice considerable movement. Throughout the night, the stars are mostly replaced – the stars visible just after sunset will mostly lie, by the end of the night, beneath the horizon. Imagine, specifically, a star that is visible just above the eastern horizon just after sunset. There are variations, depending on the position of the star to the north or to the south, but under the correct conditions of this spherical geometry, we can follow the star throughout the night, culminating near the top of the sky around midnight and then gradually setting down until finally, just as the sun rises, it will set down in the west. This, then, is when that particular star gets all the night to itself.

What constitutes that star's luck on that particular night? It then has the property that whenever it is above the horizon, the sun is beneath the horizon. We can say that throughout that night, the sun is *on the other side of the earth* from that star (this will be true from either a heliocentric or geocentric perspective). This will obviously change throughout the passage of the year. The stars that are visible throughout the night are those that are to our back throughout the day. From our perspective, as we go through the year around the sun, we have our back, during the day, to different parts of the starry sky. And so the identity of the visible stars keeps changing.

I have mentioned stars noted by several civilizations: the Mayi Mayi, the Digging Stars, the Dilyéhé. These are different names for what the Greeks (whose naming we still use) called the Pleiades. Indeed, these are quite important to the Greeks – the poet Hesiod employed the same Pleiades to mark the timing of the harvest.[2] This is typical: most civilizations refer to a fairly small and constant set of prominent stars and constellations. The Pleiades are by no means universal, but they are truly widespread. And what is noted, concerning them, is precisely the *pattern of visibility*.

Let us say that this is early in May (the precise date depends on the location of the observation and also changes, very slowly, historically; right now, I ignore such complications). Your gaze is fixed, just before sunrise, toward the eastern horizon. You notice a few last-minute stars rising, just barely preceding the sun itself. And then, in all their sisterly beauty (see Plate 4), the Pleiades (which, to your mind, might be the Mayi Mayi, the Digging Stars, or the Dilyéhé). A minute later, the sun shines, and the Pleiades are no longer visible. On the next day, the Pleiades emerge above the eastern horizon a little earlier ("earlier," in this case, means not only as measured by a clock but more fundamentally: *relative to the rising of the sun*). Thus, they will have a little longer to be visible, disappearing only when they are farther up in the sky. This goes on for the following nights until the Pleiades finally become visible throughout the night, from its beginning to end. What we notice is that in early May, the sun and the Pleiades are nearly aligned (the sun, in a sense, is nearly between the earth

[2] Hesiod, *Works and Days*, 383–387.

and the Pleiades); by November, they are opposite each other (the earth, in the same sense, is between the Pleiades and the sun).

The Pleiades were barely visible in early May because they were *so close to the sun* (in the sense that they rose only a little before the sun itself did). Just before that, then, they rose precisely together with the sun, and so their rising was invisible. When the Pleiades were above the horizon, so was the sun, and so, for all practical purposes, the Pleiades *vanished from view*. This, in fact, went on for a while: the sun's cloak of invisibility, so to speak, spreads across many nights. (The precise length of the periods of vanishing will vary, depending on the position of the observer on the surface of the earth and on the position of the star to the north or the south.) But note that before they vanished from view – in late winter into the early spring – the Pleiades were on the other side of that "cloak of invisibility"; that is, they were visible, briefly, near the western horizon, *just after sunset*. We can plot an entire sequence, then:

- The Pleiades barely visible just before sunrise (early May)
- The Pleiades visible for longer and longer stretches before sunrise
- The Pleiades visible throughout the night
- The Pleiades visible after sunset but for shorter and shorter stretches
- The Pleiades barely visible just after sunset (late March)
- The Pleiades vanishing from view entirely

At this point, the cycle repeats itself. This is once again a pendulum – appearance, growth, diminution, disappearance – that echoes the pendulum of the sun's motion north and south. Each star and constellation goes through a cycle of vanishing from view and reappearing, with its obvious metaphorical resonance of birth, death, and rebirth. Of course, it is precisely this motion of the sun that gives rise to the seasons, and so it drives the actual cycle of life and death of all living beings. We tell this as an astronomical, or geometrical, story, via diagrams and calculations, whereas the Gamilaraay people tell this as a story about the Mayi Mayi, via a particular constellation. But all humans know the same story outline, connecting life, seasons, sun, and stars.

So, to begin with, for sun and stars. One more pattern we may add, noting it very quickly (because it is still familiar to all), is that of the moon. Once again, there is a key event of "birth" – the "new moon," observed as the near-transparent circle begins to fill up again, becoming, night by night, a wider crescent. This transition also follows a pendulum pattern, the moon literally waning and waxing. (Obviously, in this case, the main phenomena of disappearance and reappearance are not related to the

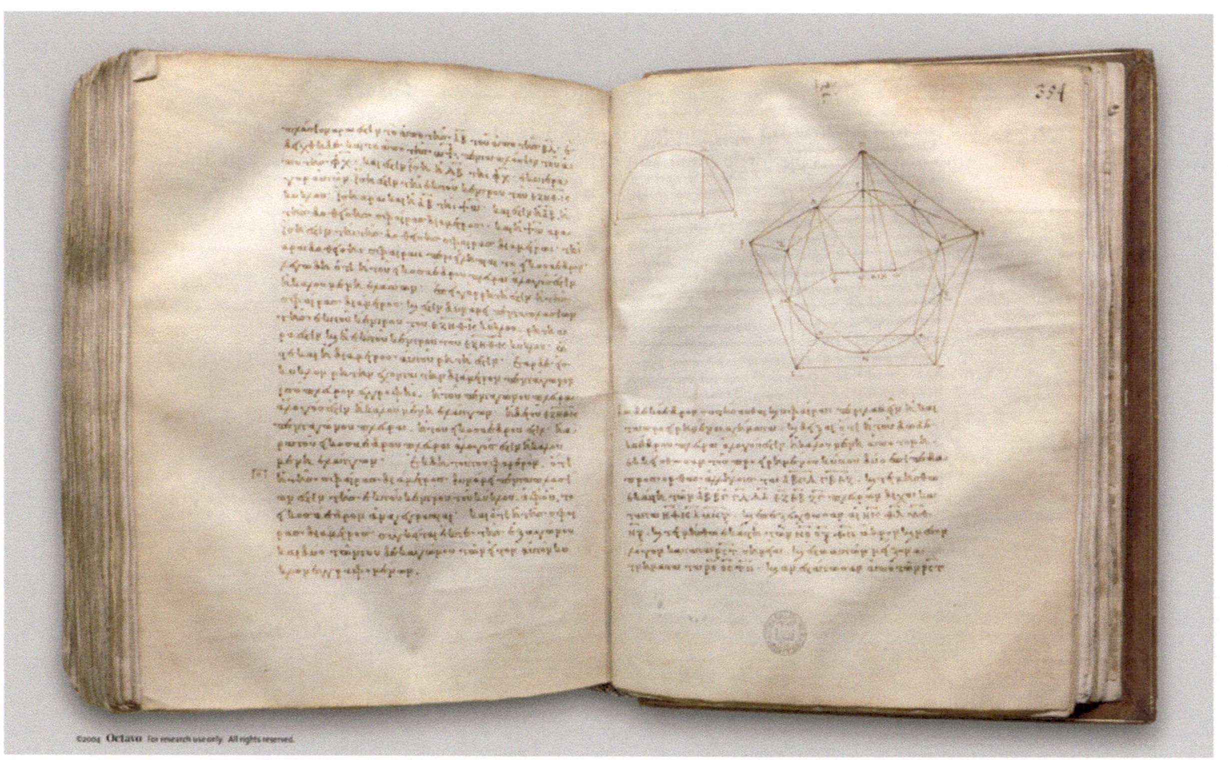


1. Pages from Euclid's *Elements* Book XIII.16, showing the construction of the regular icosahedron. MS D'Orville 301, Bodleian Library, Oxford University. Image courtesy of the Clay Mathematics Institute.

2. P. Oxy. 5299. With brief statements and rough figures from the first book of Euclid's *Elements*. Courtesy of the Egypt Exploration Society and the University of Oxford Imaging Papyri Project.

3. Doryphoros (450–400 BC). Designed by sculptor Polykleitos. Roman marble copy of Herculaneum. Naples National Archaeological Museum. Italy. Photo: PHAS/Universal Images Group/Getty Images.

4. Pleiades or Seven Sisters or the Messier 45 star cluster rising above the mountains of Leh, Himalayas. © Sukanya Ramanujan. (https://sukanyaramanujan.wordpress.com).

5. Late Babylonian clay tablet with the mul.apin. Purchased from Messrs Mann & Bishop, 1889. Museum number 86378 

6. Mosaic floor from Boscotrecase, Pompeii, showing Plato's Academy at Athens. The philosopher – sitting in the middle – teaches a group of disciples. National Archaeological Museum, Naples, Italy. Photo by Leemage/Corbis/Getty Images.

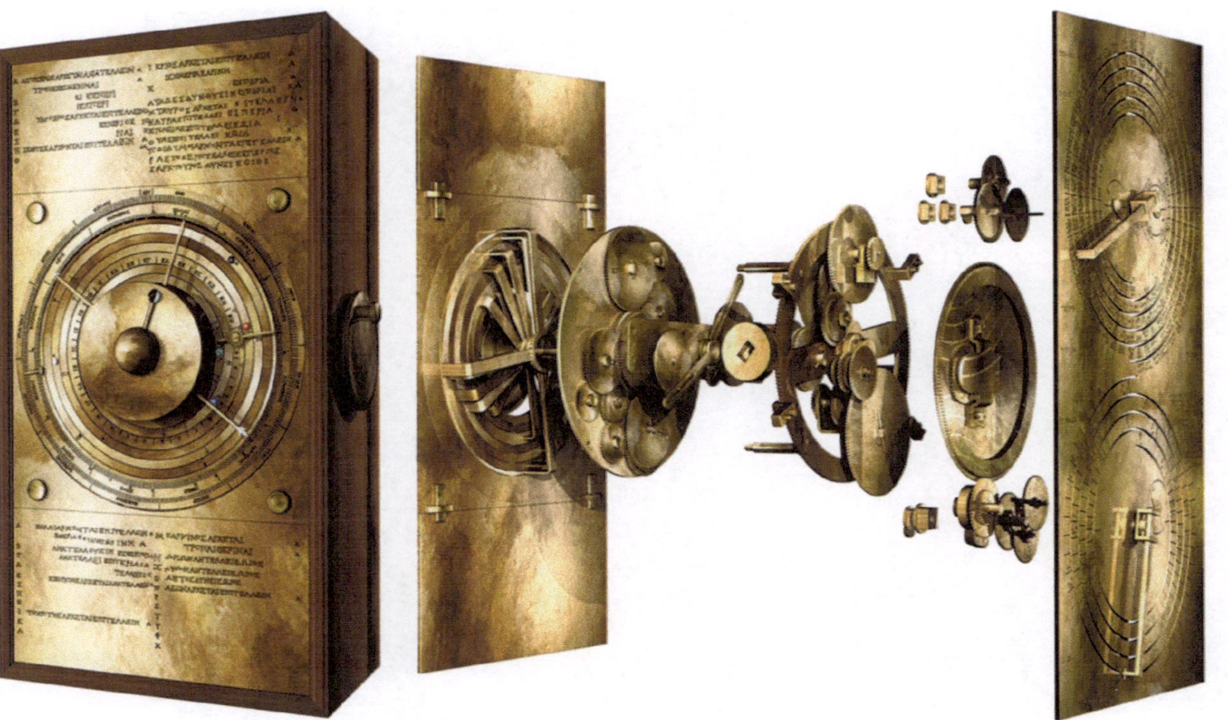

7. Exploded computer model of Antikythera mechanism. © 2020 Tony Freeth.

8. The Archimedes Palimpsest, folia 102r–98v: a diagram of a spiral, with an initial from the prayer book laid over it. Image produced by the Rochester Institute of Technology and Johns Hopkins University. Copyright courtesy of the owner of the Archimedes Palimpsest.

position relative to the horizon but to the angle between the sun, earth, and moon; we will return to this later in the chapter.)

To conclude, a few notes on the long-term significance of this widely shared type of astronomical knowledge.

First, throughout premodern times, this knowledge never goes away. It is not as if scientific astronomy replaces the star lore of the rising and setting of the sun and the stars, the waning and waxing of the moon. Star lore remains as a widely shared knowledge, taken for granted by more scientific astronomers. Sophisticated astronomy, in the premodern world, has, among other things, to provide an account of the pattern of key qualitative events: the southernmost position of the sun, the first risings of stars such as the Pleiades, the phases of the moon.

Second, such accounts are not trivial. Already, the sun and the moon create a major headache because, you see, the cycle of the month does not align with that of the year! Bringing together the patterns of key qualitative events is not merely a cultural requirement; it is also a difficult and interesting scientific puzzle. Very naturally, then, it forces itself upon premodern scientific astronomers.

We often imagine the astronomer as *the cartographer of the sky.* Full knowledge of astronomy, in this sense, would involve the ability to produce precise spherical coordinates for each of the visible objects in the sky and provide a numerically precise trajectory for each of their movements. Some Greek astronomers would, indeed, engage in a comparable project. But this is not the heart of premodern astronomy. Historically, the astronomer was not a cartographer, but instead, he was a *calendrical master.* The cartographer knows about positions and patterns in *space* – even the trajectories are essentially spatial entities that simply require time for their unfolding. But the calendrical master knows about repetitions in *time.* For him or her, even the awareness of position in the sky is necessary simply to identify the references anchoring the temporal cycles. Star lore, historically, is significant because it is the way in which time is naturalized and made part of a society's life. To understand ancient astronomy, we must understand it as a science not of space but of time.

### *Bring in the Bureaucrats*

Knowledge of the stars is not directly mathematical. It is an awareness of patterns in time. However, the basic unit is clearly defined – a single night. Clearly defined units make number come in naturally, and as states train elite scribes – used to counting taxes and inventories – it becomes natural for them to inventorize, so to speak, the events of the sky. At this point, what was once star lore becomes a calendar.

The Maya of Central America had, during the first millennium CE, a complex civilization of city-states, with – of course – a stratum of specialized priests who were also masters of a complicated literacy. I refer to them as "bureaucrats," but it should be clear that they did not approach their work with the impersonal objectivity that Weber, the German sociologist, saw as the core of a bureaucracy. Stars, to them, were objects of intense religious veneration. What marks off those priests is their highly literate, numerical, specialized approach. Ritual, in fact, is often bureaucratic-like – precise repetition in the fulfillment of one's tasks becoming a goal to itself. It is remarkable that this specific obsession with order emerges so often – together with state formation – around the stars.

There were, for instance, many books produced by the Maya, the vast majority now lost (many destroyed by uncomprehending Europeans, most destroyed by the humidity of the Central American jungle). Four are now extant and all are astronomical and, specifically, calendrical. A similar emphasis on astronomy and the calendar (although, unfortunately, with less elaboration than found in the books) is seen in Maya inscriptions on stone: clearly, then, this astronomy was central to their elite culture. This is, in fact, a fiendishly demanding system. What is at stake – we see the same across many civilizations – is the *naming of days*. The most basic format of labeling days had what to us appears, at first glance, as an arbitrary cycle with thirteen repetitions of a twenty-day sequence. Bear in mind, however, that Maya numeration used not a base-ten but a base-twenty system; hence, the twenty-day sequence came naturally. Note also that although this sequence of labeling days strikes us as arbitrary, we ourselves find it essential to keep records of a seven-day cycle – the week – a division around which we have structured a civilizational pulse, longer than the day but shorter than the month/year; the need for such pulses is shared by many complex civilizations. The Maya 260-day pattern aligns neither with the moon nor with the sun, but the Maya kept, alongside the 260-day-based year, another, more practical year with 365 days. It took considerable numerical ingenuity to align the two and assign the correct 260-day calendrical labels for given astronomical events. Masters of ritual, the Maya priests aligned together not just 260-day and 365-day cycles but many other cycles as well (for instance, a 9-day cycle – even more closely approximating our week – and on top of this, of course, the cycles of the moon). Many of the inscriptions contain a long list of the different names of a day, all the names sacred and the inscription becoming a kind of theological spreadsheet.

All this made even more sense to the Maya because they cared deeply not only for the sun and the moon but, no less, for planets: thus the 365-day cycle was just one among several that needed to be accommodated. In the Maya codices, we find detailed knowledge, for instance, of Venus and its cycle, which is 584 days long. (I will get to this later on; numerical everywhere, Maya priests must have been impressed by the fact that 584 × 5 = 365 × 8.) With this information, one can begin to build numerical systems that align a 584-day cycle, a 365-day cycle, and a 260-day cycle, which Maya priests developed and produced in richly illustrated tables, a thousand and more years ago – a complex art, now almost entirely lost. The art of the astronomer-priest is more sophisticated yet: they explicitly recognized the imprecision of the alignments and developed numerical rules for adjusting tables so as to maintain the correct naming.[3] This already becomes very challenging even for a modern reader with access to our easy-to-use writing and numbering systems. Maya writing is extremely hard – based on many glyphs, each of which strikes one, at first, as a brilliantly executed work of art. The point of Mayan scribal culture was not to make it easier to the outsiders – not in their writing, nor in their calendar.

Ancient China is better understood. Starting at 104 BCE, our grasp of official Chinese astronomy is very thorough, based on a continuous tradition kept at the courts. In that year, the imperial court of the Han dynasty brought together expert astronomers to reform the calendar (it is clear that different systems, now known only in fragmentary form, were introduced by earlier dynasties).[4] In this new system, the first day for the epoch was taken to be (what we call) December 24, 105 BCE. How is this the "first day"? It was the start of a lunar cycle (first day of the month) and solar cycle (first day of the year). For purposes of calendrical astronomy, a month was defined as 29 43/81 days; a year was defined, for the same purposes, as 365 385/1,539 days; so that 1,539 years were precisely equal to 19,035 months and to 56,150 days. One could say, then, that precisely 1,539 years after the beginning of the epoch, one would hit, once again, the same first day of the month and year. Because a separate system existed for labeling days in a somewhat arbitrary sequence of 60 days, each with its distinct label (this closely recalls the 20-day-based or 260-day-based system of the Maya), and because 56,150 does not have 3 among its factors, we need to multiply

[3] J. S. Justeson, "Ancient Maya Ethnoastronomy," in A. Aveni (ed.), *Foundations of New World Cultural Astronomy* (Boulder: University Press of Colorado, 2008), pp. 509–564.

[4] C. Cullen, "Motivations for Scientific Change in Ancient China: Emperor Wu and the Grand Inception Astronomical Reforms of 104 BC," *Journal for the History of Astronomy* 24 (1993): 185–203.

everything yet by 3 to get to 4,617 years, following which the first day of the month and year will recur, *with the very same label.* What this represented, first of all – as explicitly stated by emperors, who set such challenges for astronomers – was the serene expectation that dynasties should, and would, survive for millennia. But more than this: emperors were required to head rituals and follow omens based on the cycles of the sky; hence, one needed to be able to tell in advance, precisely, the regular repetitions and thus, also, to be able to notice (which Chinese astronomers were especially good at) when irregular events – meteors, comets, supernovae – took place, *as deviations from an expected norm.* More on this later in the chapter.

This becomes clear: as star lore becomes calendar and a group of practitioners becomes responsible for its maintenance, one does not merely keep in order the familiar observations preceding the bureaucratic state. Instead, one notices many more patterns and begins to arrange them in explicitly quantitative terms. Let us turn back and consider, then, such patterns of the sky – those that are not universally well known to all civilizations but become known as astronomy is picked up by the bureaucrats. (Once again, I apologize for going through facts that are well known to many of my readers.)

Now, we all know our solar system: just like Maya, or Han, bureaucrats, we were explicitly taught about it, beginning in elementary school. Let me point out something about the solar system that I am not sure elementary school does a good enough job explaining. I refer to the following fact: *the solar system is nearly flat.* Perhaps you think of the solar system like the model shown in Figure 76.

Figure 76 Vector model of an atom. Anna Iamanova iStock/Getty Images Plus.

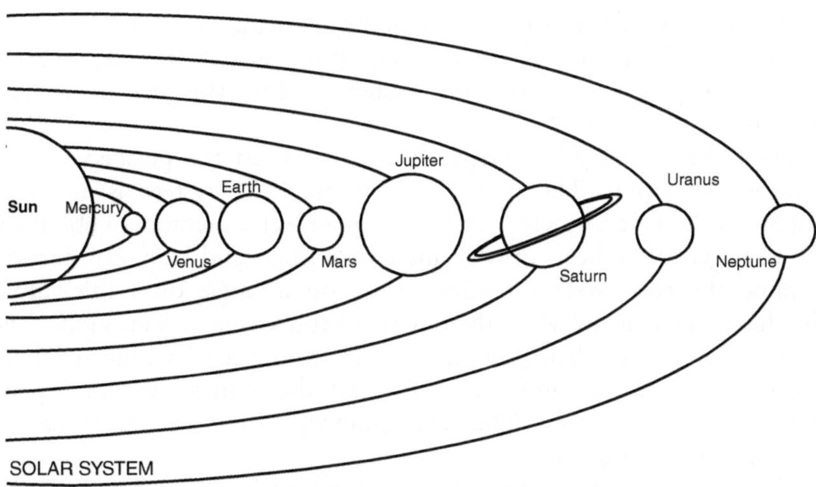


Figure 77 Solar system with sun and all planets (vector sketch). pijama61/DigitalVision Vectors/Getty Images.

That is, based on the familiar model of the orbits of the atom. This is indeed a system with a center surrounded by "planets" – and it is also a very *three-dimensional* system, the orbits occupying the entire space surrounding the center. The solar system, however, is not like this. It is often drawn as concentric rings, as shown in Figure 77.

And this is, in fact, not a schematic simplification but, instead, the correct way to draw it. The blobs of matter emitted out of the sun – planets and, eventually, our moon – were all ejected along the same path and now all reside within the confines of the same fairly thin disk. They are almost like the tracks on a vinyl record.

It also happens that the axis of the daily rotation of the earth is not orthogonal to the plane of this disk but is instead considerably oblique to it. This, indeed, you may perhaps recall from elementary school: the oblique angle at which the earth rotates has an important consequence. It means that if you are north of the equator, you spend, in summer, most of the twenty-four-hour rotation facing the sun (that is, in daylight), whereas in winter, you spend most of the twenty-four-hour rotation with your back to the sun (that is, in darkness). This is why summer (in the north) is hotter than winter and is also the reason why the sun rises at a more southern position in winter than in summer.

Had it not been for this obliquity, perhaps astronomy would not have mattered to humanity all that much. But we are, in fact, rotating obliquely to the plane of the solar system, and for this reason, we have seasons.

These are indeed all "modern" facts, told from the perspective of a heliocentric system. The near-flatness of the solar system is especially important, however, in terms of its consequences for premodern observers. Now we get to the heart of what matters to Maya and Han astronomers. Because the solar system resides mostly on a single disk, this means that from our point of view, the objects of this system – sun, moon, and planets – all change their positions (relative to the stars) within regulated boundaries. The sun and the moon and the planets do not go *just anywhere*, north or south. They are confined, instead, to a fairly narrow band, known as the *zodiac*.

To understand those terms, let us consider what is visible along this band. First of all, the sun. More specifically: it is across this band that we perceive the annual rotation of the sun. Every day, the sun rises in the east and sets in the west. And every day, it does so in a slightly different position on the band of the zodiac – changing its position upon it, each day, by roughly 1/365 the entire circle. (The daily and annual rotations are, as it happens, opposed: this annual rotation takes place from west to east.) Now, what do we mean by "the position of the sun on the zodiac"? What we mean, effectively, is the answer to the question: "What are the distant stars with which the sun is exactly aligned at a given night?" As noted earlier, the answer to this question cannot be given directly – the stars, aligning with the sun on a given night, have to be invisible. But if we gain a deep familiarity with those stars that lie along the band upon which the sun makes its annual rotation, then we can note, precisely, the *missing* stars – those that are invisible because, at this time of the year, they, so to speak, host the sun. It therefore makes sense to pay attention, in particular, to star constellations along this band – and to name them. The Babylonians ended up with twelve such named constellations, which then became the common heritage of the Near East and, slightly revised through Greek astronomy, are still with us. Based on these constellations, the Greeks ended up thinking of this band as constituted by a series of "small pictures" or *zōdia* – each constellation considered, so to speak, as a thumbnail work of art. Hence the name *zodiac.* When we say that "the sun is in Taurus," then, what we mean is that the sun aligns with the constellation known as Taurus or "passes through" Taurus – which happens today around May–June; this constellation is *not* visible at that time. (Note that those famous twelve constellations exhaust, between them, not the sky but merely the band of the zodiac; further, eventually, the meaning of a term such as *Taurus* will come to be not the star constellation but the precisely measured 1/12th of the zodiac in the general area of the constellation. In this more precise sense, the correct technical term is not *constellation* but *sign*.)

The moon, too, keeps moving along the same band – indeed, much more swiftly. If the sun returns – from our point of view – to the same position, relative to the fixed stars, every (roughly) 365 days, the moon does the same roughly every 27 days. Wait, you ask, should this rotation not take a *month*? Not quite. The universal idea of the month antedates measuring bureaucracies. It is based not on the more quantitative, measurable position of the moon along the band of its rotation but on the more qualitative phenomenon of the phases of the moon. This cycle, we now know (together with the Greeks, who figured that one out as soon as they had an astronomy), depends on the moon's optics. The moon's light is the reflection of its illumination by the sun. When the earth is near-exactly between the earth and the sun, the entire face of the moon visible from the earth is lit by the sun; hence, the moon appears full. When the moon is between the earth and the sun, it is its "back," the part turned away from us, that is illuminated, so that we barely see the moon at all. Because the sun and the moon both rotate in the same direction, this is a bit of a hare-and-tortoise situation. When the moon returns, after roughly twenty-seven days, to the same position relative to the fixed stars, the sun has already moved a little. It takes longer, then – twenty-nine days, roughly – for the moon to return to the same position *relative to the sun and the earth*. It is this three-body recurrence that we think of as a "month." It will be noted that when the moon is between the earth and the sun (that is, a new moon), it may obscure the sun from vision; when the earth is between the moon and the sun (that is, a full moon), the earth's shadow may fall upon the moon. We note the conceptual connection between the phenomena of the eclipse and the thin strip along which the annual and monthly rotations of the sun and the moon take place. The band is where – and why – eclipses take place. Hence the other name of that strip – the *ecliptic*. (Note that the terms *zodiac* and *ecliptic* are often used more technically: *ecliptic* referring to the one-dimensional line traced, throughout the year, by the center of the sun; *zodiac* referring to the two-dimensional band surrounding the ecliptic within which the moon and the planets roam.)

Now, in the foregoing I used the words *may* ("it *may* obscure the sun from vision"; "the earth's shadow *may* fall upon the moon"). This is because neither event is guaranteed because the solar system is not, in fact, exactly flat. Thus, even when the moon is just between the earth and the sun, it might be just a little too low, or a little too high, relative to the "disk" on which the earth and the moon rotate, to obscure the sun from vision. What we may conclude, then, is that a solar eclipse *may* occur during a new moon and at no other time; a lunar eclipse, during a full moon and at no other time. To be able to tell with greater precision whether or not an eclipse is likely to take place, we need a better understanding of the alignment of the overall cycles of the sun and the moon. These involve not only the cycle of the phases of the moon but also the cycle of the moon's position relative to the ecliptic: it is sometimes slightly above it, sometimes slightly beneath it, and it crosses the path of the sun – the ecliptic, strictly speaking – by regular intervals (which are, of course, much harder to detect and measure). This is already becoming quite technical and complicated. There is one pattern of motion of the moon relative to the sun, along the band of the ecliptic; this is the familiar cycle of the phases of the moon. But there is, as we now notice, a different motion, orthogonal to the ecliptic. Combine the two cycles, and you are able to more or less predict eclipses. Now, if you are a bureaucrat-astronomer, your basic duty as a calendar master is to combine cycles, and it is therefore natural that you will look for such devices.

Chinese and Babylonian astronomers came, independently, to recognize certain patterns that bring together all the motions of the moon, relative to the sun. All such cycles, stated in simple integer terms, are approximations, and so there is more than a single "correct" solution. Probably the best is the one referred to by historians of astronomy as the *Saros*: every 223 months, almost exactly 18 years elapse; more significantly (in terms of predicting eclipses), the moon returns almost to its precise same position above, below, or precisely on the ecliptic. Thus, if there was an eclipse of the moon on a certain day, it is very likely that such an eclipse will recur a Saros later.

In a sense, every calendar is a prediction. When I use a 365-day year, I make a – very rough – prediction that after 365 days, the sun will return to the same position in the pendulum of its south–north rising. This is not exactly right (the year is not exactly 365 days long; also, the motion of the sun is not uniform – although, over a 365-day cycle, average speed matters much more than its variation). But this is good enough for the purposes of a simple society making sense of its environment. Predicting eclipses differs by a matter of degree: it takes much more empirical work to verify and refine the prediction, and it is much less certain. (The prediction of solar eclipses, in particular, is far from guaranteed, even with the Saros cycle, because solar eclipses are visible only from a certain part of the earth.) Because of the imprecise alignment of the various astronomical cycles such as day, month, and year, the astronomical predictions of key qualitative events were always a matter of some uncertainty. The new moon should arrive after twenty-nine days, but sometimes it will have to wait till the thirtieth day. If you assume, year after year, that the longest day of the summer occurs every 365 days, soon enough, you will notice that you picked the wrong day. From the point of view of early observers, it is obvious that astronomy is not a matter of mere counting but also involves a certain art: a willingness to accommodate the failing of predictions.

One requires a special art – and this is just for the sun and the moon! The planets bring in yet another dimension of complexity. The human eye notices five more sources of light that keep changing their position along the zodiac: Mercury and Venus are "inferior" planets. (From the modern perspective, they are closer to the sun than the earth is. This means, in practice, they are never seen by us to be far from the sun because their circuit confines them to what, to us, appears as a region not far from the sun.) Mars, Jupiter, and Saturn are "superior" planets. (They are farther away from the sun than the earth, and so they do not necessarily appear close to the sun.)

These, too, keep changing their position along the zodiac. Look for Mars over consecutive nights, and you will likely see it surrounded by different fixed stars, eventually passing from constellation to constellation, all in the general direction of the annual rotation of the sun. However, because we understand that this apparent motion is really a composite of two distinct motions, it is not surprising for us to learn that its speed keeps changing. To be clear, this is also, in part, because the planets – just like the sun and the moon – move at nonuniform speeds and because they move along ellipses rather than circles. However, those effects are modest: the ellipses of the planets are near-circular, and their speeds do not vary dramatically. Most of the apparent nonuniformity of the speeds of the planets is due not the inherent nonuniformity of the motion but to the complex combination of the two motions, the earth's and the planet's. Sometimes the speeds of the earth and of the planet are in opposite

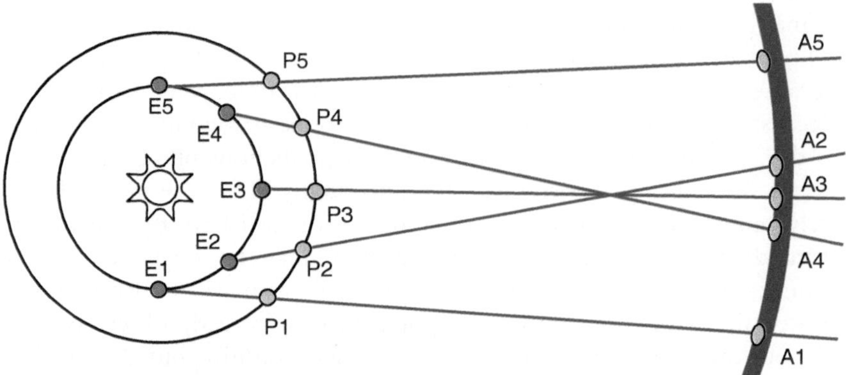


Figure 78 Retrograde motion. Brian Brondel, CC BY-SA 3.0 via Wikimedia Commons.

directions, causing the motion to appear faster; sometimes they are in the same direction, causing the motion to appear slower. In fact, sometimes the speeds cancel each other out so that the planet stops moving – on two consecutive nights, that planet appears surrounded by exactly the same distant stars (this is known technically as a *station*). Or the motion of the planet can appear to be "negative" (this is technically known as a *retrogradation*; see Figure 78). During a retrogradation, the position of the planet on consecutive nights does change, but it changes in the opposite direction to the annual rotation of the sun. Naturally enough, there is a pattern regulating such events: normal motion, followed by station, then (perhaps) retrogradation, then station again, and then normal motion resumes.

What would you make of this, as a premodern observer? Perhaps you would give up in despair and just describe the planets as "wanderers," implying that they move more or less haphazardly. (This, "wanderer," is the literal meaning of the Greek word *planet*.) But it is also natural to look for such patterns as can be found. We recall, indeed, that the Maya knew the cycle of Venus. What does that mean? The planets, in their motion, are sometimes farther away from the sun, sometimes close to it, sometimes aligned with it. This means that, just like the fixed stars, the planets have their pattern of disappearance and reappearance: there will be nights when a planet becomes invisible – followed by the first night of its rising in the east. The Maya recognized that this Venus event recurs every 584 days, and we now refer to such cycles as *synodic cycles*. Indeed, the planets go through a number of key qualitative events throughout their cycle – not only the first rising just before sunrise but, for instance, the first station (preceding retrogradation), the second station (before resuming "normal" motion), the last appearance just before sunset . . . Each planet goes, over its cycle, through a complex narrative, which ends up being recorded and predicted by star bureaucracies.

Although the average speeds of the earth and planets are – for a given astronomical era – indeed fixed, each planetary motion is nonuniform, and such combinations can quickly become very difficult. This matters little if one is interested purely in counting days between major events, for which the synodic cycle is a good enough approximation (on the lines of approximating the year by 365 days). But we have already moved to the realm of the bureaucrats, and there is often much more information that one gathers. The life cycle of planets is complicated. Some events, as usual, are related to the horizon, such as the first rising just before sunrise. But others – such as the planet's stations and retrogradations – are not related to the horizon at all. This makes it possible to begin to think of a planet's life cycle not just in terms of the horizon but also in terms of the zodiac. Thus, one naturally becomes interested in *the zodiacal position of a planet at a given event*.

Suppose, for instance, Venus was in Taurus during its most recent first rising. In which constellation will it be during its *next* first rising? Notice that this demands some kind of measurement of position not merely along the horizon but along an imaginary circle, passing through the dome of the sky. Remarkable as this is, Babylonians evidently cared for such measurements. Eventually, indeed, the Babylonians would develop sophisticated arithmetical models for the prediction of the position of planetary events. But it is clear that they were interested in such events from very early on indeed. The tablets we now call the Mul.Apin (see Plate 5) are a set of copies, clearly dated to the early seventh century BCE – that is, well before the rise of anything such as Greek science or philosophy. These tablets preserve knowledge that was developed in Mesopotamian temples over the course of many preceding centuries.

Babylonians, it turns out, recorded synodic periods for the planets – although some of the periods are very rough indeed. (For Jupiter, for instance, the Mul.Apin proposes various synodic periods: a year, a year and 20 days, or a year and a month; the correct value is 399 days. We see similarly rough measures, incidentally, in China.) But this was all good enough for the Mul.Apin bureaucrats. They were not mere observers of the sky but court astrologers: their job was, precisely, to note what was *surprising* – hence meaningful and literally ominous – about the stars. We are reminded of Chinese astronomers and their fascination with meteors, comets, and supernovae. The role of the predictive scheme, in this context, is to produce a background of expectations, *against which observations gain meaning*. You expect the planet to have its first morning appearance at a certain constellation, on a certain day – but in fact, this only happens a week later, at a constellation far, far away. And this unexpectedness constitutes an *omen*.

This seems to be true, then, whether you are in ancient Mesoamerica, China, or Mesopotamia. The events of the sky are regular enough that they are taken to constitute the very fabric of time. Hence, authority – especially as it combines political and religious dimensions – calls for knowledge of such regular patterns. But more than this: the events of the sky have sufficient apparent irregularity that, by the very markedness of the deviation of the irregular against the regular, such events are invested with powerful meaning. Rulers need to interpret this meaning, as well. Hence the astrologer – in court.

### *Prelude: The Threshold of Greek Astronomy*

The Greek trajectory leading into astronomy is – as in so much else – distinct. The Greeks had, to begin with, no powerful kingly courts or temples, no bureaucracies. Sophisticated astronomy would have to come from elsewhere. It did so, however, against the background described in the preceding two sections. The Greeks had their star lore, and the Mediterranean already had its star bureaucracies elsewhere.

Hesiod, after all, referred to the Pleiades as the sign for the harvest – and Greeks would always structure time based on such patterns. However mathematical Greek astronomy becomes, Greek scientific astronomers – like their counterparts from other early civilizations – considered the stars, above all, in terms of their key, recurring qualitative events.

Further, as we noted previously for mathematics itself, bureaucratic knowledge is leaky. The observational diaries kept by Mesopotamian astronomers were the property of temples and courts, but the output of their work could become visible in public calendrical practice and in the very language in which one referred to the stars and their phenomena. Certain calendars, the division of the zodiac into twelve equal constellations: such tools could be available, so to speak, "off the shelf."

Finally, at a later stage of the development of Greek astronomy, Greeks became aware of the working details of Mesopotamian astronomy. Most likely, this happened as Mesopotamian astronomers began to practice their craft away from the temple setting – and throughout the Levant and the Mediterranean – creating, along the way, the profession of the astrologer. The resulting science was a synthesis, then, of two traditions, Greek and Mesopotamian.

And yet this must be emphasized: Greek astronomy was also different. Let us note the earliest evidence. Aristophanes's delightful comedy the *Birds* was staged in 414 BCE. In it, two Athenians seek to escape from their city and set up Cloudcuckooland, a peaceful city in the sky (a contrast to Athens and its endless war with Sparta). Various visitors join them with advice for the foundation of the new city, including a certain Meton, who proclaims that he is famous in all of Greece and Colonos (a suburb of Athens – implying that his fame, however much he otherwise protests, is rather confined). This Meton offers to measure the city with his ruler and compass. A sky geometer! In fact, we hear more from later sources; not one but two Athenians of the time – Meton as well as a certain Euctemon – are noted for several astronomical achievements: promulgating calendars, observing the sky, and setting up public astronomical devices and inscriptions. We hear of what is known as a *parapegma*, a stone where holes are drilled, one for each day, with astronomical and seasonal observations next to some of those holes. At the most basic level, one operates with such an inscription by inserting a peg and moving it forward once a day. As one goes along, one reviews the key qualitative events of the sun and the fixed stars, as well as the more meteorological, seasonal predictions: star lore, inscribed as a public document. Note, however, that the very literacy of this monument transforms such star lore. You cover the entirety of the year, deliberately, and so, quite naturally, Greek parapegmata eventually came to include much more information than was available to, say, Hesiod. (Indeed, it now seems that parapgemata could involve much more than one peg a day, for instance, by adding in extra, fancier pegs representing, for example, the phases of the moon; this already suggests future astronomical devices, of which, much more later in the chapter.[5])

[5] For this fuller understanding of the ancient parapegma, see G. Bevan, A. Jones, and D. Lehoux, "The Miletus Parapegma and the Keskintos Astronomical Inscription: New Evidence from Reflectance Transformation Imaging," *Zeitschrift für Papyrologie und Epigraphik* 212 (2019): 137–146.

The later reports concerning Meton mention that he knew of a calendrical cycle stating the nearly precise equality of 19 years = 235 months. This, today, is referred to as the *Metonic cycle*, so that Meton, at the end, did, in fact, become famous. In all likelihood, Meton knew this simply because this was, at this point, part of the Mediterranean koine – after all, this is the basic cycle of the Babylonian calendar (and is still used in the Jewish calendar; although the Saros is especially good for predicting eclipses, the Metonic cycle is good for preserving the alignment of years and months). The Metonic cycle, – like, I believe, Pythagoras's theorem – is what many people in the Mediterranean simply *knew* back then and we ended up naming after a Greek author.

At about the same time as Meton and Euctemon, we also hear of Oenopides of Chios, perhaps producing observations of the sky and perhaps setting up sundials. (Or perhaps Meton and Euctemon set up sundials as well? Either way, the public display of scientific artifacts is a theme of Greek astronomy from its inception; more on this later in the chapter.) Astronomical or, more precisely, cosmological views are ascribed to Hippocrates of Chios and to his otherwise-obscure follower, a certain Aeschylus (to recall: not the tragedian!). Anaxagoras, a philosopher famous for many other ideas of a "materialistic" character, appears to have been the first to explain in writing the optical basis of eclipses (he did this, remarkably, with the aid of geometrical diagrams).[6] Even earlier – near the beginning of the fifth century – Parmenides argued for the claim that the earth was spherical. Greek astronomy begins to stir.

But notice where it is, at this point: continuous with the evidence for past civilizations. A parapegma is the traditional society's star lore, enshrined in stone. Otherwise, it seems that to the extent that there was any interest in the mathematical aspects of astronomy, this had to do mostly with the calendar. Those are mere leftovers from the table of past star bureaucracies.

Several things, however, are new. The most obvious is that Meton, at the end of the day, did *not* work for the state. In fact, whatever calendar he promulgated, this did not become, in his lifetime, the state calendar of Athens or of any other state. He is made, by Aristophanes, to offer his services to the nascent city of Cloudcuckooland, but the point is, precisely, that he is too impractical even for the birds. This stands in marked contrast

[6] The evidence for Anaxagoras's astronomy, including his theory of the eclipse, is Hyppolytus, *Refutation of All Heresies* I.8.6–I.8.10. Diogenes Laërtius II.11 reports that Anaxagoras was the first to write with a diagram; it is no more than a likely hypothesis that the two went together.

to everything we saw in all other civilizations that had produced sophisticated astronomy. For if not the state's, whose interests *did* Meton pursue? Clearly, his own. What he produced was *Meton's* calendar, *Meton's* parapegma. Aristophanes made the figure of Meton appear, mocked, in front of the full theater. But this was Meton's aim from the start: he sought attention. He set up a public inscription, after all. It is for this reason that we refer to "Pythagoras's theorem" and also to the "Metonic cycle": not because the Greeks invented the theorem or the cycle but because they invented the *habit of naming*. Astronomy was now pursued as a field of authorship, and one went into it with the aim of producing something new and original, something that would live in memory.

This, then, opens up the possibility of a new kind of astronomy: a field of more speculative science. Soon enough, this will be pursued by Greek mathematicians.

### *The First Generation of Greek Astronomy: The Extent of Our Knowledge*

It does not appear as if Meton broke any new ground, relative to the knowledge already available in the ancient Near East. And so, the first green shoots of Greek astronomy differ from those of Greek mathematics. Hippocrates of Chios, with his lunules, already made something entirely new. Recall our observation that, of course, *he had to*: no one would have been interested in boring, mundane mathematics. Astronomy is different. Everyone cared about it already: produce a parapegma, and you are bound to get an audience.

A uniquely Greek astronomy does emerge, eventually: within the first generation of Greek mathematics and late even by these standards, championed by the last great figure of that generation, Eudoxus. Astronomy – as mathematics' echo. We will see more of that later on in the chapter.

The evidence for Eudoxus's astronomy is early, clear, and tantalizingly insufficient. His original works are now all lost – this is typical for his generation. But we have a clear sense of a wide astronomical program.

To begin with, Eudoxus is a key figure at the background of two extant works: (1) Aratus's poem the *Phaenomena* and (2) Hipparchus's critique of that poem (this critique is often referred to, misleadingly, as Hipparchus's "commentary to Aratus"). Aratus, a major poet of the third century BCE, wrote a poem that presented the elements of the celestial globe, especially the fixed constellations (but *avoiding* the planets). You recall how Hesiod – one of the canonical poets of Archaic Greece – knew about the Pleiades. We know this from a major poem called *Works and Days* that recounts information such as, for instance, the astronomical beat of the seasons. It is typical of Hellenistic poetry to go back to Archaic poetry and to retell it in new, clever combinations, and it appears that Aratus, among other things, retold Hesiod – but based on more recent, "scientific" astronomy! Specifically, Aratus's poem was apparently based on a scientific treatise in prose by Eudoxus – or perhaps more than one – which set out to describe the fixed stars. The terms in which Aratus presents his facts are vague and almost entirely qualitative, but Hipparchus nonetheless criticizes them as if they were precise descriptions implying precise numerical values. This is sometimes put forward by Hipparchus as a critique of Aratus, sometimes of his source, Eudoxus. The overall implication is that Eudoxus – unlike past astronomers in past civilizations – did engage in that "geographical" kind of astronomy where one's goal was to produce a map of the sky. Perhaps we should imagine a three-dimensional model, described in writing. And so, once again, the astronomer as the artificer of monuments, in this case a globe?

If indeed Eudoxus tried to provide some kind of quantitative mapping of the stars, he did something quite unprecedented. But this may well have been – quite literally – merely background for his real innovation, having to do not with the fixed constellations but with the planets.

Here is where our evidence is most authoritative – and most tantalizing. We have a direct testimony from Aristotle (and also much more detailed reports – although much less reliable – by late ancient commentators to Aristotle). We also have what appear to be references to the same system from Plato and from Epicurus. The leading philosophers of the day all noted Eudoxus's astronomy. This is all excitingly early evidence (the reference from Plato, indeed, must have come when Eudoxus *was still alive*). (In later depictions, Plato could be envisaged gazing at a model of the stars – see Plate 6 – perhaps that of Eudoxus?).

In Book Λ, Chapter 8, of the *Metaphysics*, Aristotle provides us with his own account of motions in the heavens (this is quite significant to Aristotle, as such motions are in a sense, for him, a manifestation of the divine and, relatedly, the ultimate explanation for physical reality as a whole). Aristotle explicitly cites his sources: he follows an account proposed by Eudoxus with what seems like an improvement by Callippus (apparently, a follower of Eudoxus), and finally, he himself, Aristotle, adds some original detail. This is remarkable: in his astronomy, Aristotle was content to be, effectively, Eudoxus's disciple. There are no obvious parallels to Aristotle relying so directly on another person, and the implication is that for many in this generation, Eudoxus's astronomy appeared authoritative.

Here, then, is what Aristotle learned from his master:[7]

> Eudoxus held that the motion of the sun and moon involves in either case three spheres, of which the outermost is that of the fixed stars, the second revolves in the circle which bisects the zodiac, and the third revolves in a circle which is inclined across the breadth of the zodiac. But the circle in which the moon moves is inclined at a greater angle than that in which the sun moves.

We assume that the fixed stars constitute a large hollow sphere – a black shell on which bright dots are inset. This sphere rotates around the axis of the universe, which passes through the earth's North and South Poles. (The center of this sphere is the same as that of the earth; this is technically known as being *homocentric* with the earth.) This rotation takes place from east to west, at uniform speed. This is the sphere responsible for our experience of seeing the stars rotating through the night, and this is all uncontroversial: practically all Greeks, at this point, assume that, and indeed, this is a very natural way of describing the nightly progression of the fixed stars.

The idea is to derive everything else in the sky through the insertion of more homocentric spheres, each determined by the following:

- The position of its poles
- The direction of its rotation
- The speed of its rotation (assumed to be uniform)

Specifically, the idea is to rely on a combination of such spheres. One sphere is inset within another so that as the outer sphere rotates, it carries the inner one, which also has its own independent motion. The two motions end up combining to produce a distinct, more complicated pattern. The pattern of inset spheres is a physical (or mechanical) representation of what is, geometrically, a kind of parallelogram of forces: the geometrical solution to the problem of what happens when two motions combine. (We have seen this kind of curve-producing combination of motions with Archimedes's spirals; the idea may well have originated in astronomy.)

We can envisage, then, a sphere such as that of the fixed stars. Within it and attached to it is another sphere, rotating around an oblique pole, whose angle is determined by the obliquity of the ecliptic. This second

[7] The quotations here are from Aristotle's *Metaphysics* L8: H. Tredennick (trans.), *Aristotle: Metaphysics Books* X–XIV (Cambridge, MA: Harvard University Press, 1962), pp. 157–159.

rotation is not daily but annual, and it moves in the opposite direction to the daily rotation. Fix the sun on the equator of this inner sphere, and you get something like the apparent motion of the sun. Meanwhile, imagine another two-sphere combination, once again with the first sphere with the daily rotation of the fixed stars while the inner sphere, in this case, is arranged obliquely as that of the sun – but with a *monthly* rotation in the opposite direction. This gets us close to the moon. To both sets, however, we add one extra inner sphere, rotating very slowly and somewhat obliquely to the zodiac. There are several reasons why we would like to add at least one more sphere for the sun and the moon. First, the assumption of a uniform motion through the ecliptic is quite evidently false. We did not mention this yet, but it is a well-known part of basic star lore – the kind the Greeks enshrined in parapegmata – that the seasons vary in their lengths. The motion of the sun, in other words, is not exactly uniform (more on this later in the discussion). Further, the moon regularly deviates – not by much – from the ecliptic, which, as we noted, is why eclipses do not happen each month. (One should note immediately that the moon has, in fact, many other complications: the one object in the sky that is *actually* geocentric is also the hardest to reduce to simple circular motions, which we understand as the consequence of being under the influence of not one but two significant gravities.)

So far, then, for the sun and the moon. That they can be accounted for by uniform spherical motions is satisfying and ultimately not all that counterintuitive because their motion does appear, as a first approximation, like a regular rotation. One would think that the motion of the planets, with their stations and retrogradations, cannot be accounted for similarly, and Eudoxus's key insight must have been to realize that this impression is wrong. It is possible, in principle, to combine regular, uniform motions so as to derive back-and-forth patterns. I return to the same quote from Aristotle:

> And he held that the motion of the planets involved in each case four spheres; and that of these the first and second are the same as before (for the sphere of the fixed stars is that which carries round all the other spheres, and the sphere next in order, which has its motion in the circle which bisects the zodiac, is common to all the planets); the third sphere of all the planets has its poles in the circle which bisects the zodiac; and the fourth sphere moves in the circle inclined to the equator of the third.

All the planets move (at various speeds) along the zodiac, in the same direction as the sun. For this, we need, once again, the two basic spheres of the daily and annual rotations. If we had only those two spheres, however, the planets would simply imitate the motion of the sun, although on different locations on the zodiac. But in fact, the planets change their positions relative to the sun, in complicated patterns. To account for this, Eudoxus added two more spheres.

What these did is left out by Aristotle (his readers could consult Eudoxus directly). This provided an opportunity for later commentators to display their erudition, and fortunately for us, one of them did and is still extant: Simplicius, writing in the sixth century CE. (This is the same author from whom we have the lunules of Hippocrates of Chios; we rely a great deal on the prolixity, and competitive erudition, of this author.) It is clear that Simplicius did not simply make up his account, although it is also evident that he does not rely directly on Eudoxus's original work and instead gleans his own information from other commentators; it is also clear that he does not understand the underlying astronomy very well. (This need not be taken as a major criticism because the material is indeed very difficult and also *bore no relation to the astronomy known in Simplicius's own time*; more on this later in the discussion.) Still, some details are suggestive; perhaps the most useful piece of information is that Simplicius informs us that the third and fourth spheres, combined, produce a particular kind of curve, which Eudoxus called the *hippopede* ("horse-fetter"). From other references to the term, it seems probable that the hippopede is produced by the intersection of a cylinder and a sphere internally touching each other, that is, an inclined 8-shaped figure. The Greeks did not have the symbol "8," but they did tie their horses by binding their feet together, which will naturally give rise to two contiguous closed curves. So, in all likelihood, the hippopede is an 8 shape, and the combination of the two spheres specific to the planets was supposed to generate motion along an 8-shaped curve.

If the 8-shaped figure is very "narrow" and close to a single line, a point moving along it essentially moves back and forth. One way of applying this to the motion of the planets is by superimposing an 8-shaped curve, directly, on a motion similar to that of the sun. This means that you get a planet that moves generally as the sun but changes its speed, sometimes faster, sometimes slower; with the right parameters, it can be made slow enough so as not to move at all or even to move "backward" altogether. We derive stations and retrogradations. The 8 shape naturally forces the planet to deviate from the ecliptic, which the planets do, of course, even if to a minor extent. Four homocentric spheres, each rotating at its own uniform speed – and their combination gives rise to the entire range of nonuniformity of the planets!

Plato insisted on the ultimate rationality of an order of reality above our own and perhaps was attracted, for this reason, to the idea that even the planets, ultimately, were not actually "wanderers." In Plato's last dialogue, the *Laws* – so, in about the year 350 – Plato has his "Athenian Stranger" (more or less, Plato's mouthpiece) complain that it is believed that the sun, the moon, and the planets do not keep the same motion, whereas, in fact, it is now known that they do.[8] It is very hard to believe this could refer to anything other than Eudoxus's theory – which, at around 350 BCE, must have been very recent. As promised: not only Aristotle but also Plato referred to this theory! And this last reference, by Plato, is suggestive of the theory's possible motivations.

But what were the theory's mathematical details? You will note that I do not give here a geometrical account of how the combination of spheres gives rise to an 8-shaped figure. In the Suggestions for Further Reading, I refer you to articles that provide such derivations, but the fundamental point is that I can refer you to multiple papers, each deriving different motions through different *types* of combinations. This is, in fact, a new development. The first modern full reconstruction of a homocentric spheres model was offered by an Italian astronomer, Schiaparelli, in 1877. In his model, the inner two spheres characterizing each planet (or spheres three and four, counting from the outermost sphere) were arranged as follows: the pole of sphere three on the ecliptic, the pole of sphere four oblique to it, the two rotating in equal and opposite speeds; the planet was fixed on the equator of sphere four. This specific combination had two distinct advantages. First, it hewed very close to Simplicius's description. Second, it provided a model that was sufficiently simple, in geometrical terms, so that one could, in principle, calculate the positions at given times.

Unfortunately, Simplicius is not a reliable source, nor can we assume that Eudoxus's model was designed to be geometrically calculated. Here is the crux: the many possible reconstructions differ not just in their geometrical detail but also in their implications for Eudoxus's very project. Which is why, indeed, our evidence is so tantalizing: we are told so much by Aristotle and his commentators – and yet, we remain profoundly ignorant about what it even meant, for Eudoxus, to invent mathematical astronomy.

[8] Plato's *Laws* 821a–822c.

**How to Fetter a Horse?**

The horse-fetter is produced by two homocentric spheres, one nested within the other, rotating around distinct poles, at equal but opposite speeds. Why does this combination of rotations yield just this figure? We do not have Eudoxus's proof (indeed, it is a historical inference even that this was the shape of the hippopede, and that this is how it was produced). It is useful, however, to get a sense of how such a result could have been proved by Greek means. The proof that follows is the one suggested by Neugebauer. Not only is it an impressive piece of mathematics, but it is also tantalizingly reminiscent of Archytas's finding of two mean proportionals; one is tempted to say this must have been Eudoxus's original proof!

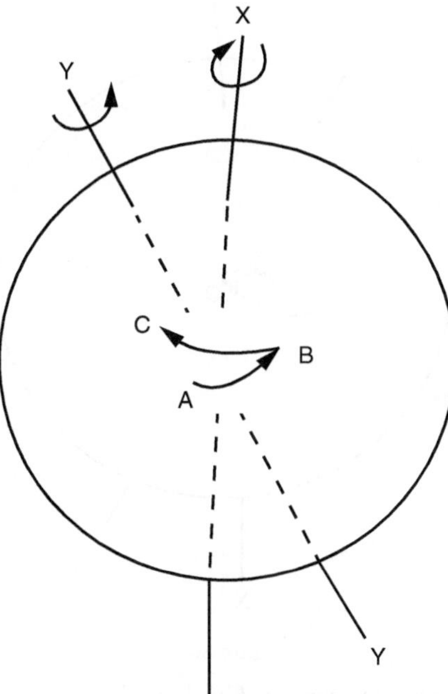


Figure 79

The two rotations happen simultaneously, but we can find the position of the planet at the end of the rotation by considering them in order. As the planet rotates around the axis Y, it moves from point A to B; as it rotates around the axis X, it moves from B to C. With the two combined, then, it ends up moving from A to C. The only thing we need to know at this point is that for each time frame chosen, because the speeds are equal, the two arcs are equal as well, each on its own circle. (We assume at this point that the circles are great circles and that the planet is located on the equator of the relevant sphere.) Arc AB, then, is equal to arc AC.

It is convenient, for the sake of the following argument, to further subdivide the motion around the axis Y into two components. First, imagine that the planet moved around the axis X, starting from A but getting not to B but to Q (once again, AQ = AB = BC); then, rotate the entire system by rotating the axis X until it coincides with the axis Y. The point Q would then rotate along a *small* circle, reaching from point Q to, necessarily, point B. The motion from Q to B is reminiscent of motions we have seen in Archytas's finding of two mean proportionals. We take point Q and draw a small circle, passing parallel to the plane that joins the two axes X and Y. This circle is also perpendicular to the equator of the sphere whose axis is X. Note that point Q is the "variable" of this construction – representing, effectively, the time frame chosen for the rotation. From now on, all points are either independently fixed or determined by the choice of Q.

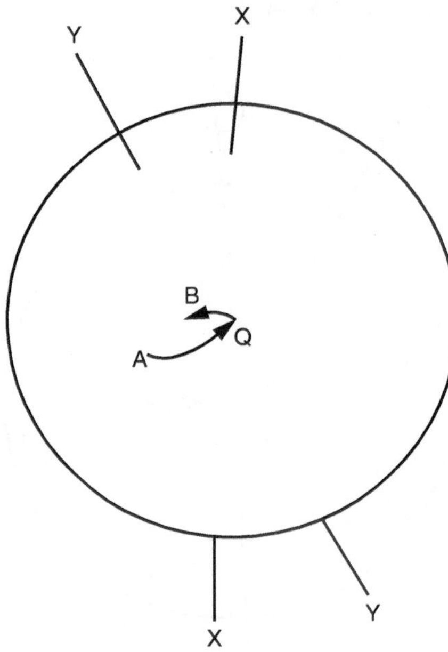


Figure 80

In what follows, we consider – as Greek mathematicians often preferred to do – plane figures, easy to diagram. One is the plane of the equator around the axis X. The other is the plane – orthogonal to this equator – that joins the two axes X and Y. In both diagrams, O is the center of the sphere (= circle in the plane section). OR is orthogonal to OA.

From now on, instead of thinking of the actual position of the planet, we think about "its projection onto the equator" – that is, the point that is right beneath where the planet is (the point where a perpendicular, drawn from the planet down to the plane of this equator, hits the equator). We will refer to this projection point – in a very modern symbolism – with the prime mark. B′ is the projection of point B; C′ is the projection of point C (points A and Q, obviously, lie on the equator itself).

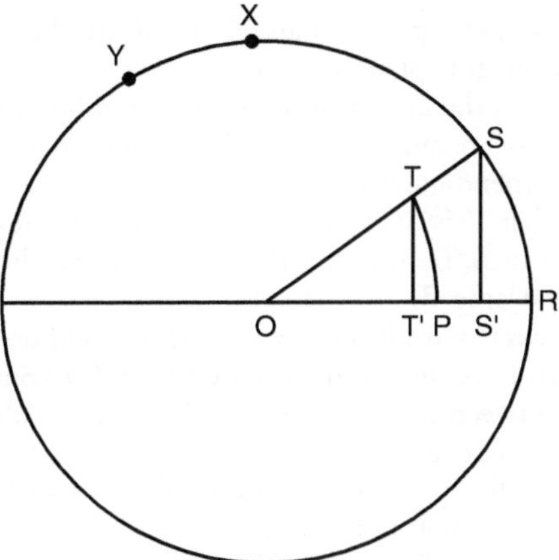


Figure 81 The plane joining the two axes.

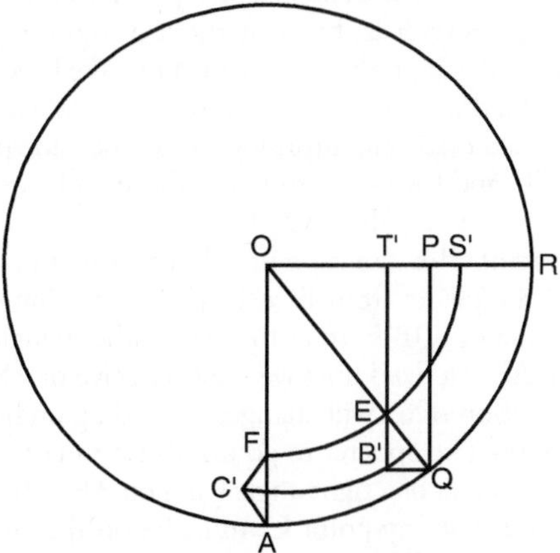


Figure 82 The plane of the equator.

You recall that we had obtained point B from point Q by rotating the entire sphere so that the axis X will become the axis Y. The same rotation takes line OR and brings it to line OS in Figure 81. In short, the angle SOR is equal to the angle between the two axes. And so, in this rotation, R becomes S; S′ is the projection of S on the plane of the equator (which is obviously seen here on the line OT′PS′R). What are the points T′, T, and P, you are asking? Well, let us look at the next diagram, of the equator, shown in Figure 82.

We know the meaning of the points A, Q, B′, and C′: they represent the journey of the planet through its rotations. (It gets from A to Q, then from

Q to B, then from B to C; points B and C are above the equator, so we only see their projection on this plane.)

Point O, to recall, is the center of the sphere, and line OR is perpendicular to OA. OR is thus also parallel to line QB′ (recall that the circle along which Q rotates to B is perpendicular to OA).

We also know what S′ is: it is the projection on the equator plane of point S, which we get, in turn, by rotating R upward, in exactly the same rotation through which Q became B.

Now we draw a circle with its center at O and its radius equal to OS′. This hits line OA at point F. Note that the rotation from R to S is fixed (the angle is equal to the angle between the two axes, which is a fixed value, independent of the frame of time we choose for our particular rotation). Thus, point S is fixed, as is point S′, and so also point F. No matter for how little or long we rotate the two spheres, point F will not budge.

We may now continue. We draw B′T′ and QP, both perpendicular to B′Q. We know now what P and T′ are, and we need reflect but a minute to realize that the rotation of P to T is the same as the rotation of Q to B. Hence, T′ is indeed the projection of T. Further, E is the point where B′T′ cuts OQ. Notice that the triangle QB′E is, by construction, right-angled.

So far, really, all we did was produce a construction. We have a long, complicated thought experiment of a three-dimensional arrangement. What we want to prove is a very straightforward, although still surprising, claim: the same point E – defined by the intersection of B′T′ and OQ – is also where the arc S′F cuts the same line OQ (that is, we need to prove that OE = OS′).

Why is this important? Because if point E lies on this arc, we may conceive of the final rotation of the planet from B to C as also carrying the entire triangle projected onto the equator, QB′E, so as to become the triangle AC′F. Recall that F is a fixed point, so AF is a fixed line; we may conceive of AF as the diameter of a fixed circle. The position of C′ will change, depending on how little or long we wish the rotations of the two spheres to go on. But no matter where we pick C′, it will always be at the vertex of a right angle lying on AF – the diameter of a fixed circle – and so we find that any point C′ must be on the circumference of the circle whose diameter is AF. This in turn means that point C must be on a cylinder, erected on top of the circle whose diameter is AF. And this means that point C – the result of combining the rotations of the two spheres – must lie on the intersection of a cylinder and a sphere, mutually tangent at a point, with the cylinder contained in the sphere – which is obviously 8-shaped.

At this point you must be eager to find out why OE = OS′. Well, we have quite a few similar triangles to play with (this is very Archytean). Obviously, by similar triangles in the plane of the equator:

$$OT' : OP :: OE : OQ.$$

But also, by similar triangles in the plane of the two axes:

$$OT' : OT :: OS' : OS.$$

And because OP = OT (the same line, rotated), we have:

$$OT' : OT :: OT' : OP.$$

So that we can establish

$$OS' : OS :: OT' : OP.$$

Which, combined with the first proportion, yields

$$OS' : OS :: OE : OQ.$$

But OS and OQ are both radii of the sphere; they are equal to each other, and therefore, so must be OS′ and OE. Thus, the arc OS′, rotated, must cut the line OQ at point E, and the horse is fettered.

Note finally that the precise position of point C′ is given by the angle AFC′, same as QEB or QOA. In other words, we find that as the radius sweeps the circle, point C′ keeps rotating on its circle, proving that the figure in fact covers the entire hippopede and also suggesting how, under the most simple interpretation of the hippopede, one may use the time (which translates directly to the angle QOA) to find the position of the planet.

### *The First Generation of Greek Astronomy: The Limits of Our Knowledge*

In the eighteenth century, as Pompeii and its environs were dug up, one of the most spectacular finds was a villa with thousands of burned shreds of papyri, miraculously still – just barely – legible.[9] The reading of these papyri is one of the most demanding skills in classical scholarship, and even now, more than two hundred years after they were first unearthed, papyri are being deciphered. In 1976 David Sedley succeeded in recovering meaning in a long passage of a papyrus that must have come from Epicurus's massive work *On Nature*. (For this and the following studies cited in this section, see Suggestions for Further Reading.) There, Epicurus inveighed against the mathematicians active in Cyzicus. This gets our attention! We are told by ancient biographies that in his later years, Eudoxus set up a school in that city. This, then, completes our trio of fourth-century philosophers, all commenting on Eudoxus (Plato, Aristotle, Epicurus); usefully, we also gain a different kind of perspective. Epicurus *hated* them all: science, astronomy, and likely, Eudoxus himself.

[9] For this incredible find, see D. Sider, *The Library of the Villa dei Papiri at Herculaneum* (Los Angeles: J. Paul Getty Museum, 2005). It is notable, and puzzling, that no other papyri are known to have been excavated from anywhere else in Pompeii and its environs. (Very few, indeed, are ever found outside of Egypt.)

So, what did Epicurus dislike about the mathematicians of Cyzicus? We learn from his comments that those mathematicians were making mechanical models of the motions of the sky, but these – so argued Epicurus – were not reliable: What makes you think that such a mechanical model is an accurate representation of what takes place among the stars?

Epicurus, in fact, wished to argue against the divinity of the sky and, indeed, against the rational order of the universe: the exact opposite of Plato. It is not surprising that he would be anti-Eudoxean. But it is remarkable that we learn that according to him, Eudoxus (or one of his followers) was, at least in part, the builder of mechanical models. This, once again, is not inherently surprising: we recall the inscriptions with parapegmata, the sundials, the many ways in which Greek astronomers *displayed* their astronomy. Why shouldn't the mathematicians of Cyzicus? But we should bear this in mind as we consider the meaning of Eudoxus's astronomy.

I noted previously that under Schiaparelli's specific reconstruction, Eudoxus's model could be calculated. That is, one could, in principle, generate numbers: on *this* day, the position of this planet should be *at that point* along the zodiac. Indeed, as we will see in the following discussion, centuries later than Eudoxus, Greek astronomical texts did include tools that could generate such precise numerical values. Historians of ancient astronomy mostly assumed, until recently, that so did Eudoxus's treatise, as well.

In an influential article from 1983, Bowen and Goldstein called this assumption – that Eudoxus offered quantitative models – into question. As we shall see from the following discussion, it is clear that from the second century BCE onward, Greek astronomers had access to Babylonian records of observations. Bowen and Goldstein suggested that until then, Greeks simply did not have the available astronomical data with which to generate theoretical predictions and, indeed, did not even have the habit of thinking about the positions of the planets in these numerical terms. If so, it could be that Eudoxus merely attempted to provide qualitative geometrical models: the point was simply that certain shapes could be derived by the combination of spherical motions – the shapes agreeing in their overall character with the somewhat halting motion of the sun and the moon, the somewhat pretzel-like motion of the planets.

Now, one possible advantage of such a deflationary account is that as a quantitative model, Eudoxus's homocentric sphere, in Schiaparelli's reconstruction, is problematic. This model generates the component of the planets' motion that goes back and forth, through a particular 8-shaped figure, and we get the variations in speed through the length and breadth of the links in the shape. This means that the model predicts a correlation between the breadth of the planet's motion (how far it deviates from a strict line) and the non-uniformity of its speed. Mars, in particular, an especially changeable planet, would also have to deviate considerably – much more than it actually does – from the ecliptic.

In fact, the problem runs deeper. Although Schiaparelli's reconstruction of Eudoxus's model does generate nonuniform, zigzagging speeds, it is not at all "topologically" correct. For instance, this model generates neat, symmetrical, precisely repeated loops, but in fact, planets do not deviate from the ecliptic symmetrically and do not repeat precisely the same pattern in each cycle. Arguably, then, *even on Bowen and Goldstein's deflationary account*, Eudoxus's model should have been seen as an obvious empirical failure.

One way to solve this would be to go back to the historical data and find reconstructions different from Schiaparelli's, with better fit to the phenomena. When all has been said and done, the few comments in Aristotle and even in Simplicius do not really constrain us so much. No one ever did this until the last generation, primarily because this is such a hard task. Personal computing made it suddenly doable, and in the late 1990s, not one but *two* new reconstructions were offered simultaneously, by Yavetz and by Mendell, for which, indeed, see the Suggestions for Further Reading. The essence of such models is that they allow much more freedom in their parameters – for instance, the planet is allowed to rest not on the equator of the fourth sphere but on some other point; or the angle at which the 8-shaped figure is aligned with the first two spheres is allowed to shift. All of this quickly becomes very hard to calculate analytically. Yavetz and Mendell can calculate their predicted planetary trajectories on their computers, but Eudoxus could not. Perhaps such reconstructions work best with the evidence found by Sedley, then. It is perfectly possible to imagine a complicated mechanical arrangement of spheres (rings, in practice?). One can follow the motion of a point set on the innermost ring and note that its trajectory in fact fits that of a particular planet. If you throw in, within your mechanism, a graduated ring representing the ecliptic, you can measure the position of the planet so that this device becomes an analog computer. Even without this, the model can be used to derive an observed trajectory, which may or may not fit qualitatively with observed planetary trajectories. So was *this* Eudoxus's project?

To many readers, the idea of a calculation based on a mechanical device appears unsatisfactory. I suppose the intuition is that to gain a scientific understanding of the behavior of an object such as a planet, one would like to have some kind of solvable equation in which the position in space is provided by the laws, the rational principles, governing the planet's motions. We look, then, for a function of the following form:

$$\text{position} = \text{function (time)}.$$

The moment in time should be given in suitable numerical terms, which we then plug in to the function, calculate it, and derive the numerically given position. For this to work, we also want the function to be practically calculable with the tools at our disposal. The point of such a calculation is not just ease of use but also a certain epistemic transparency: the process of calculation tells us *how* the passage of time causes positions to change.

Obviously, the Greeks did not use our modern equations. They did, however, have the concept of the "given": *Y* being given in terms of *X*. So, in a geometrical configuration, once the magnitude of a certain object X is known, so is that of another object Y, dependent upon it. We would therefore look for a geometrical solution where the position of the planet in space is known, once the moment in time (represented, perhaps, by the arc of rotation of the first sphere?) is known as well, based on a well-defined construction. In a sense, this epistemic function is indeed satisfied by Eudoxus's model. But it is harder to define what we mean by "practically calculable." If we assume any of the more difficult reconstructions by Yavetz or Mendell, we can no longer (using Greek tools) plug in units of time to derive, by a well-defined calculation, a position in space. We need to resort, instead, to a black-boxed mechanism such as the turning of the knob in a device made of rings riding upon rings. Such a "calculation" involves no understanding because it involves no rational access to the manner in which one number gives rise to another.

This is a reasonable critique, but I wonder if it may not, in fact, be naïve concerning scientific computation. More to the point, I wonder if this is not naïve as regards the scientific computations *ultimately admitted by the Greeks themselves*. I explain. We will see that Ptolemy will indeed be capable of doing such things as telling the position of a planet, given a point in time. However, this does not mean that Ptolemy has an equation, simple or otherwise, deriving positions from moments in time. What Ptolemy does, instead, is to refer to tables with numerical values, in turn derived from previous tables, and so on in a complicated chain of derivation. (We will see a simple example of how this works in practice on pp. 377–378.) The astronomer, calculating a position from the date, does not, in fact, directly see the rationality of the computation, and it is not unreasonable to consider the system of tables as, so to speak, a mechanical black box. It is a device that, with the appropriate input, gives out the correct output. What we do not have in Ptolemy is anything like Kepler's laws. (To be clear, in the practical application of modern astronomy, actual computations are, in fact, still dependent on devices such as tables, more recently embedded in digital computers. Science is full of black boxes!)

But if so, why would it be so shocking for Eudoxus to rely on an actual machine – so as to calculate the positions of the planets? This is certainly within the range of the possible. What is unfortunate is that this range is so wide. At the minimalist end – proposed by Bowen and Goldstein – perhaps all Eudoxus did was to offer a purely conceptual account, showing that complicated patterns can be derived from the combinations of regular spheres. Or perhaps he did offer a somewhat stronger, but still qualitative, approximation of the actual motions of (some of?) the planets. Or finally, he could have provided a quantitative model, whose calculation, in turn, we can understand in two ways: as a mechanical device or, finally, as the geometrical solution to a problem in given magnitudes.

Let me recall, finally, a few more pieces of evidence.

First, the statement by Aristotle that Callippus revised Eudoxus's model. (He did so by adding in an extra sphere for each planet and also adding two more for each of the sun and the moon.) Scholars from Schiaparelli to Mendell and Yavetz have toyed with various accounts of how such added spheres could have improved Eudoxus's initial models. Perhaps nothing may be said with certainty on such reconstructions. But this much is at least suggested: Eudoxus's project was not merely a conceptual exercise, displaying the possibility, in principle, of generating nonuniform speeds out of uniform, homocentric spheres. Likely enough, one paid *some* attention to the actual resulting motions, whether in qualitative or quantitative terms. For otherwise, what was even the meaning of trying to *improve* Eudoxus?

Second, we should recall Eudoxus's other astronomical contribution: the mapping of the constellations, reflected by Aratus's poem. At the end of the day, Hipparchus's critique would make no sense had Eudoxus not provided his map in explicit quantitative terms. And this gives rise to a strong suspicion that Eudoxus's planetary theories, too, were offered in numerical terms. Not only because it makes sense, if one begins to measure the skies, to measure the positions not only of stars but of planets, but even more to the point, because both problems are in fact conceptually related. When we measure the positions of the planets, we do not do so in terms of absolute space; we do so in terms of reference points. Nowadays, we provide everything in terms of a coordinate system, and theoretically, Eudoxus could have done so as well, but it is much easier – indeed, to begin with, it is almost inevitable – to locate a planet, first of all, in terms of its surrounding stars. This, indeed, was the primary use of the constellations of the zodiac, and mapping those constellations in precise, quantitative detail is therefore a crucial first step for numerically given planetary observations.

Finally, perhaps the most significant context is that of the mathematics of the generation of Archytas as a whole. So much of it, we recall, was about specifically numerical ratios. Indeed, our sources for this generation (above all, Plato) often pair up music and astronomy. Archytas's music must have been a major inspiration for Eudoxus – and that was not some kind of qualitative musical theory, describing, say, how pitch could *in principle* be related to the speed of motions in the air. It was, instead, a fully developed system of numbers, with specified pairs of integer ratios composed to form other ratios.

The admission of ignorance remains warranted. The entire range from a purely conceptual exercise to a fully developed quantitative geometrical model of the planets is all possible. But the more likely of all the options, in my view, is that Eudoxus produced planet-imitating devices and wrote them up, with a good amount of data given in numerical terms. The evidence, however, is far from decisive, and it is better to conclude with what can be noted, regardless of any such specific reconstructions.

We note, first of all, that Plato, as an old man, was impressed with the new astronomy – just as he was impressed, in his middle age, with Archytas's new music. I envisage mathematical authors, in this generation, seeking a wider audience than just that of mathematicians. An astronomical theory such as Eudoxus's would have been calculated for this purpose. Indeed, it was massively successful, as so many contemporary philosophers did, in fact, take notice. Eudoxus was not a card-carrying Platonist or "Pythagorean," but the overall pattern of this theory is clear and consistent with Archytas's music: a seemingly messy part of physical reality is shown to be composed of simple mathematical elements (simple integer ratios in the case of Archytas's music; uniform and homocentric spherical rotations in the case of Eudoxus's astronomy).

The most obvious analogy between the first burst of Greek mathematics, that of the generation of Archytas, and the first burst of Greek astronomy, that of Eudoxus's homocentric spheres, is that both were generational events. This generational pattern is much more obvious and much more striking in the case of astronomy. Eudoxus inspired, precisely, a generation – the philosophers around him; a handful of mathematical practitioners, such as Callippus. And then, his theory appears to have been completely dropped. This, in itself, is a major historical puzzle. But in a sense, it simply reminds us of the most basic, striking fact about this new form of science. Eudoxus did not serve the state; he did not work for a court or for a temple. Unmoored from formal institutions, Eudoxus aimed to make a name for himself and therefore came up with an original, persuasive theory, perhaps displaying it in public with his own, newly devised instruments. Being autonomous can do wonders for scientific originality! It is less helpful, however, for scientific continuity. Even as Eudoxus was proposing his theories, unbeknown to him, temple practitioners in Mesopotamia were busily improving their empirical data as well as their predictive devices, year after year, generation after generation, heirs to institutional traditions stretching back centuries. The main achievement of ancient astronomy would have to wait till the moment that the two – the science inspired by Eudoxus's example and that informed by the masters of Babylon – would come together.

## II. Conjunction: Greece, Babylon, and Hipparchus

### *The Astronomical In-Between?*

Simplicius's problem was real. Writing a commentary to Aristotle, he needed to explain the astronomical theory there – clearly based on homocentric spheres. But when he was writing his commentary, in the sixth century CE, this theory had been dead for almost a thousand years. I emphasize: there is no evidence that *anyone* other than a particular network of authors around Eudoxus himself had ever pursued this astronomical model. Which added to the puzzle faced by Simplicius: besides having to come up with some account of Eudoxus's theory, he also had to come up with an account for why it was dropped. The best he could muster was a report that Autolycus – active a little later than the generation of Eudoxus – criticized the model of homocentric spheres because it implied that the planets must always remain at the same distance from the earth. In fact, Autolycus noted, Venus does undergo variations in its brightness, which would naturally suggest changes in its distance. This is not a stupid argument against the model of homocentric spheres, but I do not think it can be seriously suggested that this is the reason for the model's demise. There are easy ways to account for such brightness phenomena within a model of homocentric spheres, and surely Eudoxus, and everyone around him, was well aware of such considerations to begin with. They didn't need Autolycus to tell them that! But an even simpler point is that in the entire history of Greek ideas, there is no parallel for people giving up a position simply because an objection was raised. Greek cultural life thrived on debate; an objection should have naturally led to counter-objections and a proliferation of alternatives. The point is not that Autolycus shut out Eudoxus; the point is that he did not appear to offer an alternative and that by the time Autolycus came on the scene, the entire model-building exercise seems to have been dropped.

Indeed, with Autolycus, our historical light shines a little more brightly. Two works can still be read: *On Rotating Spheres* and *On Risings and Settings*. They may be put side by side with the *Phaenomena*, an extant work ascribed, by our manuscripts, to Euclid. So: three extant astronomical works, possibly from around the year 300 BCE. All three are closely related. The subject of study is a two-sphere model. One sphere is stationary, representing the earth; another sphere, representing the fixed stars, rotates, daily, around a pole common to the two spheres. One attaches to the moving sphere a circle – the ecliptic – whose pole is oblique to the pole of the spheres. One assumes that the sun rotates annually through this ecliptic (in this discussion, there is no mention of the moon and the planets; also, the speed of the sun is taken to be uniform). The basic phenomena of the seasons are the risings and settings of the fixed stars; these may be deduced from the aforementioned. Theorem 1.5 of Autolycus's *On Risings and Settings* may be cited as an example:

> Among fixed stars, those which are on the zodiac get from the apparent evening rising to the apparent evening setting in half a year, while those to the north get there in more than a half a year, those to the south – in less.

This begins to look familiar! A star is at its "apparent evening rising" when it rises in the east just as the sun sets; it is at its "apparent evening setting" when it is visible, a little above the western horizon, just as the sun sets. These are the qualitative events people actually always note about the sky – the type of information that got enshrined in Hesiod's poem. Historians of astronomy often turn to Autolycus with something like a groan – after Eudoxus's brilliance, is this all we get? – but to make sense of the history of Greek astronomy, we must remember that the subject matter of Autolycus's discussion mattered much more than that of Eudoxus's.

There is a simple sociological reason, as noted previously, why a Greek intellectual project need not have progressed beyond a single generation. Eudoxus did not set up a permanent institution. But we must also bear in mind that the problem of the irregular motion of the planets was not exactly *urgent*. Of course, Greeks cared about the sky. But it was not Mars's retrogradations that made the sky paramount in one's lived experience. The one main complication to this is that certain philosophers, such as Plato, cared greatly for the idea that underneath apparent unintelligibility, there could lie a perfectly ordered cosmos, discovered by science. With this enters our other sociological observation: throughout much of the fourth century, mathematicians were aware of an interested audience of philosophically minded readers. As the fourth century turned into the third, this audience dried up. As we noted in Chapter 3: philosophical attention drifted away from science. And so, those interested in science no longer envisaged a philosophical audience. But astronomy was of genuine interest to a wider audience: the audience that cared about the fixed stars, the zodiac, risings and settings. Perhaps the survival of Autolycus's work is more than sheer accident. Perhaps it was typical of most astronomy produced throughout this era?

Astronomy, perhaps, could have been recalled to its natural position – looking at the fixed stars, from the court. As noted on page 110, in the early third century, Aristyllus and Timocharis made observations, in Alexandria, of the fixed stars, which perhaps makes sense as a project designed to create globes for the kings (this, then, is directly parallel to Aratus's poem). In the middle of the century, we have Conon, the astronomer, involved in court intrigue, finding the lock of Berenice in the sky (see p. 118). Once again, what underlies this story is an observation of the fixed stars. As we saw, even the basic spherical geometry of the fixed stars gives rise to nontrivial (even if not very difficult) geometry, and the same is true for the observations of the fixed stars. In later Greek sources, we find sophisticated discussions of the geometrical optics of instruments of observation and, in particular, a mathematical theory for the measurement of position on a sphere. (More on this later in the discussion.) It is perfectly conceivable that mathematicians of the era engaged with such questions. All of this is perhaps a little anticlimactic, and so far, I have told this as a story of a Greek retreat, the individual authorship of a daring astronomical innovation – folded into a project funded by the state. But Hellenistic courts were anything but boring. They sought the remarkable, the surprising, the gigantic. We recall Eratosthenes, measuring the size of the earth through a geometrical thought experiment. From Alexandria, through mathematics – measure all lands. Aristarchus, earlier in the third century, went further. From Alexandria, through mathematics, he went on to measure the sky. This, *On the Sizes and Distances of the Sun and the Moon*, is by far the most impressive piece of early Greek astronomy still extant. It is a brief text, stated without any discursive introduction (perhaps such introductions were made part of mathematical writing only by Archimedes himself). Based on a few postulates, Aristarchus quickly deduces – strictly geometrically – the terms mentioned in the treatise's title, so that the smallest of mathematical tools give rise to the greatest of magnitudes. Let us linger a little longer on this exercise.

Aristarchus takes for granted the model – developed, as noted, by philosophers as early as Anaxagoras – of eclipses as shadows cast by the earth or the moon. Several specific quantitative observations are postulated as well, of which I mention the following:[10]

- As a background, remember that (to a first approximation) when the moon appears halved, the line joining the center of the earth to the center of the moon is perpendicular to the line joining the center of the moon to the center of the sun. It is, in principle, possible to measure the angle created, then, by the moon–earth–sun, and Aristarchus postulates that this angle is (what we call) eighty-seven degrees. (Aristarchus himself expresses this differently: that the angle is *less than a quadrant by one-thirtieth of a quadrant.*)
- The shadow of the earth intercepts two moon-breadths (that is, a maximal lunar eclipse covers twice the breadth of the moon; this value is, in principle, measurable based on the length of eclipses).

(It should be noted immediately that these are presented as postulates, not as actual measurements, and that although both measurements are somewhat tricky, it is clear that careful naked-eye observations should yield better values. The low quality of the observations, in the context of the brilliance of the geometrical deduction, is astonishing.)

Now, the measurement of the angle created by the moon–earth–sun, when the moon appears halved, can be directly used to deduce the relative distances of the moon and the sun from the earth. Aristarchus pursues this in proposition 7. Even this fairly simple unpacking calls for some geometrical work – in particular, Aristarchus assumes (without proof) the theorem that in an arrangement such as that in Figure 83, we have $(GE:EH) > (\angle GBE:\angle HBE)$.

[10] Aristarchus's text may be read in English in T. L. Heath, *Aristarchus of Samos, the Ancient Copernicus* (Oxford: Clarendon Press, 1913). (Although the original Oxford printing may be hard to find, there are many later reprintings, by Dover and by Cambridge University Press.)

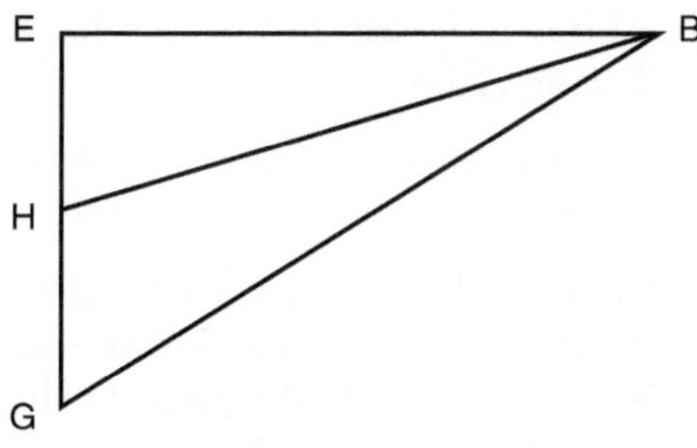


Figure 83

That is, in two right-angled triangles, having a fixed side BE, the ratio between the unfixed sides is greater than the ratio between the unfixed angles or, if you will, a trigonometric inequality of the following type, for example:

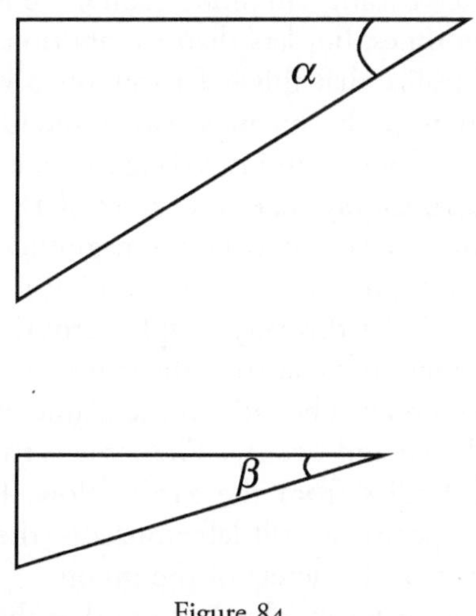


Figure 84

$$\tan(\alpha) : \tan(\beta) > \alpha : \beta.$$

This is not presented as a result in a separate field such as trigonometry (indeed, this is not presented at all and is instead simply an assumption, likely based on previous mathematical works no longer extant). In this case we appeal to trigonometry simply because we measure sides based on angles in a right-angled triangle, but for reasons that will become apparent through the chapter, we will see more and more trigonometry. It is astronomy's most lasting contribution to pure mathematics. (I used to hate trigonometry in high school, and I think it is regrettable that the discipline was taught without any sense of what any of it could be good for. Astronomy, that's what!) More of this, then, later in this subchapter: it appears that with Hipparchus, trigonometry comes into its own.

Back to Aristarchus himself, he can now show that his postulated angle translates into the following claim:

> *The distance of the sun from the earth is more than eighteen times, but less than twenty times, the distance of the moon from the earth.*

Note that the fact that we have inequalities rather than precise measurements is due not to Aristarchus's admission of some uncertainty with his observations – this is simply beside the point for him – but due to his having to rely on a trigonometric inequality.

Aristarchus now states as a theorem, number 8, that during eclipses of the sun, the sun and the moon coincide exactly (this could equally have been stated as a postulate; it is asserted as a theorem, with observations brought to support it, probably because Aristarchus, rightly, believes this is correct). Our observation concerning the relative distances of the sun and the moon now has the nice payoff that we learn – in proposition 9 – that the diameter of the sun is at least eighteen times, but less than twenty times, that of the moon. (Modern readers will realize that this is far too low a value, which suggests that the postulated moon–earth–sun angle was far too low as well; the correct angle is not eighty-seven degrees but instead eighty-nine degrees and fifty-one minutes!) Proposition 10 simply states the cubes of 18 and 20, respectively, concluding that the sun (now taken as a solid magnitude) is more than 5,832 times but less than 8,000 times the moon. Startling numbers and probably fun for courtly readers; I cite this very simple derivation of cubic numbers simply so as to give a sense of what such theorems were *for*.

Recall what we now know. The ratio of the diameters of the sun and the moon is between eighteen and twenty. The same is true of the ratio of the distances of the two bodies from the earth. Now, let us bring all that together with the one postulate still left unused – the size of the shadow cast by the earth: it is two diameters of the moon.

Now, what do we even mean when we say that the earth casts a "two-moons" shadow? To simplify Aristarchus's argument, we mean that if we interpose the earth between the sun and the moon, at the given relative distances of the three objects – then the shadow cast by the earth cuts off an arc of a circle, whose center is the center of the sun and whose circumference passes through the center of the moon, such that this arc is twice the diameter of the moon.

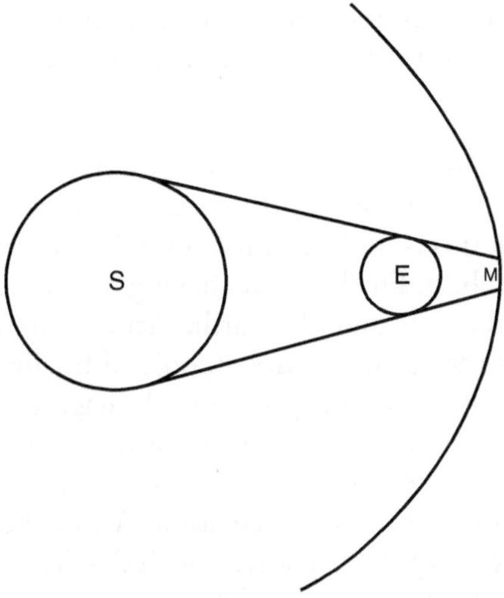


Figure 85

We have a succession of three circles with the three centers S, E, M:

S(un) – E(arth) – M(oon).

We know the ratio of the distances SE to EM (between 18:1 and 20:1), as well as the ratio of the diameters of the extreme circles, S and M (as it happens, the same ratio, ranging between 18:1 and 20:1).

The postulate that the shadow of the earth on the moon is twice the diameter of the moon carries a precise geometrical meaning, as noted previously, in terms of the tangents drawn to the two circles, S and E.

We know the ratio of the circles of S and M but not yet of E. This, in fact, is our single degree of freedom in the entire arrangement. We may think of this as a point E that we may expand at will. If it is just a tiny speck, then obviously, it will cast a tiny shadow (the tangents drawn to the two circles, S and E, will not even reach M). If we make it very big, obviously, its shadow can easily grow to more than two moons (for instance, if we make E equal to S – the earth equal to the sun – the tangents are parallel, so the shadow at the moon becomes, ignoring some complications, about eighteen to twenty moons).

But we postulate, in fact, that the shadow of E is two Ms, which determines the diameter of E. Intuitively, the earth should be bigger than the moon but smaller than the sun; in fact, the ratio of the diameters of the sun and the earth (call it S:E) is found, in proposition 15, to be

$$43:6 > S:E > 19:3.$$

Although it is clear, in principle, that once we have the ratios SE:EM and S:M, and we also know the size of the shadow of the earth, the ratio S:E is indeed determined, this is, in fact, still a complicated calculation, involving some really big numbers. Furthermore, having established the basic ratio of the diameters of the sun and the earth, Aristarchus proceeds to derive further the ratio of the earth and the moon and finally to cube everything so as to get the ratios of the geometrical magnitudes involved – so, obviously, once again, big numbers! Indeed, when one considers the boundaries of ratios, one can get to very big numbers simply by being *precise*. (At some point along the calculation, Aristarchus looks at the ratio 71,755,875:61,735,500.) The treatise concludes with the statement that the ratio of the earth to the moon, E:M (now considered as magnitudes, not as diameters), is bounded as follows:

$$216{,}000:6{,}859 > E:M > 12{,}519{,}712:79{,}507.$$

And it is clear that Aristarchus wanted to emphasize the sheer size and complexity of the numbers.

I cited Aristarchus in some detail to make two points. First of all, it would be only a small abstraction away to restate Aristarchus's claim as a functional relation. *Given* the angle of the earth/sun/moon, as well as the size of the earth's shadow cast on the moon in moon-diameter terms, it is possible to calculate the ratios of size distances of the earth, the sun, and the moon. The general validity of this geometrical relation is implicitly displayed throughout the treatise, but most of the work is dedicated to the actual calculation, with specific parameters. And this is the point: even with the specific parameters, and even with a fairly simple geometrical determination, the calculation for this single particular case takes many steps. The mere existence of a direct function does not make its calculation direct. We are content, in pure geometry, to show the solubility, in principle, of a task. In astronomical calculation, however, one is asked to go through the actual steps of the task, which is often an iterative, laborious, and ultimately, opaque exercise. Hence the black boxes of astronomy (and why machine computation is not such a bad idea after all). We will see more of this in the discussion to come.

And yet – and this is my second point – with all the laborious calculation, this treatise is a joy to read. In fact, Aristarchus is our best witness to what astronomy, pursued as a science for science's sake, could have looked like in the Hellenistic era. Astronomy did not simply revert to its role in the state, comparable to those Mesoamerican or Mesopotamian temples or Chinese courts. Quite simply, even in the world of the Hellenistic courts, the figure of the scientific author was too well established. The cultural habit was to deploy the literary genre of mathematics so as to make a name for oneself. The kings reached to the astronomers so as to set up new monuments: a globe, perhaps. But significantly, they did not introduce new, mathematically inspired calendrical reforms. And scientific authorship, itself, was not so much recruited by the courts as, so to speak, it became inflected by its field of gravity. And so we note, in Aristarchus, the emphasis on a surprising result, surprisingly told – with some big numbers – all of which being reminiscent, directly, of what we have seen of Archimedes in Chapter 3.

In fact, the line from Aristarchus to Archimedes is direct. Archimedes's own treatise the *Sand-Reckoner* is even more Aristarchean than Aristarchus, seeking even bigger spaces.[11] Archimedes sets out to name a number that will be bigger than the number of the grains of sand it would take to fill *the entire universe* – that is, to fill up the entire sphere of fixed stars. Archimedes runs through estimates of the size of the cosmos, beginning with the distance of the sun. (He improves directly on Aristarchus and also quotes other authorities; in particular, we learn that a certain Pheidias, *Archimedes's father*, also offered an estimate of the ratio of the sizes of the sun and the earth!) Going further, Archimedes suggests that we may get – to paraphrase – a much bigger universe by assuming that the earth is not stationary. If we assume that the earth goes around the sun, it is reasonable to take what was the ratio of the sphere of the earth to the sphere carrying the sun and turn it into the ratio between the sphere carrying the earth and the one carrying the fixed stars (this is because in both cases, no parallax is observed, although Archimedes does not linger on this explanation). Famously, Archimedes asserts that this model, too, was proposed by Aristarchus. And so we find – as Copernicus himself knew very well, centuries later! – that the Copernican model was on offer, by a major astronomer, already in antiquity.

This is simultaneously inspiring and disappointing. The Greeks achieved so much! And the Greeks cast *that* away! This is rather like

[11] The best English translation of this treatise is by Henry Mendell, available online at https://web.calstatela.edu/faculty/hmendel/Ancient%20Mathematics/Archimedes/SandReckoner/SandReckoner.html (accessed February 15, 2021).

Hero, passing right next to the steam engine without noticing it. For indeed, it does not seem that Aristarchus's model was picked up at all significantly in antiquity itself (one astronomer – a certain Seleucus – is said to have explored it further; the mathematics of Aristarchus's model is never cited, not even so as to be refuted, by any of our later extant astronomers).

In some ways, the question of the disappearance of Aristarchus's heliocentric model is similar to that of the disappearance of Eudoxus's homocentric spheres (in fact, had it not been for Eudoxus's brush against his contemporary philosophers, we would have known about his model no more than we know about Aristarchus's). Perhaps all we need to say is that Aristarchus was isolated and, once again, that he had no institutional continuity to perpetuate him. Surely, few cared, at the time, for whatever cosmological speculations were on offer by the mathematicians. We are told by Plutarch, a Roman-era author, that Cleanthes, the Stoic philosopher and Aristarchus's contemporary, gruffly complained that the Greeks ought to put Aristarchus on trial for disturbing the sun. Most likely, the story itself is no more than one of those apocryphal anecdotes that later Greeks loved to invent, bringing together, in vivid conflict, their past masters. At any rate, this apparent anticipation of Galileo's trial is no more than one of those little jokes planted by history: in the third century BCE itself, there was no "Aristarchus Affair." The scandal was that no one cared.

But even more to the point is the actual historical context, not of the seventeenth century CE but of the third century BCE. We are reminded, indeed, of all those sympotic pieces of mechanics. Aristarchus's own extant work and Archimedes's reception – in the *Sand-Reckoner* – are both suggestive. It is likely that Aristarchus proposed his model precisely for its shock value and, likely enough, also precisely for the sake of the kind of big numbers it could generate – such as those exploited by Archimedes. And if so, why should anyone be perturbed? There was probably always a flight-of-fancy quality to Aristarchus's project. It was made for the delight of the erudite, meeting in banquets; as the cup went around, so – for a brief instant, in the scholars' minds – went the earth, lifted by Aristarchus from under their feet in a dizzying whirl that the most potent of wines could not match. Was this what Aristarchus's astronomy was *for*?

### *A New Astronomy: Whodunit?*

As we move forward in time from the age of Eudoxus, the puzzle seems to be that not much happens. As we move even later into the third and second centuries, beyond Aristarchus himself, the puzzle has a more positive character: a new astronomy seems to emerge, and we do not know how. In general: our sources begin to fail us. We have extant, later than Aristarchus and before the coming of Roman cultural patronage two centuries later, no more than three securely dated works related to astronomy, of which two were mentioned already: Archimedes's *Sand-Reckoner* and Hipparchus's critique of Aratus's poem. Both offer no more than an oblique view of the astronomy of their time. The third – *Spherics*, by an author named Theodosius – takes up the geometry of spheres, clearly with astronomical motivation but, as it stands, mostly as a pure geometrical exercise; this, then, too, provides no evidence for new astronomical theories. And yet, as we emerge into the Roman era – on the other side of that historical lacuna – it is clear that a completely new kind of astronomy has already been proposed. Who proposed this and why? This is the "whodunit" in the title of this section.

Hipparchus is our first suspect. Extant, as noted, only in what to us seems rather absurd – the critique of a *poem* – enough is reported to make clear that he was one of the greatest creative mathematicians of antiquity. We will look at his achievement (not only the astronomical one) through the remainder of this subchapter.

And yet, it may be that some of the important events happened even deeper in the shadows. This much is known: Archimedes did make an astronomical contribution, beyond that of the *Sand-Reckoner*; Apollonius probably did too; both are almost entirely lost to us, and Apollonius's contribution – and perhaps that of Archimedes, too – could have shaped the route taken by Hipparchus himself. So, we must be even more speculative – and also rely on more tenuous evidence. Indeed, our deductive gears all turn, ultimately, on a single piece of the Antikythera mechanism.

We have already noted the Antikythera mechanism as an example of ancient mathematical mechanics, an object of marvel, perhaps in some sense related to Archimedes. Let us look at it a little more closely (see Plate 7).

First of all, we really need to look from *up close*. This is a miniature planetarium, a little more than thirty centimeters tall. (Jones's classic study of this artifact is titled *A Portable Cosmos*.) The sense of "miniature" is especially obvious because so much of its surface is inscribed – and this, with tiny letters (in current terms, it gets to a font size of less than 6 pt.: it is *this* small). So, this was not meant primarily for reading. It was meant for *turning*. Attached to the inside of the box was a knob, and by turning it either way, one could get the machine to move, essentially, back and forth in time. Notice that it is impossible to make any discontinuous jumps in time so that, in a sense, this is most like a clock, although one without a spring providing its own motion. (Not an idle comparison: such springs were first introduced in medieval planetaria – rather different in character from the Antikythera mechanism – which is how the clock got invented.)

As the knob turned, so did several dials. One dial gave the correct date label for the *month*, in the traditional calendrical system used in a certain region (specifically, in Epirus, a remote part of northwestern Greece; more on this in the following discussion). This is, of course, typical of the way in which ancient astronomical systems did not supplant traditional calendars. Another dial, on a different plate, showed whether or not eclipses were possible in that month. Yet another plate had at least two dials, this time adjusted by individual days rather than months. It is certain that two dials showed the zodiacal positions of the sun and the moon. In all probability – although a fatal lacuna prevents us from stating this with certainty – this plate had five more dials, showing the positions of the planets.

The remains of the Antikythera mechanism were found at sea near the island of Antikythera in 1901, among the (many other, often remarkable) remains of a shipwreck from the 60s BCE. All indications suggest that it was heading west. Indeed, there are several features of this instrument that suggest that it was "in transit." A particular plate on the instrument is set to begin at a date that suggests a "year zero" of the mechanism; this is surprisingly early for the date of the shipwreck – it seems that this was set in the final decade or so of the third century BCE. ("Aha!" – you are tempted to say. We will get to that.) Many features of the implied astronomy suggest the southern latitudes of the Greek-speaking world (many scholars think of Rhodes, for reasons that will become apparent), but its calendar, as noted, is that of northern Epirus. Indeed, the calendrical system refers to athletic festivals. That is: Greeks often labeled their years by referring to the games taking place, in four-year cycles, of which the most famous were, indeed, the Olympic Games, that is, those taking place in the town of Olympia. The Antikythera mechanism relies on such year names, but alongside the more famous games, it makes reference to the truly obscure games of the Naa, held in Epirus. The destination in Epirus, perhaps in the 60s BCE, is therefore clearly established – for an instrument whose chronological origins were perhaps in the late 200s, whose geographical origins were perhaps in Rhodes?

All of this is especially surprising if we think of the instrument as intended for some significant patron because Epirus was always impoverished and marginal. What are the odds, really? I personally find it much more likely that the intended recipient was not a human patron but a god. Marginal spaces are often holy, and it was in Epirus that the Greeks had one of their holiest places, the precinct of Dodona, sacred to Zeus (where, indeed, the Naa games took place). What better to dedicate to this god – the master of all – than *a portable cosmos*? I imagine, then, a device, originally devised elsewhere, copied (also, miniaturized?) for the sake of a dedication to Zeus; Poseidon intervened to prevent the dedication, hitting the ship with a storm near the island of Antikythera.

The actual device is in all likelihood *derivative*. Still, we cannot really say for certain what such a derivation may have involved. Although it is not clear that the Antikythera mechanism would have functioned very well in mechanical terms – that is, it is not clear how much one could turn its knob before it got broken – it is clear that in its underlying structure, it was really well conceived. Anyone who was capable of producing it must have had sufficient mastery over the underlying astronomy and its translation to gear structure to be able to introduce some of their own innovations. Which means that we cannot infer, from the make of this instrument, the make of its ancestor. As usual, we press against a real lacuna in our knowledge.

But this much is clear: Archimedes did write, as noted in the previous chapter, on the making of planetarium-like instruments. The key evidence is this. In the middle of the first century BCE, the Roman orator Cicero wrote his *Republic*, a dialogue in imitation of Plato. The main speaker was Scipio, whom we have met already: the protector of Polybius in the second century BCE. At one point in this dialogue, another speaker by the name of Philus recalls how Marcellus – the conqueror of Syracuse – brought out for display the sphere produced by Archimedes. As Philus tells it, this sphere was the only piece of loot taken by Marcellus (a clear mark, then, that we are reading polite fiction). Cicero goes on (*Republic* 1.22):[12]

> The invention of Archimedes deserved special admiration because he had thought out a way to represent accurately by a single turn, those various and divergent movements [of the sun, moon, and planets] with their different rates of speed.

The description here – a single knob, turned so that all the various motions of the sky are produced – fits the Antikythera mechanism almost too well. Indeed, we may suspect anachronism. Cicero was among the Romans who came to pay homage at Rhodes to Posidonius, the famous

[12] C. W. Keyes, *Cicero: De Re Publica, De Legibus* (Cambridge, MA: Harvard University Press, 1928), p. 43.

philosopher-historian-geographer, and we know that he saw the planetarium that Posidonius himself had produced. Everything suggests, then, that the Antikythera mechanism we now have is somehow indebted to the planetarium of Posidonius, and so it would have been possible that Cicero could merely ascribe (as could be suggested by the fiction of a dialogue, set in the second century BCE) a more recent device to the near-legendary Archimedes. (Indeed, it is hard not to recall here the fact that Stoics were especially close to the idea that the cosmos acted all in concert, guided by Zeus. Was it Posidonius, or one of his Stoic followers, who devised a version to be dedicated at Dodona?)

As we recall, Posidonius was the first philosopher for nearly two centuries to have been closely interested in science, and it is not surprising that he would pick up this new innovation – produced by astronomers while philosophers looked elsewhere – of the portable cosmos. But this much is clear, even from the little we saw in the preceding chapter: Posidonius was sufficiently familiar with science to be able to adapt something like the Antikythera mechanism. But he certainly was not at the mathematical level of *inventing* anything like it. We must go back further in time.

Perhaps we do not need to go far. Hipparchus was active in Rhodes in the middle of the second century, and as noted, he almost certainly was the most significant astronomer of his era. It is perfectly conceivable that he could invent from scratch something akin to the Antikythera mechanism. This is particularly plausible because of – as promised – a particular, tiny piece of the Antikythera mechanism.

Obviously, a mechanism where a knob drives gears can yield circular motions of various speeds, each uniform with itself (different gear arrangements can translate the same rotation into different fractions of a rotation). The mathematics required for this device, then, at this level, is simply that of finding the best factorization (and approximation) given the periods in use. So, for instance, the calendrical dial is based on the assumption that the moon revolves 254 times as the sun revolves 19 times. (Recall that this is not "254 months per 19 years" because a month is not a revolution of the moon but, rather, the return of the moon to the same angle with the sun: 254 moon revolutions to 19 sun revolutions translates, in fact, to the excellent calendrical, "Metonic" approximation of 235 months in 19 years.) We want 19 rotations of the axle to transform into 254 rotations of a particular gear – 254 is obviously too many teeth, whereas 19 is too few for the ideal gear (it is preferable for the teeth to be reasonably close to each other so as to latch on to other wheels). Thus, the relevant part of the mechanism deploys six gears for this purpose, with the teeth counts as follows:

> 64 that drives 38, mounted on 48 that drives 24, mounted on 127 that drives 32.

With all due cancellations, this is the same as a 254-tooth gear driving a 19-tooth gear.

This, then, is ingenious but not so different, astronomically, from the calendrical games played by many other civilizations. This is a mere star bureaucracy on wheels. As far as this goes, then, the Antikythera mechanism is (in all likelihood) *mechanically* superior to whatever arrangement of spheres or rings Eudoxus may have set up in Cyzicus, but astronomically speaking, this – as yet – is less ambitious. This is because Eudoxus's models were designed to display the *non*uniform speed of the sun, the moon, and the planets. By having spheres, each moving uniformly, oblique to each other, Eudoxus derived composite nonuniform speeds.

In fact, it is clear that the Antikythera mechanism, as well, derived nonuniform speeds. It would do no good to try to have gears oblique to each other, but another idea achieved the same result much more directly. I have not shared with you, in fact, the entirety of the train yielding the motion of the moon. The actual train followed not the route described previously but contained, instead, what might be considered "idle" parts – an added couple of gears that got canceled out. Why? Because these were not designed to add or take away from the count of rotations. They were set on top of each other, their centers slightly apart. A pin attached to the bottom gear went up through a radial slot cut through the top one. Depending on where the gears were in their cycle, the pin, like one of those on–off circles we "toggle" on our touchscreens, would change position in the slot. The motion of the gear would be constrained depending on the position of the pin in the slot. In other words, the gear would change its speed, in a predictable manner, depending on its position in the rotation. We have now obtained nonuniform speed. Incredibly, this detail – arguably, the most direct evidence for astronomical theory between Eudoxus and Ptolemy! – was first understood by modern scholars only in 2006.

When this observation was first made, scholars immediately set out to calculate the resulting nonuniformity of the speed of the moon. This was found to be close enough to the one that – according to Ptolemy – Hipparchus predicted, providing us with a possible identification of the origins of the mechanism. But of course, we have no proof that Hipparchus was the first to suggest this particular nonuniformity. In fact, we have some evidence to the contrary.

Once again, we rely on Ptolemy. Later on, we will get to consider Ptolemy's compendium of astronomy, the *Mathematical Systematic Treatise* (also known as *The Major [Collection]*, which, in Arabic transliteration of the Greek becomes *Almagest*, the name with which we will refer to this book from now on). In this book, Ptolemy accounts for all the regularities of the sky, and in Book III, he begins to tackle the apparent anomalies – to begin with, the relatively simple case of the sun (which we will return to follow in greater detail later in the discussion). Book III.3 is where he states that all astronomical motions are, in fact, circular and uniform and explains why the appearance of irregularity comes about:[13]

> The reason for the appearance of irregularity can be explained by two hypotheses, which are the most basic and simple. When [the bodies'] motion is viewed with respect to a circle imagined to be in the plane of the ecliptic, the center of which coincides with the center of the universe . . . then we can suppose, either that the uniform motion of each body takes place on a circle which is not concentric with the universe, or that they have such a concentric circle, but their uniform motion takes place, not actually on that circle, but on another circle, which is carried by the first circle, and is known as the "epicycle."

These, then, are two methods of deriving apparent irregularity from uniform circular motions. One – the eccentric – is to position the viewer away from the center of the uniform rotation. Obviously, then, the rotation would appear to be nonuniform (faster when the moving object passes closer to us; slower otherwise). The method of the epicycle is this: we start with a circle rotating around the earth, but instead of fixing the star, directly, to that circle, we fix another, smaller rotating circle (the epicycle) whose center is fixed on the original rotating circle. The star is fixed on the small circle.

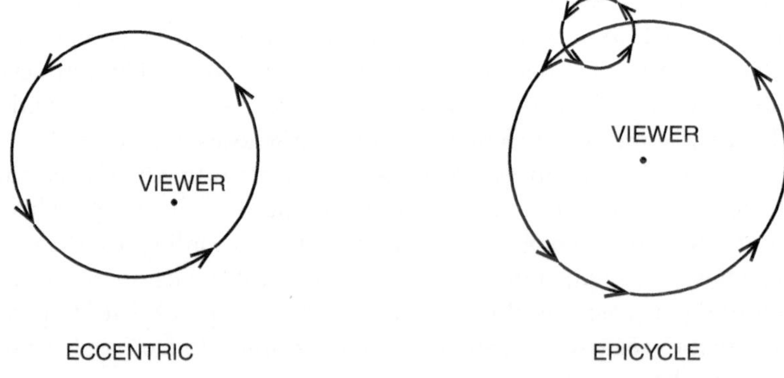


Figure 86

[13] G. J. Toomer, *Ptolemy's Almagest* (London: Duckworth, 1984), p. 141.

Once again, as with Eudoxus's homocentric spheres, we combine regular motions to produce an irregular one. It so happens that with the case of the sun, either eccentrics or epicycles will do, and in fact, the two are related – a given eccentric model can be transformed into an epicyclic model. Now, Ptolemy proceeds, in Book XII.1 – a much more difficult part of the treatise, dealing now with the theory of the planets – to quote a geometrical result concerning the determination of epicycles. Remarkably, he ascribes this result to Apollonius. Ptolemy shows repeatedly the ways in which an eccentric and an epicyclic model are related, and he does so immediately as he shows his proof in XII.1. It is at least possible, based on Ptolemy's presentation, that Apollonius himself, already, was aware of the relation between eccentrics and epicycles. It is unfortunate that we do not know this for sure, nor do we know anything about the use to which Apollonius put such theorems. This is unfortunate because epicycles and eccenters are the two main tools of Ptolemaic astronomy. (They will remain so all the way down to Kepler – Copernicus very much included.) This brief mention of Apollonius in *Almagest* XII.1 remains as our only indication that this astronomical tradition had already begun in the third century BCE.

To this, we may add a few more hints: a chatty biographical fragment informing us that Apollonius's nickname was "Epsilon" because of the similarity of the letter to the moon, his beloved subject. (This is a strange view, both of the moon and of epsilon.) Then again, there are other, very late – and not very reliable – sources also reporting specific points concerning Apollonius and the moon; all in all, it seems possible, at least, that Apollonius did indeed write on astronomy.[14] We have lost a great deal by him: Could this have included a treatise foundational for all of future astronomy?

But this is assuming that Apollonius *originated* epicycles and eccentrics. Nothing in our evidence explicitly states this – all we know is that Apollonius authored a particular theorem concerning epicycles. Our evidence beyond that runs dry. Unless, that is, we return to the Antikythera mechanism itself. Looking again at the arrangement of pin and slot, tying together two equal gears with slightly mismatched centers, it is hard now not to see there a combination of eccentric and epicycle. (The circle turns on a pivot, which, away from its actual center, keeps changing its position.)

[14] There is no systematic study of Apollonius as an astronomer (the evidence, at any rate, is too flimsy for that). Once again, the best starting point for the evidence and the discussion is the article "Apollonius of Perga" in the *Dictionary of Scientific Biography* by G. J. Toomer.

Not only does the Antikythera mechanism solve the problem of the nonuniform speed of the moon, but it does this with a mechanical model, which is the most direct implementation possible (given a system of gears), of the idea of eccentric/epicyclic rotations. To the extent that we believe that the overall features of the Antikythera mechanism go back to Archimedes, then, it becomes attractive to assume that he was in possession of something akin to such astronomical models.

Of course, nothing forces us to ascribe any particular feature of the mechanism to Archimedes. But recall that the epoch – "year zero" – of the mechanism itself seems to be set for a year near the end of the third century. There is no certainty that the design of the Antikythera mechanism goes back to that era, but is there anything to show otherwise? The point is not merely our lack of evidence. Rather, to repeat a point made earlier, without anything such as the pin-and-slot device, a device such as the Antikythera mechanism would be reduced to displaying *fixed* speeds only. It would be good enough for the basic calendrical purpose of aligning months and years in the Metonic cycle, but it would be useless for displaying the actual positions of the sun, the moon, and the planets. Specifically, in this regard, it would likely be at least theoretically inferior to the models produced by Eudoxus. Would Archimedes really accept that?

What we seem to suggest is that, on first principles, Archimedes *should* have incorporated a pin-and-slot mechanism (or some other variation of a device producing nonuniform speeds) into whatever planetarium he contrived. And if so, why not the device now attested?

This, then, is the "whodunit" of the title of this section. We need to find who first introduced a new astronomy: based not on homocentric spheres but on epicycles and eccentrics. Hipparchus certainly had such an astronomy; he is a plausible suspect. Ptolemy directly tells us that Apollonius, at the very least, investigated the geometrical properties of epicycles. He is a plausible suspect. And we know that Archimedes invented a new planetarium-like device that, by all appearances, seems rather like the Antikythera mechanism and so should have had some kind of device, other than Eudoxus's homocentric spheres, producing nonuniform motion. He, too, then, is a plausible suspect.

It is a mystery we cannot solve. But looking at our prime suspects, I like an account suggested by Evans and Carman in 2014. This is that the (mechanical equivalent of) epicycle/eccentric models could have been proposed *originally* as a mechanical solution to the problem of producing nonuniform speeds. The geometrical theory of such models could then have been devised in dialogue regarding, or even in response to, such mechanical approaches. It is obvious how this might work with our evidence. We may even find, at the end of the day, that all three suspects as responsible, in different ways, for the emergence of the new astronomy. We can envisage Archimedes as the author of a strictly mechanical treatise, describing the mechanism of a planetarium, glancing, no more, at the astronomical theory implied by the working of the machine. Apollonius, responding to Archimedes, both emulated as well as criticized him by showing some surprising geometrical properties of the motions so produced. Hipparchus, finally, could have been the first to derive all of this in an astronomically responsible way, relating the models to carefully executed observations and explaining how they work as models of the sky. The suggested sequence is mechanics, followed by geometry, followed by astronomy. This, of course, is just one story out of the many that, based on our slim evidence, we may tell. And it should be clear that the three names – Archimedes, Apollonius, Hipparchus – are but three barely visible islands, all that remain from an entire mountain range, now submerged. It is perfectly plausible that several dozen authors, now completely unknown, engaged with this kind of astronomy from the late third and throughout the second century BCE, each offering slightly different approaches.

But one advantage of our story is that it comports, in general, with our view of Archimedes, his influence, and his use of the mathematical past. Once again, Eudoxus emerges as a major inspiration. But also, once again, Archimedes reacts to Eudoxus as if Eudoxus was not a philosopher – as if he were merely interested, in this case, in the construction of marvelous machines. The cosmological significance of Eudoxus's models – their representation of the universe as a system of concentric spheres – is cast aside. Instead, one looks for the clever (and hidden) arrangement of gears that produces a surprising combination of motions.

We will return, later in the chapter, to discuss the question of "instrumentalism" in astronomy – whether an astronomical theory is meant as a description of the cosmos or merely as a device for calculation. We can note immediately that the model of epicycles and eccentrics suggests a certain retreat from cosmology in favor of mathematical ease (or perhaps, mechanical ease). This retreat from philosophy, then, fits best the mathematics of the age of Archimedes. And once again, just as with Eudoxus and his homocentric spheres, we find a new astronomy rising as a tentative, belated echo to a flourishing mathematical generation. Its acme, Hipparchus, brings us already to the middle of the second century and to an era that is, otherwise, almost entirely without attestation. The most important figure in the history of ancient astronomy must be studied almost without sources – and almost without context.

### *Hipparchus and the Babylonians*

Historians love astronomy. Herodotus tells us that a war waged by Persia against Lydia was called off because of the scary omen of a solar eclipse – and how else would we be able to date this event, so confidently, to *585 BCE*?[15] There is a certain class of astronomical observations – specific positions of planets, eclipses – such that their very description allows us to narrow down the potential dates on which they were made. It should be said, then, that up until this point in this chapter, the astronomical observations made by Greek astronomers hardly help us date them. Few astronomers engaged in systematic observations (the exception seems to be Timocharis and Aristyllus; those of Aristyllus were of the fixed stars, which once again provide only vague clues for dating, although those of Timocharis referred enough to the moon to allow better dating). Otherwise, coming closest to precise observations, allowing dating, are observations of the solstice (so, the most basic pendulum of the sun moving south and north), which, most scholars agree, imply 432 BCE for either Meton or Euctemon and 280 BCE for Aristarchus. But this is as far as it goes.

This changes with Hipparchus. This means much more than that we can date him with somewhat more confidence. It also means that, for once, with this author, systematic observations seem to have *mattered*. Ptolemy cites observations from Hipparchus made in the years 147–127 BCE; other observations cited by Ptolemy, not explicitly ascribed to Hipparchus but probably by him, bring us further back to 162. All of this suggests an author substantially removed in time from the generation of Archimedes:

15 Herodotus, *Histories* 1.74. To be fair, though, we do not date this confidently at all. We rely, in practice, on a report by Pliny the Elder (*Natural History* II.9), according to which Thales's prediction pertained to an eclipse taking place in the year 585 BCE (obviously not the date provided by Pliny himself, who uses instead the dating system based on counting Olympic Games). This in fact corresponds to an astronomically computed eclipse, and for this reason, the report is now usually trusted, but after all, there were several other solar eclipses in the surrounding years, so effectively, the date "585" expresses, at the end of the day, not the certitude of astronomical science but the feeble conjectural force of trusting Pliny the Elder. (I recommend F. Bailey, "On the Solar Eclipse Which Is Said to Have Been Predicted by Thales," *The Philosophical Magazine* 38 [1811]: 357–371.) Such anticlimactic conclusions are typical of the promise, and reality, of the use of astronomy in history.

born, perhaps, in around the same years that Apollonius and Archimedes's other interlocutors passed away.

Indeed, we didn't look that much at this period. In the previous chapter, I did mention Hipparchus's contemporary, Polybius, a harbinger of the coming of Roman patronage in the first century BCE. But otherwise, the second century BCE is surprisingly bare, coming after the great scientific florescence of the third century BCE – and before the great remaking of Greek culture under Roman influence in the first century BCE. In between is a century for which we have hardly any evidence at all – perhaps because, in general, less was achieved? We have very little new to report from philosophy and medicine, as well. I tend to suspect that this is the price the Greeks paid, ultimately, for the rise of inequality. One thing that democracy did was to establish a certain egalitarianism. This, in turn, allowed more people to participate in culture. The rise of the Hellenistic kingdoms and the later rise of the Roman empire dealt a blow to this egalitarianism. This was balanced, to a large extent, by the sheer rise of the demographic base due to the expansion of Greek ideas across the Mediterranean, but before the influx of Latin-speaking elites in the first century BCE, this geographical expansion did not match the socioeconomic collapse. The second century BCE was not yet Roman but already distant from the classical era. It enjoyed no more than a distant echo of the democracy that made that classical past so vibrant.

In this world, then, a rare genius – quite possibly, an isolated one. It is noteworthy how much of Hipparchus's work took the past as its foil. The one extant work, as noted, is the critique of Aratus (and through him, of Eudoxus). He also was considered by Strabo, the Roman-era geographer, to have been one of the leading authorities on the subject of geography; what Strabo cites from Hipparchus seems to come from one source above all: an extensive treatise by Hipparchus attacking the geography of Eratosthenes.[16] Another attested work took Chrysippus, the third-century Stoic philosopher, as its target.

The last one, in fact, calls for a brief detour. Chrysippus – who invented propositional logic – had proclaimed that from ten assertions, "millions of combinations can be made." Hipparchus then went through the trouble of showing that, in fact, there are only 103,049, or at most, 310,952 such combinations. We know very little about this episode: Plutarch, an author from the first century CE, was keen to attack Chrysippus, and so he cites the mere fact without much elaboration. It was considered a mere anecdote until a mathematician noted in 1996 that 103,049 is, in fact, what is

[16] See especially Strabo, *Geography* II.38–II.41.

known as the tenth Schröder number, or the number of different ways that ten objects in a sequence can be put into brackets. (This, then, is an obvious interpretation of "combinations of ten assertions"; I leave the higher number 310,952 aside, but with some complications this, too, is now known to have a mathematical meaning.) It is impossible to come up with such a number by simple enumeration, and so we learn that Hipparchus engaged in theoretical combinatorics.[17] Although a lot of what Archimedes, in particular, did involved the counting of a number of combinations (this is particularly true for the *Stomachion*), nothing suggests he developed pure rules of combinatorics; evidently, Hipparchus did – likely enough, the first to do so. First – and alone, facing only the foil of Chrysippus's dead authority. Something similar may well be the context for Hipparchus's measurements of the sizes and distances of the sun and the moon – of which we have only tiny reports. These had the sun and moon farther away than Aristarchus's (Hipparchus had the moon about 70 Earth radii away – essentially correct – the sun about 490 radii away, still a couple of orders of magnitude too close), and the little we do know suggests the distances were based on a multitude of considerations and observations, not just on one geometrical construction; it seems possible, then, that this, too, was in some sense directed against Aristarchus.[18]

And so: attacking Aratus, Chrysippus, Eratosthenes, perhaps Aristarchus – almost as if this author of the second century systematically went through the leading authorities of the third. Specifically, we also see a pattern in all those refutations. Hipparchus is the master of painstaking calculations. Aratus (and, behind him, Eudoxus) threw out rough approximations of the positions of the fixed stars; Eratosthenes did not calculate his geographical coordinates with precision; Chrysippus merely stated "millions," never bothering to do the actual work of counting. Aristarchus was happy, relying on a single geometrical flight of fancy. Hipparchus would prove them wrong by being *precise*.

Such an author would be interested, then, in gaining more quantitative data. Indeed, Hipparchus clearly had access to a trove of Babylonian knowledge of astronomy; as far as we can tell, no previous Greek did.

[17] R. P. Stanley, "Hipparchus, Plutarch, Schröder, and Hough," *The American Mathematical Monthly* 104, no. 4 (1997): 344–350.

[18] To be clear: is entirely speculative to suggest that this calculation was polemical against Aristarchus. We have no indications of Hipparchus's context for the calculation and no more than the bare indications for the calculation itself. See C. Carman, "On the Distances of the Sun and the Moon According to Hipparchus," in C. Carman and A. Jones (eds.), *Instruments – Observations – Theories*, 2020, https://doi.org/10.5281/zenodo.3928498, pp. 177–203.

This, then, calls for another, more prolonged detour. We last met Babylonian astronomy with the Mul.Apin, a substantial set of observations – and synodic periods – collected and relayed through many centuries down to the seventh century BCE. Such synodic periods were the basis for the Babylonian temple practice of calendars and divination. Late in this tradition, the historian David Brown notes a transformation. In the eighth and seventh centuries BCE, Mesopotamian astronomers started to *predict* eclipses. We lack precise knowledge of the historical context for such developments, but Brown notes a certain competitive professionalization of scholarship at the court. There could have been extra prestige in becoming the Sherlock Holmes diviner, the one capable of telling the omen *in advance*. At any rate, such predictions derive, at this point, merely from observations of cycles of recurrences of a calendrical nature. One can say, therefore, that the Mesopotamians were, at this point, masters of calendrical prediction. By the fifth century if not before, such knowledge was already sufficiently familiar around the Mediterranean to make possible the calendars – perhaps even eclipse predictions – of Meton and of later Greek authors.

But even as Greeks were starting on their own distinctive astronomical journey, Babylonian temple astronomers were beginning to develop a new set of techniques.

To recall, for the temple-based science of the stars – whether in Babylon or in China – the ultimate unpredictability of the planets became something of a feature rather than a bug. One could not tell precisely where the next planetary event would take place – and that made its identity all the more ominous. This unpredictability arises because the apparent speeds of a planet are nonuniform. The average speeds of the planets, however, considered over longer periods, are much more stable – hence the possibility of observing synodic patterns. It is a bit like the weather: if it is January, it is very hard to tell if tomorrow will be colder or warmer, but it is almost certain that (in the Northern Hemisphere) it will be hotter *six months from now*.

Add to this also that what motivates astronomy, to begin with, and certainly in the Mesopotamian context of divination, is the key qualitative events, and it becomes easy to understand why the Babylonian predictive planetary system was organized by those events and not by the day-to-day motions of the planets. Someone, or perhaps a small group of practitioners, came up not with one but with two schemes that did a fairly good job of predicting the location on the ecliptic of planetary events such as first rising, beginning of retrogradation, and so forth. Each scheme modified, each in its own way, the obviously false assumption that the planets move at uniform speeds. In one of the schemes, the speed was assumed to be uniform *given a portion of the zodiac*. (In certain signs, the planet was supposed to move faster; in others, slower.) In the other, speeds were assumed to vary regularly through what is known as a *zigzag function*: becoming faster until you hit a maximum, then becoming slower until you hit a minimum, and back again. This is a bit surprising to us, but the ideas are not absurd: the different-speeds-in-different-signs approach feels very natural in a context where one assumes, to begin with, that the entry of a planet into a sign is an ominous, consequential event. The zigzag hypothesis, in turn, feels very natural within an astronomical context where one keeps noticing "pendulum" events such as the sun's motion from north to south, the moon's waxing and waning. Further, my language – assuming the algebra of continuous speeds, times, and distances – is already an anachronistic modernization of a system that, for its practitioners, was based on the synodic cycle as a discrete series of qualitative events. But this, once again, is meaningful. Just as with the weather: those schemes do a poor job of predicting the position of the planet from day to day because, in fact, they do not even conceive of this problem in time, as a continuous series. Because, you see: taken in the aggregate, across the many days separating one qualitative event from the next, and provided the right parameters, those schemes can become quite effective approximations. Perfected in the fourth century or perhaps even as early the fifth, this may well have been the first fully scientific theory of the planets.

In past scholarship, this Babylonian achievement was sometimes dismissed as "merely" practical, the Babylonians unfavorably compared with the Greeks in that they did not produce a *geometrical* account of the sky, hence no physical model, so, unlike the Greeks, "not real science." This is obviously an absurd special pleading, where one defines as scientific whatever it is that the Greeks do and then reprimands the non-Greeks for failing to be Greek. The Babylonian theory is in fact directly analogous to the Greek mathematical theory of *music* – whose scientific significance no one doubts. To recall: Greek music theorists explained musical harmonies in terms of simple integer ratios. With hindsight, we understand that simple integer ratios between pitches are significant because they stand for certain geometrical relations between waves. But it is perfectly valid to look at the evidence and to consider the integer ratios, themselves, as a significant scientific discovery that could, in principle, be directly explanatory. They were taken as such by the Greeks themselves. We have no indication as to the metaphysics that Babylonian astronomers attached to their numerical schemes, but as Rochberg insists, they were dead serious about the stars. These were gods; the astronomers' business was to worship them. Can we really doubt that the Babylonian astronomers felt that in their schemes, they had discovered deep truths about the planet gods? Those truths were not related to geometrical shapes but to numerical parameters: And so what?

More than this: had Eudoxus presented Plato not with homocentric spheres but with Babylonian-like numerical schemes, would Plato really object? Not at all: in fact, it seems, from Plato's remarks in the *Republic*, that he might have preferred a more abstract, Archytas-like astronomy!

The Babylonians were scientists. The contrast with the Greeks is real, though. The Greeks started afresh, without a deep apparatus of observations. A disadvantage – but also a certain freedom. The Greek astronomical revolution involved a transition from a fairly unsophisticated Metonic, integer-based, calendrical astronomy to a different *kind* of astronomy – geometrically based modeling, whose empirical foundations include no more than a handful of observations, perhaps focused not only on treatise writing but also on public display and on mechanical models. The Babylonian astronomical revolution, in contrast, had much more to draw upon in terms of observations of patterns, and so it kept much closer to calendrical practices. Both approaches have their distinctive strengths, and astronomy was the most successful science of antiquity because it was a synthesis of the two.

This synthesis was likely produced by Hipparchus. He learned all about Babylonian astronomy through . . . Actually, we have no idea. Did he go to Babylon? Or conversely, did he meet Babylonian astronomers who visited Rhodes? Was such knowledge perhaps committed to writing, perhaps even translated into Greek, even prior to Hipparchus? One way or another, the one clear thing is that Hipparchus got his hands on Babylonian records as well as on Babylonian techniques. He now had in his possession, in particular, centuries-long data of eclipses. He knew certain periodicities and algorithms that the Babylonians deduced from their observations. Was he perhaps responsible for the transition into calculating astronomical values with sexagesimals? (The main argument that he did is that anyone writing after him in theoretical astronomy would have wished to be able to compare their numbers with Hipparchus's, and surely switching bases would be a pain! However, the little direct evidence for Hipparchus's use of numbers does not involve sexagesimals, and it is quite possible that sexagesimals trickled up under the Romans, like astronomy itself, not from any theoretical sources but, instead, from astrological practice; more on this later in the discussion.)

Let us quote from Ptolemy's *Almagest* (III.1) – the very beginning of Ptolemy's detailed model-building:[19]

> The ancients were in disagreement and confusion in their pronouncements on the [length of the year], as can be seen from their treatises, especially those of Hipparchus, who was both industrious and a lover of truth. The main cause of the confusion on this topic which even he displayed is the fact that, when one examines the apparent returns to equinox and solstice, one finds that the length of the year exceeds 365 days by less than 1/4 day, but when one examines its return to the fixed stars it is greater [than 365 1/4 days]. Hence Hipparchus comes to the idea that the sphere of the fixed stars too has a very slow motion.

We have already noticed that the "year" means, to early observers, above all, the cycle from one solstice (northernmost or southernmost position of the rising of the sun) to the next. This is now known as a *tropical year*. Hipparchus claims that this is slightly *less* than 365 1/4 days.

There is also a separate meaning for the "year," which is the return of the sun to the same position against the fixed stars (or against the zodiac). This cannot be directly observed (it is impossible to see the sun and the fixed stars together) but can be inferred, as noted earlier, from the first risings of constellations on the zodiac. This is now known as a *sidereal year*. Hipparchus claims that this is slightly *more* than 365 1/4 days.

At first glance, both the tropical year and the sidereal year seem to measure the very same process, namely (from a geocentric perspective), the sun's annual motion around the earth. One would expect solstices or equinoxes to occur at "fixed positions on the zodiac" – when the sun is against the very same fixed stars, almost as a matter of *definition*. It is remarkable that Hipparchus was certain enough in the values he obtained for those cycles that he preferred not to explain away such tiny differences but instead to use them as a basis for a theoretical claim. He was right. The fixed stars move very slowly, so the stars against which the sun is positioned during the solstice, or equinox, very gradually transform over the centuries: this is the "precession of the equinoxes." We now understand it as a tiny wobble in the earth's rotation: even while the earth rotates around its axis, the axis itself draws, slowly, a small circle or, more precisely, a cone. (All, ultimately – as Newton realized – because of the interference of the gravity of the moon.) Hipparchus thought that this precession of the equinoxes completed a circle every thirty-six thousand years, whereas in fact, it takes about twenty-six thousand years: a truly tiny error (an artifact

[19] G. J. Toomer, *Ptolemy's Almagest* (London: Duckworth, 1984), p. 131.

of rounding, perhaps?), given Hipparchus's very limited empirical basis. But indeed, what was his basis? Ptolemy gives few further details (according to him, it appears that Hipparchus relies strictly on comparing his own observations to those made by previous Greek astronomers about 150 years earlier). Modern scholars sometimes wonder if Hipparchus could have based one or even both measures of the year not directly on observations but, instead, on Babylonian parameters, but we can't know and can only marvel at Hipparchus's intuition.

In the considerations concerning the lengths of the year, the contribution of Babylonian knowledge, in this case, could be that of parameters, assuming which one may deduce underlying cycles. In the *Almagest*, we have evidence for a more direct use of Babylonian sources: in several cases, Ptolemy cites Babylonian observations, especially those of eclipses. We are specifically told that Hipparchus compiled such observations, and because we find such a compilation already on Babylonian clay tablets, the conclusion appears to be that Hipparchus, or one of his predecessors, had the Babylonian series (or some selection from it) translated into Greek. In all likelihood, Ptolemy takes his data from Hipparchus – so, Babylonian two times or more removed – and perhaps he also follows the outline of Hipparchus's own interpretation. (It is telling that this use of Babylonian sources takes place mostly around the theory of the moon, which Hipparchus apparently advanced the most: it is possible that the first iteration of Ptolemy's moon theory is essentially Hipparchus's.)

So let us read an example, once again, from Ptolemy (IV.6):[20]

> The first [of three eclipses used by Ptolemy] is recorded as occurring in the first year of Mardokempad, Thoth 29/30 in the Egyptian calendar. The eclipse began, it says, well over an hour after moonrise, and was total.

The data were indeed compiled in Babylon: an event, taking place in 720 BCE, available as the raw material for a theory promulgated in the second century BCE.

Notice how much the raw material has to be processed. The travel of astronomical data requires the greatest care! Even the reference to the Egyptian calendar involves, already, a reference to calendrical equivalences that Ptolemy (or Hipparchus before him) had to establish independently. (The Babylonian data, of course, were recorded according to Babylonian calendars.) Following that, Ptolemy transforms the data into an absolute time of the night, based on (a) the position in the solar (tropical) year of

[20] G. J. Toomer, *Ptolemy's Almagest* (London: Duckworth, 1984), p. 191.

the given calendrical event and (b) the length of the night at that point in the tropical year. He finds that mid-eclipse was 2½ hours before midnight. He then transforms this further, as he makes all such observations uniformly observed from the longitude of Alexandria (which perhaps was taken to be the same as that of Rhodes, so he might have directly followed Hipparchus in this manipulation). He thus shaves 5/6 hours (the time-zone difference between Babylon and Alexandria) to get that the mid-eclipse, as observed from Alexandria, was 3⅓ hours before midnight. He then calculates the zodiacal position of the sun at that particular moment, based on his tables. And this finally makes it possible to extract, from a Babylonian dating of an eclipse, the zodiacal position of the moon at a given time.

This is all required for only a single observation. Ptolemy repeats the process twice more so that he acquires three precisely dated zodiacal positions of the moon for three moments in time. This allows Ptolemy to calculate two values: the average speed of the moon between the first and the second moment; the average speed between the second and the third. Because Ptolemy derives independently (a much simpler exercise) the underlying average speed of the moon, each result is interpreted as a deviation from the moon's underlying average speed. We have two data points: in period 1, the deviation was of size $X$; in period 2, the deviation was of size $Y$.

In this exercise, we assume an epicyclic model, and the parameter we are seeking is the size of the epicycle (in practice, its radius) in terms of the radius of the main circle. The crux is that the two values we now have can in fact determine, geometrically, the value of this parameter.

This is all in Ptolemy, but as Ptolemy makes clear, he is following the approach of Hipparchus himself – perhaps the one represented in the Antikythera mechanism. Depending on our solution of the whodunit of the previous section, we may find him to be either entirely original or perhaps merely deducing and refining an Archimedean/Apollonian solution, based on new Babylonian materials. But in all likelihood, what we would have found, had we had access to Hipparchus's original works, was – in Ptolemy's words – *industriousness* and *love of truth.*

### *"Industrious": A Detour Regarding Trigonometry*

Let us take those in order. First, Hipparchus was industrious. That is, he did not merely develop theories but also observed and calculated in detail. It is clear that Hipparchus's calculations were based, in some instances, on Babylonian algorithms. But he had to go beyond the Babylonian methods in order to derive the parameters for his own theories, based, as they were, on circular motions that derive particular speeds.

Even this account elides much of the complexity of the calculations required for fixing parameters based on observations. Suppose one knows that the average speed of the moon, along a certain arc of the circle, is *X*, and that its average speed, along a different arc, is *Y*. This, indeed, can be derived from no more than three observations. And it is geometrically correct that if one assumes a model of an epicycle or of an eccenter, these two average speeds should be enough to determine just which epicycle to take or just where to position the eccenter. However, the geometrical argument for determining the parameter, given the values, is much more complex than you might at first think, not just in terms of the underlying geometry but in terms of practical calculation. In particular, the argument has to work, in part, through the determination of chords inside the circle – straight lines – on the evidence of arcs of the circle. I will not go through the detail of the argument but merely observe one of the several diagrams used by Ptolemy in his own derivation to illustrate the kind of difficulty we encounter.

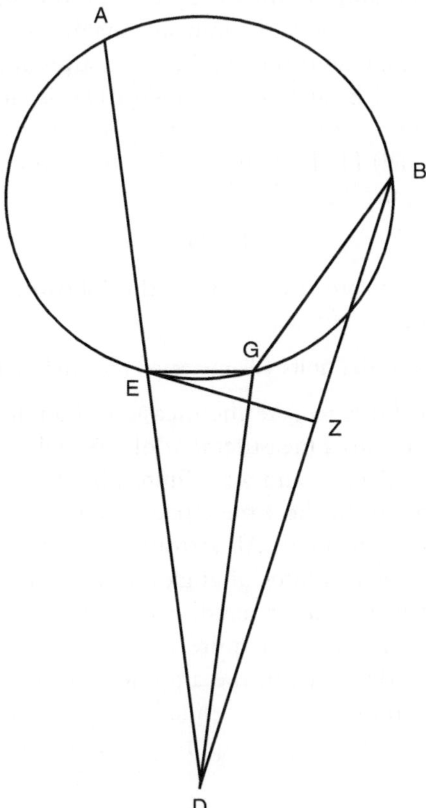


Figure 87

To recall: we assume that the nonuniform speed of the moon is produced by an epicycle and that the positions of the moon on the epicycle at the midpoint of the three observed eclipses are A, B, and G. Our evidence is the length of the arcs between the points. We need to find the size of this epicycle – that is, its radius – that will derive the observed speeds. You may begin to sense the difficulty: we start with evidence concerning lengths of arcs and derive from it a result concerning the length of a linear unit – a radius. This becomes, in fact, much more difficult because in practice, we do not get directly from the arc to the radius. The radius is derived geometrically by considerations of other geometrical relations determined by the three points A, B, G. In this case, Ptolemy draws lines from the center of the main circle (that is, the center of the earth), at D, to the three points A, B, and G, and the operation is based on manipulating the resulting straight lines. EZ, for instance, is the perpendicular dropped from point E (where line DA intersects with the circle) to line DB. The length of EZ is significant in Ptolemy's derivation, and he finds it by first producing the size that the arc EZ would be at if we were to draw an imaginary circle based on the points DEZ. (This may be produced because, after all, we see the moon at the positions A and B, and so the angle EDZ is simply the observed angle between the two observations – our eyes do not care that the moon travels on an epicycle!) He finds the arc EZ to be (in degrees, written in sexagesimal numbers):

$$15;24.$$

And then immediately proceeds to state the following claim concerning the straight line EZ:

$$\text{EZ} = 16;4,42 \text{ with the units (diameter of the circle DEZ) DE} = 120.$$

What Ptolemy just did is to take the measure of an arc – which one can observe – and turn it into a measurement of a straight line. The arc EZ of the circle DEZ is 15 degrees and 24 minutes; hence, the chord EZ stands at the ratio 16;4,42:120 to the same circle's diameter.

This is a fundamental point. All astronomical observations are understood by the Greeks as measures of angles on a circle or (equivalently) as measures of arcs. On the other hand, a typically geometrical derivation from such observations would require the finding of particular straight lines as, at least, interim steps in the argument. But how to do this? We wanted to do astronomy – and we ended up, pretty much, squaring the circle! In short, we need to rely on approximation. Indeed, Ptolemy had a table for this, containing such approximations: a table of chords. This went through the possible measurements of arcs – in increments of half a degree (what is the chord if the arc is half a degree? A degree? A degree and a half?), all the way up to 180 – and found the length of the associated chord in terms of 1/120th parts of the diameter of the same circle. The chord of the arc 180, for instance, is 120 1/120th parts of that diameter – because it *is* the diameter! (This is one of no more than a handful of measurements that are *exact*.)

We recall Aristarchus deriving conclusions from his postulate of the angle of the moon–earth–sun when the moon is bisected – supposedly, this angle was then equal to what we call (following Hipparchus and the Babylonians) three degrees. From this, Aristarchus derived the ratio between the distance of the earth and the moon, and the earth and the sun. In other words, he produced a trigonometric calculation of a particular value – three degrees. It is likely enough that at the time of Aristarchus, Greek mathematicians would indeed produce such results piecemeal, as the need arose. But for the kind of purposes required by Hipparchus, a more systematic table would become handy, and indeed, we are told by Theon – a later astronomical commentator, on whom more in the following chapter – that Hipparchus wrote an entire treatise (in twelve books, no less!) titled *On Chords*. Hipparchus must have produced a work dedicated to what we now call "trigonometry."

It is impossible to say what this book included (and the report of twelve books almost defies belief), but we can make some guesses concerning its overall strategy. The field of trigonometry lies at the intersection of geometry and astronomy, and our evidence is divided accordingly. For the geometrical inspiration, we probably need to look back to Archimedes; for the astronomical context, we will rely on Ptolemy (but we are lucky to have further evidence, in particular from Menelaus).

I have only briefly mentioned Archimedes's *Measurement of Circle*. In part, this treatise is a standard (and surprisingly easy to accomplish) Eudoxean, proof-by-contradiction measurement of a curvilinear object, showing that a circle is equal to a right-angled triangle whose two sides are the radius and the circumference of the circle. The rest of it is dedicated to an explicit calculation of boundary values on the actual length of the circumference, which is translated, in practice, to finding the ratio of the circumference to the diameter – that is, finding the value of $\pi$. Just like the Eudoxean proof, this is done by considering the ratio to the diameter not of the circle but of regular polygons, inscribed and circumscribed. For what follows, we are mostly interested in the case of the inscribed polygon. We start with the hexagon, a special case that allows direct computation (Figure 88).

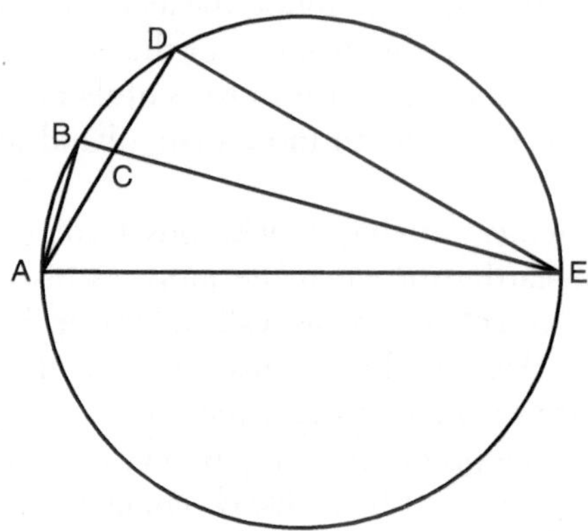


Figure 88

AE is a diameter, AD is the side of the inscribed regular hexagon, and AB is the side of the inscribed regular dodecagon. Note that the angles at D, B are both right (all of this goes back all the way to Hippocrates of Chios). Archimedes proves generally that whenever BE bisects the angle AED,

$$AE + ED : AD :: BE : BA.$$

(The proof is not self-evident, but because we are looking, right now, for the evidence for Hipparchus, not Archimedes, I skip it.) Now, the angle of a regular hexagon is what we call 120 degrees, so that its half – DAE here – is 60 degrees. If we were to draw radii from the center to A, D, we would have an equilateral triangle, and so AD is equal to the radius or to half of AE. Whatever unit AE is, AD is half that. DE, then, is what we would call $\sqrt{3}$ of AD, and Archimedes has a very close numerical approximation of that. (This gives the lie, incidentally, to the notion that the Greeks needed geometrical tools for the practical problems of calculating cubic roots for the sake of "duplicating the cube." Such problems could easily be solved numerically.) We can therefore use the previous theorem to find BE:BA and then use Pythagoras's theorem to find AE:AB, for which Archimedes finds the following:

$$AE : AB < 3{,}013\ \frac{3}{4} : 780.$$

Which provides us with a close boundary on the chord of the arc of the dodecagon, or what we call the thirty-degrees arc or fifteen-degrees angle.

If you will, the aforementioned value provides an answer to the question, "What is the cosine – hypotenuse divided by opposite side – of fifteen degrees?" However, in the Greek conception, the triangle is incidental, and the question concerns arcs and chords.

Now that we have the values for the triangle associated with the dodecagon, we may go on by applying the same operation to find the values for the triangle associated with the 24-gon, which Archimedes iterates, for both circumscribing and inscribed polygons, and then iterates again, and again, until he reaches the 96-gon. His goal in the *Measurement of the Circle* is this very elegant boundary on π:

$$3\frac{1}{7} > \pi > 3\frac{10}{71}.$$

But we have also learned, along the way, to measure a certain set of chords. There are some indications that Hipparchus may have produced such trigonometric tables for continuous bisections of what we think of as 60 degrees up to 3.75 degrees, or perhaps even smaller bisections.

It was suggested in the past that perhaps this was all Hipparchus did. I personally suspect that he should have displayed some one-upmanship, going beyond Archimedes. It remains significant that the starting point for this exercise (as for so many others!) is Archimedean and, specifically, dependent on the quest for measuring curvilinear objects. To repeat, what we now identify as ancient trigonometry never loses its connection to the circle: it never becomes a field pursued for its own sake and is always studied (aside from the *Measurement of the Circle* itself) with astronomical applications in mind. However, because a measurement of an arc can always be construed as a measurement of an acute angle in a right-angled triangle (whose hypotenuse is the diameter), every astronomical calculation of this type can become a trigonometric calculation in our own, modern sense (as noted earlier for "cosine 15"). Indeed, the ratio of the circumference of the circle to its diameter is simply a special case of the ratio of arcs to chords. An alternative geometrical universe, where there is a single formula with which a chord may be calculated – a general, straightforward construction for the general chord, given the general arc – must also be a universe furnishing a straightforward construction of the straight line equal to a given circumference of the circle. There isn't such a simple geometrical construction. (Archimedes achieves such a construction in *On Spirals*, but this is possible only because the construction of the spiral, itself, relies on the circle: this is, literally, a circular construction!) This means that trigonometry, too, cannot be done on the basis of any single formula and must instead be pursued through a grab bag of tricks, among which Archimedes's iteration of bisections was the first to be found.

What was inside Hipparchus's own grab bag? We can't really tell. Our next source of evidence is Ptolemy, whose originality cannot be ascertained. Indeed, as usual, Ptolemy is economical in his presentation, providing an overview of, effectively, not more than the tools he requires for his precise astronomical needs (*Almagest* 1.10). I find it likely that Hipparchus's discussions were more broad and systematic. Ptolemy's discussion can be taken, in my view, as a kind of lower boundary on Hipparchus's.

Ptolemy's goal, to repeat, is to produce a table of chords with ½-degree intervals. As Archimedes does in the *Sand-Reckoner*, Ptolemy starts from measurements that can be found directly by basic considerations of plane geometry (to the self-evident case of the hexagon, Ptolemy adds the somewhat more complicated starting points of the pentagon and the decagon; this is all material related to *Elements*, Book XIII). We thus have the chord measurements for seventy-two, sixty, and thirty-six degrees. It's helpful to have multiple starting points because, from this point onward, we will use a kind of combinatorics to derive further values.

The simplest tool for extending our measurements is that of supplementary angles. If we have the chord of arc 72, for instance, known in terms of parts of the diameter, we also have the chord of arc 108. This is because we can draw a right-angled triangle, one of whose sides is the chord on arc 72, another of whose sides is the chord on angle 108, its hypotenuse being the diameter.

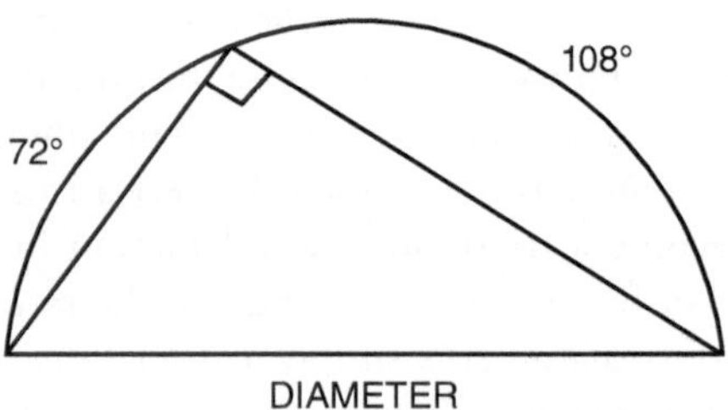


Figure 89

Again and again, we apply Pythagoras's theorem (notice that we will need to find a square root of some number, which we will most likely need to approximate; this process is based on upper and lower boundaries throughout). Ptolemy adds to this another simple and clever device. In Figure 90, AD is the diameter, and the chords AB, AG are given.

The claim is that we can then compute chord BG. This allows us to find the chord of a *difference* between chords, then. We achieve this based on the following theorem:

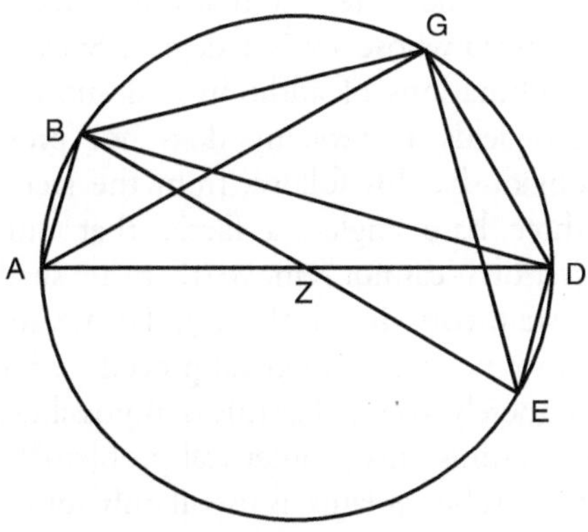


Figure 90

> The rectangle contained by AG, BD is equal to the sum of the rectangle contained by AB, DG and the rectangle contained by BG, AD.

(I will not prove this theorem; it is a simple result based on considerations of similar triangles, that become proportions, that become equalities between rectangles – as we have seen so many times before in Greek mathematics!)

Now, we have the value of the rectangle contained by AG, BD (AG is given directly, and BD may be found as the supplement to AB). We have the value of the rectangle contained by AB, DG (AB is given directly, and DG may be found as the supplement to AG). We thus can find the value of the rectangle contained by BG, AD, of which AD is the diameter, that is, given. We thus can find BG.

To this, Ptolemy adds an analogous theorem, allowing him to compute not only the difference but also the sum of solved arcs. So, to recapitulate, we have at this point the tools with which, given the chord measurements for arcs X, Y, we can find the chord measurements for the arcs

$$180 - X, \quad X - Y, \quad X + Y.$$

And also, based on Archimedes's technique, we can find the chord measurements for the continued bisections of X, Y. The entire process is bootstrapped by three geometrically given starting points – 36, 60, and 72. The smallest value we can directly derive is 72 − 60 = 12, and with continued bisections, Ptolemy gets to 1½ and ¾.

Ptolemy's desired increment is ½, which he plans to get as the difference between 1½ and 1. But he notes, at this point, that he cannot get the measurement of the chord whose arc is 1 degree. Starting from 36, 60, and 72, there are no combinations of addition and subtraction of which any continuous bisection yields 1! Ptolemy does not prove this theorem of impossibility, but obviously, this follows from the fact that the factor 3 is common to our three base angles, a factor that our tools – addition, subtraction, and bisection – cannot remove (there are shades of the numerical manipulation of music theory, in all of that). To measure the chord whose arc is 1 degree, then, we need a geometrical procedure for *trisecting the angle.* Ptolemy, once again, merely asserts that this is impossible (did Hipparchus go on a long excursus regarding this geometrical problem?) and uses, instead, a clever roundabout that, as he explains, is useful only for very small arcs. (Note throughout the attention to the detail of actual calculation, taking precedence even over ideals of generality; this, above all, strikes me as Hipparchean, if not in precise technique than in its overall spirit.)

Ptolemy proceeds by proving that, in general, given associated chords and arcs in a circle, the ratio of the greater to the smaller arc is greater than the ratio of the greater to the smaller chord:

$$\text{arc AD} : \text{arc AC} > \text{chord AD} : \text{chord AC}.$$

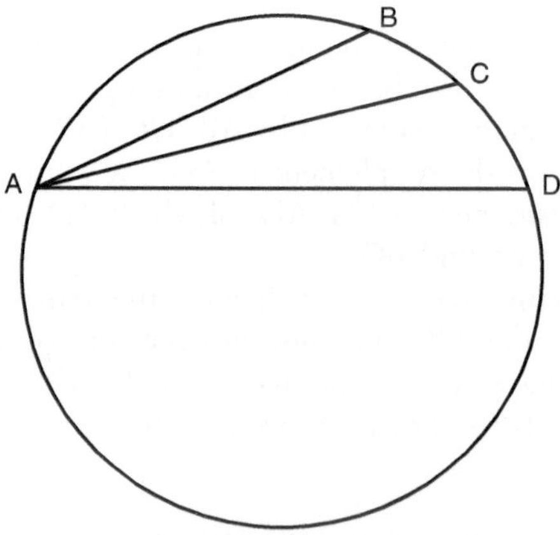


Figure 91

In Figure 91, we may now take the arc AD to be that of 1½ degrees; AC, 1 degree; and AB, ¾ degree. Ptolemy has already computed the length of the chords of the 3/4 degree – this is, in sexagesimal value, 0;47,8 units (where the radius is 60 units) – and of the 1½ degrees – this is, in sexagesimal value, 1;34,15 units. We therefore know that the chord of 1 degree is as follows (slightly modernizing):

More than 4/3 times 0;47,8 but less than 2/3 1;34,15.

At the level of precision of two sexagesimal points, this settles the value as 1;2,50.

And with this, Ptolemy can now derive his table: first, finding the chord for ½ degree based on the values for 1 and 1½; and then, applying this value for ½ degree across the board, based on addition and subtraction from the many starting points available already. Almost all of this vast operation of calculation is black-boxed so that all that we see is a table providing a value: an arc measurement and, next to it, its associated chord measurement. Ptolemy's table begins like this:

| Arc | Chord |
|---|---|
| ½ | 0;31,25 |
| 1 | 1;2,50 |
| 1½ | 1;34,15 |
| 2 | 2;5,40 |
| 2½ | 2;37,4 |
| 3 | 3;8,28 |

Even sitting down to copy these values (which I take from Toomer's translation of the *Almagest*) caused me a little headache. I only got down to 3 out of 180, and needless to say, I didn't compute anything. Ancient astronomers computed all the values, without the aid of the calculator ... Hipparchus, in all likelihood, the very first to do so.

Note the overall structure. One begins by producing theorems, showing how, given a certain value *X*, another value *Z* is given as well. Such proofs are indeed common in Greek mathematics (Euclid's *Data* is all about such relations). Elsewhere in Greek geometry, those are "proofs of concept" whose point is to show the fact of a functional relation, hence the logical *possibility* of a calculation. We do not know what proofs of this kind Hipparchus actually used. What we do know is that whatever proofs he used, he then had to use them not just as showing the possibility of a calculation but, instead, as practical tools, applied again and again – and again – until he had written down a full set of chords. This, I think, is the type of attitude Ptolemy had in mind when praising Hipparchus's "industriousness." There was real labor involved in this astronomical craft. (More on the practical aspect of such tables follows in our discussion of Ptolemy.)

The labor becomes even more forbidding once we begin to consider *spherical* trigonometry. Once again, we need to piece our evidence together, although in this case it all postdates Hipparchus. We have extant, but only in Arabic (and likely, somewhat mediated) translation, Menelaus's *Spherics*: a work from late in the first century CE. The work considers various questions in the geometry of spheres and, in particular, various arrangements with spherical "triangles" (that is, figures on the surface of the sphere bounded by three arcs). One result stands out: *Spherics* III.1, which allows the calculation of the ratio between spherical chords. The same theorem is proved in the *Almagest*, I.13. Most scholars today believe that both Menelaus and Ptolemy relied on a common source, who then, most likely, was Hipparchus; by scholarly habit, this result is still referred to as "Menelaus's theorem."[21] It is obviously useful to be able to calculate arcs on a sphere, based on other given arcs. For instance, if we know that a planet starts at a certain position on the ecliptic and then moves with a certain average speed for a given time, we can tell which position the planet will be in on the ecliptic. This, however, will not tell us directly, for instance, how far up it will be relative to the *horizon*. We seek to transform one arc measurement to another. The way this is done in practice is through proportions that hold between the chords of such arcs, and this is provided by Menelaus's theorem.

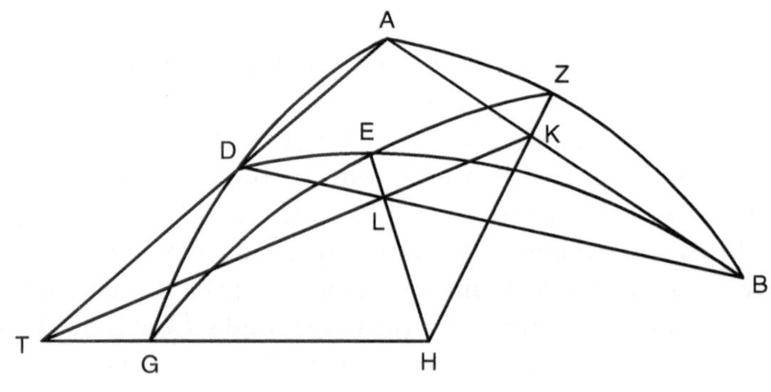


Figure 92

In the arrangement in Figure 92, GA, BA, BD, GZ are all arcs on the surface of a sphere; it is required that AG, AB are less than a semicircle and

[21] My account is informed by N. Sidoli, "The Sector Theorem Attributed to Menelaus," *SCIAMVS* 7 (2006): 43–79.

that E is taken (anywhere) inside the spherical triangle GAB, obviously determining the positions of D and Z. We also take H, the center of the sphere, and draw all the straight lines (it is proved that the straight lines actually produce the configuration shown here, in itself not a trivial observation). We associate each arc not with its own chord but with the chord on its double. In the Latin translation, this is referred to as the "nadir" of the arc, and regardless of the actual etymology, this may serve to explicate the significance of the chord-on-double-arc. In Figure 93, I concentrate just on the arc ADG, adding in the points M, N so that the arc DN is twice the arc DG, and the arc AM is twice the arc AG. I label the intersections of the chords DN, AM with the line HT as P, R, respectively.

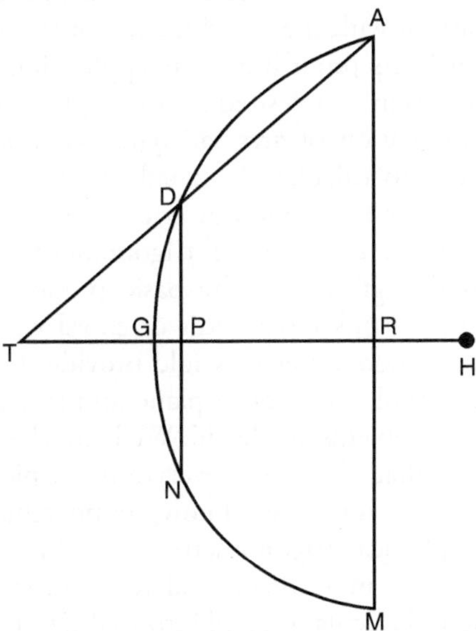


Figure 93

HG is a radius of the sphere, and so, because DG = GN, AG = GM, we find that it is orthogonal to lines DN, AM. From the point of view of line HG, point N is "directly underneath" D, and point M is "directly underneath" A – so that, at the very least, a modern reader can use "nadir" as a mnemonic to clarify the meaning of this chord-on-double-arc.[22] As

[22] To be clear: this is likely not the actual etymology of the word. In general for the theorem and the history of its transmission, see, once again, N. Sidoli, "The Sector Theorem Attributed to Menelaus," *SCIAMVS* 7 (2006): 43–79.

usual, the presence of parallel lines is suggestive, to the Greek reader, of proportions, and we immediately notice that DN:AM::DP:AR::DT:AT. The "nadirs" of DG, AG are to each other as the line segments DT:AT.

In what follows, I refer to the nadir of an arc as "nadir (arc)," and I use the expression "comp." to refer to the composition of ratios (which is what we think of as the multiplication of fractions). Based on this, it takes only a moderate manipulation of proportion to get to Menelaus's theorem (Figure 92):

$$\text{nadir}(AZ):\text{nadir}(BZ)::(\text{nadir}(AG):\text{nadir}(GD))\text{comp.}(\text{nadir}(DE):\text{nadir}(EB)).$$

The proportion manipulation underlying this theorem is not too demanding; what makes it difficult is the three-dimensional intuition required. But this pales in comparison with the real difficulty of this proposition, which lies, once again, not in its proof but in its application. For every specific problem in the calculation of a desired arc on a sphere, one would need to find the useful combination of arcs satisfying Menelaus's theorem, and then one would need to calculate in detail: in the given configuration, certain arcs are known, and so one can find the ratios of their chords or nadirs (this, itself, takes up a lot of trigonometry). One would then discover, based on this given and the basic proportion calculations of Menelaus's theorem, the unknown nadir or chord which, working backward with some more trigonometry, would provide the desired arc.

Did Hipparchus's book study both plane and spherical trigonometry? Did it work out all problems in detail? Well, maybe it did take twelve books, after all! Not that we know: everything is pieced together from much later sources, and for all we know, Hipparchus could well have treated plane and spherical trigonometry in distinct places, developing them in some works in a more theoretical vein and in others, developing the theory into practical calculation and into well-developed tables. He was a prolific author, after all. Nothing is clear, with his trigonometry as with everything else. But the label of "industriousness" seems spot-on.

### *"A Lover of Truth" and Hipparchus's Character as a Scientist*

Now, what about the "love of truth"? This, too, calls for some unpacking. I am going out on a limb a little – but just a little – here when I suggest that when Hipparchus is said to be a "lover of truth," what is really meant is that he was a "hater of falsehood." We do not need to bring any new evidence for this and instead just consider the bulk of the evidence summoned so far. We noted his one extant work: it was a heated critique of Aratus and, through him, of Eudoxus (another astronomer, Attalus, who justified Aratus in his writings, ends up, so to speak, as collateral damage). We noted Hipparchus's geography: it seems to have been a series of critiques of Eratosthenes. We noted Hipparchus's combinatorics: it was all prompted by a critique of Chrysippus. Throughout, we see the same pattern: Hipparchus leverages his industriousness (his ability to bring together not only mathematical understanding but also plenty of hard-won data and calculations) so as to refute a statement – regardless of the degree of numerical precision by which the statement was even intended by its author. Hipparchus, implicitly, demands everyone to be Hipparchus, then savagely criticizes them for failing.

We recall (p. 328) Ptolemy's statement concerning the length of the year (that is, effectively, Hipparchus's great discovery of the precession of the equinoxes):

> The ancients were in disagreement and confusion in their pronouncements on the [length of the year], as can be seen from their treatises, especially those of Hipparchus, who was both industrious and a lover of truth.

What I believe this means is that Ptolemy had at hand a book by Hipparchus on this question, in which Hipparchus went, in detail, through all past work on the length of the year (on this, more calendrical subject, claims could be found from many past authorities), refuting them all.

This, then, is an interpretation of what, I believe, might have been one of Hipparchus's major springs of intellectual action: the detailed, numerical refutation. As I said, this goes out on a limb (so little is known about this author) but not too far: the emerging picture, pugnacious and prolix, is one that we can easily fit a Greek author to.

But if so, I think this suggests a certain interpretation of the most famous of Hipparchus's silences. I gave a flavor – no more – of Hipparchus's theory of the moon, and we know that he produced a theory of the sun. It is unthinkable that an astronomer of his ambitions would not have written on the planets, and given his future influence, any such theories he might have had would surely have been cited. Ptolemy clarifies that Hipparchus did not, in fact, produce a positive theory of the planets (*Almagest* IX.2):[23]

> Hipparchus, being a great lover of truth ... did not even make a beginning in establishing theories for the five planets ... All that he did was to make a compilation of the planetary observations arranged in a more useful way, and to show by means of these that the phenomena were not in agreement with the hypotheses of the astronomers of that time.

[23] G. J. Toomer, *Ptolemy's Almagest* (London: Duckworth, 1984), p. 421.

What I think is clear enough is that the point of Hipparchus's exercise was not to say, with the wistful recognition of one's own limits, "That's it, folks: here are the facts concerning the planets, and unfortunately, even I can't produce a theory based on such data!" I am more inclined to believe that Hipparchus's main tenor was perhaps rather like, "Those are the past theories regarding the planets, and thanks to the new data I have now arranged, I can calculate in detail and refute them one by one, definitely showing that all past astronomers were wrong."

Once again: this is going out on a limb. It is, of course, *conceivable* that Hipparchus's main target was simply Eudoxus's homocentric spheres (this, indeed, would be in line with Hipparchus's extant treatise criticizing Aratus and, through him, Eudoxus's mapping of the sky). But because Hipparchus's modeling techniques were so distinct from those of Eudoxus, I find it hard to see how a critique of Eudoxus could have been viewed as the grounds for Hipparchus's refraining from pursuing his own model. A more likely account, then, is that Hipparchus criticized a previous model – or several previous models – describing the motions of the planets with epicycles or eccenters. It is indeed extremely difficult to account for the planets in such terms alone, as the history of astronomy would continue to attest from Ptolemy to Copernicus and beyond. (Ptolemy's own approach adds more ad hoc mechanisms, as I shall briefly return to note later in the discussion.) Lacking Hipparchus's Babylonian data and numerical techniques and probably lacking even Hipparchus's table of chords, it is easy to imagine how previous efforts by, say, Apollonius, could have been, to say the least, empirically crude. Would such a failure even matter, all that much, to an author such as Apollonius, who may well have been more interested in the geometrical polish of the theoretical arguments? It is easy to imagine Hipparchus pulverizing such efforts to the extent that, later on, no one took them with great seriousness. One would go on reproducing the mechanisms, inspired by Archimedes's invention, but they would be seen as mere approximations, and it was only much later – possibly, only with Ptolemy? – that new planetary theories would be on offer. Indeed, why bother? For practical purposes, all one needed was to calculate the positions of the planets – which one could do, tolerably well, with Babylonian-style calculations. (The practical purposes would become ever more practical in the generations following Hipparchus, through the explosion of astrology, to which we turn immediately – which, indeed, continued, for many centuries, to rely on the Babylonian tools.)

A negative result, then, but with perhaps a small addition to our evidence concerning the whodunit of the previous section. So much of

Hipparchus's output was critical and therefore reactive. This is especially likely for his remarks concerning planetary theory, and thus I conclude that the generation of Archimedes probably did produce some astronomical works, broadly anticipating the models of epicycles and eccenters that came to define future astronomy.

We do not reduce anything from Hipparchus's greatness by noting his use of the past. Indeed, part of his greatness lies in his intuition of which past was worth using, indeed, in some sense finding and even inventing it. Likely enough, Hipparchus created the route through which Greek astronomy learned how to use a *Babylonian* past. This was by far the most significant scientific synthesis of distinct cultural traditions, up to that point.

We have one more small piece of evidence that invites us to put Hipparchus in the context of his past – and, indeed, of his future. The evidence, as is so often the case with Hipparchus, is tenuous. All we have are a few comments from – who else? – Simplicius (*Commentary on Aristotle's* De Caelo 264.25–265.6):

> Hipparchus ... in his work entitled On Bodies Carried Down by Their Weight declares that in the case of earth thrown upward it is the projecting force that is the cause of the upward motion, so long as the projecting force overpowers the downward tendency of the projectile, and that to the extent that this projecting force predominates, the object moves more swiftly upwards; then, as this force is diminished (1) the upward motion proceeds but no longer at the same rate, (2) the body moves downward under the influence of its own internal impulse, even though the original projecting force lingers in some measure, and (3) as this force continues to diminish the object moves downward more swiftly, and most swiftly when this force is entirely lost. Now Hipparchus asserts that the same cause operates in the case of bodies let fall from above. For, he says, the force which held them back remains with them up to a certain point which accounts for the slower movement at the start of the fall.

This is a tantalizing passage, and it is infuriating that we hear no more of it (and indeed, see nothing even like it prior to yet another commentator to Aristotle – Philoponus, active even later than Simplicius in the sixth century CE! Philoponus develops a distinct theory of how the mover provides the moved body with an "impetus," which is then gradually lost through the moved body's motion, a theory that will come into its own throughout the Middle Ages[24]). But the lack of evidence is in itself significant. This was just not a typical area for Greek writing. Philosophers

[24] For Philoponus as a creative natural philosopher, see R. Sorabji, et al. (eds.), *Philoponus and the Rejection of Aristotelian Science* (London: Duckbacks, 2010).

sometimes worried about the problem of motion; mathematicians, rarely, made their geometrical claims significant in physical terms. What we do not see anywhere is any attempt to produce a fully fledged mathematical model of motion. Aristotle, indeed, comes closest, in that he discusses some of the basic equivalences of motion ("faster" means the same distance traversed in lesser time, etc.). He specifically believed that objects have a natural tendency to move in certain directions – hence, earth falls, fire rises, and so forth. (Notoriously, it seems as if Aristotle concluded that the heavier the body is, the faster it will fall.) All of this in Aristotle is not mathematics. I say this because modern scholarship (to the extent that it engages with Hipparchus's *On Bodies Carried Down by Their Weight* at all) tends to read Hipparchus specifically within an Aristotelian context. The attraction of such a reading is clear: our evidence is from a commentary to Aristotle, and other writings on the topic, from Philoponus through the Middle Ages and all the way down to Galileo, are primarily responses to Aristotle. But this much is clear: in the second century BCE itself, Aristotle was marginal even within philosophy, let alone to a scientist such as Hipparchus. It is quite likely that Hipparchus never even read Aristotle's *Physics*.

What is suggestive, however, in Hipparchus's treatment – mediated and fragmentary as it is – is the process of treating, first, the configuration where two forces are acting against each other and then, from this basis, drawing consequences for the configuration where only one force is active. This is reminiscent, specifically, of Archimedes's *On Floating Bodies*. There, one considers a floating solid, lighter than the liquid, as if it were composed of two components: one, immersed in the liquid, pushed up; the other, outside the liquid, pushing down. From the equality between the two forces, one can deduce, then, the power with which a solid, immersed in a liquid, is pushed up by it. It seems to me at least possible that Hipparchus follows a similar train of thought: identifying the motion of a projectile, pushed up, as the composite of two motions, its projectile force and its downward-heading force. From this, Hipparchus moved on to results concerning the nonprojectile, falling motion, where the downward-heading force acts alone. (As we recall from Pappus and Hero's treatment of a body falling on an inclined plane [pp. 259–260], there is precedent for trying – unsuccessfully – to extend Archimedes's approach to other physical problems.)

Simplicius states nothing on the detail of Hipparchus's argument. He probably knew it only through intermediate sources, likely philosophical in character. He probably knew nothing, then, of Hipparchus's original mathematics in this treatise. But given everything else we know of Hipparchus, I do not believe that this treatise was strictly philosophical, merely engaged with the metaphysical analysis of the forces acting on moving bodies. Further, I find it incredible – given everything else we know about him – that Hipparchus would not produce lots of calculations in support of his statements.

But if so, we come close to imagining a very Galilean Hipparchus. Let me indulge myself, as in my mind's eye Hipparchus climbs up the steps of the remains of Rhodes's Colossus – ready to drop his balls and measure the time in fall ... This, of course, is sheer romance, and I understand why past scholars have resisted engaging in such speculation. After all, had Hipparchus anticipated Galileo, how come we had to wait so many years for the real Galileo? How come we hear so little of this decisive breakthrough?

There are two complementary ways to approach this. First, everything we know of Hipparchus depended on the destinies and choices of *other* authors. He attached himself to the fame of Aratus by writing a critique of that massively popular poem, thus securing his only extant treatise. Strabo chose to cite Hipparchus extensively in his own, extant, great geographical compilation; Ptolemy chose to cite Hipparchus extensively in his own, extant, great astronomical compilation. Plutarch hated Chrysippus with such relish that he was glad to cite any attacks upon him. It is via such sources that Hipparchus's light is reflected to us. He was an extremely prolific author (it appears that among his treatises is a catalogue of his own works – surely the mark of a truly active author!). There was a lot to get reflected, and so, even from such bits and pieces, a picture does emerge. But you need later authors, engaged with their own enterprises, for such reflections to emerge. There simply was not any more expansion of the Archimedean project of the equivalence of geometry and physics – largely because there was very little by way of advanced, pure geometry at all. Hipparchus himself may be seen as a belated, critical echo of the second generation of Greek mathematics. As we will see, he is followed by a cliff. Mathematics nearly disappears and is then reconstituted quite differently. The Archimedean moment does not give rise to an Archimedean era.

And here is the second point: Hipparchus, after all, does come after Archimedes. In which case, there is nothing very counterintuitive about the idea that he should anticipate Galileo. As mentioned already in the introduction, and as I will return to discuss in a little more detail in Chapter 7, Galileo's starting point was, quite simply, Archimedes, and he did not use, in fact, any significant mathematical techniques not available already in the late third century BCE.

Taruskin, a music historian, quotes his colleague, musicologist Joseph Kerman, as fond of saying, "We live in the valley of the Ninth Symphony." What I point out is that for almost two thousand years, beginning already with Apollonius and going all the way up to Galileo and beyond, mathematicians had lived in the valley of Archimedes. Archimedes had in some cases perfected, in other cases created, the tools that would make possible the achievements of a more modern science. We will take a quick look at this in Chapter 7: how Islamic science almost got there; how long it took the European Renaissance to lift off. The story is one of centuries of false starts. Of the many not-quite-scientific revolutions, Hipparchus's was the first.

## III. Culmination: Ptolemy

### *Before Ptolemy: Signs and Tables*

This may be the story not of one but of two multicultural empires. Iran, a land of deserts, mountains, and plateaus, was marginal to the ancient Near East until, in the sixth century, it was united by the Persians. In the year 539, King Cyrus conquered Babylon itself. For the next two centuries, all of the landmass between India and Greece was ruled from an Iranian center.[25] It was an empire with a light touch, gaining its legitimacy through many acts of local delegation. Everywhere, life went on as before. The Jews were allowed back into their holy city; the Chaldeans, too, were allowed to keep on doing, in their temples, whatever it was they had been doing since time immemorial.

What they were doing was *looking for signs*. So, for instance, for many centuries, the priests were responsible – in Babylon as everywhere else – for animal sacrifice. One killed the animal, cut it open – and looked inside. It was a fraught religious event, rich with significance. There were texts to teach the priests how to read those moments:[26]

> If there are holes in the "head of the finger" [a technical identification of a part of the liver] – the destruction of the land will take place.

Our clay tablets record many statements of the form "if <an observation is made> [then] <a prediction follows>." The observations range widely across many fields of human experience. The slaughtered animal body is indeed an important site – but perhaps none was as important as the stars.

[25] A comprehensive history of this remarkable state is P. Briant, *From Cyrus to Alexander: A History of the Persian Empire* (State College, PA: Eisenbrauns, 2002/1996).

[26] I quote from the cuneiform tablet K. 3816 + K. 6777 from the British Museum, Omens 5 and 4, via I. Starr, "Omen Texts Concerning Holes in the Liver," *Archiv Für Orientforschung* 26 (1978): 45–55.

These were two places where many ancient people felt that they met the divine. Indeed, why not both? Quoting from the very same tablet:

> If there are holes in the "head of the finger," on top of the "finger" – there will be eclipse (predicting) a mourning in the country.

A prediction of a prediction! The world was a web of signs, pointing to each other and ultimately shining light, above all, on the big questions of the state: the rivers, the kings, the outcome of war. These were not mere empirical observations. As Rochberg shows, the signs were written by gods, and apparently, so were the collections of signs themselves: the books of the form "if X, then Y" were taken to have, themselves, divine authorship. How could one ignore that?

It should be clear that the star lore accumulated by the Mesopotamians was motivated entirely by such encounters with the divine. The Mul.Apin, recording events in the sky (together with indications of synodic periods), was there for the sake of the Enuma Anu Enlil, the series of astronomical omens. In the late kingdoms of Mesopotamia, in particular, observations in the form of detailed diaries were preserved throughout many generations. Toward the end of this period, as noted earlier, master diviners outdid each other by their skill in *predicting* eclipses. But this is as far as Babylonian astronomy went, before Cyrus.

Now, with the Persians arriving, with the old kings of Mesopotamia gone – what would happen to this old tradition? Well, remarkably, it went on. Persepolis, however, was distant, and the astronomers found themselves, so to speak, for the first time, on their own. The interesting thing is that it was then, precisely, that prediction took off. As we recall, schemes were now developed to predict not only eclipses but also the major planetary events. At the same time, the entire point of divination became reoriented, from the court to something like daily life. The major new invention was that of the horoscope, an astrological tool designed to tell the fortune not of the state but of an individual. The two developments are independent: the specific predictive techniques developed were focused on what was still conceptually and religiously primary – the qualitative events of the sky. They were not that good at deriving continuous predictions that allow the calculation of positions at given points of time, but this is precisely what is required for the casting of a horoscope (where one needs to calculate positions for a day defined through an extra-astronomical consideration: a person's birthday). To anticipate: one advantage of Greek geometrical techniques was precisely that they led more naturally to continuous predictions and so were, in principle, better suited to astrology!

The astronomers of Babylon definitely came down in the world. Your ancestors had the ear of the kings: no longer. On the positive side, however, when you tell the fortunes of mere individuals, your potential customers suddenly become the whole human race. Why stick around in Babylon? It is clear that many did not. Perhaps our earliest unequivocal evidence is an inscription from Thessaly – Greece itself! – from the second century BCE, commemorating Antipatros, born in Syria (so, already away from Babylon), a "Chaldean Astrologer."[27] This is already contemporary with Hipparchus. There is yet another set of archeological evidence that is considerably later but is truly abundant: the Egyptian papyri of the Roman era. Hundreds of fragments – in both Greek as well as Demotic Egyptian – record the very same kind of astronomical and astrological calculations found in Babylonian clay tablets. (The most significant group of such texts has been edited by Alexander Jones.)

The date of the papyri edited by Jones is not dispositive – most papyri, anyway, are from the Roman era, and so there is nothing surprising or meaningful about the fact that the same is true for the astrological papyri. But it seems as if this other multicultural empire – Rome – did transform the fortunes of astrology. For this, we need to turn to our literary evidence. Antipatros the Chaldean was active in Thessaly in the middle of the second century BCE. He surely had his clients, and his type of craft was certainly known to at least some Greeks. But it just did not register very much. In fact, we are hard-pressed to find even a single Greek author showing *any* awareness of astrology prior to the first century BCE. Polybius keeps emphasizing the technical skills required for leadership and keeps heaping scorn on charlatans. One way or another – as a fan or as a detractor – we would have expected him to refer to astrology if it had been a significant cultural presence. He never does. The same is true of Hipparchus: had he believed in astrology, he surely would have made a contribution to it; had he denied it, he could write splendidly scathing critiques – either way, he would be likely to be quoted by his astrology-crazed posterity. It seems like he didn't care either way. (And this, while he clearly had a superior grasp of Babylonian star lore, as a strictly astronomical technique!) And then – in the first century BCE, our evidence explodes. Cicero keeps discussing astrology as a key example of divination; our historical sources describe how Roman leaders from Sulla onward rely on astrological predictions

[27] The evidence is just the inscription, SEG 31:576, discussed in detail only in Italian: I. Savalli, "Un 'Astronomo Caldeo'nella Tessaglia tardo-ellenistica," *Annali della Scuola Normale Superiore di Pisa. Classe di Lettere e Filosofia* 15 (1985): 539–558.

(beware the Ides of March!). Sulla was active early in the first century BCE; we hear nothing similar of earlier Roman leaders or even, more remarkably, of Hellenistic kings. The transformation is decisive, quick, and Roman. (Curiously, the Babylonian tradition of astrology ceases at the same time. Clay tablets with astronomical contents become rare throughout the first century BCE and extinct in the first century CE. Did they all decamp to the Roman empire?)

It is not that the Romans were congenitally more superstitious or less critical. Rather, astrology fitted well the wider cultural project pursued by Roman patrons throughout the first century BCE. From Polybius onward but especially from the first century BCE and later, Greeks served Roman patrons with useful combinations of Greek knowledge. It was the age of eclecticism and synthesis. And so, what's better than this combination of Greek and even more exotic knowledge? Which – in contrast to much else that Greeks had on offer – was also so obviously *practical*?

Greek and more exotic knowledge – combined. Let us pause to consider this moment of astrological encounter. Just what was it that a Chaldean-style astrologer *knew*? What was involved in the Hellenization of this knowledge?

This should be made clear: most practicing astrologers were not, remotely, engaged in scientific research, however capaciously we may define the term. Already in Mesopotamia itself – already in the evidence of the clay tablets – we cease to find any innovation soon after the emergence of the new predictive systems. It is unclear whether any Babylonian astrologer, from about 300 BCE onward, ever paused to consider why he used the tools he did; they had no occasion to. It was a craft, to be mastered. Admittedly, a difficult technique. One did not have access to a single source that detailed the position of any heavenly body at any given point in time. Instead, positions had to be computed. To get this all started, one needed to acquire tables with the numerical coefficients determining the predictions – and to learn the technique of working with such tables. Only then could one produce yet another set of tables – almanacs detailing the major synodic events across a certain period of years. (The calculations are laborious and iterative; as we will return to see with Ptolemy, one uses tables as tools for the making other tables.) With these at hand, one would further extrapolate the positions of the stars at a given time. There is a great deal of technical knowledge and computation even after the bare positions are established. The position of a planet, for instance, is to be given in terms of $1/12$th of a sign – each sign, so to speak, divided into its own twelve signs: the logic of divination is always to maximize meaning by adding on yet more layers of what is considered significant. One also adds meaning by considering the relations between signs: How many signs are the planets apart? One adds up all this information and finally gets to the actual work of interpretation, where suddenly, computation halts. This is now the realm of art. Whether from training or from explicit handbooks, one knows a large number of "if, then ..." statements, which, however, by their sheer accumulation, rule out algorithms: the significance of the signs is contradictory and variable, and so one, at the end, has to choose and refine one's own narrative.

The practicing astrologer, then, is a master calculator and a narrative artist *but not a theoretical scholar*. Indeed, he need not even observe the stars. His ancestors in Mesopotamia scoured the sky for portents because their work was understood on the grand scale of court and cosmos so that events, on the spot, carried immediate meaning (the same, of course, was true in China, in the Maya, and in many other civilizations). Individual destinies, however, were understood to be made at birth and so were calculated and *not* observed. Neither a theoretical astronomer nor an observer, the practicing astrologer was most like the practicing doctor, carrying a set of books with the old lore and producing recipes – or horoscopes – out of such old knowledge. The two professions would continue to be plied, side by side, for centuries to come (in the Islamic world, in particular, often by the same person; already the Romans, late in the first century BCE, went through a quick fad of astrologically based medicine). These were, so to speak, the two main scientific service industries of the premodern world.

If we look for applied mathematics, we find it, finally, here, with the symbolic service provided by astrologers. "Symbolic," quite literally: they took signs and applied a specialized technique so as to make those signs "speak" so as to provide narrative meaning to individual life.

It is remarkable, then, that the chief applied mathematics of antiquity was not a *Greek* mathematics. The numerical techniques were, in origin, Babylonian. But this must be qualified. Babylonian astrology did become partly Greek. It happened almost by necessity. It was a traveling astrology; and astrology, in fact, has a hard time traveling unless it adapts. This is true for a technical, astronomical reason. We have seen Hipparchus's use of Babylonian eclipse observations. He had the utmost faith in the accuracy of the observations themselves (which, indeed, is remarkable from such a critical author). But he did not take them as is. He first transformed the recorded time and transformed it into a time measured on an absolute scale and then transformed this, in turn, to the time that would have been observed from Rhodes. The first thing one needs to do with astronomical data is to calibrate them for time and place, which was never required *at a given Mesopotamian temple*. Not so if your astrology is to be practiced across the Mediterranean.

We are coming close to the Kuhnian notion of "paradigm," with its attendant questions of the "reality" studied by science. Is reality truly the same across scientific paradigms? The Babylonians look at the sky and see there the sun, the moon, and the planets, producing key events such as eclipses or first risings. The Greeks look at the same sky and see there a rotating sphere. For the Babylonians, the zodiac is simply the baseline within which events are recorded. For the Greeks, the zodiac is an astronomical agent, its rotation and the angles measured upon it producing the passage of the year. And so, Greeks do not merely introduce new techniques to recalibrate astronomical events based on latitudes and longitudes. The zodiac itself is turned into an astrological sign. Specifically, at the moment of birth, one notes not only the position of the sun, moon, and planets but also the position at which the ecliptic rises above the horizon – the "ascendant."

The craft of the astrologer, then, is composite in at least two senses. First, it combines a craft of computation (for the finding of significant astronomical positions, provided for given times and places) with the hermeneutic art of turning astronomical positions into astrological predictions. Second, in the computation of the positions, it combines Babylonian arithmetic techniques, with Greek geometrical ones. All of this is reflected in a rich and varied extant literary tradition. We have the Greek didactic poetry of Anubion (from the first century BCE, extant in substantial papyrus fragments), the Latin didactic poetry of Manilius (from the early first century CE), the Greek prose of Vettius Valens (second century CE), the Latin prose of Firmicus Maternus (fourth century). Altogether, about a dozen ancient astrological authors are now extant. Not bad, for what might strike one, at first blush, as an *esoteric* field! Indeed, we would have expected the computations (which are, after all, a matter of objective fact) to be widely shared while the hermeneutic art (which is a matter of quasi-religious, exotic doctrine) was kept from wide audiences and merely transmitted from master to pupil. We find nearly the opposite: most treatises merely glance at the technical details of astronomical computations and instead engage almost entirely with the art of interpretation. (Vettius Valens gives more such details than most others, but this is because his text is enormous – in nine books – and even he concentrates on the hermeneutic aspect.) What we know of techniques of astronomical computation in the ancient world is known to us, most often, *indirectly* – pieced together from the documentary evidence of the papyri and from other, later sources, requiring painstaking deduction by modern scholars. (Perhaps the most remarkable element in this modern jigsaw puzzle is the argument – pursued, above all, by David Pingree – that we must trace back, to independent developments in the Greek Mediterranean, certain techniques of medieval *Indian* astronomy.) Likely enough, such treatises were written with the interests of the wider audience in mind. The audience did not want to know how to find the position of Jupiter. They wanted to know how long they would live and if they would be successful in love. In short, this astrological literature hardly belongs to the history of mathematics. It is a cliché of modern scholarship that until modern times, the line between "astronomy" and "astrology" was blurred (indeed, the words themselves – both, obviously, are Greek – were interchangeable). But the truth is that we find very few authors in antiquity who were noted both as practicing astrologers and mathematical astronomers in the tradition of Eudoxus or Hipparchus.

Let us pursue this briefly – so as to understand better the nature of the (qualified) yet major exception in the figure of Ptolemy.

Throughout the first century BCE, we simply do not find authors whose main identity was that of mathematical astronomers for the simple reason that throughout this century, we find no mathematical authors *at all*. This is curious in an era of eclecticism and synthesis. Was not everything mixed with everything? Indeed so: in Rhodes, some philosophers are even positively interested in mathematics. We noted Posidonius in the preceding chapter. One of his followers, Geminus, wrote even more on mathematical subjects, including an astronomical handbook that is still extant. This work is important for historians of astronomy because so little else is extant. But in all likelihood, Geminus was a Stoic philosopher who merely dabbled in mathematics. The new spirit of eclecticism of this Roman era meant that Greek mathematics could be added together with other endeavors – whether philosophy or Babylonian astrology – but it was precisely that, an *addition*, the side dish to other, more central projects. Later in the first century BCE, Vitruvius wrote on architecture – and also brought in some mathematics. Even later, in the first century CE, our evidence for activity in the exact sciences often comes from philosophers, engaging with mathematics.

Once again from Rhodes, once again early in the first century BCE, we have the rare find of an inscription. It is known as the Keskintos inscription, after the modern name of the locality where it was found.

This is an astronomical inscription (like the Antikythera mechanism itself, then, we have yet another example of astronomers exhibiting their work in public). It lists the synodic periods of the planets, culminating with a very long, overarching cycle at the end of which all cycles return again: Was the purpose of this cosmological, no less than astronomical? We do not know: the remarkable inscription – like so much else of the astronomy of the time – remains isolated.[28] Geminus's astronomical handbook is followed, perhaps two centuries later, by another astronomical handbook by Cleomedes; from the first century CE we have extant a work by one Theon of Smyrna, titled *On the Mathematics Useful for Understanding Plato*. Even Nicomachus, an important mathematical author of the first century CE, appears to have a more philosophical set of interests (we shall revisit him, and the rise of a new mathematical philosophy, in the following chapter). These are all extant authors, but it is not as if we have a plethora of merely attested authors whose works were somehow lost; on the contrary, even though the amount of literature extant from the Roman era is significant (and certainly much bigger than that for the third and second centuries BCE), it mentions mathematicians only very rarely. We noted that the audience for astrology was interested more in its predictions for life and love than in its computational underpinnings. And similarly, the new audience for eclectic works, combining science and philosophy, was interested in philosophy more than in science.

It is remarkable that Ptolemy cites astronomical observations that can be dated to 98 CE, probably made by Menelaus. As mentioned on pages 340–342, the *Spherics* of the same author is extant in Arabic. A work on chords (certainly including a table of chords, as well) is also attested, and so the impression is of a Hipparchus-like figure, finally: an author who made contributions to astronomical theory – and also to the advanced mathematics one requires to underpin it. But this already brings us very close to the second century CE and to the time of Ptolemy himself. It is certainly not the case that theoretical astronomy went from Hipparchus to, say, Menelaus, simply untouched. We have lost so much that our view must be distorted. But the overall impression is clear: not only astronomy but the exact sciences as a whole went, at the very least, into decline. For generations, the most widely practiced form of mathematics was simply that of astrology. Little wonder that the Latin word *mathematicus*, at this point, comes to mean "astrologer."

[28] See A. Jones, "The Astronomical Inscription from Keskintos, Rhodes," *Mediterranean Archaeology and Archaeometry, Special Issue* (2006): 213–220.

However much they could be described as mere "craftsmen," the craft of the astrologers did carry a new meaning – and only now, following its (fairly recent) transformation into a horoscopic science, based on the position of the planets at given moments in time. Throughout history, across civilizations, people looked to the sky and cared, most, about the key qualitative events: the year, the moon, the vanishings and reappearances of stars. I noted, in connection with the loss of interest in Eudoxus's homocentric spheres, that, among other things, his research question (What is the trajectory of the planets?) just was not all that urgent. The stars were at the center of cultural attention, of course, but not for *this*. That is – not as long as star lore was defined by the Hesiod-like measures of the passage of the year. But now, following the triumph of horoscopic astrology, the position of the planets became absolutely crucial. The Babylonians, in the last flourish of their divinatory science, had changed the very meaning of the sky. And it was in this, above all, that all future astronomy would be inspired by Babylon.

Indeed, even more technically. Babylonian astronomy became Hellenized, but at the same time, Greek astronomy became Babylonized. This is apparent in the use of sexagesimals and, more generally, in the practice of sustained computations. But there is more than this: the Greeks did not merely inherit the Babylonian practice of calculating with sexagesimals; they inherited, with it, the method of computation reliant upon, and productive of, tables.

The use of tables was perhaps less remarkable on the Babylonian clay tablet. It was, after all, a *tablet*! The written surface of Babylonian writing was (typically) a unique, handheld store of information related to a particular task (a handy-tablet, so to speak). The simplest arrangement (found in most tablets) was to divide the surface of the tablet into irregularly shaped "cells," each of which included its own distinct information. (Many tablets look as if you had to take notes and had only one sheet of paper so that you ended up utilizing each piece of its surface, finding creative ways to fit in more and more notes.) Early in the second millennium BCE, some tablets began to be arranged in table form, the cells arranged vertically and horizontally and each row and column having its specified meaning. Almost all Babylonian tablets are accounts, and many, beginning with the second millennium BCE, take the form of a spreadsheet. These administrative tables become less common in later Mesopotamian history, during the first millennium BCE, but as Eleanor Robson explains, this is an effect of language. The old kingdoms used, for their records, Akkadian-sprinkled-with-Sumerian (Sumerian was a dead language, but the scribes prided themselves on their mastery of this esoteric culture). The new kingdoms used Aramaic. Now, the cuneiform script of Akkadian-sprinkled-with-Sumerian goes very well with the table form because Sumerian is written in logograms (single signs standing for entire words). This correspondence of visual and verbal symbolism naturally suggests how the layout of information on the page can represent, visually, the structure of the information. Hence, the habit of the table went out of fashion in the late period of Mesopotamian administration. The temple scribal practices, however, went on as before, and the Akkadian-sprinkled-with-Sumerian, less central to the affairs of the state, remained the vehicle for the practices of the priests. Now, there are certain cognitive tasks for which tables are good – such as the compilation of numerical information and its use for reference purposes. And so, Babylonians had the means to produce an effective computational astronomy. The evidence for Babylonian astronomy is, in fact, entirely tabular. To be a Babylonian astronomer was to be the maker and user of astronomical tables, and so as the Chaldeans began to move elsewhere, they took their tabular practices with them.

Aramaic was not naturally put into table form, much less Greek. We have noted earlier how Greek writing was produced in a literary culture organized around *performance*. What mattered most was poetry, drama, oratory. You read a text – and imagined it performed. It would have been completely wrong to use the spatial layout of the papyrus to convey information because the only information you were interested in was *speech* itself. Thus, ordinary writing in Greek was entirely linear and unadorned, like that shown in Figure 94.

As the Chaldeans bring their tables to Greek papyrus, they create something unprecedented. Alexander Jones produced his edition of astronomical papyri by going through the trove of unedited papyri from Oxyrhynchus. In many cases, all he had to do was to look for those with tables on them. Nobody, other than the astrologers, possessed *such* papyrus (Figure 95).

As modern scholars, we are never quite sure what to make of ancient astrology. It is important to point out that it was nothing like the quack nonsense we read in today's tabloids, and so we sometimes go overboard, stating that ancient astrology was actually a real science, that it actually was of a piece with ancient astronomy. Neither is quite precise. It was a (difficult, honest) craft, one that involved a great deal of numeracy. But it was socially and intellectually distinct from the practice of theoretical astronomy, and beyond the original impetus provided by its Babylonian

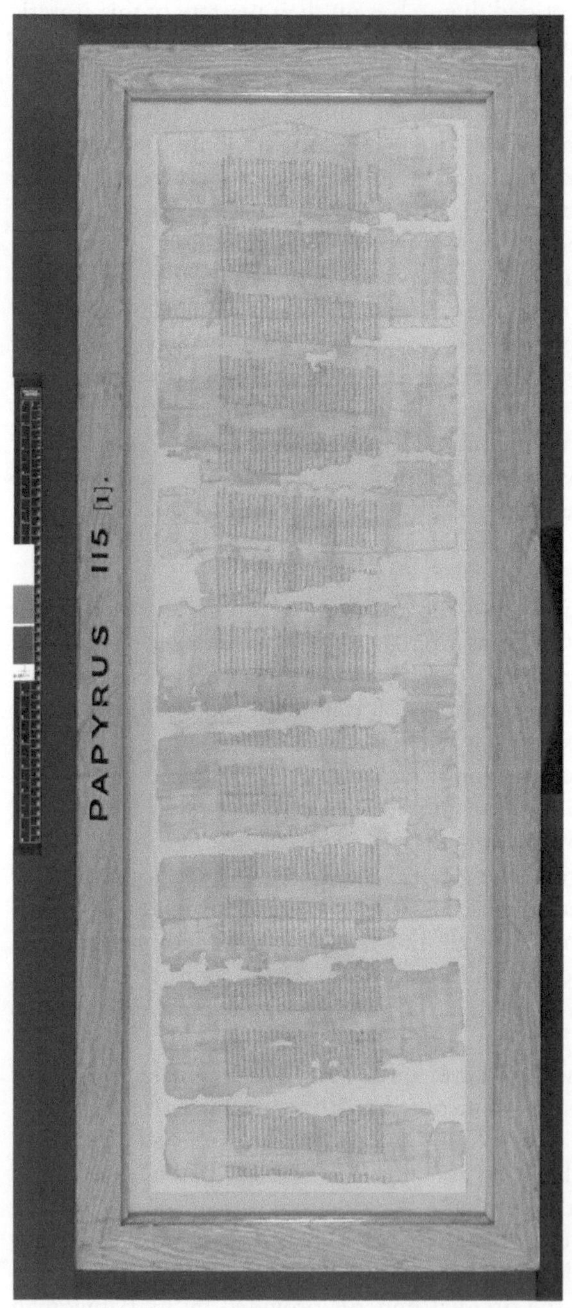


Figure 94 Papyrus 115 from Hyperides's *Pro Lycophrone*. Egyptian (first–second century). © British Library Board. All Rights Reserved/ Bridgeman Images.

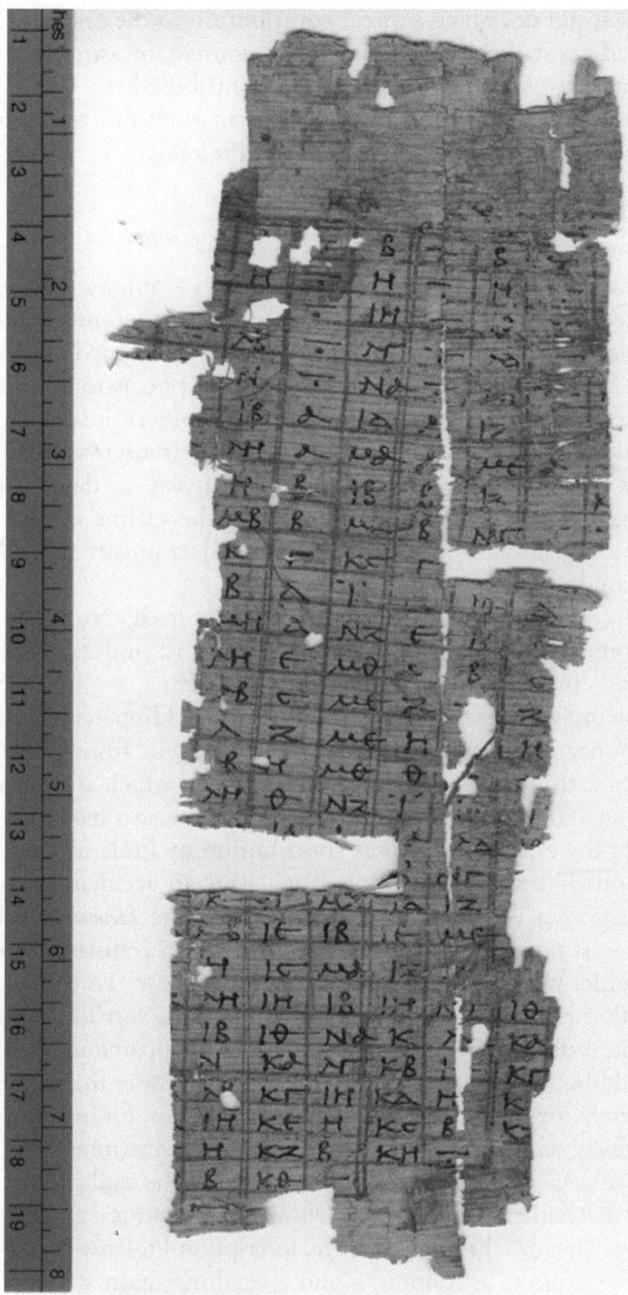

Figure 95 Fragment from P. Oxy. 61.4217.
Courtesy of the Egypt Exploration Society and the University of Oxford Imaging Papyri Project.

origins, it did not make a direct contribution to the science. Perhaps what mattered most, however, was not the content of astrology but its form. Through astrology, the Babylonians contributed, to Greek science, the tabular arrangement of quantitative information; this would blossom with print but is very significant, already, in Ptolemy.

### *Ptolemy: Toward the Canon*

Ptolemy followed Hipparchus in many ways – among other things, in his practice of producing and recording his own astronomical observations. In this case, a very large corpus is still extant. The recorded observations range from 127 to 141 CE. He also set up at least one astronomical inscription (almost a professional requirement for astronomers, it seems), which is lost but was sufficiently well known to have been transcribed and preserved in ancient manuscripts. The inscription carried a date (146/147 CE). Ptolemy's extant works revised some of the claims of that inscription. We find that Ptolemy's extant work must date mostly from the middle of the second century CE and later.

The comparison to Hipparchus suggests itself. Not only is the older astronomer continuously referred to – it is, indeed, mostly through Ptolemy that we know about Hipparchus – but the contours of Ptolemy's project partly reflect those of Hipparchus. In a nutshell: "astronomy – but much more." We have extant from Ptolemy an entire series of astronomical works, at the center of which is a huge and highly ambitious composition, the *Almagest*. But we also have, for many other fields of the exact sciences, *one* contribution by Ptolemy (indeed, exactly a single one – was this intentional or is this an accident of our survival?). His *Geography* is extant in Greek, and so is the *Harmonics*. His *Optics* is extant – as noted on page 205 – only in Latin translation made through the Arabic, with the beginning and end now lost. To be clear: we do not know that Hipparchus wrote on optics, and it is very likely that he did not write on music. Is the project, then, one of "Hipparchus-plus"? At any rate, it should be noted that optics relates in a direct way to astronomy. (This is not merely my own, outsider observation; Pappus, for instance, included his own survey of optics as a part of his survey of his astronomy.) Geography as a science is largely an extension of astronomy; the explicit goal of Ptolemy's *Harmonics* is the finding of parallelism between musical notes and the stars (this was already a key theme of the inscription Ptolemy set up in 146/147). So is the project "astronomy – and everything upon which it touches"? It comes as no surprise, then, that Ptolemy also wrote, as noted, a book on astrology, known as the *Tetrabiblos*, discussing mostly – as was typical in the genre – the hermeneutics of astrological signs. As noted previously, this treatise seems to be a compilation by a sympathetic outsider rather than the work of a practicing astrologer.

Two more works in the exact sciences are merely attested. We recall Simplicius providing us a tantalizing glimpse of Hipparchus with the treatise *On Bodies Carried Down by Their Weight*. Simplicius notes all this in his commentary to Aristotle's *On the Heavens*. It is from here that we also learn of a work by Ptolemy titled *On Balancing Forces* or the *Elements* (this must mean, in context, the elements of matter such as earth and fire, not "elements" in Euclid's sense). The argument in that work was explicitly anti-Aristotelian (for once, the reference by our sources to Aristotle is probably from Ptolemy's original: by Ptolemy's time, Aristotle was already once again a major figure). Bodies move linearly when they are away from their natural place (so, earth down, fire up). In their natural place, however, they may stand still – or rotate. The point of interest to us is, once again, the direct analogy with Hipparchus, down to the very theoretical claims – Ptolemy explicitly stated, in fact, that bodies in their natural position have zero weight (the pull up and down exactly balances, then). Further, for Hipparchus, we have the record of the strange combinatoric work against Chrysippus, counting the number of combinations made from the propositions. This is continuous with Hipparchus's insistence on precise calculations (and his penchant for polemic) but is also one piece of evidence for some contributions to "pure" mathematics. We do not have the same for Ptolemy, which is probably significant: Had he written on, say, the conics, surely Pappus, writing not that much later, would have known and at least referred to such works? Proclus, however, cites an attempt by Ptolemy to prove Euclid's parallels' postulate, and this suggests that he authored some kind of study of the *Elements*. The simplest hypothesis, in my view, is that he wrote a treatise introducing students to the serious study of Euclid's *Elements*: – much like Khayyam, nearly a millennium later, as we will see in Chapter 7. The study of the nature of matter in motion – ultimately explaining why the stars rotate – as well as the introduction to the study of Euclid's *Elements*, can belong to a project of "everything to do with astronomy." (Although Ptolemy never cites any advanced mathematics – he does, however, develop some of it for his trigonometric calculations and for his derivations of astronomical parameters – he does assume, throughout, elementary geometry: something like an *Introduction to the Reading of Euclid's Elements* could be seen, then, as ancillary to Ptolemy's astronomy.) With some stretch, we can even add in here a brief study, *On the Criterion* – now extant – that is purely philosophical. The subject matter is epistemology and so could plausibly be seen as an introduction to any scientific knowledge, including, obviously, astronomy. But this is already tenuous. If indeed Ptolemy's project was so strictly organized around astronomy, however, it is strange that we do not have from him more abstract works in spherics and trigonometry. Perhaps the succinct treatment of such issues in Books I and II of the *Almagest* was all he required. Perhaps he did write them and those are simply not attested: even with such a well-preserved author, it is clear that there are significant lacunae in our knowledge. But enough is known to suggest not only the continuity with Hipparchus but also one obvious contrast. Hipparchus's project was often polemic and largely piecemeal: "this thing, about which that one is wrong; and that other thing, about which yet another one is wronger." Ptolemy is almost never piecemeal. Everything that comes down from him is a summa of its field, organized by its own internal logic, polemic becoming fairly rare and subservient to the internal needs of the theoretical field. For which reason – with all the bulk of his extant corpus – there is a residue of an enigma about Ptolemy. More than most Greek authors, he disappears into his subject matter, coming close to being the mute expositor of an impersonal logic.

Let us take the *Harmonics*. A compendium on a small scale – merely three books – it achieves a sense of comprehensive treatment. Ptolemy begins by noting that the two criteria for judging harmony are listening and reason. The remainder of Book I is organized around this duality. Ptolemy first sets out the purely Pythagorean theory, according to which harmony is judged purely as a ratio between integers, and then the alternative (put forward by Aristoxenus, Aristotle's pupil), that harmony is to be judged empirically, by the ear. Both positions are developed with what (in the Greek context) can be described as sympathy, but both are ultimately rejected, Ptolemy claiming to offer some kind of synthesis. As a matter of fact, however, it is clear that Ptolemy is much more of a Pythagorean, and he does not, in fact, follow any specific Aristoxenian doctrine, taking from the Aristoxenians merely the overall attitude of respect for empirical evidence. Thus, the mathematical route is pursued fairly rapidly, but it is decisive.

Ptolemy takes over the assumptions that the octave is 2:1, divided into 3:2 (the fifth) and 4:3 (the fourth); that overall it is composed of two fourths with a single tone added in between; that relations within the fourth should be integer ratios of the form $(n + 1):n$; and that the octave is 2:1. Note that the question "What is the structure of the octave?" resolves into the question "What is the structure of the fourth?" Now, to recall (p. 73), the Greeks did not even aim to have the octave evenly spaced out. Let me explain this further: such even spacing is important if one wishes to modulate or harmonize the same melody across different keys, so that if one modulates by a third, say – or two tones – the melody should be recognizably the same but simply "two tones down." This is specifically a concern of modern tonal music, where modulation of keys is the main organizing principle – a unique development of modern European music. This modulation is easily achieved on the modern piano, which is indeed "even-tempered." To a Greek ear, such a tuning was less motivated because modulation mattered less (indeed, sometimes was frowned upon), and at any rate, such tuning would sound lacking because it did not provide for the very close pitch differences – nearly quarter tones – that were important for the Greek musical experience. The fourth was divided, instead, unequally, allowing for greater and smaller gaps. The question then becomes, to Ptolemy, very well defined: how to divide 4:3 (the main ratio of the fourth) into two smaller ratios, such that their own ratios take the form $(n + 1){:}n$. Ptolemy's originality, in fact, seems to be that he insisted that all ratios have such form. (It appears that past numerical musical theorists were more interested in developing a variety of systems with other interesting numerical relations.) Ptolemy's insistence on $(n + 1){:}n$ as a unique criterion within the fourth creates a constraint that allows a sense of finality. At this point, Ptolemy simply asserts (without proof: this will be simple) that there are only three ways to divide the ratio 4:3 into two ratios of the form $(n + 1){:}n$:

$$16 : 15 \times 5 : 4, \quad 10 : 9 \times 6 : 5, \quad 8 : 7 \times 7 : 6.$$

Which determines the three allowed tunings.

In the second book (and reaching into the beginning of the third), Ptolemy puts forward his instruments for the empirical judgment of harmony. This gives rise to a discussion of a musical/physical device, either a single-string monochord with a moveable bridge or a many-stringed "harmonic canon" (with strings of varying length). If the argument concerning integers is mathematical through its calculations, the argument concerning the monochord and the harmonic canon is mathematical through its geometrical diagrams of the instruments. Lo and behold, experiments with the instruments confirm the theoretical tunings, and Ptolemy's system is complete. This takes over two books, as Ptolemy describes, in detail, the proposals of past authors and as he calculates, in detail, the entire systems resulting from all such proposals. These are summed up in table form, with the ratios between different notes in the different systems all laid side by side. (Book III remains; I return to it later in the discussion.) Ptolemy ends up with completion at several levels: completing a theory of harmonics that is necessary and proved, that covers both theory and experience, and surveys all the significant past proposals. The sense, indeed, is of *finality*.

Even in its fragmentary, mediated form, it is evident that this finality was the goal of Ptolemy's *Optics*. The extant work is made of three surveys in sequence. First, a long array of optical illusions, surveyed and resolved one by one (these are not very rationally ordered – it is hard to structure a list of, essentially, odd visual phenomena – but their very great number provides a sense of exhaustion). Second, a mathematical theory of mirror illusions with both convex and concave mirrors, in this case going systematically through all the possible combinations of the geometry of the mirror as well as the distance of the viewer. Third and finally, a study of refraction, which once again joins together reason (in the form of geometrical theory) and experience (in the form of observations of angles of refractions in the transitions between three concrete media: air, water, and glass; those values are provided in table form). We see a crescendo of complexity as more difficult problems are tackled, reaching the totality of Greek optical knowledge and probably adding, originally, new observations and, especially, a new, more systematic presentation. (The basic theory of light and perspective was undoubtedly treated in Book I, now lost.) The *Geography* is even more explicitly about the project of synthesis. One has to combine mathematical reason and ordinary experience (one has to rely primarily, we learn, on astronomical principles, but the distances reported by travelers should be taken into account as well). The main contribution is in the systematic presentation: Ptolemy is trying to provide the precise information required for a universal *world* map. He discusses the mathematics of planar projection, but the original work itself (unlike many of its later copies) did not include a world map and instead was constituted – beyond its relatively brief discussion of theoretical principles – by a massive set of tables of coordinates. More than any other work by Ptolemy, the *Geography* is organized around a foil – Marinus of Tyre – and so it is striking that even here, the tone of engagement is respectful. (I.6: "[Marinus wrote] with absolute diligence . . . treated nearly all his predecessors with care." So Ptolemy, implying his own values – "Since, however, even he turns out to have given assent to certain things that have not been creditably established" – and Ptolemy's own approach ensues.)

Ptolemy's astronomical project was much more extensive than in any of the other sciences. It includes a number of more specialized works (the *Tetrabiblos*, as noted, on astrology; the *Phaseis*, a calendar or, in the Greek precise sense, a parapegma). Two more technical works are the *Planispherium*, the projection of a celestial sphere on a plane, required for the production of the star-observation device called the astrolabe; as well as the *Analemma*, a theory of the sundial. With both works, we see an engagement with the mathematical theory underlying astronomical instruments, which, once again, appears to be a repeated feature of the ancient astronomical genre. It is mildly surprising, perhaps, that Ptolemy did not write a comprehensive survey of astronomical devices, instead. But at any rate, it is clear that he must have been even more prolific than his extant corpus suggests, and his lost works could very well have been mostly astronomical and mostly dedicated to more isolated problems.

Yet the core of Ptolemy's life must have been the creation of that major astronomical summa, the *Mathematical Composition* or the *Almagest*. Two other major works, in fact, are essentially derivative from it.

One, the *Handy Tables*, marks a transformation in the encounter of Babylonian and Greek astronomies. As noted in the previous section, Chaldean-inspired astronomers were active, certainly from the first century BCE onward, all throughout the Mediterranean. Their craft was the making of numerical tables, based on algorithms developed centuries earlier by Babylonians: calculating key events such as first risings of planets and, based on these, calculating the positions of planets at given times. (To clarify again: one could not just have the tables set once and for all because, after all, the position of the stars kept changing! The "fixed" tables provided the tools with which one could calculate tables for one's own time and – in the Greek tradition – place.) Techniques developed between Apollonius and Hipparchus could make it possible, in principle, to calculate such positions relying on geometrical models, together with some trigonometry. Even apart from the normal reluctance of craftsmen to give up the tools of their trade, it is clear why practicing astrologers went on using, in part, the Babylonian tables: these, after all, were already on the shelf. It is not certain who was the first to offer an alternative set of geometry-based tables, completely doing away with Babylonian techniques. At any rate, this was achieved, in the definitive form, by Ptolemy himself. *Handy Tables* was basically an extract from the *Almagest*, collecting the tables useful for astronomical practice and setting them out with some improvements and with a new, dedicated introduction. (To professors in the tenure track, this is all very relatable.) The remarkable thing is that these tables turn up in significant numbers – almost uniquely, for the corpus of advanced Greek mathematics – in the papyrological evidence. It appears that almost immediately, astrologers started using Ptolemy's tables, largely replacing the old techniques. Active Babylonian astronomy ends with Ptolemy – at the moment that the Greeks learned, from the Babylonians, to use *tables*.

Another significant contribution, essentially derived from the *Almagest*, is the *Planetary Hypotheses*. *Handy Tables* was written for the practicing astronomer and, already since Ptolemy's time, was widely circulated. *Planetary Hypotheses* was much less copied, and in fact, only the beginning is still extant in Greek, whereas much of the work is known only in Arabic translation (and some of it lost entirely). This, in a sense, is surprising because this is the ultimate payoff of the *Almagest*: Ptolemy finally tells us *what the universe is like*. Ptolemy provides a systematic description of the structure of the cosmos in terms of nested spheres or rings, within which are located the epicycles (all derived from values calculated in the *Almagest*). The hypothesis of epicycles and eccenters (unlike that of homocentric spheres) implies a variation in a given star's distance as it rotates around the earth. Ptolemy calculates this variation in the *Planetary Hypotheses* and uses it to derive a minimal workable set of sizes and distances for the entire cosmos – not just for the sun and the moon, as we already saw with Aristarchus (and the many later refinements by other Greek astronomers), but for the planets as well. All of this is framed, however, not purely as an exercise in cosmology. Instead, Ptolemy presents the treatise as his contribution to the genre of planetarium-making (an echo, then, of the original Archimedean *Sphere-Making*). Unlike past devices, Ptolemy claims that his should be able to represent not just the visible effect of the apparent positions of the heavenly bodies but, instead, their actual spatial arrangement. (Was any such planetarium ever built by Ptolemy? Or was he a merely theoretical maker of planetaria, just as he was a merely theoretical caster of horoscopes?)

At the core of all of this stands the *Almagest*. This, quite simply, is a massive book – almost a quarter million Greek words, in thirteen books. This is the same as the number of books in Euclid's *Elements* – certainly established in Ptolemy's time, as noted on pages 232–233, as part of normal education. Clearly, Ptolemy aimed to provide a full survey of the sky, just as Euclid did of the elements of mathematics. At any rate, the scale of the work emerges fairly naturally from the material itself. Nothing here is superfluous, although everything is prepared in precise detail. Book 1 provides the basic presuppositions and tools – a geocentric model (for which Ptolemy explicitly argues, mentioning potential alternatives; this will matter to future debates), the elements of trigonometry. Book II provides the elementary results of spherical astronomy: the obliquity of the ecliptic and the consequences for different latitudes. This is all the same as one found already, centuries before, in Autolycus or Euclid – but now with rules for the derivation of specific numerical values, based on spherical trigonometry (which, in turn, could well derive from Hipparchus; see pp. 340–342). Book III provides the theory of the sun, and Book IV provides that of the moon, to the extent that it follows the same logic as the sun. The motion of the moon is, in fact, rather more complicated, and Book V deals with such added complications. Obviously, all the calculations regarding the sun assume the tools of Books I (general trigonometry) and II (spherics and the ecliptic). Less obviously, the calculations for the moon already use measurements that derive from the position of the sun; hence, Books IV and V, too, rely on Book III. (You recall how we relied on eclipse observations to determine the positions of the moon: an eclipse is the conjunction of the moon and the sun – and the position of the sun is the one that is known already.) This nearly axiomatic structure continues into Book VI, which combines the theories of the sun and the moon into a theory of eclipse prediction. Books VII and VIII are something of a fresh start in terms of their subject matter and overall spirit. They discuss the practice and theory of observations of fixed stars and end up building a fixed-star catalogue set out as a massive table (1,022 stars!). This is not really a hiatus in the deductive sequence, however, because one requires Books I–III for the overall structure of the sphere, and because the position of the moon (from Books IV and V) does come in because angular distances between a star and the moon are used for determining the positions of stars. The fixed stars are at the center of the *Almagest* as a whole; following them, we get to the main business of geometrical modeling with, it so happens, five books for the five planets (not, however, on a one-for-one basis) – Books IX–XIII. These keep referring to positions of the sun, moon, and stars, relative to which the positions of planets are calculated – keeping with sense of a continuous deductive progression. Book IX offers the general approach, followed by the theory of Mercury; Venus and Mars are studied in Book X; Jupiter and Saturn in Book XI. All of those theories involve serious complications – it is never the case that a simple epicycle or eccenter suffices to account for the observed motions, and Ptolemy instead uses several ad hoc extra assumptions. (For instance, that the motion is around an eccenter but that its speed is such that equal angles are swept over angle times, relative to a point that is neither the center of the circle nor its eccenter. For almost two thousand years, original contributions to astronomy will engage not with the rather empty question of heliocentrism versus geocentrism but, rather, with the problem of how to make Ptolemy's special tools for dealing with planets less ad hoc.) Even so, the presentation is logically compelling: Ptolemy's clear aim is to convince us that the tools emerge as a mathematical consequence of the observations.

Only based on the individual theories for each planet does Ptolemy proceed to calculate particular outcomes: the retrogradations and other features of the motions of the planets (Book XII), the motion in latitude (that is, departing from the ecliptic), and finally, the observations that are at the core of the entire endeavor – the "phases" or such key events as first risings. The major recurrences, out of which it all began – now, the outcome of a massive, thirteen-book-long deduction.

Modern readers sometimes learn about, say, Galileo and the scientific revolution, about geocentrism and heliocentrism, and they then become impatient with the past. *How could people have been so naïve? Why follow Ptolemy?* In truth, how could you not? There is nothing quite like the *Almagest* in the extant scientific literature from antiquity, and in truth, there are not many books that compare with it since. Newton's *Principia Mathematica* comes to mind, or perhaps Darwin's *Origin of Species*. (But in a sense, Newton modeled himself on Ptolemy, Darwin on Newton.) Here is a book that thoroughly derives the totality of a realm of nature, going through both a lucid explanation of the process as well as the painstaking detail of all the derivations. The Antikythera mechanism was "a portable cosmos." But so was the *Almagest* – somewhat more bulky, with its thirteen thick rolls – a rational, ordered, massive system – a counterpart to the system of no less than the universe itself.

A work of this ambition is unique in the extant literature from antiquity. Was Ptolemy, in fact, the first to have written a work of this type? This is a difficult question but one we must answer so as to try to understand Ptolemy's historical context.

As noted, it appears that later than Hipparchus, the very figure of the author in the exact sciences became, at least, very rare. It is conceivable that for generations, no significant new works in mathematical astronomy were produced at all – let alone works of Ptolemy's ambition. In fact, between Hipparchus and Ptolemy, the only astronomer of note emerging from our evidence is Menelaus – now known to us (as mentioned earlier) mostly through his very competent spherical trigonometry. Ptolemy does not cite Menelaus for any astronomical theory, merely for a couple of astronomical observations. Now, among the astronomical/astrological papyri, Jones notes a few that seem more theoretical in character. With one exception, these might be more "cosmological" in character (and so, perhaps, from a more "philosophical" as opposed to a mathematical origin?), but one clearly belongs to mathematical and observational astronomy and is a report of an observation close in time, and similar in format, to those Ptolemy cites from Menelaus: this involves Jupiter. Jones concludes that this may well be, in fact, a fragment of Menelaus.[29]

Perhaps this small shred of papyrus could have come from a basket of rolls as imposing as that of the *Almagest*. Now, it is true that Ptolemy is the opposite of a name-dropper: he mentions authorities sparingly, looking for a handful of representative authors he needs for specific purposes. He does use such authorities, however, and cites them as necessary. For the cases of the sun, the moon, and the fixed stars, his foil is largely Hipparchus, who did work constructively on those very topics (albeit, it appears, in piecemeal works). Such foils, as noted, structure the *Harmonics* and the *Geography*, and I suspect that we have lost Ptolemy's account of previous work, in the lost first book of the *Optics*. However, at the beginning of Book IX, as Ptolemy turns to the daunting task of the theory of the planets, he explicitly states his originality – "there is a good reason why no-one before us has yet succeeded in it" (IX.2) – and continues to explain (as mentioned earlier) that Hipparchus decided *not* to produce his own planetary theories, citing, beyond that, no past authorities. I suggested previously that Hipparchus's refusal to issue his own theories of the planets could well have come from a polemic against past theories (from the generation, say, of Apollonius?). If so, Ptolemy could have felt justified in referring to Hipparchus alone: any pre-Hipparchean theories, or their later descendants, would have been sufficiently covered by Ptolemy's reference to Hipparchus's own antitheoretical statements.

Menelaus remains, and it is true that he was unlikely to have recorded a position of Jupiter purely for the pleasure of observation. Indeed, the brief passage now extant from the papyrus refers not only to the new observation of Jupiter but also to a much older one, dated to 344 years before. This is the sure sign of a theoretical deduction (two observations of the same object, given in terms of position and date of observation, are most

[29] This papyrus is P. Oxy. 4133, discussed in A. Jones, *Astronomical Papyri from Oxyrhinchus* (Philadelphia: Amer Philosophical Society, 1999), I.73–I.75.

likely there for the sake of a calculation of the mean speed: change in position divided by time elapsed). But this should be qualified in two ways.

First, Menelaus is surprisingly close in time to Ptolemy. His recorded observations are now dated to 98–105 CE, as against Ptolemy's, from 126 to 141 CE, so that he is a generation older than Ptolemy. Many Greek authors show a preference not to cite living authorities by name, and if so, Menelaus could well have been at the absolute latest moment of citability. As Ptolemy cites authorities later than Hipparchus, he seems to find an author standing right behind his shoulder – and nearly no one else farther back.

Second, Jones's relative confidence in the identification of Menelaus is remarkable. After all, we know very little about the author of the papyrus fragment. However, the dates, place, and format of observation are all precisely aligned, and – incredible luck – the author uses the first person in referring to the observation. But notice the implication: Of course this is a sample of one, but is it not striking that when we find a "randomly selected" non-Ptolemy astronomer, this happens to be, likely, the *one* astronomer of his time that Ptolemy cares to cite? The implication of all of this is that major astronomers, in Ptolemy's time, were recent and rare.

There is little astronomical context within which to put Ptolemy, which means that to get some of his setting, we need to look at a broader context: the historical setting as a whole and the evidence for the science of Ptolemy's era.

As we widen the zoom, our picture suddenly fills with life. This, after all, is the height of the Roman Empire! The archeological sites now admired by modern tourists all over the Mediterranean – theaters, temples, statues, inscriptions, stone and marble everywhere – are mostly the product of this era. Money is lavishly spent as patrons display their power and their generosity. Perhaps the Mediterranean was somewhat richer, at this moment of Roman Peace. It certainly was much more unequal, and the patrons displaying their generosity had a lot to put on display. Little wonder, then, that we also see, in this era, a mad rush to reach the elite and to enjoy its goodwill. Everyone seeks "paideia," the good education that marks one as a true gentleman. This is represented by knowledge of the Greek past and its literature, above all in one's rhetorical skills. There are orators everywhere, as well as teachers of rhetoric offering their services. But of more relevance to the history of science, the two service industries – astrology and medicine – are also in demand. For the case of medicine, our evidence is the most direct. We have extant the huge corpus of Galen, who fills, in the history of medicine, a role analogous to that of Ptolemy in astronomy. A much more rhetorical author (as was always the case for Greek medicine – a discipline more "literary" than the exact sciences), we hear from Galen a lot – I mean, a *lot* – about his interactions with his many interlocutors. Born in Pergamon and reaching the heights of Roman society (at times, a physician to the emperor!), he is completely marked by the rush for patronage. He was a great scientist – the greatest anatomist of antiquity and the father of experimental methods in physiology. All of this, with him, was pursued through vicious polemic, indiscriminately attacking physicians past and present. We hear through Galen of many dozens of physicians, many of them contemporaries or near-contemporaries (if we include those who are merely reported in passing for contributing a pharmacological recipe, the numbers rise to the hundreds). The clear impression of his corpus is of a bustling and growing field of medical literature. Indeed, it is hard not to explain Galen's great prolixity – his extant works now occupy twenty volumes, but many have been lost – other than through this rush of competition. Rome was swarming with Greek medical authors, vying for attention. Galen sought to triumph over them all by becoming the best experimentalist but also, quite simply, by out-writing everyone else.

It is hard to say exactly how much astronomy was different from that. Ptolemy is simply not as chatty, which suggests, perhaps, that he has less of a need to prove himself – so, perhaps, a less severely competitive field? – but that also means we do not have the chance to learn about his immediate context. The simplest account is that, indeed, there was less activity in astronomy than in medicine, but that, once again, things did begin to stir up somewhat by Ptolemy's time. Or again: we do not see in Ptolemy anything like the rush for patronage evident in Galen. But it is noteworthy that most of his astronomical works are addressed to one Syrus. The language of the address is laconic, and so we cannot tell who this Syrus was, but Cristian Tolsa has recently proposed the plausible identification with the Roman governor of Egypt, Sura (*Syrus* being, then, a Hellenized version of the Roman name).[30] Perhaps something structurally similar to the case of medicine needs to be inferred for astronomy, after all: on a more modest scale, authors in the exact sciences were seeking patronage, were competing for it; a logic of inflation ensued, and

[30] C. Tolsa, "Claudius Ptolemy and Self-Promotion. A Study on Ptolemy's Intellectual Milieu in Roman Alexandria" (doctoral dissertation, University of Barcelona, 2013), Chapter 5.

something such as Ptolemy's project – seeking the monumentality of a complete system – was its culmination.

A Galen-like chattiness, a constant bickering against past and present authors, would have been alien to Ptolemy's endeavor. A key goal of Ptolemy's project, we said, was its finality: such were the musical harmonies; these were all the optical illusions produced by all variety of mirrors; such were the motions of the planets. The constant back-and-forth of debate, ultimately, undercuts such a sense of finality. To punch is to suggest the possibility of a counterpunch. Or perhaps – and I wonder if this more parsimonious explanation might not suffice – Ptolemy set out to emulate Hipparchus but also to differ from him, and so, among other things, he sought to mark himself as the non-Hipparchus. Perhaps, simply, by being *polite*?

### *A View of the Monument*

The fact is, Ptolemy got his finality. I start the following chapter with a commentary to Ptolemy's *Harmonics*, and it appears that not long after his death, Ptolemy had already obtained canonical status: recognized as *the final word*. He would continue to enjoy this for almost fifteen hundred years. The same, indeed, was true of Galen: it was a good time to get canonized. We will return to see, in the following chapter, why and how this canonization took place, but one fundamental reason should be clear already: both Galen and Ptolemy were great scientists.

In such a rapid overview, it is hard to do the greatness of Ptolemy any justice. I did give a sense of the monumentality of the work and of its logical character. What is harder to convey is the mechanism through which the logic is conveyed. Let me try to do so, very briefly, for the simplest of Ptolemy's theories – that of the sun. A glancing overview, then, of Ptolemy's Book III.[31]

The book begins with an exposition of the year, a concept that (as Ptolemy explains) has two meanings – relative to the zodiac and relative to the fixed stars – which, as Hipparchus discovered, are slightly different (this is the precession of the equinox that we have mentioned on pp. 328–329). The calculation of the length of the year is observationally and conceptually difficult, but once this is achieved, one can deduce the

[31] For an English translation of the *Almagest*, we have G. J. Toomer, *Ptolemy's Almagest* (London: Duckworth, 1984). A monumental translation, worthy of its source.

mean position of the sun at a given point in time ("mean" position – because the speed of the sun is not uniform). This is produced by Ptolemy explicitly, not as a function (as we would do) but as a table; indeed, as Sidoli notes, tables are often Ptolemy's way of conveying a function.[32] A set of tables provides the mean position of the sun in eighteen-year periods (starting at a defined moment, of course, based on a Babylonian epoch), in a given year within the period, in a month within the year, in a day within the month, and in an hour within the day. Notice that all of this is not merely scientifically precise but also supremely practical.

Had the speed of the sun been uniform, Ptolemy could have stopped there. It is not. (The reasons for that are evident from a modern perspective: the earth proceeds not along a circle around the sun but along an ellipse. Further, the earth does not cover equal *angles* at given times – which would have been perceived as a fixed speed – but rather, it is *the area of the sector of the ellipse swept by the earth through its motion* that is fixed, given the time.) The nonuniform speed of the sun is not some arcane observation but is instead a very familiar one to ancient observers, known as the "inequality of the seasons." Our next task is therefore to find tables that set out the deviation from such mean motion or the tables of the "anomaly of the sun." This is typical for the general task of premodern astronomy: given an object, construct the tables for its mean motion, followed by the tables for anomalies.

Ptolemy approaches this problem, as he typically does, from the more general and abstract to the more empirical. He begins by setting out the type of techniques available for deriving anomalies from regular circular motions, which are shown to be the eccenter and the epicycle. He then proceeds to show their equivalence for the problem at hand, as well as further general properties (for instance, at what point of the circle we find the greatest and least speeds). This discussion goes beyond the needs of the immediate theory of the sun but not beyond the needs of what would, in fact, be applied through the treatise, all of which is typical to Ptolemy. This passage, then, is a kind of "The Elements of Epicycles and Eccenters Required for the *Almagest*." Following that, Ptolemy simply states that he will produce the theory of the sun through the eccenter (equivalent to the epicycle, and simpler). The following proof is explicitly said to follow Hipparchus.

[32] N. Sidoli, "Mathematical Tables in Ptolemy's *Almagest*," *Historia Mathematica* 41 (2014): 13–37.

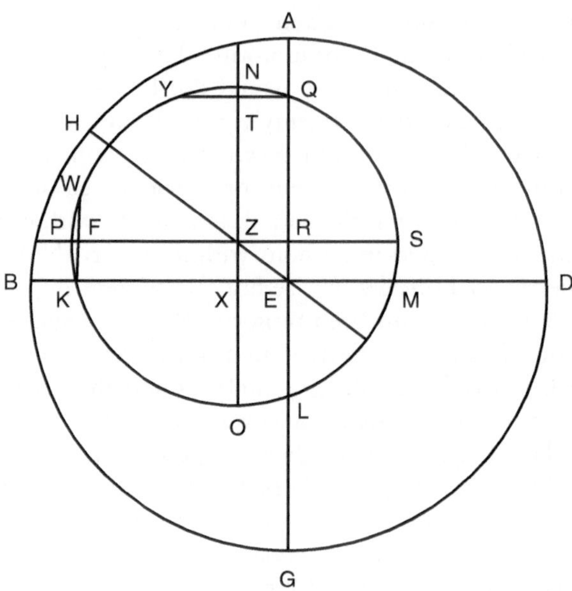


Figure 96

In Figure 96, the sun rotates along the circle ABGD, and the earth is at Z. A is the position of the sun at the spring equinox (when the day and night are exactly equal); B, at the summer solstice; G, at the autumn equinox; D, at the winter solstice. Had the earth been at the center, E, the length of the seasons would have been equal. The eccenter hypothesis is that the earth is at some point other than E, and we can fully determine the theory of the sun by finding the position Z where the produced angles agree with the observed facts. (In practical terms, we need to find the length EZ as a fraction of the diameter.) Assume Z found, and draw an arbitrary circle around Z as the center. Our data for the length of the seasons can be given as follows (this is an approximation, but it is a valid one, given the limits on our observations): that the arc QK, equivalent to the length of the spring, is traversed in 94½ days, whereas the arc KL, equivalent to the length of the summer, is traversed in 92½ days. Divide the days by the length of the year to get the length of the arcs in degrees. At this point, we have the arc measurement of QK, KL and also NO (semicircle), NP, and PO (quadrant). It is, at this point, easy to find the arc measurement of NQ, PK (differences between the aforementioned) and so of the arcs QY, KW. Applying the trigonometric tables of Book 1, we can find the length of QY – hence of QT, that is, RZ – as a fraction of the diameter; similarly, we can find the length of FK or RE as a fraction of the diameter. Once we have both, we apply Pythagoras's theorem and find the length of EZ as a fraction of the diameter (2;29½ parts where the diameter is 120 – not too far from the center!). I merely outline the geometric logic: for Ptolemy, everything has to be calculated in detail. Following that, however, it is noteworthy that Ptolemy merely corroborates this calculation in passing, based on those two seasons, with reference to the lengths of the autumn and the winter (instead of redoing the calculation from scratch, based on the remainder of the data). This is typical of his approach: calculations are thorough for the cases chosen, but there is no effort to expand the evidence base beyond those few cases. We remember, for instance, how the changes in the speed of the moon were estimated, to begin with, based on three observations, or two time intervals (i.e., the absolute minimum). Later on, Ptolemy does check his theory of the moon, obtained based on one set, against a single, other set of three observations; but that's it. It is true that Ptolemy does not yet have a statistical technique with which he could find the position of the eccenter as the best fit, given a number of measurements; however, he could still have pursued several separate measurements and noted their agreement, an option that he avoids, preferring the simplicity of relying on minimal tools.

The distance between the earth and the center of the rotation of the sun determines all we need to know in the *cosmological* theory of the sun – answering the question of "what the universe is like" – and if that was our goal, the study of the sun could have ended right there. Ptolemy, however, is an astronomer, not a cosmologist, and so his goal is not just to say "what the universe is like" but, rather, the much more ambitious and more technical goal of *computing* the position of objects. Hence, the position of the eccenter is just the beginning. The main part comes now.

Suppose that in Figure 97, Q is the position of the earth, and ABG is a circle concentric with the circular trajectory of the sun or the ecliptic – with its center at D. What we have established once and for all is the position of the points D, Q, so we know the distance DQ – as a fraction of the radius of the ecliptic – and we can also join the two points D, Q and extend them to cut the circles so that the line EAQD forms a natural starting point from which to measure angles. Now, here is what we will need in astronomical practice. We will observe the sun at a certain angle, which we measure from line EAQD. For instance, let us say that we observe the sun at point Z, at "thirty degrees from point E." This means that we found the angle EQZ to be equal to thirty degrees. What we are interested in, however, is the real position of the sun, which is the angle it is covered on the ecliptic. This is the angle BDA or ZDQ. Now, from *Elements* 1.32, we know that

the exterior angle ZQE = interior angle QZD + interior angle ZDQ.

And so, to be able to find, from the angle ZQE, the angle ZDQ, we need to find the angle DZQ first.

Ptolemy approaches this neither as a purely geometrical problem nor (as we would do) as a statement of an algebraic formula but through a worked-out numerical example. This is indeed close to the experience of education we are familiar with.

We take the example of the angle ZQE being thirty degrees.

We now extend ZQ to K where DK is perpendicular to ZK. Thus, we have a right-angled triangle DKQ with a right angle at K and DQK being also equal to thirty degrees.

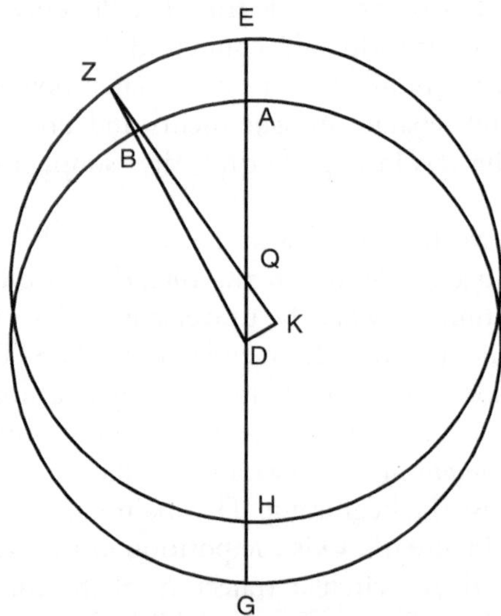


Figure 97

Now, the measure of QD is known: we found that it is 2;29½ parts – Ptolemy now rounds this up to 2;30 – where the diameter of the ecliptic is 120. Recall how sexagesimals operate: 2;30 is what we call 2½. Recall how the table of chords works. A right-angled triangle, for Ptolemy, is really a site for a semicircle. We imagine a semicircle around DQK with DQ as its diameter, DK as the chord on the arc of 60 degrees (as the arc is measured from the circumference), and KQ as the chord on the arc of 120 degrees. We turn to the table of chords and find, rounded up, that the chord of arc

120 is 103;55, whereas that of arc 60 is, of course, 60 (this is the side of the hexagon, which is, of course, equal to the radius), all when DQ – the diameter – is 120 units. Now, in the case at hand, DQ is considered to be not 120 but 2;30 units, and so we need to transform the values. DK obviously becomes half of 2;30, so 1;15. Ptolemy merely states the result. I personally would go about this by first dividing 103;55, which gets me to the very same: ;103,55 – these are sexagesimals! This gets me to the value I would have had DQ been 2. It is, in fact, 2;30, or a quarter more. A quarter of ;103,55 is almost ;26, and adding this to ;103,55, we get almost ;129,55 or almost 2;9,55, which Ptolemy rounds up, once again, to 2;10.

Now, QZ = QE, which is equal to 60 parts of the same units; thus, KZ is 62;10 of the same units, whereas KD is 1;15. The triangle KZD is right-angled, so we can apply Pythagoras's theorem to find the value of ZD, which Ptolemy asserts to be 62;11. We may now once again conceive of the triangle ZDK as a site for a semicircle, with ZD as its diameter. Once again, we transform everything to make ZD equal to 120 units so that we can apply the table of chords. DK has now become 2;25. We now go back to the table of chords and consider it with the inverted query: For the chord length 2;25, what is the arc? This turns out to be the arc 2;18 degrees. This, once again, is the arc as measured from the center and the arc as measured from the circumference – that is, the angle DZK – now found to be 1;9 degrees. This is what we were seeking, and now we can state that when the sun is seen to cover 30 degrees from point E, it actually covers 1;9 degrees less than this.

In a table of "anomaly" – one that takes every measured angle and tells us how much to subtract – we need to put, next to "30 degrees measured," the value "1;9 degrees to subtract."

Book III, Chapter 6, of the *Almagest* is the table of the sun's anomaly, which means that for each value, Ptolemy had to find the chords of the given arc and of its complement; transforming both values from 120 parts to 2;30 parts; adding 60 to one of those values; applying Pythagoras's theorem to find a third value, and then transforming that third value into 120 parts; then picking up the other previously found value and plugging it into the inverse chord table to find an arc – and finally to halve it so as to find the anomaly for the given arc. There are forty-five values (we skip by six degrees each time through the quadrant 0–90, and by three degrees through the quadrant 90–180; the remaining quadrants are obviously symmetrical). These are forty-five iterations of going through the same procedure, each time with different numbers, dividing, multiplying, adding, squaring . . . The table presents itself as a pristine two-dimensional array of numbers, and it is, obviously, rather slim, a mere inset between longer discursive and geometrical chapters. It should be clear, however, that this table represents the more substantial mathematical investment – and value.

It would seem as if at this point, one could simply take the two tables – the table of mean motion together with the table of the anomalies – and find the position of the sun at a given time. However, as Ptolemy now explains, our work is not yet done! In order to know which anomaly to apply, one also needs to "calibrate" the two tables together: the first day of the table of mean motion is (what we call) February 26, 747 BCE, but this is not necessarily "degree zero" of the table of anomaly. We need to find absolute dates on which to anchor the table of anomaly, and fortunately, our result for the position of the maximal anomaly allows us to do just that. Ptolemy records a precise date for an autumn equinox that he observed – what we call September 25, 132 CE (once again, relying for this operation on the bare minimum – a single observation!). He calculates the number of years and days passing between "day zero" and that measured equinox. On this, he finally applies the table of anomaly to find the position of the sun on "day zero" and, significantly, the implied position, for that day, of the sun on the table of anomaly. Only once this is achieved does it become possible to take the two tables – of mean motion and of anomaly – and calculate, with no need for any more observational or geometrical input, the position of the sun on the zodiac for any given date.

From this point in the *Almagest* onward, this is how positions of the sun – necessary references for almost any other astronomical calculation – will be found. Ptolemy's discussion of the sun, however, does not end there because besides calculating the position of the sun on a given day, the astronomer also needs to define times in terms of hours of the day. Besides the familiar fact that the length of the night changes from day to day based on the time of the year and the latitude (as thoroughly discussed already in Book II), one now needs to take into account that the "twenty-four hours" of the day change somewhat according to the changing speed of the sun, and Ptolemy ends the book with the calculations required to take care of this deviation.

This, then, is Book III – nearly the simplest of the *Almagest*. The whole, thirteen books long, may well have been intended to echo, as noted, the *Elements*. The comparison is significant not only for the similarities but also for the differences. Both works have, as their skeletons, long chains of deduction. They differ in the function of the chain.

First, the similarity. When Euclid proves a problem, such as I.I ("on a given line, to set up an equilateral triangle"), he relies on auxiliary constructions (in this case, there are circles set up on the original line). Then, when such a problem is applied again, the construction of the auxiliary circles is ignored. Now, a Ptolemaic table serves an analogous function. There is tremendous labor going into the calculation of a given table (such as the anomaly of the sun). This then is applied later: say, when a position of the moon at a given time is correlated, later in the *Almagest*, to a position of the sun computed to the same time. Similarly, the computation of the anomaly of the sun sent Ptolemy back, time and again, to the table of chords. Once again, as with the Euclidean problem: all the labor originally invested in finding the table of chords is black-boxed relative to finding the anomaly of the sun; all the labor invested in finding the anomaly of the sun is black-boxed in finding the position of the moon. One needs merely to find and copy the numbers stated in the table. These, then, are the deductive skeletons of the *Almagest*: a mathematical treatise constructed not as a sequence of proved propositions but as a sequence of argued-for tables.

With both Euclid's geometry as well as Ptolemy's astronomy, one sets up tools that are then black-boxed for future applications. But here is the difference. In geometry, no one actually constructs equilateral triangles. When the need arises for one's diagram, one simply assumes that the equilateral triangle has been magically constructed on the given line because this can be done *in principle*. The proof of a problem in Euclid acts as a *conceptual* labor-saving device. A Ptolemaic table is different. The goal is not the mere statement that one could, in principle, determine the position of the moon: the goal is actually to compute it. The labor saved by the table is not conceptual but real (and it is related to a work produced habitually by actual craftsmen, the astrologers).

This means that the *Almagest*, in a sense, is more "necessary" than the *Elements*. It is nice to have all the results one usually needs in place, arranged in logical order, but in normal mathematical practice, Greek mathematicians merely take the results of the *Elements* for granted. Indeed, many works start "from the middle," not from first principles, and merely postulate what they need along the way. Surely, this is how the Greeks had started – Hippocrates of Chios's squaring of the lunules must have postulated the required proportions. What's more, the same is true of the little we see in Hellenistic astronomy – Aristarchus's *Sizes and Distances* or Archimedes's *Sand-Reckoner*. There are calculations, but it is less significant that the calculations be *right*; the emphasis is instead on their interesting conceptual structure.

By the time we get to the *Almagest* – partly, perhaps, because of the example of Hipparchus, partly, perhaps, because of the habits of working astrology – the emphasis shifted. Astronomy became a field where actual computation is paramount, and so its logical structure changed its meaning.

Each actual computation in astronomy does in fact depend on previous calculations. It is necessary to refer to the position of the sun in the computation of the positions of the moon and the planets, and all of them do depend on the length of the year. An astronomy that is interested in the *conceptual* computability of certain values can "start from the middle" and handle, piecemeal, isolated topics. But once one is truly committed to a valid computation of, say, the positions of the planets, one is naturally led to consider the set of tools in its entirety.

This, finally, helps us understand why the *Almagest* gets written. This is, quite simply, the thing to do – once one is doing astronomy and is serious about it. The reason people before Ptolemy did not is simply that the task is very hard. But if so, we can begin to understand why when the field revives and engages more authors, perhaps in the second half of the first century, it would not take more than a couple of generations to reach its summa.

As a practical matter, then, the *Almagest* is a device for calculating astronomical positions, and in some sense, however big, it is the "correct" size for this purpose. This purpose, of course, does not exhaust the range of theoretical questions that an astronomy could, in principle, address. In the twentieth century, one of the most influential accounts of Ptolemy – that of Pierre Duhem – emphasizes the constrained, limited remit of the *Almagest* and of its entire astronomical tradition. Citing an Aristotelian phrase, Duhem thought that ancient astronomers were interested merely in "Saving the Phenomena." The goal was not to find what the universe was like but simply, instead, to predict astronomical positions. The astronomical models (such as epicycles and eccenters) were seen, according to Duhem, purely as scaffolding, mathematical instruments of no inherent meaning. The Ptolemaic response to the question of "what the universe is like" would be mere shoulder-shrugging: such questions are metaphysical but not scientific. For authors such as Duhem (and many modern philosophers following him), this was to be lauded because the shoulder-shrugging expresses a correct, "instrumentalist" (as opposed to "realist") philosophy of science. Theories, according to instrumentalism, are merely tools for predicting observable results; they are not a window into the hidden reality that lies behind those observables. Whether or not such a hidden reality behind the observables exists at all, we have no access to it, and so, at best, reality is a matter for metaphysical speculation, not for scientific argument.

As Lloyd made absolutely clear, this is terribly wrong for the ancients. No one in antiquity engaged with astronomy as merely "instrumental": every author in the field took it for granted that astronomy could, in principle, describe the real structure of the universe. That different astronomers took this task at different levels of seriousness is a different matter. But this is the point: when Aristarchus, perhaps half in jest, proposed that the world could be heliocentric, the piquant power of the proposal lay precisely in that it would be a remarkable, real phenomenon – and not a mere instrumental device – if the earth did revolve around the sun. Ptolemy is the worst possible instrumentalist, in that in his *Planetary Hypotheses* (mostly unknown, however, to Duhem, writing when its Arabic text was barely published), he patiently spelled out the implications of the *Almagest* as a realist model!

And still, Duhem's intuition was not silly. There is, after all, a difference of emphasis between the *Almagest* and the *Planetary Hypotheses*, and it is, after all, the *Almagest* that matters most. This difference of emphasis, however, belongs not to the history of ideas but to the history of scientific practices. At issue is not whether ancient astronomers were realists or instrumentalists (of course they were realists) but whether or not they engaged primarily with the question of the constitution of the universe or with the question of astronomical computation. And once the cultural choice was made to take astronomical numbers seriously, it is clear that the practical questions of computation must occupy the bulk of astronomical work.

Computation anchors the astronomical practice – to become its own goal. This is the most surprising from a modern point of view – so that not only "realism" but also "instrumentalism" become misleading labels. Today, we normally take the structure of a scientific theory as a set of hypotheses out of which predictions are deduced and then tested against empirical observations. Realists think that the point of the observations is to establish the truth of the hypotheses; instrumentalists think that the point of the hypotheses is to derive, in an economical way, the totality of observations. But it is extremely rare that Ptolemy engages with any of that. We noted that even with the length of the seasons, Ptolemy relied on the length of two seasons – which are geometrically sufficient to determine the parameter of the theory – merely noting in passing that the result comports with the length of the remaining seasons, even though (what we consider) the empirical test of the theory is so readily available. And so, in general: having produced the theory of an astronomical body, Ptolemy displays little interest in testing this theory against a series of observations. He could easily assert that having produced the theory of the sun, he then observed the sun on many different occasions, and lo and behold, applying the table of mean motion together with the table of anomaly allowed him to predict precisely where the sun would be. He does not say this because, to him, this is just *obvious*. That the heavenly bodies recur precisely, once a pattern has been deduced, is just a given, well known to all astronomers. One needs to deduce the patterns, not to test them.

Ptolemy is not as chatty as many other ancient authors; he does not speak about himself and only rarely mentions past astronomers other than Hipparchus. What he does do, rather more than most other authors in the exact sciences, is to engage, explicitly, with mathematical methodology, not merely producing each particular derivation but also explaining why it is called for. His goal was to achieve objective truth, and it seems that for him, the true guarantor of truth was the rationality of one's method of deduction. Observational prediction becomes, then, almost beside the point, and in a way, Ptolemy hardly predicts *observations*. The output of his tables is a position at a given time, and the normal use of those tables is not necessarily to find the position at the present moment (which is what we require for the empirical testing of a hypothesis). Arguably, the normal use for which such tables were intended was to find the position at some point in the *past*. This, in fact, comes up already in the example we just saw, in the theory of the sun. Ptolemy there made an observation of the present-day position of the sun on September 25, 132 CE – so as to be able to deduce the position of the sun on February 26, 747 BCE. In this theoretical structure, the present observation was an input, and the prediction was the Babylonian-era position. Which is actually not too far from the astronomical practice against which the *Almagest* took place: astrologers computed, primarily, the positions of the stars at the (past, no longer observable) moment of birth.

This is related to a key dilemma in our interpretation of astronomy in the Ptolemaic tradition. Theoretical computation is the core of the practice and is routinely applied as the input for further computations. It is functionally equivalent, as a source of information, to observation. Hence, it becomes impossible to tell, for any particular case, whether a statement is based on observation or on computation. What did they actually see; what did they merely extrapolate? Can we trust them? Some scholars in the past have been scathing. To them, Ptolemy was a fraud. To a modern, the confusion of empirical data and theoretical constructs is a grave sin: it is only through the clean separation of the two that one can test hypothetico-deductive theories. What we find is that Ptolemy was not a modern hypothetico-deductive scientist, and separating observation from extrapolation simply did not matter that much.

But then again, as noted, neither was he an instrumentalist. Not only because he ultimately did use his models so as to deduce a physical reality but also because this physical reality actually carried, for him, a meaning. His daily lived practice was the making of tables, but setting things side by side – the great table of all things, as it were – was, to Ptolemy, not just a tool but also a metaphysical statement. This is hinted throughout his oeuvre but comes to the fore at the conclusion of the treatise that, at first glance, appears the most removed from astronomy – the *Harmonics*. We recall how Ptolemy produces, there, a synthesis of Pythagorean and Aristoxenian positions, ultimately producing empirically grounded arithmetical tables of the division of the octave, superseding (politely!) those of his predecessors. These are the first two books. The final book is the payoff for this musical excursion. Having pursued the arithmetical theory of harmony in the context of music, Ptolemy finally observes that harmony can be found elsewhere – in souls, and in stars. He then pursues a systematic analogy, first between musical harmonies and ethical categories (some ethical positions are more "balanced," you see) and finally, concluding the entire work, between music and astronomical phenomena. The division of the octave is like divisions of the zodiac; particular notes may be likened to particular planets. The discussion, throughout, is closer to astrology than to astronomy, engaged with the more symbolic values of the stars. There is no effort to connect the precise numerical values of Ptolemy's astronomical theories with those of his music. (Ptolemy did pursue such more direct analogies in his Canobic inscription, perhaps representing an earlier stage of his thought.) Modern readers might be embarrassed by such metaphysical speculation, but in fact, if anything, this is where Ptolemy precisely anticipates the scientific revolution. Kepler and Newton – to mention just two examples – were attracted to the construction of a grand scientific model of the universe, precisely as the key to the finding of the universe's rational, divine structure, a pursuit that, in all likelihood, goes back at least to Eudoxus. We need to understand Ptolemy as under the influence of two unequal forces: the pull of a Pythagoreanizing metaphysics of the universe as a harmonious, mathematical structure and the pull of the mathematical technique of the construction and computation of astronomical tables. In the practice of Ptolemy's astronomy, the pull of technique is the more powerful, and so technique becomes nearly its own independent pursuit, the metaphysics revealing itself merely intermittently, through the cracks, as it were, of the system. The tide of science will recede and rise again for many centuries to come, but generally speaking – as we will see in the remaining chapters – the pull of metaphysics tends to get stronger. With authors such as Kepler, the pull of metaphysics is overpowering even as they seek to maintain the precision of Ptolemy's mathematical technique: it is then that modern science takes off.

## Suggestions for Further Reading

The history of astronomy is a specialized scientific expertise: one needs to understand the mathematics of a theory – but then also to calculate its consequences and compare them against currently known astronomical observations. This is a piecemeal exercise of puzzle-solving, very much like normal scientific practice. This was most brilliantly pursued by Otto Neugebauer, whose three-volume monument must be mentioned first: O. Neugebauer, *A History of Ancient Mathematical Astronomy* (New York: Springer, 1975). (Known in the profession simply as *HAMA*.) Neugebauer was a genius. Focused on getting the details of the science right, Neugebauer was less interested in questions of historical context, whether in society as a whole or even within the scientific practice itself. To him, it was all parameters. Once you get into the thick of a particular technical question, his writings become compelling, even gripping, but I am not sure I can recommend Neugebauer to most of my readers: his writings might strike you as impenetrable and, frankly, unmotivated.

Alexander Jones is, more recently, the foremost author trained in Neugebauer's tradition, and unlike him, he is as much a historian as a scientist, not only recovering the detail of past astronomical systems but also accounting for their meanings and motivations to their original authors. His exposition of the Antikythera mechanism – A. Jones, *A Portable Cosmos* (Oxford: Oxford University Press, 2017) – touches on most topics of ancient astronomy. (Jones's claim is that the mechanism was in fact designed to serve as an introduction to ancient astronomy, and in Jones's hands, it certainly becomes so.) This is also, in its own way, as much a work of genius as HAMA and is, quite simply, the best introduction to ancient astronomy we now have. Another excellent introduction –

structured around examples and exercises and so almost a textbook – is J. Evans, *The History and Practice of Ancient Astronomy* (Oxford: Oxford University Press, 1998).

F. Rochberg, *The Heavenly Writing* (Cambridge: Cambridge University Press, 2004) focuses on Babylonian astral divination, but (because this was, indeed, the motivation to Babylonian astral science as a whole) it is the best introduction to Babylonian astronomy as a cultural phenomenon. Particular points I discuss include the Mesopotamian invention of astronomical prediction, the spread of astrological practice, and the invention of the table. For the first, see D. Brown, *Mesopotamian Planetary Astronomy-Astrology* (Groningen: Brill, 2000). For the second, see A. Jones, *Astronomical Papyri from Oxyrhynchus* (Philadelphia: American Philosophical Society, 1999). For the third, see E.Robson, "Tables and Tabular Formatting in Sumer, Babylonia and Assyria, 2500 BCE–50 CE," in M. Campbell-Kelly, M. Croarken, R. Flood, and E. Robson (eds.), *The History of Mathematical Tables: From Sumer to Spreadsheets* (Oxford: Oxford University Press, 2003).

I refer in particular to four individual studies on early Greek astronomy: H. Mendell, "Reflections on Eudoxus, Callippus and Their Curves: Hippopedes and Callippopedes," *Centaurus* 40 (1998): 177–275, and I. Yavetz, "On the Homocentric Spheres of Eudoxus," *Archive for History of Exact Sciences* 52 (1998): 221–278 – the two articles that came out, independently, in the same year, reopening the study of Eudoxus's astronomy. I suggest this should be read against the background of the evidence of Epicurus's reaction: D. N. Sedley, "Epicurus and the Mathematicians of Cyzicus," *Cronache Ercolanesi* 6 (1976): 23–54. An attempt to debunk the entire project is B. R. Goldstein and A. C. Bowen, "A New View of Early Greek Astronomy," *Isis* (1983): 330–340, now mostly seen as an extreme but salutary reminder of our limited evidence. For the whodunit – the question of how a new astronomy based on epicycles and eccentrics was offered in the Hellenistic era – I refer to J. Evans and C. C. Carman, "Mechanical Astronomy: A Route to the Ancient Discovery of Epicycles and Eccentrics," in N. Sidoli and G. Van Brummelen (eds.), *From Alexandria, through Baghdad: Surveys and Studies in the Ancient Greek and Medieval Islamic Mathematical Sciences in Honor of J. L. Berggren* (New York: Springer, 2014), pp. 145–174.

The major figure of ancient astronomy must have been Hipparchus. There is a real need for a monograph dedicated to him, but incredibly, there is none (admittedly, there are relatively few sources). We do a little better with Ptolemy, although once again usually focusing on the technical detail of the astronomy. O. Pedersen, *A Survey of the* Almagest*: With Annotation and New Commentary by Alexander Jones* (New York: Springer, 2010), is the best guide to the contents of this work, hence to the final developed form of ancient astronomy. Finally, the realism (as against instrumentalism) of ancient astronomy is the topic of G. E. R. Lloyd, "Saving the Appearances," *Classical Quarterly* 28 (1978): 202–222.

CHAPTER 6

# *The Canonization of Greek Mathematics*

## Plan of the Chapter

Time for a survey, for a look back. What we may call "the Greek achievement" is all in: Archytas, Archimedes, Ptolemy . . . Which does not mean, of course, that history stopped. At around the third century CE, however, there is a marked shift. "Porphyry and a New Start" is where we begin – an author of the second half of the third century whose works include a Neoplatonist commentary on past mathematical works. To understand this, we need to bring in two contexts. One is "Platonism and the Return of Number," where we survey the history of Platonism until its remaking as Neoplatonism. This philosophy has mathematics – and especially, number – at its center. From the third century CE onward, practically all pagan philosophy becomes Neoplatonist. (To understand this process, we also note several mathematical works of the imperial era, more focused on number.) The other context, discussed in "Teachers, Commentaries, Books," is the rise of commentary as the major form of creativity, starting in the third century CE.

With this in place, we survey our evidence for the mathematics of the fourth to the sixth centuries, roughly in chronological order. "Pappus and the Mathematicians of Alexandria" brings in three teachers of mathematics – Pappus, Theon, and Hypatia – active in Alexandria across a century, from the early fourth century to early fifth century CE. Especially for Pappus, the mathematical remains are significant and will exercise an important influence on the future growth of science. From mathematicians, we move back to philosophers. "Proclus and the Philosophical Schools" concentrates on the figure of Proclus, the leading Neoplatonist philosopher of the fifth century CE and author of an influential commentary to Euclid. This brings in the wider phenomenon of the reception of mathematics, in this era, by philosophers. "Eutocius – and a Coda" looks, finally, at an author trained in Neoplatonist philosophy but visible to us only as an author of commentaries on advanced mathematics. Eutocius, of the sixth century CE, already lived in a fully Christian Byzantium, and it is thanks to the authors of his milieu that so much of Greek mathematics survived into the Middle Ages – preparing the way for the many renaissances of Greek mathematics to come.

## Porphyry and a New Start

> "There are many schools of thought in musical science, Eudoxius, on the subject of attunement, but there are two that one might reckon pre-eminent, those of the Pythagoreans and the Aristoxenians."

This is, in Barker's translation, the beginning of a work from the late third century CE.[1] Which work, you ask? Well, let me cite, now, the title:

> *Porphyry's Commentary on Ptolemy's* Harmonics.

Right there, a revolution. Several generations after Ptolemy's death, his works were not merely read: they were the subject of *commentary*. Surely Ptolemy never anticipated this. But he would have been pleased by Porphyry's explanation of why he, Ptolemy, was chosen:

> Before Ptolemy, Didymus, the writer on music, described the distinctions between [Pythagoreans and Aristoxenians] very adequately in his pioneering treatise about them, and Ptolemy examined them in his Harmonics, making clear what is useful in each of them and reconciling the apparent conflict between them. I have therefore decided to attempt an explication of Ptolemy's Harmonics, since . . . I know of no one up to this time who has done it, and also because I recognize that reading it is no easy task for people who have not made a detailed study of these schools of thought or have been trained in the mathematical sciences. . . . I was led to undertake this exposition also by the fact that only Ptolemy, or he above all, brought the science of attunement to perfection, not so much by what he added . . . as by his critical assessment. . . . He rejected in their entirety his predecessors' quarrelsome polemics . . . bringing out instead what had been well said, and showing that it is in concord with the facts and with the criteria by which they should be judged.

Porphyry singles out Ptolemy's attitude – of finding common ground – as a positive, indeed, as the reason for taking Ptolemy as his starting point.

[1] A. Barker, *Porphyry's Commentary on Ptolemy's Harmonics* (Cambridge: Cambridge University Press, 2015), p. 63.

This is not a matter of tone alone – Porphyry, in his own writings, could often become a sharp, indeed vicious, polemicist. This is rather a matter of intellectual goals. Porphyry's stated aim is the finding of a lasting, overarching synthesis: it is therefore useful to take, as one's starting point, an author who covers the past well and aims, himself, at possible areas of agreement.

A few more observations. First, Porphyry's main audience, we learn, was those "not trained in the mathematical sciences." Porphyry himself was no mathematician. He does comment competently on the more arithmetical passages in Ptolemy's text, but he concentrates mostly on the methodological passages. The imagined audience is surely that of intellectuals who, like Porphyry himself, seek to be erudite – and philosophically trained (the "Eudoxius," addressed in the first sentence, is otherwise unknown). Here, then, is where the mathematical sciences now fit: within a large project of philosophical erudition.

We also learn that Porphyry knew of no previous commentary to Ptolemy's *Harmonics*. This is, in fact, decisive. For Porphyry's generation, the writing of commentaries came naturally, but this may not have been true of any previous generation. The one work mentioned so far in this book, earlier than Porphyry, that was referred to as a "commentary" was the one Hipparchus wrote on Aratus – and that was no commentary at all but, instead, a critique. Until the third century CE, the writing of commentary was restricted to a few domains (mostly the literary canon; we return to survey the history of commentary later in the chapter). Commentary was never the main vehicle of intellectual production and was never produced for the exact sciences. From the middle of the third century CE onward, however, commentary suddenly becomes the main form of intellectual production, across all fields. Even noncommentary works take up some of the characteristics of commentary, often interested in the production of overarching synthesis, often organized around a canonical corpus that is excerpted and rearranged. And so it remains. The triumph of the commentary took but a generation or two in the third century, but from then on: commentaries, encyclopedias, and various other compendia remain the mainstay of cultural life, in the various Mediterranean languages – Greek, Latin, Arabic, Hebrew, and more – for over a millennium. We transition into Late Antiquity and, with it, into the Middle Ages.

Porphyry was there right at the beginning, and as it happens, we know him rather well.[2] Commentators, by the nature of their project, are

[2] I provide a fuller account of Porphyry's scholarly practice in R. Netz, *Scale, Space and Canon in Ancient Literary Practice* (Cambridge: Cambridge University Press, 2020), pp. 745–753.

prolific. And although they tend to be self-effacing, they often do show through the cracks. Porphyry's life mission was the dissemination and elucidation of the works of his philosophical master, Plotinus. The centerpiece of all that was an edition of Plotinus's works, the *Enneads*, introduced with a "Life of Plotinus" – a hagiography of the master with the pupil, Porphyry himself, in prominent display. We learn that Porphyry was a Phoenician, born in Tyre in the year 233/234. He went to study in Athens, which – in this age of rhetorical display and pageantry of the past – had become once again something of a liberal arts college. He became recognized for his good education in Greek literature and philosophy and, at age thirty, got to Rome. Soon after, he became distinguished – in the very last years of Plotinus's life – as Plotinus's star pupil. Plotinus died in 270; over the following decades, Porphyry wrote millions of words, editing, reintroducing, reproducing Plotinus's philosophy.

Many works implore their readers to a life of learning (a short extant work addressed to his wife, Marcella, encourages her to spend the time, when Porphyry is away on travel, in philosophical study). This is often polemical: you should study *our* philosophy, not another's. Plotinus was a Platonist, and late in the third century CE, the polemic was mostly against Christianity. Once safely pagan, one was expected to study Plato as well as Aristotle, seeking their synthesis. Indeed, a lot of Porphyry's effort was invested in the commentary to Aristotle, including no fewer than three known commentaries to one of Aristotle's most elementary works – taken, at this point, as a kind of introduction to philosophy as such – the *Categories*. This then led to a great many commentaries on all of Aristotle's logical works and many others and then to commentaries on many Platonic dialogues. But there are many further indications of Porphyry's erudition: works commenting upon Homer (the traditional subject of ancient literary commentary). He went to study in Athens, so he was obviously a competent rhetorician, and lo and behold, we find that he wrote a commentary on the works of Minucianus, a past theoretician of rhetoric (as far as we know, this is the first such commentary to rhetorical handbooks). And he did indeed write commentaries and introductions to the exact sciences: not only to Ptolemy's *Harmonics* but also to his *Tetrabiblos* (that is, Ptolemy's astrological work) and also other introductions to astronomy.

Porphyry's life is all about the teaching, within a close circle, of like-minded erudites: above all, they become immersed in a philosophy, which they learn in succession through Aristotle, Plato, Plotinus, and the many commentaries along the way; they maintain their literary and rhetorical erudition; they add to this, also, a certain knowledge of music, and of the stars. They are Platonists, after all. This Platonism is in many ways true to the spirit of the original engagement between Plato and Archytas, with a vision of a true world, hidden behind the visible one, made of abstract structures: so, mathematics as an intimation of hidden realities. This Platonism is also much more rigorous in its view of such hidden realities, constructing specific cosmologies of the nonmaterial reality. It is a philosophy that verges into a kind of cartography of the divine – even more so as it comes to confront the outright blasphemy of newly assertive religions, such as Christianity, that deny the gods.

All of this sounds incredibly specific. A particular variety of Platonism, an exposition of the structure of the hidden divine; the writing of commentaries to selected mathematical texts – surely this would be just a small part of the history of mathematics? But in fact, from the third to the sixth centuries, almost everything we see in mathematics is at least within the field of influence of this distinctive project, and indeed, the very transmission of mathematics into the Middle Ages seems to have depended on it. We now have Ptolemy's *Harmonics* to read, for which we should be grateful, above all, to the tradition represented by Porphyry. Why did this variety of Platonism become so significant? How did culture come to be ruled by Neoplatonist commentary?

## Platonism and the Return of Number

Already in his lifetime, Plato made it into the canon. He was among the small group of authors – made mostly of the leading cultural figures of classical Athens – who were read and admired everywhere. This small group included tragedians such as Euripides; public speakers such as Demosthenes; historians such as Thucydides. Greek readers long continued to be able to read Athens' dialect (even while they were speaking very different versions of the language), and they continued to feel at home reading works that took place in that city (even while they were living far away from it and under very different regimes). Such is the power of literary memory. It was through Plato, above all, that Greeks came to know one of the most beloved characters of antiquity, Socrates: clever, questioning, iconoclastic. Plato's prose is supremely elegant, witty, and precise. Greeks were schooled, from childhood on, to write like him. He talked about love, about death: his readers came to think about the most intimate parts of their lives in Platonic terms. The dialogues depict debate at its best – powerful theses, logically demolished. This mattered a lot to

Greek readers, who liked to imagine themselves reenacting Socrates's wit in their own conversations. Indeed, Plato had a memorable scene of a symposium and many other scenes of social gatherings, all important rituals in the lives of the elite. In short, Plato was an author to live by; he still remains.

He died in 347 BCE. His immediate associates continued to meet near Plato's villa, just outside the city walls, in the gardens dedicated to the hero Akademos: hence, they were "Academic."[3] This informal study group was first led by Plato's nephew Speusippus (leader of the school, 347–339), then by his much younger associate Xenocrates (leader of the school, 339–313). Xenocrates – no longer a blood relation to Plato – replaced Plato's charisma with the routine of something akin to a formal school. (Indeed, at the same time, another younger associate of Plato – Aristotle – set up his own institution of teaching at Athens; the city was beginning to build its apparatus of a college town.) Whereas readers of Plato elsewhere emphasized his elegant prose, his wit, Socrates's own irony, the Academic group led by Speusippus and Xenocrates emphasized Plato's doctrines, especially those he held toward the end of his life. The emphasis was on the hidden mathematical structure of the universe and the truth of proportion, seen in music and the stars. In particular, there was a specific doctrine, developed perhaps more by Plato's followers than by Plato himself, of numbers as foundational. Plato's *Timaeus* describes the universe in quasi-mythical terms, as constructed by a divine maker. There was a debate in antiquity over how this was to be taken (as referring to a temporal sequence of events in which the world is gradually made or as a conceptual analysis of the structure of reality). At any rate, faithful readers of Plato's *Timaeus* came to think of the world as a well-ordered succession. There were the ideal forms, among which "one" and "two" were paramount; combined, they gave rise to the entire succession of numbers, which gave rise, in turn, to geometrical figures, of which all the rest was made. It might strike us as a flight of fancy, but in fact, at this point, among this group – heir to the encounter between Plato and Archytas – the idea that numbers and their ratios underlay music, hence the stars, hence reality, was quite uncontroversial. The Italian, that is, Pythagorean, pedigree of this view of

[3] In the following couple of pages, I offer a brief history of Plato's ancient legacy up to the High Empire. There is no single monograph surveying this tradition in its totality; essential resources include J. M. Dillon, *The Middle Platonists, 80 BC to 220 AD* (Ithaca, NY: Cornell University Press, 2003); J. Dillon, *The Heirs of Plato: A Study of the Old Academy (347–274 BC)* (Oxford: Clarendon Press, 1977); J. Glucker, *Antiochus and the Late Academy* (Göttingen: Vandenhoeck & Ruprecht, 1978).

reality was noted (already by Plato: the eponymous, main speaker of the dialogue *Timaeus* is a visitor from southern Italy), and so, because authors such as Speusippus and Xenocrates thought of themselves as Platonists, this also carried, for them, some measure of allegiance to Pythagoreanism. Numbers, Plato, Pythagoras: an important alliance was formed.

We may imagine Arcesilaus, born in Pitane (modern-day Turkey) in about 315 BCE, thinking, at the very last years of the century, of his future study at Plato's Academy; little wonder that he first studied mathematics with the local astronomer, Autolycus. Coming to Athens, he soon became one of the most popular members of the Academy. At about fifty years old, he became head of the school. His mathematical education might have been impeccable – but he threw it away. He was clearly more a public dialectician, in the old Socratic manner, than the maker of doctrines (Xenocrates, who, more than anyone, else created the doctrine of Platonism/Pythagoreanism, wrote at least seventy books; Arcesilaus wrote none). This dialectical prowess is probably why Arcesilaus was so popular and also why he turned away from the doctrines of Speusippus and Xenocrates to what must have appeared as a more "Socratic" version of Plato, one where the emphasis was on dialectics and debate as such. Technically, Arcesilaus was a skeptic. One should never assent to any doctrine; instead, one should always continue "looking" (so, in the original Greek, he was a *skeptikos*, a "looker"). Such was the force of Arcesilaus's stamp – and such was the stability of Athenian schools, at that point – that for the remaining two hundred years of the school's existence, it would keep close to Arcesilaus's skepticism (elaborated and made even more radical by Carneades, in the second century BCE).

Like everything else, such stability was shattered by the coming of the Romans. We recall Posidonius, pursuing Stoic philosophy away from the Athenian school – but in the presence of Roman patronage from the likes of Pompey and Cicero. Philosophy, we recall, moved away from Athens and looked for various syntheses, combining the old philosophies with other tendencies of Hellenism – always crafting original, eclectic packages of "what the Greeks knew." Science, literature, and philosophy all got combined, in many distinct ways. Posidonius was not only a philosopher but also a historian and a geographer. He even wrote on military tactics; he even commissioned an astronomical device! The Stoa was not alone in this new parade of novel combinations. We find Antiochus of Ascalon, a Platonist trained in Athens (so, obviously, trained in Carneades's skepticism), in Alexandria in the year 87 BCE, under the protection of the Roman general Lucullus. The time and place are extremely specific because Cicero (our witness to the event) recalled it as a famous scandal: here was an "Academic" – follower of the school of Plato – claiming that knowledge – gasp – was possible! This was explicitly advertised by Antiochus himself as a return to original Platonism. It is indeed most notable that Antiochus also returned to the form of the master – writing in dialogue form (the first philosopher to do so, it appears, after many generations). In fact, many scholars think that Antiochus's new philosophy was not more than a Platonic variation of Stoicism. The main theme is not of originality but of a combination of variety – and now the floodgates are open. Labels can be fluid. Potamon of Alexandria – a couple of generations later than Antiochus – even called his philosophy "eclecticism." We know almost nothing of the contents of his philosophy, but what we do find – attested, almost as usual, by Simplicius – is a mathematical diagram explicating Plato's constructions in the *Timaeus*!

Clearly, throughout the first century, there is a revival of interest in the mathematizing philosophy of the *Timaeus* and of Plato's immediate followers. This philosophy is all about obscure realities: a world, behind a world. It is natural for its adherents to look for secrets, lost theories, hidden meanings. One has to tease out Plato's allegories to look for new sources. Throughout the first century BCE emerges a host of works purporting to come from old Pythagorean masters, above all from Archytas. Those new apocrypha state Pythagorean theses as the keys with which to understand Plato's hints. Other authors, such as Eudorus of Alexandria, argued for similar theses in their own names. It is hard to know how much of this is original, how much goes back to Speusippus and Xenocrates (whose works are all lost). Broadly speaking, the emphasis in all these philosophies is the same: the world is a layered emanation from numerical principles. This abstract approach is provided with concrete meaning because beyond the bare assertion that "one" and "two" beget the other numbers, one also finds particular qualitative meanings of individual numbers, and through them, one can produce significant interpretations. Philo of Alexandria, for instance, early in the first century CE, produced his own eclectic mix of Pythagoreanism with *Judaism*. Thus, Genesis has the world created in six days. Of course! This is because 2 is the first female number, 3 the first male (because even numbers are female, and odd are male, obviously), so six is their combination. Also – a somewhat more serious point – six is the first *perfect* number (as defined in Euclid's *Elements*: a number equal to the sum of its factors, in this case, $1 + 2 + 3 = 6$). Taking numbers as possessing qualitative significance may strike us as nonsense, but in a way, that was the simple scientific consensus concerning music theory. And indeed, right at the same time that a new, Pythagorean and numerological version of Platonism emerges, we see a revival of mathematical music. This, of course, was a central area of research for the generation of Archytas, documented down to the Euclidean *Section of the Canon* (if indeed this is a work of the late fourth century BCE). From that point in time onward, right until the first century BCE – with but one exception – no mathematical interest in music is attested at all. Nothing in Archimedes, nothing in Apollonius, nothing in Hipparchus: an entire field that simply fell away. (The single exception is intriguing: this is Eratosthenes, who was also – among many other things – a philosopher, almost the only one attested for Alexandria of the third century BCE. To engage with music, then, was the mark of true eccentricity!) Perhaps late in the first century BCE, we come across Ptolemais, a daring new author – a woman, also, writing on mathematics (interestingly, we will have more of this). Known now only through her quotations by Porphyry – in the commentary just mentioned to Ptolemy's *Harmonics* – she seems to have reinvented the mathematical study of music. It is clear that writings on music quickly burgeon. The field resonates with new, Pythagoreanizing philosophies and is also obviously accessible to anyone with general education (music, after all, is song, which is poetry: so everybody with literary education also had some musical education). Late in the first century CE – we are now not that far away from the time of Ptolemy himself – we have extant Nicomachus's *Manual of Harmonics*, where this sense of an appeal to a wider audience is especially palpable. Here is how the work is addressed to its patron:[4]

> Though [*Harmonics*] is in itself complex and difficult ... and though I especially, because of the restlessness and hurry of a traveller's life, am unable to devote myself ... nevertheless, best and noblest of women, I must arouse my greatest efforts, since it was you who have bidden me at least to set out the major propositions for you in simple form.

The woman is not named (was it the empress?). It is in fact rare to see such groveling in ancient dedications. (Ptolemy, as we recall, merely put the name of the dedicatee – Syrus, perhaps the patron Sura – in between commas.) The overall sense with Nicomachus is of an author seeking patronage, on the power of his skill as a *teacher*. Elementary mathematical education always involved, above all, such things as the manipulation of number: actual counting and calculation. In this, it differed from Euclid's

[4] Translation adapted. Context can be read in F. R. Levin, *The Manual of Harmonics of Nicomachus the Pythagorean* (Grand Rapids, MI: Phanes Press, 1994), p. 33.

*Elements*, where the three arithmetical books are much more abstract and almost never refer to any actual numbers. Finally, arithmetic – in the sense of manipulation and calculation with simple numbers, especially with their ratios – is the basic tool required for mathematical music theory. In short, there is little wonder that we have from Nicomachus not only a manual of harmonics but also a manual of arithmetic. This is clearly a teacherly work, an exposition of terms and a series of classifications, above all of *ratios* and of *means* (arithmetic means, geometric means, harmonic means, etc. – the terms are still familiar to us), all illustrated by concrete examples of actual calculations. There are no proofs, and at first glance, such works strike modern readers hardly as mathematics at all. Nicomachus did fill a gap, and for his many readers of a philosophical bent, for centuries to come, he was seen as the equal to Euclid if not greater than his predecessor. The very meaning of mathematics, it appears, has changed: usurped, so to speak, by the new Pythagoreanizing philosophy.

Of course, a new focus on number – and teaching – need not mean the relatively superficial treatment of Nicomachus. Nor was this era in any way a nadir of scientific activity. It is possible that we should date to about this era another of our extant authors – Diophantus – whose works take numerical problems, essentially of the kind taught in elementary education, and transforms them into sophisticated mathematics.

In fact, there's very little evidence for the date of Diophantus. The one work securely ascribed to him is addressed to one "Dionysius," perhaps the most common of all Greek names. There is nothing further in the work concerning the person Diophantus. He is first mentioned by Theon, a mid-fourth-century author, and there was a commentary on him written by Theon's daughter Hypatia (on both of whom, see the following discussion). He could have been active as late as the early fourth century, then, but the late reference does not prove a late date: there was – as we will note in the following section – very little literary, let alone mathematical, activity all throughout the third century. Diophantus could easily be a second-century author, perhaps to be considered side by side with Ptolemy. The extant work, the *Arithmetic*, is perhaps to be considered side by side, in some sense, with the *Almagest* itself. It was a gigantic – specifically, a thirteen-book – work. Unlike the *Almagest*, not all books are now extant. In the Greek, we have six books, known in Europe since the fifteenth century: these formed an important impetus to the rise of European arithmetic (Fermat's Last Theorem was originally too long for Fermat's margins . . . of his printed text of *Diophantus*). Modern scholarship came to know of an Arabic version thanks to a sensational find by

Sesiano, published in 1982:[5] a manuscript with seven nonoverlapping books. As it turns out, in the Greek, we had Books I–III, then (probably) Books VIII–X; from the new Arabic manuscript, we can now recover the complementary Books IV–VII (XI–XIII appear to have been entirely lost). That we did not even know that the Greek text was nonconsecutive goes to show that Diophantus's work, in fact, does not have a tight deductive structure. It is, instead, a long list of example problems, arranged not by axiomatic dependency but by growing complexity. To take one of the simplest examples (I.10):

> To two given numbers: to add to the smaller of them, and to take away from the greater, and to make the resulting [number] have a given ratio to the remainder.
>
> Let it be set forth to add to 20, and to take away from 100 the same number, and to make the greater 4-times the smaller.
>
> Let the [number] which is added and taken away from each number [sc. of the two given numbers] be set down, [namely] ς, one. And if it is added to twenty, results: ς1 Mo20. And if it is taken away from 100, results: Mo100 lacking ς1. And it shall be required that the greater be 4-times the smaller. Therefore four-times the smaller is equal to the greater; (5) but four-times the smaller results: Mo400 lacking ς4; these equal ς1 Mo20.
>
> Let the subtraction be added [as] common, and let similar [kinds] be taken away from similar [kinds]. Remaining: ς5, equal Mo380. And the ς results: Mo, 76.
>
> To the positions. I put the added and the taken away on each number, ς1; it shall be Mo76. And if Mo76 is added to 20, result: Mo 96; and if it is taken away from 100, remaining: Mo 24. And the greater shall stand being 4-times the smaller.

This passage is opaque only because of two strange symbols: "ς" is an ad hoc abbreviation for the Greek word *number*; "Mo" is a more transparent abbreviation for the Greek word *monas*, or "unit." In this problem, we are asked to find a number – the same in both cases – that, added to a certain value, removed from another, makes the greater resulting number a certain multiple of the smaller (four times is taken as the example). We would put it, perhaps, as follows:

> given a $p$, finding, for a pair of $m$, $n$, the value of $x$ that satisfies:
> $$(m + x) = p(n - x).$$

[5] J. Sesiano, *Books IV to VII of Diophantus' Arithmetica* (Berlin: Springer-Verlag, 1982).

And here, the fact that we use the same symbol, $x$, on both sides makes it easier for us to see that it is the same number we subtract and add. The same function is provided by the Greek by using not simply the word *number* but by abbreviating this expression. Instead of $p$, $m$, and $n$, Diophantus takes some token numbers that act, in this problem, rather like the way that an individual diagram acts in a geometrical argument. Those are obviously selected for ease of manipulation – in this case, they are 4, 100, and 20. Diophantus refers to these as "monads," which is, in practice, a way of signaling that this is a problem in the first degree and that Diophantus does not refer here to "squares" or other exponents. Other problems might demand that Diophantus find, for instance, two square numbers equal to a third one – the problem that inspired Fermat's Last Theorem – and this would be expressed by referring the token numbers chosen not as "monads" but as "powers," meaning the second power. Other problems might refer to cubes, to power-powers, to power-cubes, to cube-cubes . . . In all such cases, Diophantus would use a similar system of abbreviation: the Greek word for "power" is *dunamis*, and so the abbreviation is Du; for "cube," one has Ku (*kubos*), and so forth. Not only does Diophantus use these abbreviations: he explains them in his introduction to Book 1, which, otherwise, lacks any more detail, neither a personal introduction (so, not helpful for the date) nor any axiomatic material (other than, effectively, a set of definitions of the powers). The reliance on explicit symbolism is remarkable and could be compared with Ptolemy's reliance on the textual tool of the table.

As noted, Diophantus's *Arithmetic* does not have Ptolemy's or Euclid's logical structure of a chain. The various problems instead apply, to different configurations, the same techniques. Thus, a most basic technique is, in fact, explained in the introduction to Book 1:

> If, from the terms of a problem, there arise some kinds equal to the same kinds, but of different multitudes, one should take away from either part, similar [kinds] from similar [kinds], until one kind will be equal to one [other] kind.

This elementary algebraic technique is clearly implemented here. Diophantus starts from what we would write down as

$$20 + x = 4(100 - x).$$

Unpacked as

$$20 + x = 400 - 4x.$$

Which is turned, explicitly relying on "taking away similar [kinds] from similar [kinds]," into

$$5x = 380, \text{or } x = 76.$$

This is finally verified by plugging the number back into the terms of the problem. Diophantus never proves anything, but it is significant that the problem is stated in general terms, and indeed, it is not difficult to see how this particular solution could be generalized: whatever terms one is given, simply follow the same route, with the same algebraic technique. Technique – replacing proof. There is no doubt that this is the attitude of the schoolroom, and problems at this level of complexity were indeed encountered in ordinary education, since the times of the ancient Near East. Soon thereafter in Diophantus's book, the problems begin to require much more dexterity: one brings in higher powers; most problems are, technically, "indeterminate" (they allow multiple solutions); and adding to the complexity of sheer calculation, fractions are allowed. Here, for instance, is VI.22:

> To find a right-angled triangle so that the [number] in its perimeter is a cube but adding on the [number] in its area, it makes a square.

We look for a Pythagorean triplet whose sum is a cube number; its sum, together with half the multiple of the first two terms, is a square number. I leave this as an exercise (as noted already, there are multiple solutions; also, "square" and "cube" numbers mean here the square or cube of what we would now call a *rational* number). Who in the second century – or later – was interested in such an exercise? It is hard to tell. As we have seen, the work was very little cited or studied – in this, for sure, it differed from Euclid, Ptolemy, and from Nicomachus. Diophantus was a fiendishly clever setter of arithmetical problems, but to most readers of his time, what mattered about numbers was the much simpler classifications of "ratio" and "mean." If Diophantus did survive, it was thanks to Hypatia, late in the fourth century, and in all likelihood thanks to her sense that numbers – even in the level of complexity required by Diophantus – had deep ontological significance. For by the time we get to Hypatia, it appears that even the practicing mathematicians, such as her, were always motivated by Platonist and Pythagoreanizing philosophies. Perhaps the same was true already for Diophantus himself.

Indeed, Platonizing, in one way or another, seems to be everywhere already in the imperial era. We see it with scientists who are not even mathematicians at all. Galen is explicit that he is *not* a Platonist, meaning simply that he does not see himself as belonging to any philosophical school. And yet he cites Plato abundantly and dedicates entire works to him (so, for instance, he is the author of summaries of dialogues by Plato, which are now extant in Arabic). *On the Doctrines of Hippocrates and Plato* is one of Galen's important works, showing – naturally enough – that the two are in agreement (in practice, Galen tries to prove that both anticipate Galen's own positions: the title refers, of course, to Hippocrates of Cos, the father of medicine). Why would a physician care so much to find the philosopher in agreement with him? Why would he write summaries of Plato's dialogues? Closer to home, let us consider Ptolemy himself. Ptolemy is never as talkative as Galen, and his underlying metaphysics needs to be extracted from hints left by his science. But those are clear enough. We have already noted the obvious Platonizing thrust of Book III of the *Harmonics*. (Indeed, in that book, Ptolemy analogizes the harmonies not only to the stars but also to the soul. This soul is made to be tripartite, which is possibly the clearest case of Ptolemy holding a specifically Platonic position.) Wietzke suggests that we pay attention to the precise words in which Ptolemy introduces this discussion in the *Harmonics*, in III.3:[6]

> Since it may follow for a person who has theorized on [θεωρήσαντι] these matters to be filled with wonder immediately – if he wonders also at other things of exceptional beauty – at the extreme rationality of the harmonic power, and at the fact that it finds and creates with perfect precision the differences of its own forms; and, on the other hand, [since it may follow for him], owing to some divine love, to desire, to behold, as it were, the nature of [the harmonic power].

I quote this at length because this is the clearest statement any ancient mathematician provides, anywhere, for the goal of his pursuit. But this, as Wietzke points out, is not quite an original statement. Essentially, Ptolemy's statement of the soul's desire to learn about musical harmony is a clever reweaving of words and themes from Plato's narrative, in his dialogue the *Phaedrus*, on the soul's desire to perceive beauty, whether material or (preferable, to Plato) abstract. This is, of course, apposite to the *Harmonics* (whose business is the passage from the perceived beauty of musical harmonies to their underlying mathematical structures, thence to such structures in the soul as well as in the stars). But the main lesson from Wietzke's reading of this passage is that Ptolemy could produce a clever allusion to Plato – and count on his readers to get it. There is a very simple reason for this: Plato, quite simply, was everywhere. Throughout the imperial era, we see Greek authors trafficking in a variety of eclectic sciences, literary practices, and philosophies, but the one core assumption, throughout, remains the literary canon. Everyone knows Euripides,

[6] I follow the discussion in J. M. Wietzke, "A Fashionable Curiosity: Claudius Ptolemy's 'Desire for Knowledge' in Literary Context," in S. Ju, B. Löwe, T. Müller, and Y. Xie (eds.), *Cultures of Mathematics and Logic* (Cham: Birkhäuser, 2016), pp. 81–105.

Homer, Demosthenes – and everyone knows Plato. (Indeed – this is why Wietzke can be certain of his interpretation – allusions to the *Phaedrus* are an absolutely standard trope of imperial-era literature.)

It is not precisely that everyone, all of a sudden, became a number mysticist. But the revival of the old Platonism was, in fact, very successful. Plato differed from all other philosophers, in that he was part of the literary canon. His old "Pythagorean" interpretation by Speusippus and Xenocrates differed from the later, skeptical interpretation by Arcesilaus and Carneades, in that the Pythagorean interpretation – unlike the skeptical one – could readily be synthesized with other sciences as well as other philosophies. In short, Pythagorean-like Platonism provided a natural point of intersection between science, philosophy, and literature. By Late Antiquity, it would remain nearly alone – as the very practices of the literary canon, finally, came to define science itself.

## Teachers, Commentaries, Books[7]

The literary canon survived because ancient readers liked a particular kind of literature. They liked public performances with a sharp sense of conflict and debate. Homer, and then, the classics of democratic Athens: the tragic and comic stage, the orator's podium, Socrates's refutations. It also survived because ancient readers were trained, generation after generation, through the same literary corpus. There are a handful of papyri of Euclid, deriving from the mathematical classroom. There are also about a thousand schoolroom literary papyri, providing the entire gamut of learning – from the alphabet to sophisticated rhetoric – through the close study of a very narrow range of works from the literary canon. Already in the early third century BCE, especially in Alexandria, a new field of knowledge emerges, "grammar" – literally, the knowledge of letters. This practice will always remain close to that of elementary education. I have mentioned medicine and astrology earlier as "service industries," fields that, especially under the Roman Empire, thrived under elite patronage. The same is true of grammar: members of the elite would have tutors at hand to teach their children and to serve as clever company: one needed to have the right quotation ready at hand.

Grammarians always showed their mettle by the edition and commentary of the literary canon. Homer was the primary subject, but one

[7] As noted in the Suggestions for Further Reading, this section is based on R. Netz, *Scale, Space and Canon in Ancient Literary Culture* (Cambridge: Cambridge University Press, 2020), Chapter 6.

commented upon many other poets as well. Early on – perhaps already by the middle of the third century BCE – several medical authors started to distinguish themselves by writing commentaries on the works ascribed to Hippocrates. In this way, medicine created for itself something like a canon, although physicians would continue to propose their own original, radically conflicting medical theories. Other fields did not acquire any comparable sense of a canon and definitely no tradition of commentary. As noted already, it is not even clear that any commentaries were produced in the exact sciences throughout the Hellenistic and imperial eras.

Philosophical commentaries were written from time to time. Early on in the academy – while it was still Pythagoreanizing – Crantor, a pupil of Xenocrates, wrote a commentary on the *Timaeus*. Was this the first of its kind? There were not many following that. Even in the Hellenistic philosophical schools of Athens – structured around the works of the masters – one expressed one's admiration of the master's words through the writing of one's own original works.

Let us now go back, once again, to Athens in the fourth century BCE. Alongside the Academic school led by Speusippus and Xenocrates, there was yet another school led by a former pupil of Plato – that of Aristotle. Aristotle's school was even more structured, research based, than that of Xenocrates. As the Academics sought to codify the master's doctrine, Aristotle and his followers set up something of a research seminar, with working documents that went through the various problems related to a given field. Aristotle, of course, was a prolific author besides that – how else would he become famous? – and his writings (many, imitating Plato, in the form of dialogue) were famous for their elegance. Thus the Aristotelian corpus had two components: a widely reproduced one, elegant and accessible; another, much less frequently copied and much harder to read (because it was not exactly meant for publication): in a sense, no more than the school's internal documents. It is clear that after Aristotle's death, both types of works were available to the public, but it is also clear that Aristotle was not very widely read, and his internal school works were in some cases nearly forgotten. (Indeed, even his own school – just like the Academy – lost interest in science, hence in the great bulk of the works of the master; it is for this reason that we should assume that Hipparchus did *not*, in fact, respond to Aristotle in his physics.)

As with so much else, this would change in the first century BCE. It suddenly became important to be eclectic, encyclopedic. Well, Aristotle, already, produced works on an encyclopedic range of topics! The school's internal works gained a new interest – but they were by no means accessible. To make them useful, one needed a re-edition and a rearrangement. Andronicus of Rhodes – active perhaps in the middle of the first century BCE? – accomplished just that. Besides editing the school's documents and republishing them as literary works, Andronicus went on to add commentaries on works by Aristotle – certainly on the *Categories*, the most basic of Aristotle's works in logic, but likely on more works as well. This transition now sticks: from this point onward, it seems that nearly anyone who considered himself an Aristotelian philosopher wrote commentaries, in particular on the more introductory, logical works. Platonism, from the first century BCE, came to be bundled together with a Pythagoreanizing, arithmetical science – of a sort. At the same time, Aristotelianism came to be bundled with the grammarian practice of editions and commentary.

If we survey the literary landscape in the age of the High Empire – the Pax Romana of the late second century CE, when Galen is still alive, the ink still fresh on Ptolemy's *Almagest* – we see a world dominated, first of all, by rhetoric. There are many authors vying for position in what I call the "service industries," above all, medicine. There are novels (a recent innovation!), even new epics, all sorts of eclectic philosophies, all striving for unique effect, clamoring to make a name on a competitive, crowded literary stage. The writing of commentaries is there on the literary scene, as well – as a genre produced by grammarians, yet another important service industry, displaying their mastery over the literary classics. It is also used on occasion by a few medical authors, as well as by the followers of Aristotle. But the writing of commentary on works outside the literary canon represents, at around the year 165 CE, no more than a tiny fraction of total literary production.

At this point, many things begin to change at once.

Already Galen comments, in his writings, on repeated outbreaks of a virulent illness that forced him to escape out of town. This might not be clear from Galen's descriptions – and it is likely enough that he himself did not understand it – but what struck the empire roughly from the years 165 to 180 CE was no ordinary disease. The empires have been growing and pacifying their neighbors for quite some time: China, reaching west, and Rome, reaching east. Finally, the germs met. Plagues hit both empires – in Rome, was it smallpox? – killing millions. Even under the conditions of extreme Roman inequality, we suddenly see wages rising for ordinary working people: there were so few of them left, you see. The elite, and the state, had to manage with less: fewer taxes collected, fewer resources to redistribute as patronage. The empire became less of a well-oiled machine, and soon enough, early in the third century, dynastic troubles began. As soon as a new emperor would be installed, others would conspire and rebel against him. The cycle of civil wars went on to drain the coffers of the emperors, forcing them to debase the currency, which, in turn, took its toll on the economy and so on tax collection. We begin to see, in the second and third decades of the third century, a rapid collapse in the building of new monuments. By the fourth decade, the currency and the state nearly collapse. Civil war becomes essentially uninterrupted, the empire often divided into several competing regions; the reality of endless civil war means that the emperors, themselves, are now recruited directly from the military. It is no longer soldiers in the service of competing dynastic factions. Instead, the soldiers take control. Little wonder that cultural patronage nearly disappears. It is not only that one does not endow new temples, baths, theaters. It seems that the entourages of patronage – doctors, astrologers, rhetors, grammarians – have drastically thinned, as well. Intellectuals fall back to a few centers: mostly, indeed, Athens. Then again, it is not as if the new soldier-emperors make the borders any more secure. Incursions by nomads become ever more frequent, and in the year 267, one tribe – the Heruli – reached as far as Athens and burned it to the ground. In the archaeological record, this appears as the most violent moment in the history of the city. It was essentially gone. To be clear: Athens would remain, as cultural memory, and so it would become inhabited again, and years later, it would briefly become a college town for one final time. (Indeed, millennia later, it would be re-created as the capital of a Greek state.) But for now, the last site of Greek civilization was lost. The elderly Longinus – Porphyry's teacher – found refuge in faraway Palmyra, in the Syrian desert, protégé of the queen Zenobia (who carved out this part of the empire as her own). Such was Greek culture in the age of Porphyry: dispersed and small. I estimate that there were, as an order of magnitude, perhaps 1,500 authors active at about the year 170 CE and perhaps about 150 active a century later.

Now, the picture was definitely dire right at the top. It was a bad time to be an emperor (you would expect to last a year or two) – and so a bad time to depend on the goodwill of emperors. Hence, the drying up of patronage with the consequent drying up of cultural life – because, you see, for the last couple of centuries, this cultural life had been sucked dry by the corrupting influence of inequality and its consequence, cultural dependence on the top elite. But below that, things look normal. At some level, does it even matter if the emperor's name keeps changing? The basic contours of life remained, and the sons of the elite went on going through the same education, learning the same Homer, the same Euripides, eventually marking themselves for their rhetorical abilities. Which means that even as elite authorship collapsed, the experience of teaching and education hardly changed. Indeed, if anything, we find an *expansion* of paideia. The Mediterranean began to fragment, and so local centers emerge: one consequence being that one looked for elite status (which still implied elite education) even in remote places such as the Syrian desert.

The very few remaining authors were swamped by a rising sea of teachers. And so, the very identity of the intellectual subtly changed, perhaps without the authors even noticing it. Quite simply, teaching became the natural model of culture. Philosophers came to think of themselves *primarily* as teachers of philosophy. And what was teaching, after all? This was enshrined in the old practice of the grammarians: teaching was the brokering of the texts of the canon, leading pupils through the reading of Homer, of Euripides, by the reading of a commentary. Philosophers would do the same, then, for their own field. Porphyry was a philosopher, in the tradition of Plotinus, which means that he saw himself as a teacher of the philosophy of Plotinus, which means that his core project as an author was the collection, edition, and commentary to the works of his master. He was also a polymath, erudite in many other fields as well. And so, he saw himself as teacher in those fields, too, and hence authored more commentaries: rhetoric, astrology, harmonics . . .

To be clear: I do not mean that Porphyry opened up a university with that many classrooms. Indeed, the relation between a commentary and "teaching" took many forms and was only rarely the commentary as the actual write-up of lecture notes. A comparison might help. We recall, perhaps, the ancient genre – made popular especially from the imperial era onward – of sympotic literature, depicting elite members in polite conversation around their wine cups. Of course, any literary work titled a "symposium" is not the transcript of an actual banquet. It is the literary, idealized version of what social gatherings *aimed* to be. This is the way we should think about the ancient commentary. It was, generally speaking, not actual lecture notes but rather a literary, idealized version of what education aimed at. Indeed, the mention of the "symposium" is directly relevant here. Hear the word *lecture*, and you think of a professor speaking to a large auditorium full of eager eighteen-year-olds (actually, even this image is a cultural construct, more *Legally Blonde* than genuine Harvard). Advanced teaching in antiquity – confined to a small slice of the elite – was closer in nature to tutoring rather than to a lecture course. Even if several learners were involved, the experience was closer to that of a social gathering of elite members, seeking to better themselves under the guidance of a teacher. This is ancient teaching, what the commentary was about: the idealized picture of the perfect erudite teacher, ready to lead his tutees into a master text. In some cases, perhaps the commentary did no more than project this image – no further teaching even taking place. Perhaps Porphyry's "teaching" of harmonics consisted only in the authorship of that commentary to Ptolemy.

A new way of writing, then. And so, together with it, we find a new *medium* of writing. Ancient books were written on rolls of papyrus: you took the stalks of papyrus reeds – grown in the marshes of Egypt – crushed their pulp, and produced thin layers that, glued together, made for reliable and surprisingly sturdy writing surfaces. They were, indeed, rolls – think of a toilet roll, for instance – that one had to wind and rewind for every reading. But then again, reading was understood on the model of performance (most rolls had copied on them the works of the canon: an epic poem, a play, a speech, a dialogue): they were meant to be read in sequence. The roll made sense. So did the size. A play takes about an hour or two to perform, perhaps ten thousand Greek words or ten yards' length of papyrus roll. The typical roll was about that size, the size of a normal performance; longer works demanded more than a single roll or "book" (which is why Homer has twenty-four "books," for instance); but that was just taken for granted.

There were other ways of packaging writing, and for various reasons, the Christians, in particular, preferred their books bound together in a different way. Instead of a roll, the Christians adopted the form of a codex. There, one takes individual leaves and sews them together. This is more laborious, but the book produced in such a way can become much bigger, especially if instead of sewing all leaves together in one single pile, one produces individual "gatherings" (the typical size is four double-leaves, making eight pages or sixteen sides, known technically as a "quire"), sewing each separately and then sewing the many quires to each other. With this arrangement, there is almost no limitation to a book's size. For various reasons – does papyrus not respond well to such complex handling, or were there simply more animals slaughtered, more skins to use? – we also find, early on after the adoption of the codex, the rise of a new material, the parchment. The skins of animals may be hardened to become leather, or they can be stretched and scraped to become a very sturdy writing surface. From the fourth century and for about a thousand years, almost all European books would be made on such skins: our knowledge of antiquity passed, literally, on the backs of cows.

Perhaps the Christians liked the size of the codex; perhaps they liked the fact that it defined a group of works as belonging together in a meaningful way (the collection of the four gospels might have been significant). Perhaps they liked the fact that one could leaf through such a book, identifying, with ease, chapter and verse. Perhaps they liked the very complexity, even the expense, of the new kind of book: its production (especially with parchment) involved some planning and elaboration; it became much more natural to embellish and elaborate the spaces of the writing. In short, this was a book one clearly valued; it was a good tool for the religion of scripture, one organized, more and more, around the learning of a canonical book.

But so, indeed, was the culture, everywhere! Christian books, already in the second century, are codices, and late in the third century, they seem to have become dominant everywhere. Book culture shifts: from the many, ubiquitous small rolls, so closely associated with the old literary performances to the bulky, expensive, rare, parchment codices, to be cherished, sometimes worshiped, always – studied. The very make of the book meant that from now on, literary culture was to be a culture of *study*. Perhaps this – the material transformation of media – is one of the reasons why the transformation of the third century was made to last.

For indeed, for a while at least, the empire did recover. A few soldier-emperors were more successful than others. Diocletian rose to the throne in 284, and through a sequence of successful campaigns and shrewd co-optations of his rivals, he managed to stitch the empire back together and to hold it – even more remarkably – for twenty-one years (always in concert with other co-emperors). The usual dynastic crisis ensued when Diocletian abdicated, but once again, an able soldier-emperor – Constantine – managed to retain control of the entire empire until his own death in 337 CE. For several decades, the empire was mostly stable (a major loss to the barbarians, in the Battle of Adrianople of 378, would shatter that, but this is yet in the future). In many ways, this would be an empire that any Roman, of earlier times, would easily identify. Sure, the capital was moved, by Constantine, to the new, Greek-speaking city of Constantinople: but then again, ancient culture was always predominantly Greek speaking, anyway. Sure, the emperor converted to Christianity. But throughout much of the fourth century, many members of the elite remained pagan, nor was Christianity that alien by now: one could cherish both Homer and the Bible. And so, some measure of the old patronage, the old learning, returned. Alexandria had its scholars, and so, once again, did re-founded Athens.

And yet, this was a distinctive culture, among other things, stamped by the many transformations of the imperial era and of the third century. There was perhaps almost no mathematical authorship to speak of in the later parts of the third century itself – Porphyry's commentary to the *Harmonics* is, indeed, the closest we have to a mathematical text extant from that era. From the early fourth century to the early sixth century, we have a significant body of writings. All of this body of writing is shaped by these two forces: Pythagoreanizing Platonism, on the one hand, and the practices of collection and commentary, on the other. Under the influence of these forces, Greek mathematics was canonized.

## Pappus and the Mathematicians of Alexandria

We have met Pappus on several occasions already throughout the book. Indeed now, nearing the end of this book, we have something of a grand reunion of the main characters. You recall Theaetetus producing some meaningful classification of irrationals – which we learn from Pappus's commentary to Euclid's *Elements*, Book x (partially extant, in Arabic). Or you may recall several of the more surprising things we know about Archimedes and Apollonius: Archimedes's classification of semiregular solids or his alternative, nonrigorous proof for the area of the spiral; Apollonius's treatise on finding the value of a line of hexameter, as the letters multiplying each other . . . All of this is from Pappus's *Collection*. It is from there that we have the handful of truly sophisticated results – or attempted results – in mathematical mechanics (in Book VIII of the *Collection*); this is also our main witness for the position of optics as a subsidiary science to astronomy (which is how Pappus positions it in Book VI of the *Collection*). These are but a few examples, but the fundamental point is that it is through the efforts of memory and collection of late antiquity, throughout the fourth to sixth centuries CE, that Greek mathematics survives. This gives us a sense of the significance of the period – but also raises a question about the period itself: Why did its authors care so much about the mathematical past?

Pappus stands out in several ways, perhaps above all in that he had more past than most at his disposal. Working in Alexandria, early in the fourth century CE – in a city largely unscathed by the upheaval of the third century – Pappus had access to very considerable collections. The erudition he wished to flaunt (unlike most of his contemporaries) was primarily not philosophical but mathematical. This means that Pappus was interested, specifically, in displaying *difficult* mathematics. It is through him, then, that we learn of the more advanced mathematics of the past.

Who remembers the rememberer? Pappus's own works survive in a multiplicity of mediated forms. It is likely enough that his most important work, as far as his immediate readers were concerned, was a commentary to Ptolemy's *Almagest* – a massive undertaking. Only Books V and VI are now extant. Similarly, the commentary to Euclid is now known only through an Arabic fragment. It is certain that Pappus also wrote a geography, which, once again, would appear to be a reworking of Ptolemy's work. Our source for this is almost a parody of an indirect survival: an Armenian abridgment and translation from the sixth century CE! (Because this is an abridgment, we have lost what interests us most, and the extant text is almost entirely descriptive geography without any references to mathematics.) A few other works are merely implied by the extant works: the most significant is Pappus's mention, in passing – 1.246 – of a commentary by him to the *Analemma* by one Diodorus. So, yet another commentary, concerning the sundial? Pappus was by no means a shy author (he is often polemic with his contemporaries), but – always surveying the works of others – he almost never refers to his own works. In all likelihood, he authored many works, and it is quite possible that many are even extant. We think of him as the author of a single work – the *Collection* – and it is perfectly possible that he himself did the collecting. It is also possible that the work was put together posthumously, cobbling together eight of his works, of which but a single medieval copy remains. This is most of what we know of Pappus, then: the Vatican Greek manuscript number 219, a ninth-century codex, perhaps a direct copy of the early codex produced not long after Pappus's death. This is one of our most precious witnesses to the Greek mathematical past. The collector was collected – even if the rememberer was not that often remembered.

What was in this collection? The first book is now lost, and with Pappus's lack of self-referential comments, we have no direct evidence suggesting its contents. The second half of the second book is extant: it is a formal commentary to Apollonius's work on the counting of the hexameter. Book III is a critique on a solution to a geometrical problem by the pupil of a rival teacher; this becomes, implicitly, a reflection on problem solving as well as on teaching. Book IV is a medley of past problems and solutions, arranged in an ascending order of complexity: so, once again, a kind of implicit reflection on the methodology of problem solving. Book V engages with a variety of theorems emerging out of the comparison of figures, such as finding the largest area among plane figures with the same perimeter. These problems give rise to a study of polygons and polyhedra that quite naturally evokes Plato's solids – the comparison between figures morphing, then, into a comparison between mathematics and philosophy. Books VI–VIII, finally, are more systematic surveys of entire fields: astronomy (introduced, as noted previously, by optics) in Book VI, "analysis" (meaning, effectively, geometrical problem solving) in Book VII, and mechanics in Book VIII. The entirety touches on a great variety of recognized mathematical topics. (One wonders: Was Book I dedicated to music theory?) It seems natural enough that astronomy would have mattered most to Pappus's readers, which means that, in practice, he would have been, in his lived experience, primarily the teacher of Ptolemy's *Almagest.* But his *Collection* surveys, above all – even in Book VI – the mathematics of the generation of Archimedes. Early in the fourth century CE, Pappus's eye was keenly focused on a specific version of ancient mathematics. Why that?

Thanks to Cuomo (see the Suggestions for Further Reading), we now recognize that in surveying his past, Pappus made statements that mattered to his present. We may follow her analysis of Book V. She points out that Pappus assumes surprisingly little of his reader (the book is actually addressed to one Megethion, probably a stand-in for the intended audience). In other books in the *Collection*, one makes complex mathematical assumptions; in this book, one spells out even the assumptions of Euclid's *Elements.* Much of the book revolves around the regular solids: obviously, to a nonspecialist, known primarily through Plato's *Timaeus.* Those five are upended by Pappus through Archimedes's much more complicated thirteen semiregular solids. Pappus is quite clear about his ultimate point:

> The philosophers say that the first god, maker of all things, has opportunely given to the cosmos the shape of a sphere, choosing the most beautiful among things that are, and declare the natural properties inherent to the sphere and add that the sphere is the greatest among solid figures with the same area. They also say other things inherent to the sphere which are more than clear and do not require great persuasion, whereas that the sphere is greater than the other figures the philosophers do not prove, but only assert, and it is not easy to be persuaded of it without further investigation.

Pappus presents to us, as "the philosophers'" position, the Pythagoreanizing–Platonist project where the physical world is derived from more abstract mathematical forms. That this is what philosophers believe, indeed, even that this is true, does not seem in doubt: Pappus – like everyone else we will see in the remainder of this chapter – simply takes for granted this conflation of true philosophy with Neoplatonism. However true, however, the Neoplatonists fail in their project because, this being a mathematical project, they ought to become true experts in mathematics – which Pappus is, and they are not. Why is Pappus the expert? Because he can recruit to his side the mathematical authority missing from the philosophical teaching. And this is, ultimately, I would say, because the Neoplatonists – bound by the tradition of their master – envisage the early mathematics, literally that of the age of Archytas. Which is why the generation of Archimedes meant so much to Pappus: it was his advantage, vis-à-vis the philosophers.

Pappus positioned himself against the teachers of philosophy, as he did against rival mathematical teachers. In Book III, Pappus chides Pandrosion, a teacher of mathematics, by telling her (yes, Pandrosion was a woman!) that her pupil made a grave mathematical error. Trying to offer an original solution to the problem of finding four lines in continuous proportion – the "duplication of the cube." which we have seen so many times, going all the way back to Hippocrates of Chios – the pupil (if we are to trust Pappus's report) effectively assumed the required construction. How does Pappus respond to the reported error? Essentially, by two moves. First, he pursues the falsehood of the construction, not so much from the abstract perspective of a general mathematical proof but, rather, through specific numerical examples. He then proceeds to offer correct solutions by providing a historical survey of past solutions, culminating with his own. I believe that the concrete numerical examples provide a hint to what Pappus is doing: although these are less compelling than pure mathematics, they are indeed effective in the actual experience of teaching. Pappus, chiding a rival teacher of mathematics, shows her *how to teach*. From this, Pappus moved on to add the history. These, then, are the two tools of teaching: concrete examples – that is, the normal techniques of mathematical pedagogy – *and ancient texts*. Because, after all, this is how education was, at the time: the master, brokering a canonical corpus to his followers.

This, then, is the purpose of the *Collection* as a whole but also of the commentaries to the *Elements* and to the *Almagest*. Everywhere, Pappus constructs a corpus and presents himself as the teacher most competent to introduce it. It is perhaps this attitude of the teacher that makes Pappus – with all his obvious polemic zest – somewhat self-effacing. In the game he played, the ability to cite ancient results was valued even more than the ability to produce new ones; hence, no more than a few results are explicitly presented by Pappus as his own original contribution. The same is true, even, for the opinions offered: these are often put forward as the outcome of an ancient debate rather than as Pappus's own. Perhaps the most "methodological" book is IV, with its crescendo of complexity.[8] Pappus specifies that problems should be classified according to three types – "planar" (based on the tools of Euclid's *Elements* alone), "solid" (requiring conic sections), and "linear" (requiring special curves, not producible from either the *Elements* or the *Conics*). Now, one of the key results surveyed throughout Book IV is Archimedes's *Spiral Lines*, and so at the end of this book, Pappus adds a final discussion, introduced as a kind of addendum:

> I have inserted the analysis of the neusis that was taken by Archimedes in the book on the Spiral Lines for you, so that you will not get into difficulties when you go through the book.

This is, in fact, the most typical way in which Pappus introduces a text. Books VI and VII, in particular, are constituted by long surveys where Pappus picks a work and then produces a series of required results or lemmas that are helpful for this treatise. The finale of Book IV can therefore be seen as a mini-introduction to *Spiral Lines*: here is a difficult construction by Archimedes; here is how it should be achieved. However, this serves an extra purpose, as Pappus explains:[9]

> They [the solutions produced by Pappus] are useful for many other solid problems as well.

When one has a task that requires conic sections, this particular tool – a certain construction making possible a neusis – is *generally* helpful. A brief note: and we learn that our passage fits a larger meta-mathematical project.

What is a neusis? In *Spiral Lines*, Archimedes has a series of propositions where, for instance (SL 5), given the line AX, a point B on the circle ABG, and the length E, one finds the line BH (H on the line AX) so that the segment QH – intercepted between the line and the circle (and "verging" toward point B, "verging" being the literal meaning of *neusis*) – is equal to the length E. It is obvious why such an intercept *can* be found (start the line passing through B at BG, and the value of the intercept is zero; move this line out toward H and beyond, and it extends indefinitely; so, obviously, its length will, at some point, pass through equality with E). Archimedes does not provide a construction – I believe simply because this

[8] I will look at this book a little more closely. It is translated into English in H. Sefrin-Weis, *Pappus of Alexandria: Book 4 of the Collection* (New York: Springer, 2010).

[9] The following citations from the final stretch of Book IV are from Sefrin-Weis (n. 8 sup.), pp. 163–167.

would have put too much weight on this part of the treatise (which is intended as a mere prelude).

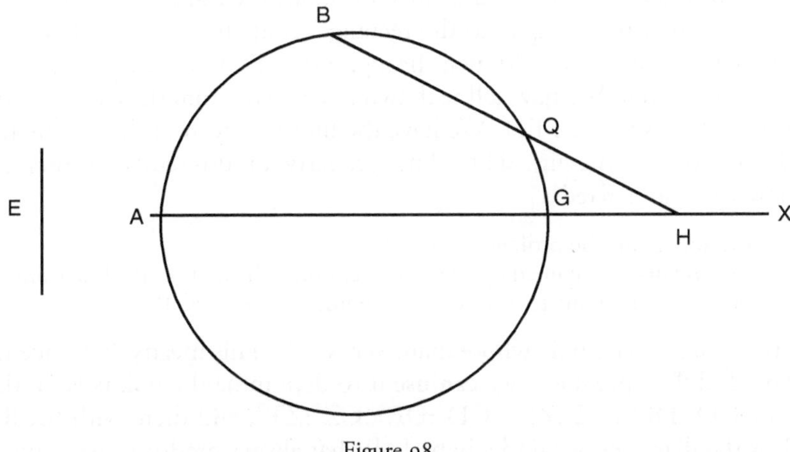


Figure 98

Pappus, introducing the text, sees an opportunity to expand on this gap, but more than this, he senses that this is required so as to make Archimedes's solution whole. From his perspective – that of the commentator – to imply "here is a mere prelude, and we allow ourselves to cover this ground more quickly" is just not an option. And so he provides a more general set of tools to neusis-type problems.

A couple of specific cases are taken, rather different, in fact, from those required by Archimedes! I will go through the first. We are given the line AB and the point C. We need to find a point on line AB, such as D, so that the line joined from D to C will have a given ratio to the line drawn perpendicularly from D, up to some otherwise-defined position.

In the typical move of Greek geometrical analysis – one especially typical to Pappus – we imagine the construction already produced. We then follow the consequences. The ratio of CD to DE is given; now draw CKZTH, perpendicular to line AB. Line CZ is given (point C and line AB are given). Points K, T are chosen so that

CZ : ZT :: CD : DE, CZ : ZK :: CD : DE (this is doable because CD : DE is given).

The clever move is that we now consider the same ratio, squared:

$$CD^2 : DE^2 :: CZ^2 : ZT^2.$$

$CD^2$:$DE^2$ is still a given, and so is the second ratio. So, then, will be the ratio of the differences:

$$CD^2 - CZ^2 : DE^2 - ZT^2.$$

Now, the difference between $CD^2$ and $CZ^2$ is clear from Pythagoras's theorem: this is $ZD^2$ or $EH^2$. The difference between $DE^2$ and $ZT^2$ is slightly less obvious, but with the kind of manipulations used in *Elements*, Book II, we can see that this is equal to the rectangle contained by KH, TH.

Now imagine the curve drawn through point E, T, with CH as its axis; point K is a given. We have a line drawn as ordinate from the curve, on the axis CH: this is the line EH. We have the line cut off from the axis by this ordinate: this is the line HT. The property of this curve is that the following ratio is fixed:

> the square on the ordinate (= $EH^2$), to
> the rectangle contained by (i) the line cut off, HT, and (ii) the line from the same cutting point to a fixed, given point (= HT × KH).

This defines a hyperbola whose diameter is KT. This means that once the ratio CD:DE is provided, we can use it to determine the points K, T that satisfy $CD^2:DE^2::CZ^2:ZK^2$, $CD^2:DE^2::CZ^2:ZT^2$, and then, with the line KT as the diameter, we find a hyperbola that always produces the required ratio: the line joined from any point such as D to C will have the desired ratio to the line joined perpendicularly to the hyperbola. With a conic section, then, we produce a construction that is rather like a neusis.

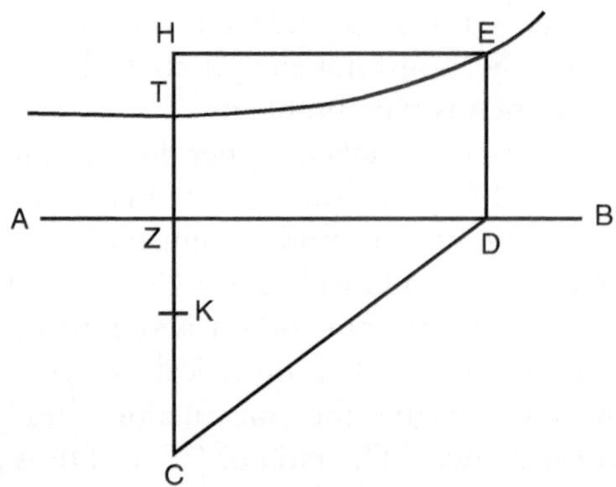


Figure 99

I quote and explain all this in detail because it is otherwise impossible to understand what Pappus's text was like: sophisticated, rich in many details; also, exasperating. It is often the case that we can barely see the *point*. How does this relate precisely to the neusis required by Archimedes? Why provide this?

Pappus extends this discussion a little further, and then, as the book breaks off, he concludes by asserting, apparently, something along the lines of the following (the end of the book – it is now meaningful that we have but a single manuscript – is in parts illegible):

> Archimedes used this neusis . . . some, however, reproach him, alleging that he did make use of a solid problem in an inappropriate way . . . [lacuna] they show that it is possible by planar . . . [lacuna].

Which is where our text effectively breaks off. Pappus seems to say that there were those who criticized Archimedes for applying a neusis instead of applying simple techniques from Euclid's *Elements*. My guess is that Pappus would have ended up defending Archimedes (which is perhaps another reason why Pappus has shown how neuses are found through conic sections, that is, are "solid" – ever the teacher, he was showing *through examples*). But this guess, as to Pappus's end point, is less significant than the terrain through which Pappus was making his moves – which is far clearer. Pappus does not really tell us all that much, in the end, about *Spiral Lines*. The turn to the past is motivated by present concerns: a methodological, prescriptive discussion as to how mathematics *should* be pursued. These are the tools, Pappus tells us, throughout – planar, solid, and linear; analysis and synthesis; these are the disciplines; these are the authorities – mathematics, made into a well-regulated endeavor. All of this forms the attitude of the commentator-teacher, surveying the mathematics available to him as a whole. The very breadth of the survey, its historical detail but, in particular, its insistence on questions of methodology, would be extremely significant as soon as the work was read with an eye to advancing its mathematics. We will return to this in the following chapter.

Pappus is not alone: cultural life did pick up in the fourth century, and we see a few glimpses, even, of other teachers of mathematics. Still in Alexandria of the fourth century, we have the evidence for yet another mathematical school, that of Theon and Hypatia. This was, literally, a two-generation event: a school led by a father, then by his daughter. Astronomical observations date the father; political events date the daughter. The two were teaching for about fifty years, from the 360s until Hypatia's death in 415. For each of these three Alexandrian teachers – Pappus, Theon, Hypatia – our sources preserve a different, complementary slice. Perhaps the three were different, but perhaps it is mostly our sources that are different, and we should use all three to get a fuller picture.

For Pappus, we have, above all, *the Collection*, a more sophisticated work expounding, in a more personal and discursive voice, a wide range of mathematical sciences, everything grounded in historical erudition. This is to some extent misleading: Pappus did write more limited commentaries and editions as well, but those are less well preserved.

For Theon, explicit commentary and edition are all we have. These are three safely attributed works: a commentary to the *Almagest* and two – different – commentaries to *Handy Tables*. There are some indications that he could have also produced editions of works by Euclid, which we will briefly revisit in the following chapter (we already broach the field of *later reception*). Did he write more ambitious works, comparable to Pappus's *Collection*? Perhaps not: the extant works are much more technical and even practical in character. It is telling, after all, that he wrote *two* commentaries to *Handy Tables*: the *Little Commentary*, almost an instruction manual for the operation of the tables, and the *Great Commentary*, a more detailed commentary with an account of the derivation of the tables. A curious comparison between Pappus's and Theon's astronomical commentaries is that we find many computational errors in Pappus and perhaps none in Theon. Perhaps Hypatia helped.

Even in the more theoretical commentaries, it is remarkable that Theon's extant works tend to avoid historiography. The beginning of Ptolemy's *Almagest* is very theoretical and reflective, which elicits from Theon a few more general statements (in particular, though, he says quite clearly that he will not write expansively, like a philosopher, but briefly, like a mathematician). He discusses the shape of the earth, which does bring out a brief mention of Archimedes's *Sphere and Cylinder*, and so forth; it is from Theon that we hear that Hipparchus wrote a treatise in twelve books on trigonometry (once again, this is for the sake of praising Ptolemy's relative brevity). What Theon mainly does is, like Pappus, provide theorems that will be helpful for Ptolemy's geometrical arguments. (The commentary to Book II of Ptolemy's *Almagest*, for instance, begins with a very general mathematical discussion of the meaning of "composition of ratios" – taking the ratio A:C as composed of A:B and B:C. This is very far from astronomy as such and instead is a basic mathematical technique applied more generally.) In all of this, erudition is downplayed, the emphasis put instead on mathematical skill and exposition. Perhaps all of this is simply comparable to Pappus's more commentary-like works. (Pappus's own commentary to the *Almagest*, in particular, is quite comparable to Theon's.)

For Hypatia, our evidence is different yet again. To be clear: her own works are entirely lost. We hear from late (and not very reliable) reports that Hypatia wrote commentaries on Diophantus's *Arithmetic* and on Apollonius's *Conics.*[10] It is possible that her commentary is reflected by

[10] On the practice of commentary in Hypatia's milieu, read A. Cameron, "Isidore of Miletus and Hypatia: On the Editing of Mathematical Texts," *Greek, Roman, and Byzantine Studies* 31 (1990): 103–127.

the Arabic translation of Diophantus (in which case we learn that, as commentator, she added more calculations and verifications: this is in the style of her father's work). We do not have any indications of more independent work, although the range of commentaries suggests a more ambitious mathematical program than that of her father.

Many of Hypatia's contemporaries, however, *are* extant. They wrote directly to parchment codices, after all. In particular, we have several contemporary collections of epistles. One of these collections is by Bishop Synesius of Cyrene – who had been (how lucky can we be?) Hypatia's pupil. Synesius's circle of correspondence included several other former pupils, and they sometimes waxed lyrical about their learning. A few letters are even addressed by the former pupil to his teacher. This, then, complements our sense of education, adding the missing angle: What would it mean for the *students*? Here we come across something of a surprise. The Hypatia of Synesius's letters is indeed a mathematician and astronomer – so, for instance, in one of his letters, Synesius presents his addressee, Paeonus, with a gift, an astrolabe, that Synesius knows how to make thanks to his studies with Hypatia. But what is cherished about the education seems to be the general philosophy or, indeed, even more, the general good education in letters and in Greek style. Hypatia is referred to as a philosopher, not as a mathematician; Synesius asks for her advice concerning the literary quality of his writing. Even the letter accompanying the astrolabe engages mostly not with astronomical details but with more general erudition. The main theme of its introduction is that, in this time and age, learning is not respected (Synesius, *On the Astrolabe* 5):

> For when Italy in bygone days possessed the same men as pupils of Pythagoras and governors of cities, it was called Magna Graecia, and rightly so. Of these men Charondas and Zaleucus gave laws, the Archytae and the Philolai were masters of the field. The greatest of astronomers ruled over a city, and was an ambassador, besides taking other parts in political life, Timaeus himself, on whose authority Plato discourses to us concerning the universe.[11]

Synesius was a bishop; Paeonius, a high-ranking member of the court in Constantinople. When Synesius addresses a gift with a scientific instrument to Paeonius, saying, with a sigh, that there used to be a time that rulers were erudite in fields such as astronomy (those were the days!), he is clearly making a statement about the bonds that tie Synesius and Paeonius together, in the present. These intellectual bonds are the main theme running through the

[11] A. Fitzgerald, *The Letters of Synesius of Cyrene* (Oxford: Oxford University Press, 1926), p. 266.

correspondence. The members of Synesius's circle recall their learning not as a mere gesture of nostalgia but as a statement of the *distinction* that marks them as members of the ruling class of the empire.

Everything suggests that Hypatia's family, too, was noble. Indeed, it appears that she was a political force to reckon with in the city, and so when a conflict emerged between the two main figures of Alexandria – its governor, Orestes, and its bishop, Cyril – it was perhaps natural that Hypatia would get sucked in. As it happens, she sided with Orestes, and in 415 CE, in one of those spasms of violence that regularly punctuate the lives of late ancient cities, Hypatia was killed by Cyril's supporters. This image – a pagan female mathematician, torn to pieces by a mob incited by the bishop – is rich in a symbolic freight that perhaps it should not be made to carry. Perhaps her being a pagan, a mathematician, or a female played some role in her death. Or perhaps it was just the bad luck of picking the wrong side of the wrong dispute.

And yet, the implications of Synesius's correspondence seem real enough: Hypatia played a role in the elite precisely through her specific brand of philosophy. Surely, Hypatia wrote technical commentaries to the likes of Diophantus and Apollonius. Surely, she taught astronomy, like her father. All this is technical mathematics, but surely, as well, all of this carried a *symbolic* meaning. The glorious references evoked by Synesius's letters are not chosen by accident: Pythagoras, Archytas, Philolaus, Timaeus, Plato . . . This is the Pythagorean pantheon. Platonism was everywhere: a teacher of mathematics did not need to spell this out. For a student such as Synesius, part of the attraction of mathematics must always have been that it carried an added aura of philosophical, and therefore, cultural significance.

The evidence for our three authors – Theon, Pappus, Hypatia (as seen by Synesius) – encompasses the same mathematics in wider layers. In the evidence for Theon, we mostly see the unvarnished technical details of computation; in the evidence for Pappus, the technicalities are conveyed so as to make the study of mathematics, itself, into a kind of erudite endeavor, historiographic and methodological; Synesius seems to understand Hypatia's mathematics as part of a much larger project of Platonizing philosophy and general erudition. Perhaps the authors did differ between them, Pappus the more prolix, Theon the more gruff, Hypatia the most humane. But should we really imagine the school of Theon turned upside down by his daughter Hypatia? This is, of course, possible, but I would prefer to place the three along a continuum. Theon's and Pappus's pupils, no less than Hypatia's, would come out of such education with the sense that they had learned something deep, with a closer sense of belonging to a

Greek tradition where mathematics mattered, partly because Platonism did. Even Pappus's competition against the philosophers should not be read, after all, as a break away from philosophy: his point, likely enough, was not that one should *ignore* the *Timaeus* but, rather, that to learn the *Timaeus* well, one should be aware of Archimedes no less than of Euclid. The mathematician, Pappus implied, was a better Platonist than even the Platonists; Synesius seems still to have believed that.

The Greek past mattered to everyone. It was the vehicle of one's identity and the mark of one's merit. In different places, one could express it in different ways. But Late Antiquity differed from earlier eras because now – in the age of the teacher – every field was treated in the manner of the canon. One could belong to the Greek past, in Alexandria of the fourth century CE, not just through Homer and Euripides and Demosthenes, but no less – if one found the right teacher – through Archimedes, Apollonius, Diophantus, and Ptolemy. The canon has been expanded to encompass mathematics.

It is very likely that from about 375 to 415, Hypatia was the best mathematician alive anywhere in the world. This is exhilarating – and makes it so much more galling that none of her works survive.

Throughout this book, we see how social and historical contexts determine the production of mathematics. People do the mathematics they like – but they end up liking this or that based on the cues available to them. It is obvious, then, that the kinds of people who will engage with mathematics will also be determined by such social and historical forces. We have seen the mathematics of scribal elites, working for courts; of free citizens; of authors competing against each other for patronage. These were all elite members and, nearly without exception, male. And now – something new and, remarkably, not even unique. We recall Pandrosion, Pappus's rival teacher of mathematics; we should mention Sosipatra, a Neoplatonist philosopher active in present-day Turkey in the first half of the fourth century. (Earlier, we recall Ptolemais, perhaps the re-founder of mathematical music.) The cluster is meaningful, in time and in intellectual affinity. How to account for this? Perhaps, in terms of the recently changed meaning of the canon and at the intersection of gender and public space in antiquity. One theme running through the ancient treatment of gender was the exclusion, in various ways, of women from the public sphere. Now, in early Archaic times, Greek literature was understood primarily through the intimate spaces of the symposium, and it is perhaps through such spaces that we can understand the qualified phenomenon of female poets (Sappho is the greatest among them but is far from alone). As the democratic ideal of public performance came to define the canon – in drama, in speech, in Socratic performance – women were excluded. For a woman to write a drama, for instance, would be for her, implicitly, to be exposed in public. This was frowned upon, and so the public nature of literature created a further barrier to female literary activity.

It was only in Late Antiquity that the canon was expanded to encompass all fields, not only the performative ones. The significance of literature changed to become, as it were, more bookish and so, also, to some extent, disembodied. This was most clearly the case for Neoplatonists who followed their master in the disdain of the body (Plotinus – the first thing Porphyry tells us about him – was *ashamed* about his body). To engage in the literary field by commenting on the scientific canon, from a Neoplatonist perspective, was – at long last! – no longer an exposure of one's body. The field became open, in relative terms, to women.

There were all sorts of highly visible barriers to female participation in ancient culture: women had less access to education and less control over their lives. But even apart from such visible barriers, there were the more subtle barriers, where the expected expression of one's gender clashed, implicitly, with the requirements of culture. Even if one was literate and wealthy, still . . . writing drama was just not something that could figure in a life lived in the female way. Writing mathematics, in the fourth century, all of a sudden, could coincide with being a woman. This is exhilarating. This is also an opportunity to pause and reflect: What subtle barriers do we put, now, in the way of those who wish to participate in science?

## Proclus and the Philosophical Schools

Throughout most of the third century, mathematics could have been nearly dead. The fourth century presents a remarkable revival. Already with Pappus, we see full mastery of its entire array, in control of both technical detail and the history of the discipline. Ptolemy's *Almagest* is one of the more demanding works of science ever produced: we see that many readers in the fourth century wished to learn it, and a few were perfectly capable of expounding it. The fourth century also marks a clear transformation. Historians of astronomy often observe that whereas astronomers, throughout antiquity, always wished to improve the astronomical models of their predecessors, Pappus and Theon do not even broach this as a possibility. The *Almagest* is seen as final; the goal is not to change it but to understand it. A commentator can be more or less competent, but no matter how sophisticated, the very attitude of the commentator tends in the direction of conservatism. If you pride yourself on being the master, capable of leading your fellow learners through the difficulties of the master text, surely you do not wish to overturn it. What is Theon worth if the *Almagest* is wrong?

There is a strong cultural bias to channel one's work into the making of commentaries and to affirm, rather than overturn, the masterworks. This has consequences in philosophy, as well. One philosophy had established itself, already in the imperial era, as based on the genre of commentary: Aristotelianism. Another was always entrenched in the canon as a set of master texts: Platonism. These two philosophies survived through the transformation into the culture of commentary, and the rest simply do not come back from the upheaval of the third century. There are no more Skeptics, no more Epicureans, even no more Stoics. All philosophy will be either Platonist or Aristotelian, and because the cultural bias is toward an affirmation and synthesis of the past, the division loses its meaning. In the late fourth century BCE, the two schools, that of Xenocrates and that of Aristotle, were in direct competition, but now a synthesis had to be found. Porphyry already perfects this combination in works such as his commentary to Ptolemy's *Harmonics*; for three centuries, it will continue largely uninterrupted.

As usual, this is not a matter of institutional continuity. We find many local events flickering through the Mediterranean. Porphyry himself does not establish a school, but while he is still alive, toward the end of the fourth century, Iamblichus attracts a following in Asia Minor. "When Iamblichus had departed from this world, his disciples were dispersed in different directions, and not one of them failed to win fame and reputation," Eunapius tells us (*Lives of the Philosophers*, 461–462), a late member of this tradition, writing its history – as a series of hagiographies – early in the fifth century. Sosipatra, mentioned already, belonged to this group. Aedesius was another of Iamblichus's followers who became his own master, now in Pergamon: Emperor Julian (who reigned from 361 to 363) saw Aedesius as his master. Julian was a principled pagan who wished to undo the Christianization of the empire. I am not sure if Hypatia's fate was really determined by religious animosities (her circle of pupils included many pious Christians, including Synesius himself, a bishop). What is clear is that beginning with Porphyry himself, but surely by the time of Julian, Neoplatonism does not merely evoke the Greek past: it is, quite simply, *the* ideology of the pagan elite. It was the surest go-to position of the likes of Emperor Julian.

The best place to commemorate the past was Athens. Perhaps the first of the Athenian Neoplatonists was Plutarch (not the same as the author of the imperial era, famous for his biographies), active early in the fifth century. The power of the place was such that a local succession did form: Plutarch was followed by Syrianus, and in 437, a twenty-five-year-old Proclus was so precocious as to become head of the school – which he continued to lead for almost fifty years until his death in 485. Throughout those forty-eight years, Proclus wrote the major corpus of Neoplatonism.

Already Porphyry wrote mathematical commentaries, the choice of works to study (harmonics, astrology) closely related to the potential function of mathematics within Neoplatonist religion and ontology. Iamblichus's philosophy verged much more powerfully into fully fledged number mysticism. His biography of Pythagoras was hugely influential; in general, he invented a mythical Pythagorean past, which he tried to imitate as a model for his own philosophical conduct and that of other Neoplatonists. So, Iamblichus's treatise *On the Common Mathematical Science* (Chapter 18):[12]

> One of their methods was the one that proceeds by symbols in mathematics; for instance, showing that the number 5 is the symbol of justice, because it signifies symbolically all the possible forms of what is just. This kind of instruction was useful to them for philosophy as a whole. . . . Nevertheless, the fact that they also taught the first principles and the discoveries made by mathematics, is clearly shown from the various branches of mathematics, and is made apparent in particular by the methods of arithmetic.

The two ways of approaching number – as a doctrine of grand symbolic emanation and as a theoretical study – are reconciled by Iamblichus as if they were the two faces of a single project. Thus, of all the mathematical sciences, the most significant one for Iamblichus is that of arithmetic, and his most sober work is a commentary on Nicomachus's *Arithmetic*. Nicomachus became perhaps the central mathematical figure to this tradition. Proclus believed that he was Nicomachus's reincarnation, a belief held in utter sincerity. Strident pagans, the philosophers of Neoplatonist Athens held most devoutly to the religious aspects of Platonism – but also with evident pride in their capacity to handle Platonism's mathematics. To be an expositor of numbers was the highest a man could aspire to.

Alongside the many commentaries to Plato (many extant) and Aristotle (all lost), Proclus was also the author of a commentary to Euclid and also of

12 L. Brisson, "Chapter 18 of the *De Communi Mathematica Scientia*: Translation and Commentary," in E. V. Afonasin, J. M. Dillon, and J. Finamore (eds.), *Iamblichus and the Foundations of Late Platonism* (Leiden: Brill, 2012), pp. 41–42.

a number of works that display original, interesting mathematical erudition. Proclus's *Commentary on the First Book of Euclid's* Elements stands out, among other Neoplatonist works, in its engagement with (what we would see as) strictly mathematical questions. It is from Proclus (as noted on p. 361) that we hear of Ptolemy's attempted proof of Euclid's parallels' postulate – which Proclus competently refutes. Proclus recounts to us the details of an exchange between an Epicurean, Zeno of Sidon, and Posidonius: the Epicurean argued that Euclid's first proposition, constructing an equilateral triangle, is false because one has to *assume* that the sides of the triangle do not partly coincide; Posidonius claimed this could be proved. Once again, Proclus competently surveys the arguments. He *understands* his mathematics. As can be surmised from the examples – and as is to be expected after all in this commentary – the emphasis is on the most foundational aspects. (About half of the treatise is the general introduction, followed by Proclus's commentary on Euclid's axiomatic introduction, while the commentary of the forty-eight propositions is handled much more rapidly.) Proclus is reasonably aware of more advanced mathematics: when he cites *Elements* 1.45, finding a parallelogram equal to a given rectilinear figure, he suggests that this must have prompted the ancient to look for the extension of this problem into the squaring of the circle, and so Proclus mentions – barely – Archimedes's *Measurement of the Circle*. When he cites *Elements* 1.9 – to bisect a given angle – he notes that the problem of cutting an angle into three equal parts is much harder, and then surveys – once again, very rapidly – some past solutions, barely mentioning the names of past authors (Nicomedes, Hippias, and – once again – Archimedes). But this is about as far as Proclus goes. It is especially remarkable that he never mentions any result having to do with conic sections. This is not just a matter of Proclus concentrating on the Euclidean, elementary mathematics at hand, avoiding mere displays of erudition. To the contrary – as we have seen – Proclus often brings up the past. But he tends to concentrate on a deeper past.

My predecessor Heath and many historians – up until the last generation – gave credence to the view according to which Thales, and then Pythagoras, made lasting contributions to mathematics. This derives almost entirely from Proclus's commentary, which, because of its overall sobriety, was taken seriously even for such obviously unfounded assertions. Proclus does, however, also discuss later and more historical mathematicians. There is one extended historical passage in Proclus, near the beginning. Introducing his topic, Euclid (for whom Proclus evidently had no biographical information), Proclus provides a brief history of the writing of mathematics up to his time. This is most likely based (if perhaps indirectly), as noted on page 57, on the history of geometry written by Eudemus, Aristotle's pupil. So, in some sense, this is as far as Proclus needed his history to go, and it is also possible that this is all Proclus had: There were no histories later than Eudemus, perhaps? But perhaps the correct account is that for rather obvious reasons, having to do with his Platonism, Proclus chose to emphasize – to use the terms of this book – the generation of Archytas. (A generation that I reconstruct, in part, based on Proclus's own summary! Such is the charmed circle of historiography.) Proclus and Pappus both belong to the same world, of Greek erudites of a philosophically minded mathematics or a mathematically minded philosophy. The opposition between them is that of a philosophical as opposed to a mathematical emphasis, which is also the emphasis on the generation of Archytas – or on the generation of Archimedes.

Proclus was clearly fascinated with the axiomatic structure of Euclid's *Elements* (two works, *Elements of Physics* and *Elements and Theology*, attempt to present philosophical doctrine *more geometrico*, as a series of proved theorems: from our perspective, the derivations can hardly be seen as logically rigorous). This is where his heart was. He mentions actual, advanced mathematics almost reluctantly, as if such considerations take him away from the more elementary mathematics, which, through its proximity to mathematical foundations, carries more philosophical weight. This reluctance becomes explicit in the introduction to Proclus's major astronomical work, the *Hypotyposis*:[13]

> The great Plato, my friend, expects the true philosopher at least to say goodbye to the senses and the whole of wandering substance and to transfer astronomy above the heavens and to study there slowness-itself and speed-itself in true number. But you seem to me to lead us down from those contemplations to these periods in the heavens and to the observations of those clever at astronomy and to the hypotheses they devised from these, [hypotheses] which Aristarchuses and Hipparchuses and Ptolemies and such-like people are used to babbling about.

Proclus's (fictional?) reader asks him to comment on astronomical theory, and Proclus obliges, but with the (valid) caveat that true Platonists engage not with the apparent astronomy of material stars but, rather, with the more abstract contemplation of pure mathematical form. The ensuing discussion takes off – of course – from Ptolemaic astronomy but is

[13] G. E. R. Lloyd, "Saving the Appearances," *Classical Quarterly* 28 (1978): 207.

fascinating precisely because of Proclus's relative distance from the field. Approaching Ptolemaic astronomy not as an astronomical commentator but as a philosopher, Proclus's main observation is that Ptolemy's tools – epicycles and eccenters – are problematic for a philosophy that truly aims to recover all the motions in the heavens as combinations of perfect spherical rotations. (Does an epicycle not introduce a discontinuity within the sphere that carries it?) Proclus's critique shows that the philosophers in Athens understood Ptolemy. But also: that even though the astronomical commentators canonized Ptolemy, the canonized system, itself, was rich with internal contradictions, suggesting potential avenues for future critique and development. There will be much more of that in the later history of astronomy.

Proclus's commentaries mark the high point of Neoplatonism and of Neoplatonist reflection upon mathematics. Still, Neoplatonist commentary literature is massive and very well preserved. There is a good deal more. From Domninus – Proclus's co-eval – we have a handbook summarizing Nicomachus's *Arithmetic* (this is essentially a stripping down of Nicomachus into mere definitions, with numerical examples). Marinus – Proclus's heir in the school of Athens, by now a genuine institution – wrote an introduction to Euclid's *Data*, in Marinus's case, a slight work, not at all comparable to Proclus's commentary to the *Elements*. The only real significance of such works is to show that it was important to Neoplatonist philosophers to teach not only the works of Plato and Aristotle but works of mathematics as well. This is indeed evident from the extant philosophical commentaries themselves. Without this, we would have known so much less! For instance: a passage from Olympiodorus is our key source for Archimedes's major contribution to optics. In the middle of the sixth century, this Neoplatonist – active in Alexandria – wrote commentaries to Plato and Aristotle. In the commentary to Aristotle's *Meteorology*, Olympiodorus gets to the rainbow; this gives rise to a discussion of the general phenomenon of refraction and, soon enough – a substantial treatment of a nine-hundred-year-old theory (pp. 205–206).

Boethius was an early-sixth-century Neoplatonist active in Italy (and so, writing in Latin, and in many other ways somewhat distinct from the Greek commentators discussed here). He writes a survey of music – and that includes a brief extract from Archytas, the most suggestive piece of information we have for the early interaction between music and arithmetic (pp. 71–72).

Philoponus – another Alexandrian, contemporary with Olympiodorus – is in many ways a sui generis commentator, and more than other

Neoplatonists, he occasionally uses the commentary form not simply to affirm the positions of the master but to criticize them (he is famous, above all, for suggesting a revision of Aristotle's theory of motion, perhaps akin to that of Hipparchus; see pp. 345–346). Perhaps his unique stance has to do with his identity as a Christian, rather than a pagan, Neoplatonist. Still: he, too, produces the same standard arrays of commentaries to Aristotle, occupies the same cultural space of philosophy and science, taken together – and, on occasion, displays his historical erudition. A methodological passage in Aristotle's *Posterior Analytics* raises the question of the place of geometry within the system of sciences. Philoponus launches into a historical excursus: he tells the legend according to which Plato's Academy had inscribed on its entrance the words "Let no one ageometrical enter." He tells another legend – that of how the problem of duplicating the cube arose of an oracle posited to the Delians, who brought it to Plato – and then proceeds, remarkably, to provide a solution for the problem of finding two mean proportions, as found by Apollonius!

Proclus never engages in such sheer mathematical name-dropping. But no one name-dropped like Simplicius. Simplicius was educated in Alexandria, then in Athens – until the year 529, when the Christian emperor finally resolved to close down that pagan school. Paganism was becoming untenable. After a sojourn to Persia, he returned to the Greek-speaking world; we do not know where. His massive, extant works are witness to a successful career as a philosophical teacher. Much more than other Neoplatonist philosophers, Simplicius brings to the fore, somewhat in the manner of Pappus, his scientific erudition. Was this a safer alternative to the religious emphasis of past philosophers, such as Proclus? Whatever the reason, we are very much in Simplicius's debt. There is a fundamental tension in the Neoplatonist marriage of mathematics and philosophy: the mathematics they cite most naturally – the one associated with Plato's time – was much earlier and so rather different from the science that eventually became canonical: that of Archimedes and Ptolemy. Proclus was somewhat reluctant to address this discrepancy. Simplicius, perhaps less philosophically sensitive to the meaning of this gap, tackled it directly in his writing. Aristotle has his theory of falling bodies; Simplicius provides us with the unique evidence that an alternative theory was offered by Hipparchus. Aristotle has a cosmology based on nested spheres; Simplicius addresses the fact that this is an old, discarded theory – and provides us with our most detailed account of how, originally, this theory was produced. Aristotle mentions, in passing, the measurement of the circle; Simplicius cites in full the relevant work, which Aristotle had in mind – the *Quadrature of Lunules* by Hippocrates. It is thus, in the sixth century CE, that we have our witness to what may be, perhaps, the earliest Greek mathematical work. In our end is our beginning.

## Eutocius – and a Coda

Eutocius of Ascalon appears, in our sources, strictly as a mathematical commentator. Extant are commentaries to three to four works by Archimedes (*Sphere and Cylinder* I, *Sphere and Cylinder* II, and *Measurement of Circle*; a commentary to *On Balancing Planes* is ascribed to Eutocius, as well, but might be by someone else from his milieu), as well as the first four books of Apollonius's *Conics* (books that he also edited). He is also known to have written on the *Almagest*. (This is no longer extant under Eutocius's name, but it is likely that some of this is now preserved among the substantial corpus of anonymous scholia.) The commentaries on *Sphere and Cylinder* are dedicated to Ammonius the philosopher, and the terms of the dedication make it likely that Eutocius was the philosopher's pupil. It is hard not to identify this Ammonius with Proclus's pupil who, from the end of the fifth century, was the head of a Neoplatonist school in Alexandria. Yet there are no philosophical commentaries attested, let alone extant, by Eutocius – indeed, no philosophical references in his writings. We recall Simplicius, thrown out from Athens in 529: perhaps it was really beginning to be unsafe to flaunt a pagan philosophy, perhaps indeed, mathematics became a safer alternative, the severable component of the Neoplatonist agenda. Philosophy nearly discarded, what replaces it is a revived mathematical ambition. Hypatia preceded Eutocius in her commentary to Apollonius, but it is not clear that anyone wrote on Archimedes before him. (There is a very late report of a commentary by one Theodosius on Archimedes's *Method*, and indeed, there are small traces of scholiastic intervention in the text of the treatise as it now stands. My own interpretation is that this Theodosius could have been active in the generation immediately following Eutocius, on which, more in the following discussion.)

Indeed, we see something of the mathematical ambition – in both technique as well as erudition – we last saw with Pappus. On a couple of occasions, Eutocius provides lengthy historical excursus, citing the authors' names and trying to provide a modicum of historical provenance.

Archimedes takes for granted, as one point of *Sphere and Cylinder* II, a solution to the problem of four lines in mean proportion ("duplicating the cube"). This sends Eutocius on the longest of his detours, citing in sequence solutions by "Plato" (this is certainly misascribed), Hero, Philo, Apollonius, Diocles, Pappus, Sporus, Menaechmus, Archytas, Eratosthenes, and Nicomedes: a who's who of Greek mathematics, providing a good deal of our knowledge of the history of the discipline. The citation from Eratosthenes contains, among other things, information about Hippocrates of Chios's original contribution to the problem; the citation from Archytas is said to be on the authority of Eudemus. Most of the solutions, however, are from the generation of Archimedes, and the fact that they extend to Late Antiquity itself suggests that Eutocius, at the very least, could have revised whatever was available to him from his source and did not simply plagiarize some previous scholar. This is indeed a more general question that arises with all ancient authors but especially in Late Antiquity: When they cite a work, does it mean they have read it, or did they rely, instead, on its citation within a previous work? What were the books still available to be read at the time? Another passage in Eutocius brings this problem into focus.

In Book II of *Sphere and Cylinder*, Archimedes divides a sphere and requires for this purpose a solution to a more general problem: given a line to be cut such as AB, a given length C, and a given area D, to divide the line AB at point E so that

$$AE : C :: BE^2 : D.$$

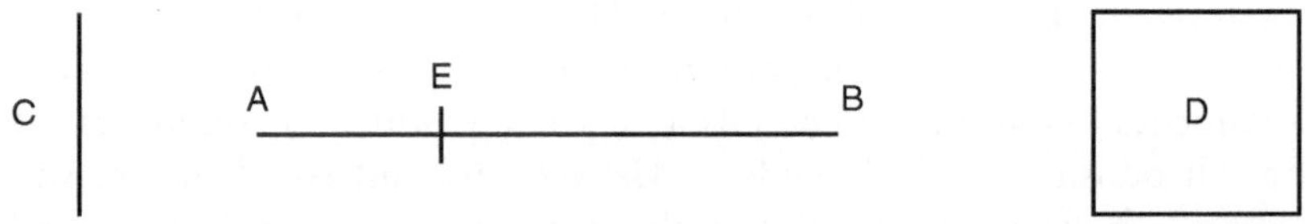


Figure 100

This, just like the problem of duplicating the cube, is rather like a cubic equation. It was not solved prior to Archimedes, and so Archimedes promised to provide it separately. Probably this was not intended as an appendix, attached to the original roll of *Sphere and Cylinder* II, but rather as a challenge, to which there were a number of ancient responses – and so

Eutocius cites the solutions by Dionysodorus and Diocles. More important, however, Eutocius managed to find Archimedes's own solution:[14]

> But – in a certain old book (for we did not cease from the search for many books), we have read theorems written very unclearly (because of the many errors), and in many ways mistaken about the diagrams. But they had to do with the same subject matter we were looking for, and they preserved in part the Doric language Archimedes liked using, written with the ancient names of things: the parabola called "Section of a right angled cone," the hyperbola "section of an obtuse angled cone."

And so – Archimedes! The text identified, Eutocius tells us he cleared its errors and proceeds to cite it. Paideia was mostly about the display of one's acquired erudition, and it is rare for us to hear anything about the labor involved in this acquisition. Were old books becoming harder to come by? Pappus was active in the early fourth century; Eutocius, early in the sixth. The difference is important: Pappus's world was still awash in imperial-era papyri, but by Eutocius's time, everything was already transformed into parchment codices; the papyri were already being discarded. Much of antiquity had been tossed away, and those who sought the past had to go out of their way to look for its traces. The commentator begins to transform into the philologer. Eutocius is an "ancient" author, for sure, but in some ways, he is already contemporary with the humanists of the Renaissance.

The lead-up to Archimedes's proof shows Eutocius as a philologer; I believe the proof, as cited, shows Eutocius as a mathematician. Having solved the original problem, Archimedes shows that it can be solved only under certain conditions, which ends up equivalent to a statement on a maximum: taking a line such as AB and cutting it at a variable point E, we may consider the value of the figure whose base is the square on BE and whose height is the length AB. Archimedes proves that the maximum value of this figure (which, once again, to us, means the maximum on a cubic equation) is when BE is twice AB, or the line AB is cut at one-third of the way. He does so by setting E at one-third of the way, raising a perpendicular, and drawing a parabola and a hyperbola passing through a certain point through that perpendicular. Archimedes then takes an arbitrary point, such as S, on the segment EB. A complicated demonstration shows that the fact that the hyperbola "above" S is contained within the parabola yields the result that the three-dimensional figure produced by

[14] R. Netz, *The Works of Archimedes: Translation and Commentary*, vol. 1 (Cambridge: Cambridge University Press, 2004), p. 318.

the cut at point S is smaller than the three-dimensional figure produced by the cut "above" E (where the parabola and hyperbola are tangent to each other).

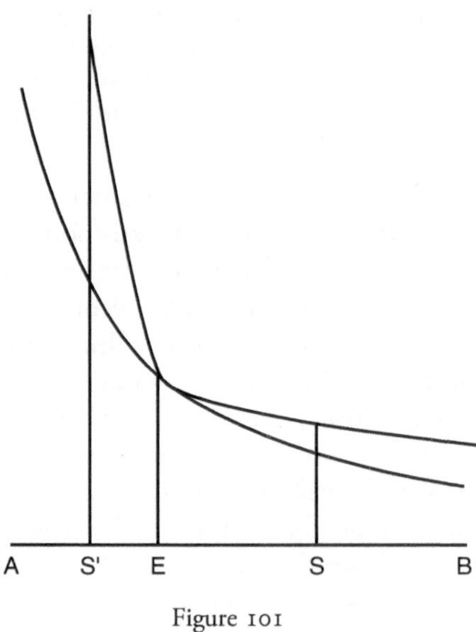


Figure 101

This is as beautiful and complex a result based on conic sections as anywhere in Greek mathematics, and it is remarkable that Eutocius was up to the task of identifying it and cleaning it up. (For several generations before Eutocius, we do not even have the trace of anyone seriously reading Apollonius's *Conics*!) But then, I believe that Eutocius adds something of his own. The proof continues, and there are textual indications that lead me to believe that its continuation is by Eutocius's own hand. Having proved the result for the arbitrary point S, to the right of E, the text goes on to pick another arbitrary point (I label it S′), to the left of E, and to produce the same conclusion. This is redundant – it is clear already that the proof works thanks to the unique tangency of the conic sections at point E and so is automatically valid for both sides of E. Eutocius's addition is motivated, perhaps, by some urge at completion. This is the same as a tendency we see in his edition of Apollonius's *Conic Sections*. In the text we now possess, Apollonius's propositions are often accompanied by several diagrams for the several possible cases that the proof allows, even if the proof itself requires no more than a single case.

As Decorps-Foulquier has shown, this is most likely the addition of Eutocius, who, in this case, expanded not the text but the figures.[15]

In my previous works on mathematical commentary, I have referred to the "deuteronomic" character of late ancient mathematics: it takes texts and transforms them into new texts. Commentary texts are always meta-textual, and this, in mathematics, carries a special meaning. As you become meta-textual, you might become meta-mathematical. Thus, in a case such as Archimedes's complicated proof for the maximum of a cubic equation, the thrust of Eutocius's contribution – perhaps, not even intentionally – is to display the *symmetry* of solutions around the maximum. Authors such as Apollonius and Archimedes, ludic, competitive, striving for that one single, remarkable result, often produced isolated theorems. Late Antiquity with its commentary – just by virtue of the practice of commentary! – tended to make such isolated results more well regimented: so, Pappus's classification of types of problems and their allowed solutions; Eutocius's surveying of possible solutions and cases, arranged side by side.

An even more worrisome question is that of Eutocius's intervention in the text. I suggested that he might have added, of his own, to one solution by Archimedes; Decorps-Foulquier suggested that he might have added, of his own, to Apollonius's figures. Knorr has argued convincingly for an even more radical intervention. Obviously, one of the requirements of Archimedes's solution is the construction of a hyperbola (we are reminded of Pappus's own solution, related to Archimedes's neusis in *Spiral Lines*, which also gave rise to the construction of a hyperbola). Now, what we are given is a point through which the hyperbola is to be drawn, as well as its two asymptotes. At the end of his discussion of this problem, Eutocius provides a construction with these givens, asserting that he needs to do so "as this is not a self-evident outcome of the Conic Elements." Indeed, Apollonius's *Conics* I.55 shows how to construct a hyperbola, given its diameter and parameter. What Eutocius does is to show (a relatively easy task, although still a tribute to his good understanding of conic sections) that from a given point and the two asymptotes, one can find the diameter and the parameter, so that the problem required by Archimedes easily reduces to the already available construction of Conics I.55. However, the very same reduction provided by Eutocius in his commentary to *Sphere and Cylinder* is now extant inside the transmitted text of Apollonius's *Conics*, as proposition II.4! Clearly – although no one noticed that prior

[15] M. Decorps-Foulquier, "Sur les figures du traité des Coniques d'Apollonios de Pergé édité par Eutocius d'Ascalon," *Revue d'histoire des mathématiques* 5 (1999): 55–76.

to Knorr – Eutocius inserted his own solution, an extension of a result by Apollonius, into Apollonius's text.[16]

Should we be alarmed? Can we even trust our texts of the ancient authors? I would say that we should be *on our guard*. One must always be aware of the mediated state of our evidence. But then again, let us remember: *Conics* II.4 was clearly a special case. Eutocius must have been especially proud of his realization, within the commentary to Archimedes, that a certain result, required by Archimedes, was not, in fact, directly furnished by Apollonius – and by his ability to furnish it himself. This is a unique case, and I do not think Eutocius, in general, interfered with his text. But we must remember: if he shows something of the attitude of the philologer, he is definitely not a modern textual editor. The point of his project was not the establishment of the text, as such.

What was the point, then? Indeed, our sense of context becomes elusive. We have very many writings from the world of Neoplatonist commentators, and we may recover their social context and agendas. But Eutocius has perhaps already burst out of this milieu, without us knowing, precisely, which milieu he joined instead. We have a handful of traces. The early commentaries to *Sphere and Cylinder* are addressed, as noted, to the teacher Ammonius; the later commentary to Apollonius's *Conics* is addressed to Anthemius, seen as something of an equal, perhaps almost a colleague.

In fact, we know this Anthemius rather well. Several histories of the sixth century are extant, and from them we learn of Anthemius of Tralles, famous in the capital, Constantinople, as a master of *mechanics*. We hear, specifically, of wonders produced with steam and mirrors. Indeed, there are even traces of a treatise – comparable in scope and sophistication to that by Diocles – on burning mirrors, and it appears that he wrote on mechanical wonders in general. He was held to such prestige that the emperor set him to plan the major church of the city – the famous Hagia Sophia – an undertaking completed, after Anthemius's death, by Isidore of Miletus. And this Isidore, too, may be connected – albeit indirectly – with Eutocius's milieu. The text we now have of Eutocius's commentaries contains four interpolations inserted by a follower of Isidore; we learn that Isidore wrote a commentary to a work by Hero (once again, we see an interest in mechanics, then) and that he, Isidore, also proofread Eutocius's commentaries.

[16] W. R. Knorr, "The Hyperbola-Construction in the *Conics*, Book II: Ancient Variations on a Theorem of Apollonius," *Centaurus* 25 (1981): 253–291.

Another extant work, attached in our manuscripts to Euclid's *Elements* and titled in them "Book xv," is an elementary study of the regular solids – and this, too, has a subscription by Isidore's follower. Finally, I note again that the commentary on Archimedes's *On Balancing Planes*, ascribed in our manuscript tradition to Eutocius, is of a far lower quality, mathematically speaking, than Eutocius's other commentaries. I personally suspect that this was a work of, perhaps, a follower of Eutocius? Perhaps here might be fitted, too, the report that there was a commentary on Archimedes's *Method*, by one Theodosius. So, maybe, a mid-sixth-century mathematical school?

Eutocius emerged from the Neoplatonist schools and then situated himself in a strictly mathematical milieu, one that was associated with the building projects of Constantinople in the sixth century. As we contemplate the spheres and cylinders of the Hagia Sophia, it is easy to imagine its construction as inspired by mathematical ideals. Is there an echo, in there, even of Neoplatonism, mathematics as the divine? Did Isidore's follower – author of a study of the five regular solids – still care about Plato? We do not know, and this, in itself, might be significant. Whether or not Eutocius or any of his followers, Anthemius, or Isidore harbored any pagan sympathies, this did not manifest itself in their mathematics. They could have been crypto-pagan, merely pretending to adore the Christian god, or they could have been genuinely pious Christians, simply carrying with them the mathematical souvenirs of a pagan past. Either way, they maintained the traditional values of a certain slice of the Greek elite: priding themselves on a particular kind of erudition, invested in the mathematical past, seeing in it the values of an otherworldly, superior form of being.

The Marranos – originally, a pejorative term, "pigs" – are the Jews who were forced to convert to Christianity in Spain and Portugal in the fourteenth and fifteenth centuries. Sometimes they just pretended to be Christian, but most of them, after the passage of several generations, became genuine Christians. And yet, habits persist. Generations ago, the entrance to the house was sanctified by the presence of the ritual Mezuzah. The Mezuzah was now long gone – but when sweeping the floor, one still made the extra effort to not let any dust settle near that once-holy ground. Does one still remember the reason for doing so? Perhaps, perhaps not: the fundamental point is the very persistence. And so: by the second half of the sixth century, Neoplatonism has nearly disappeared. But mathematical erudition persisted, the final habit, so to speak, kept by the pagan

Marranos. And so, they preserve and copy the works of Archimedes, Apollonius, Hero, Euclid . . .

There are about 270 extant Greek authors: this is what survives of an ancient civilization of perhaps tens of thousands of authors. Of these 270 authors, about 30 belong to the exact sciences (broadly defined). This is a remarkably high fraction, and I am quite confident that much fewer than 10 percent of all ancient authors were mathematicians. In other words, Greek mathematical works have a surprising tendency to survive through our manuscripts. Someone made an effort to preserve them, and I suspect that we see here why: at the moment that mattered most – in the sixth century, when Constantinople consolidated its holdings of ancient literature – a certain trace of the Neoplatonist past was still alive. And so, the works of Greek mathematics were carried inside the ark.

Not a moment too soon: the deluge was coming. The empire, stabilized with great effort at the beginning of the fourth century, was nearly shattered, once again, by the terrible defeat in the battle of Adrianople in 378. There was no option but to admit the victorious Goths as de facto rulers of parts of the empire. The ensuing litany of calamities is familiar to all: Vandals, Attila, the fall of Rome . . . A new pandemic – "the plague of Justinian" – hit first in the years 541–542. It was a virus strain similar to that of the Black Death, and just as deadly: killing perhaps 20 percent of the population of the cities in its very first year, returning again and again for decades and even centuries to come. As ever, it is hard to apportion historical blame. We do not have the statistics. But something shifts, irretrievably: from the second half of the sixth century onward, the empire no longer has the same authority. This is certainly obvious in cultural history. Anthemius and Isidore pursued science under the tutelage of the state – the last to do so, in the Mediterranean, for centuries to come.

And yet, the books remained. We have seen hiatuses already. I suggested that throughout much of the first century BCE, and then again throughout much of the third century CE, there was very little original work in the exact sciences. There was certainly no institutional continuity. But this was not the continuity that mattered in ancient culture. Greek mathematics was a literary genre, persisting through its books. As long as the books could be read – and then inspire others to imitate their genre – the legacy could be revived.

## Suggestions for Further Reading

In the past, studies of Greek mathematics largely avoided the later, less "creative" mathematics of Late Antiquity. An article of mine, R. Netz, "Deuteronomic Texts: Late Antiquity and the History of Mathematics," *Revue d'histoire des mathématiques* (1998): 261–288, was among the first seeking to understand the original contribution of this era. (I recently returned to the question of why commentary was so significant in Late Antiquity, in Chapter 6 of R. Netz, *Scale, Space and Canon in Ancient Literary Practice* [Cambridge: Cambridge University Press, 2019]).

Much more has been written since, and on Pappus, in particular, one should read S. Cuomo, *Pappus of Alexandria and the Mathematics of Late Antiquity* (Cambridge: Cambridge University Press, 2000), as well as A. Bernard, "Ancient Rhetoric and Greek Mathematics: A Response to a Modern Historiographical Dilemma," *Science in Context* 16 (2003): 391–412. (For Porphyry, we now have A. Barker, *Porphyry's Commentary on Ptolemy's* Harmonics [Cambridge: Cambridge University Press, 2015].)

It is hard to single out further contributions on other individual authors from Late Antiquity. The field is emerging and is now mostly in the form of individual article contributions or editions. An example of the recent interest is the special volume of *Historia Mathematica*: M. Sialaros, J. Christianidis, and A. Megremi (eds.), "On Mathemata: Commenting on Greek and Arabic Mathematical Texts," special issue, *Historia Mathematica* 47 (2019).

CHAPTER 7

# *Into Modern Science: The Legacy of Greek Mathematics*

## Plan of the Chapter

This chapter moves into a much faster clip. We still look at individual authors, but we take each briefly and consider the authors as tokens for entire civilizations (I also reduce, even further, my use of footnotes: other than direct quotations, I refer in general to the Suggestions for Further Reading). More than this: the chapter is decidedly *not* "the history of mathematics after the Greeks." Its subject matter is much narrower: the way in which Mediterranean civilizations, from the Middle Ages onward, responded to the legacy of ancient Greek mathematics. Specifically, our subject matter is the reception of Greek mathematics in three cultures, which we survey in sequence, one section each. "Byzantium and the Making of 'Greek Mathematics'" looks at the Greek-language tradition of mathematics in the Middle Ages, mostly in the city of Constantinople. Its main theme is manuscript transmission, and so, finally, we get a closer look at some of the primary sources informing the book so far. The following two sections engage with more original developments. "The World Made from Baghdad" concerns the Arab-language tradition of mathematics in the Middle Ages (and glances at its satellites in other languages – which encompass medieval Latin). We mention some essential new contributions – such as the rise of algebra – as we note the continuity of Arab and Greek mathematics. The same is true, to some extent, even of the last major episode – "The Renaissance to End all Renaissances" – the final section, dedicated to the return of Greek mathematics in early modern Europe. The makers of what is now known as the scientific revolution saw themselves not as revolutionaries but as restorers, and it is impossible to understand their project apart from the history, surveyed in this book, of Greek mathematics.

## Byzantium and the Making of "Greek Mathematics"

The plague was just the beginning. A century later, almost instantaneously, Arab conquerors rolled over both big empires, Byzantium and Sassanid Persia. Byzantium's holdings in Africa and the Near East were now gone. Constantinople always subsisted on the grain shipped from Egypt, and without it, the empire no longer had the resources with which to support much of a capital, let alone a centralized state. Authority was delegated to small fiefdoms nearer the borders. Cities became smaller, almost rural – even within the walls of Constantinople itself. (Its population shrank in size, perhaps from hundreds of thousands to mere tens of thousands: a postapocalyptic landscape.) Ancient states had a much weaker footprint than their modern counterparts possess, and the historical maps – displaying contiguous landmasses as "the Roman Empire" – are misleading. It was always more of a Swiss cheese: the state controlled, in a limited sense, only parts of its land, the cities, the roads. But now it was more holes, less cheese, and it did matter that the borders were so much nearer. From now on, a major defeat on the battlefield meant that the enemies could easily reach the city walls – a terror that would revisit the city of Constantinople every few decades. The state: under siege. Little wonder that ambitions, cultural or otherwise, were diminished. The remarkable fact is the very survival of a Byzantine state, in a large part subsisting on the inertia of the past: the physical capital of the walls, the symbolic capital of the emperor and the church.

These were not happy times, and the history of Byzantium's early centuries is of continuous conflict, made even more vicious – and ideological – in the eighth century. This spasm of internal conflict was marked by *iconoclasm*, the prohibition of the veneration of religious images and their active destruction. We have no direct record of the motivations of the iconoclasts (history was written, ultimately, by their enemies), but it seems likely that iconoclasm was an attack on the echoes of pagan culture, still embedded within elite behavior. We have seen the pagans gradually giving up their Neoplatonism, perhaps clinging – some of them – to mathematics, the last relic of a lost culture. Perhaps we should see iconoclasm as a final radical wave of de-paganization. At any rate, it was a kind of cultural revolution, pitting the emperors – together with much of the populace – against the cultured elite. In the end, by the early ninth century, the cultural revolution passed, Byzantine Christianity was stabilized, and the images were allowed again. Elites usually win. But the elite was chastened, and there is little wonder that not much culture was produced throughout the seventh and especially the eighth centuries.

Who could produce it? The state, indeed, had long retreated from the patronage of culture. The old traditions of learning – the mark of high status – did not disappear, but their site was transformed. By the sixth century, Christian education displaced the pagan one: the key was *biblical* study. This education did produce an institutional setting: not just the church but, specifically, a certain type of monasticism. Not all intellectuals, of course, were monks, but monasteries were especially important for keeping alive the practice of book-making. Books, now, required scriptoria because the parchment codex calls for a considerable division of labor. The parchment needs to be procured (often produced on location from slaughtered animals). The leaves are arranged into gatherings and then, often, divided between the scribes, each assigned different portions of the text. The leaves are also ruled in advance, in preparation for inserting the text. Specialist scribes might work on the rubrics (titles and other marks added in a contrasting color such as red); others still come in, on occasion, to insert more lavish art; the production of covers is a separate skill yet. These are often big books, folio-size, thick, with hundreds of leaves. Big and elaborate, a codex is best conceived as a small library unto itself, often containing multiple works as well as commentary: the attitude of book learning, folded into a single artifact.

In the ninth century, things were on the up again: monks went back to their scriptoria. As more books were made, new practices were introduced. The monumentality of the book was further enhanced by a new kind of script: minuscule. Up until then, Greek writing was usually produced in "majuscule," similar to our own "uppercase" Greek. In the ninth century, Greeks started writing in "lowercase." It now became possible to cram even more works into the same size of codex. Soon enough – like the papyrus roll preceding it – majuscule became obsolete. To survive, books had to be in minuscule. Uppercase books are now extremely rare, not only because they are so old but also because, even to their owners, they appeared obsolete: one wanted to read in minuscule. But this would be the last destructive disruption. Even the coming of paper into the Christian world, from the thirteenth century onward, simply expanded the availability of materials but did not make old parchment books obsolete; the coming of print, in the fifteenth century, made the old codices, if anything, more valuable, glowing now with the aura of a unique artifact. After the hiatus of the seventh and eighth centuries, books would be made again – and from now on, they would be kept. A book made in Byzantium in the ninth century certainly would not be guaranteed survival (in particular, it needed to be lucky with worms and fires – as testified by the browned edges and tunneled interiors of many surviving old books). But it had a sporting chance. Perhaps 5 to 10 percent of all the books made in Byzantium in the ninth century can still be found and read in modern libraries. It is thanks to the monks of Byzantium that we can access Greek antiquity, and now – at long last – we can look at the thing itself, unmediated: here, in its physical presence, is Greek mathematics.

And so, we may begin at Oxford, with Ms. D'Orville 301. Now we can see what a minuscule codex can do: it assembles the entirety of Euclid's *Elements* in a single volume; tucked in for good measure are also the so-called "Book XIV" (Hypsicles; see p. 160) and the so-called "Book XV" by someone in the milieu of Isidore of Miletus (p. 433). The manuscript is from 888 CE, which we know because this is the first artifact in our entire journey to be dated by its scribe. The scribe should well have inserted a colophon, identifying himself by name and date – he had reason to be proud! This is probably the most beautiful specimen of Euclid anywhere.

Although there is in this case only a single scribe – one "Stephen the Cleric" – the manuscript remains composite, containing several layers of scholia, many of which were introduced by the first reader, the one who commissioned the book. We can identify that reader, as well. This is Arethas – later made bishop of Caesarea – the most learned Greek of his generation.

Arethas studied with Photius, patriarch of Constantinople throughout much of the second half of the ninth century. Photius himself is a striking figure. He was a strong iconophile or "lover of images"; books, he loved even more. He is now famous chiefly for his massive compilation the *Bibliotheke*, consisting of 268 surveys of books he had read and found especially noteworthy. Some are with Christian contents; many others represent the entire range of pagan literature (we notice: the Byzantine elite would always go back to mark itself by its ability to read its pagan past). Several dozens of those authors are now not even extant. (The *Bibliotheke*, in such cases, is our best witness to an otherwise-lost book, a true treasure trove.) Photius read all the classical Greek orators; he read Greek novels; he read many histories. The goal of wide learning, for its own sake, is evident. We witness an explicit project of renaissance.

With Arethas, we see this renaissance reaching even wider. Several commissions are known besides the Euclid volume. We know about works of grammar and of rhetoric. (Books from the entire Photius range, then: all copied and preserved thanks to Arethas.) Crucially, he also commissioned copies of Plato and Aristotle. He wrote his own commentary to *Revelation*, so, a biblical scholar. But he also wrote commentaries to Aristotle's *Categories* and to Porphyry's *Isagoge*. In the relay race of Greek commentary culture, the baton was surely dropped to the ground sometime late in the sixth century. But as we noted, ancient commentaries are, above all, a literary representation of how idealized teaching should look. One can use them not just to learn but also so as to learn how to learn – or more precisely, to learn how to write new commentaries. The generation of Arethas dusted off the old books, with their commentaries – lifted the discarded baton from the ground – and started running again.

The scholia introduced by Arethas's hand into his manuscript of Euclid were certainly not his own original invention. We find similar notes, independently, in other manuscripts of Euclid, and the entire medley must have been put together in a manuscript that Arethas consulted for his own scholia. Many of these notes – but not all of them – derive from Proclus's commentary. Thus, both the source text (fifteen books of the *Elements*) and the commentary are compilations, probably of the sixth century, now put together, in the ninth century, in a single synthesis.

The milieu in which those books were made was conscious of its scholarly project. We may consider another manuscript, this time found in the Vatican – codex Vat. Gr. 190. Unlike the one found in Oxford, this manuscript does not carry a colophon with names or dates (although the handwriting suggests, perhaps, the first half of the ninth century: the kind of manuscripts Photius would have studied). As a compilation, this is even more ambitious than the one commissioned by Arethas: following the thirteen books of Euclid's *Elements* – with scholia – it has the *Data* (with Marinus's introduction); the so-called Books XIV and XV; and finally, Theon's great commentary to *Handy Tables*. The manuscript is so large that it has broken, in modern times, in two. The scribes who made such books, and their patrons, must have had access to fine libraries, from which such compositions could be compiled. For once, this is directly attested in a small comment, at the bottom of page 233, just below what is now transmitted as the sixth proposition of Book XIII:

> This theorem is not carried in most [books] of the new edition, but is found in the old one.

The scribe must have been familiar with several manuscripts of Euclid, some of which explicitly referred to themselves as a "new edition." In fact, it is easy to imagine how such a self-proclaimed "new edition" could have looked because, in fact, most of the medieval copies we now have of Euclid are exactly of this kind. Their title page usually carries some kind of variation of the phrase "based on the edition by Theon." This, in turn, is significant – and is in turn consistent with another, single passage in the text now transmitted of Theon's commentary to the *Almagest*. In Book I of this commentary, as Theon surveys basic tools required for the study of chords, he cites (Rome II 492.6–492.8) the result that sectors of equal circles are to each other as their angles, adding "as proved by us, in the edition of the Elements, at the end of the sixth book." It seems, then, that the scribe of the Vat. Gr. 190 knew several books with a title such as "of the edition of Theon" and one without such a title. Vat. Gr. 190 itself does not carry such a title, and – unlike most other manuscripts – it did not originally include the result concerning the sectors. (This result was, however, inserted into its margins by a later Byzantine hand.)

I keep saying that it is thanks to the scholars of Late Antiquity and of Byzantium that we even possess Greek mathematics. But this does not mean that we simply go to the extant medieval books, read them, and say, "Yes, this is what Euclid wrote." We read, instead, modern editions, which are not merely more practical (printed in a script we are familiar with) but also aim to be *more correct*. Each manuscript, after all, contains its own errors, makes its own additions and omissions, and modern editors try to work out what was the original form, removing the interventions of past scribes and editors. The first hints of such a philological approach are found already in authors such as Eutocius (who discovered and edited Archimedes's appendix to *Sphere and Cylinder*) and, indeed, in intermediaries such as the scribe of Vat. Gr. 190. In the nineteenth century, philology became a science. Francois Peyrard was the first librarian of the École Polytechnique, filled by the reforming zeal of the Napoleonic era. Napoleon looted many of the greatest treasures of the Vatican and brought them into Paris. Peyrard studied this Vatican manuscript of Euclid during its Parisian stay (commendably, after the fall of Napoleon, most of the treasures were returned). He noticed the marginal comment next to XIII.6, the absence of the title "based on the edition of Theon," and concluded that this codex – alone of all extant copies – was based on an old version of the *Elements*, preceding Theon's edition.

In the course of the nineteenth century, philologers gradually produced scientific editions of the great bulk of ancient literature – collecting all manuscripts, transcribing them precisely, establishing whether or not some of the extant manuscripts were copied from other extant ones, discovering the affinities between those that were independent of each other and thus constructing a family tree, at the end of this process applying linguistic, stylistic, and contextual reasoning to reconstruct lost archetypes. It took a while before philologers turned to scientific Greek texts, and we are lucky that one of the best philologers of the nineteenth century – the Danish Johan Ludvig Heiberg – edited so many mathematical works. He produced his first edition of Archimedes (we will soon get to see why a second, and much better one, was deemed necessary) in 1880, when he was merely twenty-six years old. A year later, he started on his Euclid, and his principle was clear: to rely on the Vat. Gr. 190 (called by Heiberg – and ever since – "Codex Peyrard").

Heiberg's edition was necessarily imperfect. Euclid was recopied more than most other ancient authors (already from the ninth century, we note two celebrated copies – but there are many more to come from later centuries). There are dozens of Greek copies of Euclid's *Elements*, which, as you recall, is a big book. The task of collating all of them and establishing their family tree is simply impractical. Little wonder that Heiberg was not keen to make this task even harder. Now, in Arabic, too, there exists an exceedingly large number of manuscripts with translations of Euclid's *Elements*. In 1881, another young scholar – Martin Klamroth, born 1855 and with an Orientalist PhD from Göttingen from 1878 – argued that some of those Arabic manuscripts were very faithful translations, based on manuscripts that are superior, in fact, to those now extant in Greek. That is: Klamroth implied that a scientific edition of Euclid should proceed through an edition, first, of the Arabic translations, which then should serve as the basis for reconstructing the Greek itself. The sociology of all of this is painfully evident, the Hellenist and the Arabist, each arguing that their own intellectual resources are the ones most valuable for studying Euclid. It is also painfully obvious that Heiberg did not want to *wait*. In an article from 1884, he objected, politely and firmly, to Klamroth – and proceeded to produce his own edition of Euclid, on which we still rely. Knorr argued in 1996 that Heiberg was wrong and that the Arabic tradition is, in fact, superior. Perhaps, perhaps not: many scholars today think Knorr might have overstated his case. But it is certainly true that Heiberg was wrong simply to *dismiss* the Arabic. We should be grateful to Heiberg, thanks to whom we have *an* edition of Euclid, one that is readily available and provides the information we need for several key Greek manuscripts. But the fact is that we still do not have finality on the text. We are contemporary with the scribe of Vat. Gr. 190 – faced with multiple sources and, as yet, unclear as to how to choose from them.

Should we even "choose"? Vitrac and his collaborators are skeptical: the text of Euclid is just too open-ended, and perhaps we need to think of it as a multiplicity (see the Suggestions for Further Reading). Perhaps there were *always* many Euclids? This is valid, but we should pause to consider the sense in which the multiplicity arises. Looked at from the outside, it might be considered that the text of Euclid should have been in flux because – so one would think – it was not perceived as authorial literature but instead was a much more practical, technical document that one could feel at liberty to transform for practical purposes. At the extreme – so one would think – the text of Euclid could have been rather like the notes one keeps from one's friends along with their recipes. If you copy such a collection of recipes, you do not bother with their textual fidelity, order, orthography. The text is a mere tool. And because Euclid was indeed subsumed, early on, into classroom education, we can see how such a reduction could have taken place.

In fact, this is not what is suggested by the actual pattern of variation. Vitrac and his collaborators find 150 points of significant variation between the main traditions of Euclid, which is a substantial amount even for a book as massive as the *Elements*. However, most of these are not in the nature of some kind of practical simplification. Rather, it seems that in a great many cases, some traditions (but not others) added extra explications, alternative proofs, lemmas, and corollaries. This is scholarly addition, and – what is perhaps the most significant – practically all of this variation is found in the more advanced parts of the text. (Indeed, the great bulk of the variation involves the theory of irrationality, in Book X and then in its application early in Book XIII: by far the most abstruse part of the *Elements*. Such, indeed, is XIII.6, the cause of the marginal comment in Vat. Gr. 190!)

What makes Euclid's *Elements* unique, then, is not that it was made part of the classroom and so made less than literary. It is that, more than most other scientific works, it was at the focus of scholarly works from Late Antiquity. The many branches and layers of scholarly contribution each produced their own lemmas and extra explications. These were originally intended as accompaniments to the text, not as interventions – as Theon clearly intended for the proposition concerning sectors, marked out as such as the end of Book VI and proudly stated to be his own contribution.

The task for us, scholars of the text, is complicated and, indeed, as Vitrac emphasizes, perhaps not solvable because the very earliest sources we may reconstruct – tracing the text back to Late Antiquity – are multiple. But what becomes clear is that the text was never merely technical, and indeed, the very accumulation of commentary is precisely an expression of the canonization of Euclid. This was recognized as a supremely authorial text – as, in fact, it should have been: the works of Greek mathematics were never some kind of collection of recipes, always the self-conscious contributions to a literary genre.

Our difficulty stems, indeed, from the fact that there were many copies of Euclid made, even as early as the ninth century. The mathematical past was still venerated; the habits of late paganism did not entirely disappear. More difficult authors were collected as well, and we have not one, but three separate collections made, at about this time, of the works of Archimedes. It is thanks to this effort of recopying that we now have a substantial survival of this, the most difficult and important of the works of ancient science.

Heiberg's PhD dissertation from 1879 was a study in the textual history of Archimedes, and like so many other PhD dissertations, it was mostly wrong. Heiberg relied on the eighteenth-century printed catalogue of the Laurentian Library in Florence, according to which the library's copy of Archimedes – Laur. 28.4 – was from the thirteenth century and, so, much earlier than all other manuscripts known at the time. Soon after submitting his dissertation, Heiberg visited Florence and must have immediately realized that the catalogue was wrong: the script was nothing like that of the thirteenth century and instead was an obviously late, Renaissance copy attempting to imitate what a ninth- or tenth-century handwriting should look like (and so half fooling the original cataloguer). Instead of being the original, the Florence manuscript must have been one of several copies, made during the Renaissance, out of a single – now lost – manuscript from the ninth or tenth century.

That old ninth- or tenth-century Byzantine manuscript was famous in the Renaissance. Many copies made from that manuscript are extant (five, counting only those made directly from the lost early manuscript; this, without counting Latin translations). These were all produced throughout the fifteenth and sixteenth centuries, all containing the same works, in the same order, with a close pattern of errors. One of those – the one that tripped up Heiberg – is still in Florence; another one is in Venice, one in the Vatican, and two in Paris.

Incredibly, Heiberg published his edition already in 1880 – comparing the various extant copies and thus reconstructing the lost, original, ninth- or tenth-century Byzantine manuscript. For Heiberg in 1880, this reconstruction was tantamount to the recovery of the original text: it was as far as he could go. Indeed, because there are so many extant copies, and because one of those is an apparent effort at producing a visual facsimile (which is, in fact, extremely rare), we may say that this manuscript is almost as good as extant. This lost-but-reconstructable manuscript is now known as Codex A. It is hard to judge from the "facsimile" of the Florence copy, but if anything, the handwriting it imitates suggests an early manuscript, perhaps more ninth than tenth century. Indeed, the same Florence manuscript also carries, at the end of its text of the *Quadrature of the Parabola*, an epigram celebrating one, "Leo the Geometer." This is a well-known historical figure, another member of Photius's circle who became famous especially for his knowledge of mathematical texts. (But he is not known – this is hardly surprising – for any positive contributions to mathematics. To be a scholar, in this time and place, was mostly to collect books.) And so: Archimedes's Codex A, known to originate from Byzantium's first renaissance.

There were more. Just like the dissertation preceding it, Heiberg's first edition of Archimedes soon crumbled. In 1881, Valentin Rose – a classicist who worked at the Royal Library in Berlin – visited the Vatican library and realized that one of the Latin codices – Let. Ottobon. 1850 – was in fact a translation of the works of Archimedes, some of which were not among those of Codex A. This translation had a colophon: it was made in the papal court in Viterbo, in 1269; the translator is not named, but it is certain that he is William of Moerbeke, a Latin cleric trained in Greek and a prolific translator, especially of Aristotle (traces of Arethas, there). Rose published his result in 1884, and it was clear that Heiberg's edition required some revision. In 1891, Athanasios Papadopoulos-Kerameus visited Istanbul as part of his massive catalogue – funded by the Russian emperors – of Jerusalemite libraries (either those in Jerusalem or subsidiary to the Jerusalem churches). And so, in the Istanbul subsidiary of the library of the holy sepulcher, Papadopoulos-Kerameus noted a palimpsest, and when he published his catalogue in 1899, he provided a few of the lines still visible. Heiberg came across this catalogue in 1906 and rushed to Istanbul: here was yet another Archimedes manuscript!

You may have heard of it: this was the famous Archimedes Palimpsest (see Plate 8). It turned out to be independent, once again, from Codex A. And so, in the years 1910–1915, Heiberg published his second edition of Archimedes, based on three separate Byzantine traditions. The previously known tradition – that of the 1880 edition – was now, as noted, called "A." The independent source of Moerbeke's translation came to be called "B" (Moerbeke's translation, it turns out, was a composite, partly based on A, partly on B; there is also another fifteenth-century Latin translation, probably based on A); the Palimpsest is C.

A palimpsest: that is (in Greek) "scraped-twice-over." You see: when parchment is made into a writing surface, the skin of the animal is scraped (you don't want to write on rough, hairy skin). Surprisingly frequently, ancient books could then be recycled: the codex taken apart, its leaves re-scraped and then written over with a new text. The original Archimedes Codex C was made in about 975 CE (judging by features of the handwriting). Almost certainly, this happened in Constantinople, but we have no colophon or, indeed, any indications of the context of the manuscript. We do have a colophon in Codex C as it now stands, but this is a colophon of the secondary use of the same pages, recycled. We find that the Archimedes book was taken apart in 1229, almost certainly in Jerusalem, and that a prayer book was then written over the original text. The words of Archimedes were hidden from view, scraped, faded, folded vertically into the new sewing of the prayer book. And yet, Heiberg managed to read most of them, finding new works by Archimedes and improving the text of those previously known. Heiberg died in 1928, having done for the philology of Greek science more than anyone before him or since. His editions are extremely precise and well thought-out. They were designed as tools for future scholars, and Heiberg no doubt died expecting others to follow him to Istanbul and revisit the palimpsest, filling in the lacunae. In fact, by 1928, and unbeknownst to Heiberg, the manuscript was already stolen from Istanbul – in all likelihood, amid the troubles that followed World War I. A Parisian art dealer tried to sell it, without success, between the two world wars. The art dealer knew that as an Archimedes manuscript, this was extremely valuable – but he could find no one willing to part with the thousands of British pounds he wanted. During World War II, seven forgeries of Byzantine art were laid over the manuscript, apparently in a desperate attempt – the dealer was Jewish; it was 1942 – to sell at least a few of its pages (he needed fast cash, and old art sells quickly even if mathematics does not). His heirs, finally, sold the manuscript in New York in 1998 for two million dollars. It was then that I had the immense luck to receive a phone call from William Noel – then curator of manuscripts at the Walters Art Museum. I was invited to edit the palimpsest together with the great Byzantinist and textual critic Nigel Wilson. Digital technology could lead us both where even Heiberg did not reach. A team of scientists led by Roger Easton produced digitally enhanced images of the manuscript (the most useful technique was multispectral imaging, although further readings were enabled, by Uwe Bergmann, with X-ray fluorescence). We now have the full transcription of the Palimpsest, and a critical edition is forthcoming. The journey of philology, for Archimedes just as for Euclid, is ongoing.

The number of Archimedes manuscripts is not at all comparable to that of Euclid's. But we do have witnesses to three early Byzantine traditions, and the degree of their independence is noteworthy.

We do not have here anything like Arethas's notes, the scribe's comments in Vat. Gr. 190. But the very choices of what to include in a manuscript may be telling. All the copies made from Codex A contain the following works, in the following order:

- *Sphere and Cylinder* I
- *Sphere and Cylinder* II
- *Measurement of the Circle*
- *Conoids and Spheroids*
- *Spiral Lines*
- *On Balancing Planes* I and II
- *Sand-Reckoner*
- *Quadrature of the Parabola*

Followed by Eutocius's commentaries:

- *Sphere and Cylinder* I
- *Sphere and Cylinder* II
- *Measurement of the Circle*
- *On Balancing Planes* I and II

(And finally, a small treatise by Hero, *On Weights*, thrown in for good measure.)

Now, more can be said of Codex B, the source of Moerbeke's Latin translation. After Rose identified it in 1884, an entry in the papal catalogue of 1311 allowed us to determine that this manuscript contained:

- *On Balancing Planes* I and II
- *Quadrature of the Parabola*
- *On Floating Bodies* I and II

(And also Ptolemy's *Analemma*, as well as the catoptrical works by Hero and, perhaps, Euclid.)

As for Codex C, its contents (before it was palimpsested) were, in order:

- *On Balancing Planes* I and II
- *On Floating Bodies* I and II

- *Method*
- *Spiral Lines*
- *Sphere and Cylinder* I
- *Sphere and Cylinder* II
- *Measurement of the Circle*
- *Stomachion*

Each of these codices divides the works of Archimedes between the physical and purely geometrical. Codex A has five purely geometrical books (*Sphere and Cylinder* I and II, *Measurement of the Circle*, *Conoids and Spheroids*, *Spiral Lines*), followed by four physical books (*On Balancing Planes* I and II, *Arenarius*, *Quadrature of the Parabola*), and then five books of commentary by Eutocius (*Sphere and Cylinder* I and II, *Measurement of the Circle*, *On Balancing Planes* I and II). Codex C inverts the order: five physical books (*On Balancing Planes* I and II, *On Floating Bodies* I and II, *Method*), then five geometrical ones (*Spiral Lines*, *Sphere and Cylinder* I and II, *Measurement of the Circle*, *Stomachion*). Codex B, finally, is clearly conceived as a collection of physical works with a core of five books by Archimedes (*On Balancing Planes* I and II, *Quadrature of the Parabola*, *On Floating Bodies* I and II). The repeated scale – four to five books – is perhaps meaningful: it is, in fact, the size of a majuscule codex. Are we, perhaps, glimpsing collections made already in the sixth century?

It is indeed noteworthy that Eutocius still thought that works by Archimedes could be identified by their use of Doric dialect – but that this dialect has been removed, and transformed into the forms of Greek more familiar to later readers, in many of the transmitted works. Indeed, a few of the works, but above all, *Sphere and Cylinder* I, carry many explications interpolated into the text, suggesting a scholiastic intervention – and finally, we recall the handful of mentions of Isidore of Miletus, "our teacher," inserted into the text of Archimedes.

And so, perhaps, the text of Archimedes was put through the standardization of the school sometime late in the sixth century? But here is the point: as with Euclid, the standardization did not standardize but, instead, created a certain plurality. Some works by Archimedes are now in Doric; some are not. And the three collections of Archimedes of which we know, made in the ninth and tenth centuries, are all different from each other. There was no established corpus, no established order.

Indeed, it is clear that even if the Archimedes texts did go, once, through the school, they did not primarily subsist there. The Euclid manuscripts are typically accompanied by many marginal scholia. When the text flows from one side of the leaf to the next, we often find a late hand copying, in rough outline, the diagram so that one need not flip the page back and forth. Errors are typically corrected. All of this is missing in the Archimedes manuscripts. In Codex A, the text of the *Arenarius* was missing its diagrams: the original owners of the manuscript did not bother to insert any new ones. The same is true for an entire stretch of the diagrams in Codex C (from the end of *Spiral Lines* till near the end of *Sphere and Cylinder* 1). Codex C, in particular, is rife with errors of mathematics and simple Greek, and in many places, it plainly makes no sense. There are no signs of any effort to correct any of that. Archimedes, probably, was just too hard to read anyway. Why bother, then, to collect Archimedes in the first place? The question answers itself: one collected those works because they were *by Archimedes*. In the ninth and tenth centuries, Byzantine patrons, who wished to mark themselves by their familiarity with the classical tradition of Greek literature, were heirs to values shaped in Late Antiquity and so continued to value, in particular, the exact sciences. There, Archimedes's fame was unrivaled. And so, here and there, the most sophisticated of patrons searched for old manuscripts containing Archimedes's works and had them transcribed into new, massive minuscule collections. These were not the possessions of scholars but of patrons. They served merely to adorn libraries. In fact, some of them were worse than useless, and when, in 1229, a scribe picked up the Archimedes book, the dust told him all he needed to know. For 250 years, no one did anything whatsoever with that volume: what further proof do you need that it better be recycled?

And so, Archimedes was cast away, in Jerusalem, 1229. But what was Archimedes even doing there? What was it doing in Viterbo in 1269?

Very briefly, there are two things to note concerning Byzantine learning from, say, 975 to 1229, one internal to the progress of Greek learning, the other external, concerning the vicissitudes of the state.

First, then, following the initial impetus of the revival of learning in the ninth century, Byzantine learning tended in a more strictly literary direction. We find plenty of rhetorical productions, poems, histories, literary commentaries. (The towering achievement of this era is Eustathius's commentaries, late in the twelfth century, to Homer's epics – about two million words of commentary!) The key background to all of this is that Greek – like all languages – keeps changing. To go back in time: throughout the Roman era, the language spoken by ordinary Greeks was still quite similar to that of the classics (and continuous, intensive training made sure that educated Greeks would feel at home when reading and writing in that archaizing and yet familiar language). But throughout Late Antiquity and beyond, Greek started to change more profoundly. To the Greeks of the Middle Ages, the classical language was effectively foreign, and so it was inevitable that once classical learning was reasserted as a mark of elite status, the emphasis would be on the – remarkably difficult – task of not only reading the classics but imitating them. Anna Commena was a twelfth-century princess and heir to the throne who, in political disgrace, exiled to a monastery, spent her last years writing the *Alexiad* – the history of the reign of her father, Alexios I (1081–1118). Spending one's exile in relitigating the past was a classical Greek tradition (Thucydides, Xenophon, and Polybius were all famous for this), and Anna Commena, sure enough, writes in direct imitation of the masters (although her rhetorical prose is even more purple than that of the masters: the standard peril of imitation). Such was the goal of Byzantine learning.

What histories such as Commena's reveal – and this is the second external reality – is a state in decline. The stability achieved early in the ninth century was maintained for several generations, and the historical maps reveal an expanded Byzantium, in control of much of present-day Turkey and the Balkans, early in the eleventh century. In fact, the quasi-feudal institutions created throughout Byzantium's crisis meant that the state never did resume its imperial power. Meanwhile, the continued pressure from Asian nomads kept destabilizing all powers in the Middle East. Throughout much of the eleventh century, the Byzantines fought a losing war against Seljuk Turks (in 1078, in one of the worst defeats, the emperor himself – Romanos Diogenes – was taken captive). I keep referring in this book to "present-day Turkey," referring to a part of the world that, until now, was mostly Greek. From now on, it is indeed mostly Turkish, and Constantinople becomes a frontier town. Little wonder that Christians, everywhere, became alarmed, and the Crusades (launched in 1096) did, for a while, provide Byzantium with some kind of a buffer. This, however, was, to the Greeks, a terrible bargain. The band of warriors assembled by Venice for the Fourth Crusade anchored in Byzantium in 1203. Financing was short; the riches of Byzantium were alluring. On April 12, 1204, the crusaders achieved what no barbarian did before: the walls of Constantinople fell. The city was sacked, and the crusaders found a better use of their energies in the foundation of new, Frankish kingdoms through the once-Byzantine lands.

This, then, is how Codices A and B ended up in the papal library in Viterbo in 1269. Indeed, this is how William of Moerbeke learned his Greek: born in today's Belgium, the monk was sent to the

Peloponnese. It is not far-fetched to see the Crusades as Europe's first taste of colonialism, and it was as an early-day colonialist that Moerbeke translated Archimedes. And it is not far-fetched to see Codex C – flung to Jerusalem, at the far end of the Greek-speaking world, and in this hour of need, turned into prayer book – as a mark of the depth of the Byzantine crisis.

Perhaps surprisingly, the empire had one last hurrah in it. The year 1204 ensued with a fragmented Aegean, mostly ruled by Franks but also including a Greek statelet around Nicaea (which we recall as the birthplace of Hipparchus), not far from Constantinople. In 1261, Michael VIII Palaiologos, king of Nicaea, took Constantinople itself. There was now a new line of emperors. It is obvious how important it now was, for the city's elite, to reassert their Greek identity. It was now not much more than a city-state, a rump of the old Byzantine state. But the intensity of cultural identity mattered more than the size of the city's domains, and Byzantium now went through yet another – "Palaiologan" – renaissance. George Pachymeres – born in exile in Nicaea in 1241, died in Constantinople early in the fourteenth century – is a leading light of this revival, prolific and encyclopedic. The urge was to recover *everything*. His *Philosophia* – in twelve books – is an epitome of all of Aristotle. He wrote widely in philosophy, history, and literary commentary, but also – more significantly for us – he wrote a "quadrivium," a survey of the four Platonic mathematical disciplines. Remarkably, his survey of arithmetic included not only Nicomachus but also Diophantus; the somewhat younger Maximus Planudes even wrote scholia on Diophantus's first two books. Maximus Planudes is also famous as one of the collectors of ancient epigrams, through whose effort we now possess the *Greek Anthology*. Epigrams or arithmetic, it didn't matter: if it was part of the Greek past, it was collected, recopied, preserved. Outdoing the monumentality of past collections, we have several manuscripts that aim to collect something like the totality of mathematics. A manuscript now in Paris, Gr. 2342, written in the middle of the fourteenth century, is right at the extreme of this new type of cultural production, and the only way of getting a sense of what this entails is by surveying the stupefying contents. In order, these are as follows: Euclid's *Elements*, Euclid's *Data* with Marinus's introduction, Euclid's *Optics*, Damian's *Optics* (a late ancient summary of the field), Geminus's excerpts on *Optics*, Euclid's *Catoptrics*, Theodosius's *Spherics*, Autolycus's *On the Moving Sphere*, Euclid's *Phaenomena*, Theodosius's *On Habitations*, Theodosius's *On Days and Nights*, Aristarchus's *On the Sizes and Distances of the Sun and the Moon*, Autolycus's *On Risings and Settings*,

Hypsicles's *Anaphoricus*, Apollonius's *Conics* with Eutocius's commentary, and Serenus's *Cutting of the Cone* and *Cutting of the Cylinder* . . . The same scribe went on to produce yet another massive tome (now in the Vatican, Gr. 198) with arithmetic and astronomical works.

Indeed, a tremendous amount of our sources for Greek civilization are from the fourteenth century. (The coming of paper helps: this writing surface – produced by the crushing of used rags into pulp – was ultimately cheaper and more plentiful than parchment. Following the long hiatus of meat-based scarcity – that of parchment – writing resumed, through paper, its long-lost, papyrus-like, plant-based plenty.) Theon's *Little Commentary on Ptolemy's* Handy Tables is among the more "practical" of ancient mathematical works, of use for astrologers (although at this point, it was often recognized that more recent Islamic tables were superior to those of Ptolemy). It thus has an especially large number of copies, and I count, in the modern edition produced by Tihon, thirty-two copies of the Palaiologan era – and a single fragment (six pages) from the ninth or tenth century. In the order in which they are listed by Tihon, those thirty-two Palaiologan copies are now found in Florence, Leiden, Milan, Oxford, Paris, the Vatican, Venice, and Vienna. It will be noticed that none of these are now in Istanbul, but it should be emphasized that books entered and left Constantinople from the beginning. It was a small state, surrounded by Latin and Turkish realms, depending on diplomacy for its very survival. Cardinal Bessarion – one of its last great intellectual representatives – was born early in the fifteenth century in the Greek city of Trebizond. He learned his Platonism with Gemistus Plethon, a Greek living in one of the Latin states of the Peloponnese. As a Catholic cardinal, he spent his life shuttling between Constantinople and Italy as he was negotiating between Greeks and Latins. He is remembered most for his enormous bequest – 468 Greek manuscripts! – left to the city of Venice, the basis of one of the major collections for classical scholarship today – including two of the Palaiologan copies of Theon's *Little Commentary* and one of the copies made of Archimedes's Codex A. Where else would these books go? By the time Bessarion died – in 1472 – Constantinople was already under Turkish rule. The city fell in 1453, but the Palaiologan renaissance had already become, well before that, one of the streams feeding the Renaissance in Italy.

In this section, I offered a very rapid tour through learning in Byzantium – but I ended up saying more on scholarship than on positive learning. It is perhaps unfair that we treat Byzantium in such a way, as a mere passive recipient of antiquity – the locker at the station where Greek civilization is left for safekeeping, awaiting the arrival of the train of modern Europe. Then again, I am not so sure that Arethas or Bessarion would have been upset to be remembered as the preservers of classical Greek legacy. This was an identity they actively chose. Reverence toward the totality of the Greek past is one of the fixed points of Greek culture from Late Antiquity onward, and it helps explain why so much remains and why, ultimately, it remains in reasonably good shape. Nonclassicists are often alarmed to learn that our knowledge of antiquity depends almost entirely on copies of copies of copies produced throughout the Middle Ages. It is worrying that there are so many stages along the way, and it is worrying that they depend so heavily on the Middle Ages. If you are interested mostly in the history of science, this is doubly alarming. Medieval monks: Were they not *against science*? Well, not really, and more to the point, classical antiquity always endowed their copied manuscripts with more value than any views expressed could take away from it. The church could, and did, burn books it considered blasphemous (thus, the central text of Jewish book learning was frequently burned in the later Middle Ages, which is probably why, in spite of its ubiquity in the Jewish world, the number of extant medieval manuscripts of the Talmud is quite small). But the church never burned Porphyry or Proclus, whose writings were so anti-Christian. On the contrary, Byzantine clerics, from Arethas to Bessarion, took the Neoplatonist synthesis as the cornerstone of their own intellectual life – precisely because it brought together so much of the Greek classical tradition.

The problem, if anything, is not one of censorship but of overeager embrace. Anxious to show their erudition and their mastery of the texts, scholars would often introduce their own scholia and interpretations. This, then, is a clearly defined editorial problem (although occasionally it may be hard to solve): identifying the layers of the text, original and scholia. There are, of course, even simpler problems. As texts are copied, they certainly become garbled, and once again, many nonclassicists are stunned to hear that we depend on a transmission of, often, more than a thousand years for the recovery of our original texts. For instance, 1,250 years separate Archimedes from his earliest (mutilated, palimpsested) extant manuscript. How can any signal survive through the noise of over twelve centuries? For this, it matters, then, to realize the attitude of reverence toward the past that prompted, to begin with, the making of the manuscripts. From the very first moment, Archimedes's works were conceived as literary, polished works, valued for their authorial status. It was no one's business to alter them, and scribes did their best not to. But even more significant, for us, is the realization that, in fact, this game of telephone did not have *all* that many stages. The relay race of this book is made of generational events, all the way down, with many hiatuses. And this, from the perspective of the reliability of the transmission, is *a good thing*. More likely than not, a scientific manuscript produced in the generation of Arethas would be based on majuscule copies from Late Antiquity. There were simply too few scientific books produced in between. Even when our earliest manuscripts are Palaiologan, it is quite likely that they, in turn, are based directly on early minuscule from the ninth or tenth centuries (it does seem that more scientific manuscripts were produced then than in the following eleventh and twelfth centuries). In short, in the typical case – where we have several manuscripts, at least one or two of which are from the ninth or tenth centuries – the archetype we reconstruct is morally certain to be a late ancient book. In this sense, Late Antiquity is quasi-extant. And then again: it is likely that authors of Late Antiquity would have relied on sources not far removed from the imperial era (for this, we are helped by the hiatus of the third century CE). It is likely that authors of the imperial era would have relied on sources not far removed from the Hellenistic era (for this, we are helped by the hiatus of the first century BCE). And so, our circle closes. I began this book reconstructing the genre of Greek mathematics – reading the texts we now have so as to reconstruct what writing looked like, language and diagrams and all, more than two thousand years ago. Which we can do, thanks to the fidelity – and, paradoxically, thanks to the scarcity – of our tradition.

## The World Made from Baghdad

Even as the Roman Empire was struggling through the crisis of the third century, habits of classical education – the mark of the elite – did not die down but, in fact, expanded: in a more fragmentary empire, local elites mattered more. It is well known how this happened in the western parts of the empire. This area produced its own, Latin, classical tradition, issuing commentaries to Virgil and producing its own separate writings on all subjects, literary as well as technical. (We recall Boethius's Latin-language *Introduction to Music*, which is also our unique source for Archytas's arithmetical proof: it was produced in the early sixth century, in an Italy no longer ruled by Byzantine emperors.) The journey of Latin Europe begins here. We may also recall Longinus decamping, after the sack of Athens in 267, to Palmyra in the Syrian desert. Indeed, many rhetors and scholars of Late Antiquity come from this desert borderland. Just as Latin emerged in the West, local languages – especially Syriac – emerged in the East. Greek works began to be translated: there was Greek-style philosophy and (especially) Greek-style medicine, now, in multiple languages.

The Byzantine Empire was in no condition to venture into the Arabian Peninsula. Ripples of Greek reached there, but when Muhammad brought forth his own variation of metropolitan learning – importing, above all, its unique, transcendent God – he did so in Arabic. It was Muhammad's words that brought together the Arabs, and when, in the 640s, they conquered all of Persia and most of Byzantium, they were in possession of the late Muhammad's sacred words – attributed directly to the divine – the Quran. They were also heirs to the traditions of Late Antiquity, and so almost immediately, there emerged an Arabic, book-centered learning of Muhammad's words, transcribed into a holy book. Right from the beginning, then, Arab conquest differed from that of all the many other tribes that descended on the Mediterranean – the Goths and the Huns and the Vikings and the Magyars and the Mongols and all the rest. In possession of a *book*, the Arabs came with a language and a civilization that were there to stay.

To win an ancient empire, one needed military victories. To keep it, one needed collaboration from local elites. Muhammad came to the world to assert monotheism against the pagans of Arabia, and it was easy for the conquerors to accommodate other believers – Christians and Jews – even if those did not subscribe to the specific version of Quranic monotheism. A small minority in a huge swath of conquered lands, the Arabs kept themselves to a network of garrison towns, many of them new foundations (Cairo has its beginning now). This gives rise to the real problem of maintaining an empire: How to hold it together? The various garrison towns, spread across an enormous empire, had the supreme self-confidence of their religion and their victories. Why bow down to others? The first century of Muslim empire is one of nearly uninterrupted civil war between competing leaders of armies and garrisons. Damascus seemed to be in charge most often – this was the Umayyad dynasty – but in the middle of the eighth century, another family, the Abbasids, based in the Persian East, gained power. Indeed, the Persian-region origins seem significant. Conversion to Islam seems to have picked up in the Persian part of the empire before it did in the Byzantine part (perhaps because the local religion in the Persian region was not monotheistic; hence, there was more of an incentive to convert). The demographic center of gravity tended East, and it was only natural that, eventually, the Easterners would prevail in the contest for control over the Islamic lands. In 762, Al-Mansur, with firm control over the entire Islamic world (only far-flung Iberia remained Umayyad), founded his new capital – Baghdad – in Mesopotamia. The city proclaimed the determination of the new dynasty to rule, at long last, *from the top* – no more competing garrison towns. The city was set apart from all other cities – its walls built, extravagantly, as a huge, perfect circle, enclosing a mighty palace for the ruler. Of course, this did not last all that much longer. As most empires do, the Abbasids had to delegate power. The bigger the empire, the more delegation it required. This empire was bigger than most (from the borders of India as far as the western Mediterranean; only the Mongols would rule, briefly, a larger domain). And so, it had to delegate more. Already by the tenth century, local rulers could be effectively independent. The model of Baghdad remained significant, however: each local ruler, setting up his own dynasty, aimed to establish at least the aura of Baghdad's absolute power. Courts, from now on, aimed to be radically removed, elevated. A Roman emperor was always, in theory, no more than the top figure of the senatorial elite, his court dispensing a patronage not different, qualitatively, from that of all other senators. The caliph, however – and its many sultan and vizier imitators – was meant to be radically apart: his patronage, marked by unparalleled consumption – and unparalleled forms of culture. Hence Baghdad's own Renaissance – preceding, indeed, that of Arethas in Byzantium.

Baghdad was an imposing metropolis – and within it, an even more imposing palace, set apart from the world. Its culture was similarly structured, as an outside and an inside, one part more widely available – and another, more exotic. On the outside was the study of the Quran – with its allied traditions of Arabic grammar and literature – which could not be monopolized by the court. Of course: the central assumption of Islam was always the direct availability of the word of God. The royal courts did have the best Quranic and literary scholars, but there would always remain an important tradition of scholarly study in the fields most directly related to religious tradition – Muslim law, Arabic language – away from the palaces. Starting in Baghdad, the sultans set themselves apart by the possession of a different, added layer of knowledge – the one imported from *outside* the Arabic sphere. This was, above all, the knowledge brought from the Greeks.

To begin, this reached the Arab-speaking world via the Syriac translations, and soon enough, via translations directly from the original. The first translations were done late in the eighth century; by the tenth, Arabic readers mostly achieved the translated corpus with which they would be content from that point on. The two main branches of Greek science were medicine, on the one hand, and the exact sciences, on the other hand, and in Greek antiquity itself, the two branches never cross. A few physicians (such as Galen) were proud of their modicum of mathematical knowledge, but no physician authored mathematical texts, while no one whose main identity was that of an author in the exact sciences made any contributions to medicine. Arab court scientists, more often than not, contributed to *both* mathematics and medicine. Specifically, many Arabic scholars possessed knowledge in the key service industries – astrology or medicine – but they typically were also competent philosophers (besides contributing, very often, to the more standard Quranic and grammarian sciences, as well; sometimes, their writing reflected more political or bureaucratic functions). It was rare for an Islamic astrologer not to contribute to more theoretical astronomy and rare for Islamic astronomers not to show wider mathematical skills. Encyclopedism was expected. Encyclopedic, these authors were often prolific – we now come across many authors of dozens of treatises, of millions of words. Around the court of each dynasty, there were now its scholars, so many of them massively prolific (from early on, this was a paper culture), so many of them still extant. I have surveyed Byzantium extremely rapidly. I shall not even try to "survey" the exact sciences written in Arabic. The corpus is larger than that of Greek mathematics itself. I write this *History of Greek Mathematics* to replace that of Heath, written a century ago. The "History of Islamic Mathematics" is yet to be written. We will, instead, merely glimpse at this history – consider a few examples of the way in which Islamic mathematics went on to extend, and transform, the Greek legacy.

The case of optics is typical. The field was relatively marginal inside classical mathematics. In particular, while there clearly were "authors in mechanics," "authors in music," and of course, "authors in astronomy," there were no "authors in optics." Archytas, Archimedes, or Ptolemy might have contributed to the field – in a single work, each. It was the kind of field to which one would contribute only as part of a larger project of writing in the exact sciences – and so checking, so to speak, the optical box as well. Ibn Al-Haytham (965–1040?) was the author of fourteen different works in optics that are still *extant* (he certainly authored many more). There are many particular studies – those, for instance, on burning mirrors (spherical, as well as paraboloid). Those on refraction or on the optics of the observation of eclipses are still within the familiar realm of the geometrical studies of Greek mathematics (although with a much more pronounced emphasis on actual experiment). But we are struck to find

*A Treatise on Light* – an empirical, philosophical study – and in particular, we are struck by Al-Haytham's major opus, *The Optics*, in seven books, covering all aspects of the discipline as developed by Ptolemy but based on a solid empirical basis in the study of the eye. Al-Haytham describes the anatomy and physiology of the eye with Galen's attention and experimental method – and then uses this as a basis for a geometrical model. This combination has real scientific consequences: whereas most Greek writings in optics assume that vision is affected by an optical ray extending out of the eye, Al-Haytham's physiological observations persuaded him that vision is based on an impact of light, refracted through the eye's lenses. Al-Haytham was right, but what is even more important, he hit upon the truth because he rigorously combined reason and experiment, specifically because he combined mathematics and medicine. His influence, in this respect, will be important.

The science is encyclopedic and directly critical of the Greek legacy. In fact, another of Al-Haytham's treatises is titled *Doubts on Ptolemy*, going through what Al-Haytham perceives as Ptolemy's inconsistencies and errors in optics – and in astronomy (on which, more in the following discussion). It is impossible to imagine a "Doubts on Ptolemy" written in medieval Byzantium (Proclus's *Hypotyposis*, however, could have been so titled). Al-Haytham was definitely not slavish in his attitude toward the past. But it is noteworthy that in the eleventh century – after three hundred years of Islamic science – Ptolemy should still matter so much. Al-Haytham's science is deeply original, but the terms against which such originality is to be measured are the achievements bequeathed from the Greeks. Al-Haytham's *Completion of the Conics* is a perfect example. To recall, only the first four books of Apollonius's *Conics* are now extant in Greek – those transmitted with Eutocius's commentary. Arabic translators in Baghdad managed to find the first seven books, but we know that the work originally had eight. For this, we have two indications: Apollonius's introduction to Book VII and Pappus's commentary (which does not go through the *Conics* proposition by proposition but instead provides lemmas helpful for the reading of the books). Pappus is known through a single Greek manuscript in the Vatican that apparently left Byzantium early on. There is no indication that it was ever known to Arabic readers. Al-Haytham read, however, Apollonius's introduction to Book VII and was thus moved to complete the *Conics* by the addition of his own Book VIII, based on the slight indications provided by Apollonius. Thanks to Pappus, we know that Al-Haytham's proposal was almost certainly historically wrong. Hogendijk, the modern editor of this treatise, argues that the text we now possess must have been Al-Haytham's unfinished manuscript. Even so, this is a wonderful book. Al-Haytham had little to work with and so stated explicitly that his assumption is that Apollonius was looking for *beautiful* results. And Al-Haytham got them! Let me sketch the thought of just one example (Hogendijk's "proposition 10").[1]

Al Haytham indeed got the *spirit* of Greek mathematics. Proportions are key, and so he decided, in his "Book VIII," to solve problems defined in terms of given ratios. In proposition 10, we are given a parabola, its tangent DG, and a ratio E:Z. The task is to find another tangent, AD, cutting the tangent DG, so that

$$AD : DG :: E : Z.$$

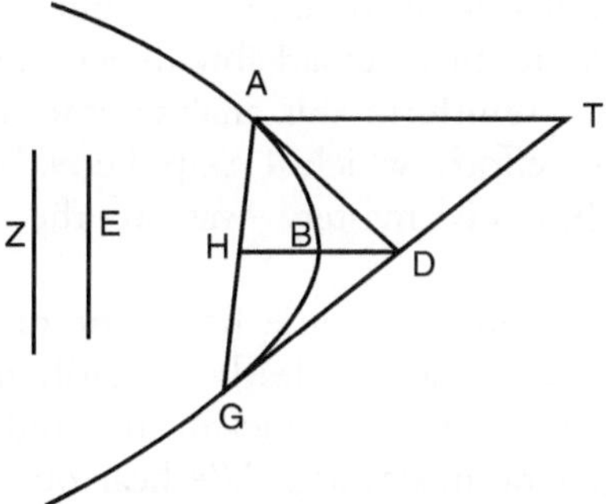


Figure 102

We follow the method of analysis and synthesis (Al-Haytham does so systematically in this treatise; in the typical metamathematical, philosophical tendency of Islamic mathematics, he even wrote a specialized treatise dedicated to the method). So, we first assume that the solution has been found. Join and bisect AG. We know that the line joining a point such as H (the bisection of the chord between the touching points of the two tangents) to a point such as D (where the two tangents meet) is a diameter of that parabola. Al-Haytham interprets the sense in which the parabola and the tangent GD are "given" to mean that we know the absolute position of the original axis of the parabola, and of course, we also know the position of the line GD. We therefore know the angle they form, and because all diameters in a parabola are parallel to the axis, we know the angle between the line GD and the diameter DH, that is, the

[1] J. P. Hogendijk, *Ibn Al-Haytham's Completion of the Conics* (New York: Springer, 1985), pp. 186–188.

angle GDH. We now draw AT parallel to HD and join GDT. Because the angle GDH is known, so is the angle GTA. Because GH = HA and HD is parallel to AT, we also have GD = DT. The ratio AD:DG is a given (because we assume the problem solved), so the ratio AD:DT is a given as well.

Thus, in the triangle ADT, we know one angle, DTA, and the ratio between two sides: AD:DT. In Euclid's *Data*, it is shown that under these conditions, all angles of the triangle can, in principle, be determined. We thus know the angle ADT, and hence also the angle ADG. And so, once again, in the triangle ADG, we know one angle and also the ratio between two sides, AD:DG. Hence, we know all its angles, and thus we can draw GA from the known point G and determine point A.

I cited this in detail because the mere statement that Al-Haytham spoke as an equal to the generation of Archimedes would not suffice; we need to see this in action. The result is remarkably direct and yet surprising. It is also still very hard (to transform this analysis into a direct construction required much more effort, which I skip here). Al-Haytham's scope was encyclopedic, but his mastery was at the level of individual mathematical detail.

I started with Al-Haytham because he is one of the greatest Islamic authors whose contributions are so clearly primarily mathematical. But of course, he was not just a mathematician, and indeed, the impression conveyed so far might be misleading. We hear of Al-Haytham feigning insanity to escape the persecution of the caliph in Cairo. It is hard to extract any history out of such obvious, Hamlet-like novelistic embellishment, but it is clear that we are dealing with a functionary of the court. We know little about his activities as a bureaucrat, but we can reconstruct Al-Haytham's writing with great confidence: we have extant a bibliography he composed of his own writings, dated by him to (what we call) February 10, 1027 – seventy treatises – supplemented by an addendum, dated July 24, 1028 (twenty-one more treatises – he was really working on his résumé!), and then by another bibliography, compiled by others, after his death, adding ninety-two more works. Even this is far from exhausting his oeuvre – the *Completion of the Conics*, as noted, is not even mentioned in the bibliographies, probably because it was never "published." Approximately seventy-five treatises by Al-Haytham are still *extant*. This is a remarkable rate of survival and a tribute to the high degree to which he was held as a scientist. These, however, are almost all strictly scientific works, and it is clear from his bibliographies that he was also the author of dozens of works of a more strictly philosophical character, for instance, studies and commentaries on Aristotle. The *Treatise on Place* – largely a critique of

Aristotle's *Physics*, Book IV – is still extant, as are several offshoots of his astronomy that are of more purely cosmological character; it is typical that later tradition conserved, among his philosophical works, those that were more closely aligned with the exact sciences. He never became recognized for the study of philosophy, which meant, above all, here as in Byzantium, the study of Aristotle: there were other masters for that.

Right at the very beginning of the translation project, in Baghdad of the ninth century, the first major Arabic scientist was also the first major Arabic philosopher – Al-Kindi. Once again, we see the same phenomenon as in Byzantium – picking up where the tradition stopped, with Neoplatonism. The philosophical corpus copied and commented upon by Late Antiquity was Aristotle, through a Neoplatonist interpretation, and it was this that Al-Kindi learned, commented upon (in his manifold writings on Aristotle), and produced himself. A prolific scientific author – of course – he also wrote works touching on astronomy, astrology, music, optics, geometry, arithmetic, *and* medicine – but what is most striking from our perspective is the manner in which the various technical fields are brought together. He writes on pharmacology – and in his recipes, he applies a system of calculation that is explicitly based on a mathematical theory of proportion. There is no known precedent for this in the Greek medical tradition. His grand metaphysical system continued the standard Neoplatonist system of emanation from the supreme astral beings down to the earthly and mundane. His contribution was to interpret this as a quasi-physicalist model of rays carrying influences down the chain of being. Metaphysics – turned into an optics of a kind. (This metaphysics of light – which is merely a metaphor in Platonic and Greek Neoplatonic writings – becomes an important influence in Arabic science and philosophy, partly accounting for the rise of optics as a central scientific field under Islam.) Al-Kindi's specific variant of Neoplatonist metaphysics would not become standard, but most later philosophers would continue to work within Neoplatonist Aristotelianism. Muhammad was not a philosopher, but his anti-pagan zeal stamped Islam with an overriding insistence on the unbridgeable gap between the mundane and the divine. Islamic philosophy could easily accommodate the emphasis on the transcendent and the dismissal of the earthly, pervading Neoplatonism. Aristotle had a number of troubling notions: that the world was not created, that the individual soul was not everlasting. This gave Islamic philosophers something to grapple with. The huge corpora of Al-Kindi or even Al-Haytham pale next to those of some of the major philosophers. Ibn-Sina, a Persian active early in the eleventh century – Al-Haytham's contemporary – wrote over four hundred works, including the major Islamic syntheses of both philosophy and medicine (it will come as no surprise that he also wrote on some mathematical issues). Ibn-Rushd, active in the twelfth century in Andalus – present-day Spain – was the major commentator on Aristotle.

We recall Moerbeke translating Archimedes from Greek into Latin in 1269. This comes at the end of a long tradition of translations into Latin made mostly throughout the twelfth century and almost entirely from Arabic. The site for most of those translations was northern Andalus, or Spain, newly reconquered by Latin Christians. Andalus, captured, captured the wild captors: Latin Europe now had a scientific-philosophical culture that was essentially that of medieval Islam. Indeed, although Islamic scientific and philosophical civilization was written in Arabic and always had the Islamic flavor of radical anti-paganism, it was not a strictly religious, "Muslim" science. (There was a science of the Quran – but this was a pursuit distinct from the more courtly combination of science and philosophy.) Most Islamic scholars were indeed Muslim, but many were Jews or Christians. Maimonides's great synthesis of Aristotelianism – and of medicine – was a twelfth-century Jewish variation of Avicenna; Aquinas's own great synthesis of Aristotelianism was a thirteenth-century Christian variation on the same. We typically do not think of Latin Europe as constituting part of Islamic science, but this is perhaps a mistake. Today, we take it for granted that there is a global science, and it makes sense to call it "Western science" – it is a practice developed mostly in western Europe throughout the nineteenth century. In the same sense, we should conceive of Islamic science as the global science of the High Middle Ages. We may consider the geographical bookends: say, on the East, take Al-Biruni, an eleventh-century author – the most prolific, perhaps, of all Islamic authors in the exact sciences, born in present-day Uzbekistan, active mostly in present-day Afghanistan but especially well traveled in India (on which he wrote extensively). On the West, say, take Duns Scotus, a thirteenth-century author, born, as his name states, in Scotland, who, in his commentaries to Aristotle (and to Aristotle's commentators), was perhaps the best philosopher of his time. Scotus would probably have difficulties following the details of Biruni's astronomy, just as Biruni would probably lose patience with the subtle distinctions Scotus introduced to his ontology. But their discourses would have been, in essence, mutually intelligible. They both worked within the same global, Islamic science.

This global civilization absorbed global traditions. Islam, as noted, was a culture based on paper – that is, based on a Chinese innovation, learned from the encounter of Muslims and Chinese, already in the eighth century, in Central Asia. India was even closer, a few of its writings translated in Baghdad alongside those from Greek sources. The most lasting influence coming this way was, of course, the Indian numerals, transformed and now familiar as Arabic numerals (confusingly, Europeans have further transformed the numerals they learned from the Arabs so that the symbols now universally used are different from those used now only in Arabic script). The significance of this innovation perhaps should not be exaggerated. People in the Mediterranean were always able to calculate with ease in a decimal, positional system, by using their abacus (pp. 226–227); perhaps the origin of Arabic/Indian numerals should be seen precisely as a variation on abacus practices. Numerals are essentially a kind of orthography, and although orthographies do have cognitive payoffs (it is easier to learn, say, Italian writing than Chinese), the mind is elastic and can manage with different tools (educated Chinese can read just as well as educated Italians). For us, it is easier to calculate with the decimal system using Arabic numerals, but Ptolemy had no problem calculating with sexagesimals and with the Greek-letter symbols for numbers. What is much more surprising is that the same was true for all Islamic astronomers: right until the end of the Middle Ages, they continued to produce their astronomical calculations (many, and subtle – as we will note in the upcoming discussion), always using the same numerical techniques as Ptolemy, merely substituting Arabic for Greek letters. Indeed, even now, sexagesimals are often used in the presentation of astronomical and geographical data: my latitude as I write this is 37° 25′ 27″ North. In some ways, sexagesimals are even better than decimals (it does help that 3 is a factor of 60). In other ways, it just takes getting used to. So: the very introduction of Arabic/Indian numerals did not constitute a revolution. What was significant, in the long run, was the way in which those numerals were manipulated. The first book to explain this was, once again, by one of the first scholars of Baghdad – Al-Khwarizmi, that is, native of the land of Khwarizm, in present-day Uzbekistan. His book, known in Latin translation as *The Book of Algorithm on Indian Numbers*, taught, well, algorithms (obviously, "algorithm" is merely how medieval speakers of Latin rendered "Al-Khwarizmi"). The trick was to arrange numbers on a surface and manipulate them (converting, that is, the manipulations previously produced on small tokens on the abacus to a manipulation of symbols). This was done now on written surfaces, and the book itself would present this with two-dimensional inserts into the written text, showing how calculations are effected. The culture of the codex made the book, itself, into the subject of study. With innovations such as the algorithm, scientific work could become a written practice in the most direct sense: to make an argument, you would manipulate symbols on the page. Such symbolic manipulation would become paramount only in modern science, but its roots, indeed, are in Al-Khwarizmi.

For symbolic manipulation to flourish, mathematics will first need to become algebraic – and this, too, has its roots in early Baghdad and, indeed, in yet another of Al-Khwarizmi's works, *The Compendious Book on Calculation by Completion [Al-Jabr] and Balancing [Al-Muqabala]*. The title – abbreviated to *Al-Jabr*, or in Latin, *Algebra* – refers, roughly, to the technique of solving a problem as an equality to which one adds and subtracts on both sides. Problems are built from three components – roots, squares, and numbers – and Al-Khwarizmi goes systematically through their combinations, showing how to solve them with concrete examples:[2]

> When you say: a square plus ten roots are equal to thirty nine dirhams [type of coins]. . . . You have the number of the roots which, in this problem, yields five; you multiply it by itself; the result is twenty five; you add it to thirty nine; the result is sixty four; you take the root, which is eight, from which you subtract half of the number of the roots, which is five. The remainder is three, that is the root of the square you want, and the square is nine.

Translating Al-Khwarizmi's algebra into the symbolic language (the distant descendant of Al-Khwarizmi's algorithm), this says that given

$$x^2 + ax = b,$$

we may find $x$ as

$$x = \sqrt{}\left((a/2)^2 + \mathrm{b}\right) - \mathrm{a}/2.$$

Which you may quickly verify if you remember the algebraic formula for a quadratic equation from school. Al-Khwarizmi himself does provide his own justification, through a geometrical construction, which provides this operation with a much deeper meaning. Effectively, Al-Khwarizmi took the step to consider geometrical magnitudes as numerical values. This epochal event is simply the unintended by-product of the typical Arabic project of synthesis. Al-Khwarizmi worked from many sources: Indian and ancient Near Eastern procedures of calculation, Greek geometry. They were all, for him, valuable as the legacy that needed to be brought into Arabic and put together inside his own, early, encyclopedic project.

[2] R. Rashed, *Al-Khwarizmi: The Beginnings of Algebra* (London: Saqi Books, 2009), p. 100.

To many past Greek scholars, perhaps, such exercises as those dealt by Al-Khwarizmi would appear as lowbrow, even unscientific (that déclassé reference to Dirhams!). The mixing of numerical and geometrical magnitudes would appear as a stylistic or perhaps even a theoretical error. From Al-Khwarizmi's vantage point, such distinctions mattered less. Once again, as happened in astronomy from Hipparchus to Ptolemy: Greek mathematics brought together with the calculation techniques of the ancient Near East – in the process, turned into something new and rich in potential.

In a sense, this potential was fully achieved in another book of algebra, by one of the greatest of all Muslim mathematicians: Omar Khayyam (1048–1131). The name is familiar thanks to the poetry – he was the author of short, epigrammatic poetry (rubaiyat – four lines, all rhymed other than the third), whose main theme is humanity's frailty and limitations:[3]

> My mind has never lacked learning,
> Few mysteries remained unsolved;
> I have meditated for seventy-two years night and day,
> To learn that nothing has been learned at all.

A deeply Muslim sentiment: human knowledge as nothing as against the divine. The response, however, is elegiac rather than tragic:

> To escape from the learned schools is best,
> And best to hang on the sweetheart's tresses:
> Before fortune drains our blood away,
> It is best to empty the bottle into your glass.

This is probably what Khayyam sang in learned symposia, perhaps in the presence of one of the sultans he served (in Isfahan and in Marv). Elegant conversation would move back and forth between Persian, the language of Khayyam's poetry, and Arabic, the language of his science. If the poetry admitted human limits, the science pushed right to those boundaries. Al-Khwarizmi solved six types of problems with roots, squares, and numbers. Khayyam's own *Algebra* started from Aristotelian philosophy and deduced, from first principles, that there are three possible types of magnitudes: lines, areas, and solids. Once all problems with these magnitudes – as well as numbers – are solved, then the field of algebra is complete (note how the field has become completely identified with that of geometry). Khayyam

[3] P. Avery and J. Heath-Stubbs, *The Ruba'iyat of Omar Khayyam* (Harmondsworth: Penguin, 1981). I cite from pp. 94, 78.

then proceeds systematically through all the combinations of the four terms – what we would call equations up to the third degree – and solves them all geometrically. Just as Pappus did, he explicitly notes the division according to the types of geometrical tools required: those that can be solved with the tools of Euclid's *Elements*; those that require, further, Apollonius's *Conic Sections*. He makes special note of how his solution fulfills the gap in Archimedes's *Sphere and Cylinder*, Book II – the one we have noted on pages 428–430, completed by Eutocius. This was the star example for my book from 2004, *The Transformation of Mathematics in the Early Mediterranean: From Problems to Equations*. Archimedes's problem has become Khayyam's equation – above all, by being inserted as one of a *set* of equations.

Khayyam's *Algebra* is surprisingly brief for such an all-encompassing project – seventy pages in Rashed's translation – which seems to be part of Khayyam's point. The completeness of the project is underlined by its quality of synoptic survey. We have lost much by Khayyam, but it seems that he never aimed to produce very big, or very many, books, aiming instead at succinct precision. His introduction to Euclid's *Elements* is very brief – a couple of observations, each a few pages long. One of these is on the theory of proportion. The other, on the parallels' postulate, carries great historical significance. We recall how Euclid postulated explicitly that "if a straight line falling on two straight lines make the interior angles on the same side less than two right angles, the two straight lines, if produced indefinitely, meet on that side on which are the angles less than the two right angles." I suggested that the need to make this into an explicit postulate arose from Euclid's choice to present elementary geometry without the help of proportion theory, which made it necessary to develop, in Book I, the properties of parallelograms. Mathematicians always felt uneasy with this postulate: this seems like the kind of thing that should be provable. Euclid, to his glory, realized that he could not prove it. Ptolemy – so we are told by Proclus – made an attempt to provide a proof, which, however, is such a weak begging of the question that we must wonder if Proclus really understood Ptolemy's purpose. (I suspect Ptolemy could simply have explained the intuition underlying the postulate, without considering this as a valid proof.)

Khayyam, too, believed he had produced a proof, which means he did commit an error. But in this case, this was a fruitful error. His idea was to pursue an entire line of reasoning, ensuing from the hypothesis that the parallels' postulate need not necessarily hold. He organized this line of reasoning around a single geometrical object, a quadrilateral, on the given line AB, AC and BD both perpendicular to AB and equal to each other. The parallels' postulate is equivalent to the statement that this quadrilateral has to be a rectangle. Khayyam pursues the consequences that one may deduce for this quadrilateral independently: for instance, even if we cannot prove that the angles at C and D are both right, we can still prove that they are equal to each other. If they are both acute, the same will be true of all the corresponding angles in all other such quadrilaterals; if they are both obtuse, the same will also be true universally.

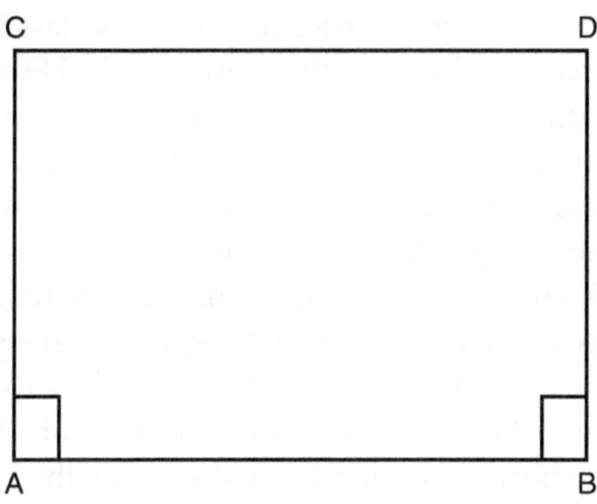


Figure 103

Khayyam takes this as a starting point for a series of geometrical deductions, from which he concludes – on what are essentially metaphysical rather than mathematical grounds (and based mostly on Aristotle) – that the nonrectangular option is impossible. In this, he is wrong. But the idea of pursuing the negation of the parallels' postulate in terms of global geometrical consequences – and the specific technique of studying this with that quadrilateral – will be significant. Khayyam's commentary to Euclid was not known in Europe, but Tusi – on whom, more in the following discussion – took up the same idea in his own work (hence the name by which this is often known, *Tusi's Quadrilateral*). Saccheri, in the middle of the eighteenth century, would produce an entire study of the quadrilateral, with several propositions resulting for each assumption of the acute and the obtuse angle; it would be down to several mathematicians in the early nineteenth century – Bolyai, Lobachevsky, and Gauss – to show that Saccheri's alternative geometries, that is, Khayyam's alternative geometries, expanded, were in fact valid alternatives. Besides Euclidean geometry, there were other, non-Euclidean geometries. It took more than two thousand years for Euclid's realization – that he needed to turn this assumption into a postulate – to reach its logical conclusion and so to give rise to a new conception of mathematics. The decisive turn taken, nearly a thousand years ago, somewhere between Nishapur, Marv, and Isfahan.

It was there – so far away from Syracuse, Athens, or Alexandria – that Greek mathematics was now advanced. Space and language were transformed, but the cultural project remained. Indeed, the continuity between Islamic science and that of Late Antiquity is obvious. Khayyam ended up opening the way to non-Euclidean geometries, but his purpose was not to open but to close. Everywhere, we see the project of grand survey, synthesis, completion. Khayyam tries to find those last few theorems that make Euclid's *Elements* axiomatically complete; Al-Haytham tries to find the last few theorems missing from the transmitted *Conics*. Al-Haytham also produces a synthesis of mathematics and medicine in his study of the eye, just as Al-Kindi synthesizes geometry and philosophy – and as Al-Khwarizmi synthesizes geometry and calculation. Aristotelianism, in its various Islamic treatments, becomes a unified system, just as do the algebraic equations of Khayyam, treated together, surveyed and classified. This might compare, perhaps, to Pappus, and perhaps of all Greek mathematical authors, he comes closest to the spirit of Arabic mathematics (even while falling short of the intellectual level of the best among Arabic mathematicians). The Greek authors of Late Antiquity, as well as the Islamic authors of the Middle Ages, took up the same corpus – the extant Greek mathematics, from its beginnings to the imperial era – and the same responsibility, of *custodians* to that corpus. In Late Antiquity, this custodianship took the form of more teacher-like commentary; in the Islamic world, this custodianship took the form of more creative, encyclopedic synthesis. In terms of the divisions proposed in this book, Islamic authors took late ancient mathematics and pursued it with a sophistication equal to that of the generation of Archimedes. This is an important experiment, then, by revealing the possibilities, and limits, of the attitude of Late Antiquity. However sophisticated, there is one thing Islamic mathematics was not: it was not modern. Through many centuries, it kept itself, largely speaking, to the conceptual boundaries of Greek mathematics itself. It was custodianship, after all. This is a science, aware of its position, indebted to a classical past. Khayyam, after all, was all too aware of the limits of human knowledge, already largely achieved – and forever bounded by the divine:

> The sphere upon which mortals come and go,
> Has no end nor beginning that we know;
> And none there is to tell us in plain truth:
> Whence do we come and whither do we go.

Knowledge of the spheres could, in fact, have been Khayyam's livelihood (it is possible that in the courts at which he served, his main function was that of an astrologer). The core of the exact sciences in Islam was astronomy: the mastery they displayed, above all, was that over Ptolemy – so much so that his major work, the *Almagest*, is known to us, now, in its Arabic title! In terms of his influence, Ptolemy was Arabic.

A stubborn historiographical myth imagines astronomers, following Ptolemy and faced with the inevitable empirical shortcoming of the model, gradually introducing more complications to the models so as to produce a better fit ("epicycles on epicycles"!) – until finally Copernicus is done with all of that. I hope it is clear that this myth is wrong from the start. It suffers from the anachronism, as if ancient astronomy – like that of Newton – was deduced from *laws*, in this case expressed in the form of, say, "epicycles." The geometrical models used by Ptolemy were important, but they provided no more than the *shape* of the theory, and one could not use them to calculate astronomical positions. Positions were calculated, instead, from *tables*. The geometrical model was significant as physical theory, but above all, it provided the mechanism whereby a handful of observations (generally speaking, the absolute minimum required) could be used to derive the parameters from which the tables ensued. To revise Ptolemy, then, was not seen primarily as a matter of revising "Ptolemy's laws" (what was that?) but of revising his tables. Furthermore, Ptolemy's system was quite precise for the main concern of his astronomy – angular position – and if one confronted the predictions, computed from the tables, with observations, one would normally find errors of less than ten minutes, occasionally a little more for the recalcitrant case of Mars. (This precision should not be seen as surprising: whether one chooses a heliocentric or geocentric reference matters little for predictions; Ptolemy's chosen combinations of circles approximate Kepler's ellipses fairly well, and anyway, the planetary orbits in our solar system are rather circular ellipses.) Finally, no one ever expected precision. Everyone – beginning with Ptolemy himself but going back to Hipparchus and, indeed, even to Archimedes – was clear that all observations contain error. This was, if anything, a reason to be conservative. (Suppose one finds discrepancies from Ptolemy's computed positions: Why should these be ascribed to Ptolemy's errors and not to one's own?) In short, it is entirely reasonable that Late Ancient authors – and, of course, their Byzantine followers – were content, on the whole, simply to comment upon Ptolemy. The remarkable thing is that Islamic astronomers were not.

Here as elsewhere, Islamic authors displayed not merely their knowledge of a field but their mastery over it. The goal was, then, to be able to remake it. This means, above all, remaking the tables. There are, of course, many specialized Islamic astronomical treatises, and some of these do obtain the monumental scale of Ptolemy himself. Al-Biruni's *Canon* does not merely derive new tables but also explains the theoretical principles with which they are achieved: a new *Almagest*, in short. But this was the exception. The standard way to remake Ptolemy was simply to produce a new set of tables, or "Zij" – the central genre of Islamic astronomy. Kennedy counted, in 1956, more than a hundred types of Islamic Zijes (!), and more have been noted since; about twenty, by his reckoning, seemed to have been based, at least in part, on original campaigns of observation. It goes without saying that this is a state project. This is also team effort – a theme that is very alien to Greek astronomical practice but that seems typical of Islamic astronomy. Our evidence consists, mostly, of fairly novelistic anecdotes, but those refer systematically to *groups* of astronomers working together. Abu'l Wafa and Al-Biruni conspired to observe the same eclipse, far apart from each other, in 997 – finding a difference of one hour between Baghdad and Khwarizm: astronomy, coordinated across an imperial domain. Much earlier, it is said that Al-Mamun – one of the first Abbasids – sent a group of astronomers to make observations in the Syrian desert, so as to measure precisely the length, on the earth's surface, of a single degree of latitude. As noted on pages 238–239, this is the kind of thing that the Chinese did but the Romans did not: this was state science.

Islamic tables could be, in fact, very precise. Many types of astronomical tables are purely mathematical (such as those for trigonometry and spherics), whereas those of the fixed stars are directly observational. Such tables can be judged on their own merits. They are, on the whole, superior to Ptolemy, and they progressively improve. Whereas precision in astronomical observation is limited by the power of the naked eye, there is no such theoretical limit in trigonometry, and the latest Islamic tables provide values to the fifth sexagesimal position (this involves not only a great effort of computation but also progress in numerical and trigonometric technique). As ever, the great bulk of the work produced does not deviate from the terms established by the Greek legacy, and indeed, most tables, in most Zijes, are simply copied over from Ptolemy. There are exceptions, however. To begin with, this is global science, with global influences. The earliest among Islamic Zijes seem to have been produced based, at least in part, on tables distinct from Ptolemy's, apparently taken over from Persian or Indian sources; awareness of Indian astronomy never ceases. This is of historical interest because it is now believed that the Indian tables, themselves, ultimately are adaptations of pre-Ptolemaic Greek techniques; to some extent, they even preserve Babylonian methods of calculation (this hybrid of Greek and Babylonian seems to have been typical of pre-Ptolemaic astronomy already in the Greek). And so: Indian, Persian, even Babylonian influences. And then, of course: the ripples of Islamic astronomy, outward. Through the Mongols, Islamic Zijes would reach into China itself; Christian Europe got its own Zij through Spain – the famous *Toledan Tables*, produced in the eleventh century, translated in the twelfth, and finally revised in the thirteenth as the *Alfonsine Tables*, produced through the patronage of King Alfonso X of Castille. (Latin Europe shared in the global Islamic science, and its kings, too, came to sponsor state science.)

The project of Islamic science involves mastery – and synthesis. In fact, Ptolemy poses a difficulty for such synthesis, as was clear already to Proclus. At the most basic level, his science is much later than Aristotle's, which will create difficulties for anyone trying to bring the two together. We recall: Aristotle explicitly assumed Eudoxus's homocentric spheres, and so a synthesis of the astronomer and the philosopher would necessarily require modifying at least one of the two. Now, this is not hopeless: Ptolemy himself was, broadly speaking, an author in the Pythagoreanizing–Platonist tradition, who wished to show the rationality of the heavens based on circular motions. The technical difficulties of astronomical motions stood, however, in the way. The myth of epicycles on epicycles has it right that with an indefinite number of circles, it is, in principle, possible to approximate curves to the desired degree. This, however, would have become obviously ad hoc and, worse, hard to calculate, and in fact, no medieval astronomer went down this route. More to the point, astronomers did not conceive of their problem fundamentally as the fitting of curves. What needed to be explained, above all, was the pattern of changing *speeds*. Ptolemy found a powerful approximation to that in the tool of the equant. Let the planet move on its single epicycle, around its eccenter, producing the rough outline of the movement required, but then let it move in such a way that in equal times it would cover equal angles, not relative to its center of revolution (itself eccentric, that is, not the center of the earth) but relative to yet another point, the equant. Obviously, this is a good way to approximate Kepler-like motions on an ellipse, so we can understand why Ptolemy's model is, after all, nearly precise. And it was nearly precise in a computable manner, based on circle-like motions alone.

Here's the rub: these were circle-like motions – but no more. Circles, if taken seriously as circles, move around their centers, not around some equants. And if one follows Aristotle, one should then wish to have the astronomical circles to be that – real circles. For if not, whence the rational account for their motion? From Al-Haytham on, Islamic authors repeatedly criticized the irrational character of this part of Ptolemy's system. So, an invitation for other astronomers – to come up with their own alternative tools.

Some Islamic authors approached the tension between Ptolemy and Aristotelianism from the vantage point of philosophy. This was especially common in Andalus: Ibn Rushd's commentaries to Aristotle are skeptical of the *Almagest*; Al-Bitruji, a twelfth-century philosopher, went much further and issued a new cosmology, trying to revive a model of homocentric spheres. Such philosophical contributions do not amount yet to a variation on Ptolemy's mathematical astronomy, but Al-Bitruji was not some kind of marginal crank, and it is important to notice this: Islamic authors felt free to deny all or any of the claims made by an author as canonical as Ptolemy. They aimed to be the masters of a Greek legacy; they were not servile to it. And indeed, if one's problem was strictly with the equant, this would imply a well-defined mathematical problem – take the equant, reduce it to circular motions alone. Can this be done?

By the middle of the thirteenth century, Mongols were firmly in control of the entire Persian-speaking world (the site, it will be noted, of much of Islamic science). Hulagu – Genghis Khan's grandson – sought to transform the brute force of military conquest into the status of a civilized Muslim monarch. And so, a court, doing what courts do. In 1259, astronomers were summoned from all over; an observatory was set up at Maragha (in today's northwestern Iran). Chief in this group was Nasir Al-Din Al-Tusi (the author through whom Saccheri learned of Khayyam's quadrilateral), but we can trace an entire group of astronomers, working at Maragha with Al-Tusi or elsewhere, immediately afterward, taught and inspired by Maragha masters. The Zij produced based on the Maragha observations was entirely traditional, but the group as a whole was characterized by the quest for innovative problems and techniques in mathematical astronomy.

Ibn Al-Shatir, a fourteenth-century astronomer, noted an empirical problem that emerged as a kind of epiphenomenon to Ptolemy's model. Focused on the angular position, Ptolemy's epicycles and eccenters also can be translated as predictions of planetary distance. Those predictions are not borne out for the sun (it varies in size a little more than Ptolemy's eccentric alone predicts), and they are especially problematic for the moon (Ptolemy's epicycle would bring the moon much closer to the earth than it gets in reality). It is incredible – and a mark of the Ptolemy-centered way in which Ptolemy was even criticized – that no one brought those discrepancies up before Ibn Al-Shatir himself. But perhaps this is because Ibn Al-Shatir, now, had the mathematical tools necessary not merely to criticize Ptolemy but to reform him. Heir to the Maragha tradition, Ibn Al-Shatir knew that precise combinations of two epicycles (known as the *Tusi couple*) could give rise to an effect equivalent to an equant. And so, Ibn Al-Shatir removed Ptolemy's eccentric and instead had the angular positions of the sun deduced by a combination of epicycles. An entire series of publications by several authors in Central Asia and the Near East, from the late thirteenth century to the mid-fourteenth century, produced variations on this Tusi theme, producing new models, preserving Ptolemy's empirical results but removing the equant. This is as far away as Islamic astronomy ever got from the Ptolemaic legacy – based, however, on the very Greek route of geometrical analysis and the demonstration of equivalence between alternative combinations of curves. What Apollonius did to epicycles and eccentrics, Tusi now did for the double epicycle and the equant. Greek mathematics – alive and *creative*, in the fourteenth century!

Ptolemaic astronomy was perfected throughout the fourteenth century, in ripples extending out of northwestern Iran. How far did those ripples extend through global Islamic science? Did they reach Andalus – by now, no more than a rump near the southern tip of the Iberian Peninsula? (In 1492, the kings of Spain would expel the last of the Moors.) Did they reach Latin Europe? This is of decisive significance to historians of the scientific revolution, and I will briefly return to this question in the following section. But before I get to the last section of this history, a final turn inward – to my own profession of history – and West, to the Arab Maghreb. Ibn Khaldun was one of the leading political figures of the fourteenth century in what is now Tunisia, at one point its effective political leader; upon his retirement – just how voluntary was it? – he turned to write a history. So did many others before him; nothing new. But Ibn Khaldun did something quite unlike Anna Commena, Polybius, Xenophon ... The first volume (out of seven!) – the *Muqaddimah*, or *Introduction* – is like nothing written before Ibn Khaldun. I said that I turn to my own profession, and the remarkable thing about the *Muqaddimah* is that Ibn Khaldun is, essentially, my colleague. He reads like a modern historian, and indeed, his *Introduction* is precisely a *sociology* of history. Ibn Khaldun seeks the general rules of Islamic civilization, and so he notes a particular pattern. I referred earlier to the way in which Hulagu, in the thirteenth century, sought to transform military conquest into the status of a civilized monarch. In this interpretation, I simply follow Ibn Khaldun's approach: he explained the trajectory of such transformations and their cyclical pattern – a civilized court falling at the hand of newcomers (often from the nomadic periphery), who establish themselves in turn as civilized court . . . The sociology is, I would say, correct. Ibn Khaldun invented a new kind of science, and he did it well. In the fourteenth century, Islamic civilization attained supreme achievements in disciplines as far apart as history and astronomy.

What are we to do with that? What are we to do, indeed, with the evidence of the legacy of Greek science in Islam?

To a historian of science, early-twenty-first-century global politics are bewildering. Many seemingly knowledgeable observers look at the tensions between present-day Islam and Western powers and see there an essential clash of civilizations. This – when, one era back, their ancestors formed part of the *same* civilization! What is so wrong with the twenty-first century? I am primarily a historian of Greek mathematics, after all, and this question is not for me to answer. Perhaps I can make my tiny contribution by pointing out how historically contingent – how far from essential – this purported clash of civilizations really is.

It is, however, within my purview – as I survey the legacy of Greek mathematics – to point out that the scientific revolution, that is, the final transformation of Greek mathematics, took place not in the Islamic lands but in Latin Europe. The Needham question – framing much of twentieth-century scholarship concerning Chinese civilization – was why the scientific revolution took place in Europe and not in China. I think this is the easier question to answer: it is hard to get to the scientific revolution without the Greek legacy, and so China – which had its own brilliant contributions but had nothing like the generation of Archytas, let alone that of Archimedes – was at a major disadvantage. But the same is no longer true for Islam. In the High Middle Ages, the Greek scientific legacy was much better known and advanced in the Islamic lands than in its poor relation, Latin Europe. And so: How did the last become first?

The question divides into two. The first part has to do with what happened in Europe, to which I shall immediately turn. The second part has to do with the Islamic world itself, and here I shall no more than outline my proposed account. (No Ibn Khaldun myself, I speak with little authority on the sociology of early modern Islam: but the question is naturally raised at this point of the argument, and I am not aware of satisfactory explanations.) The one observation that stands out concerning Islamic science is the centrality of the court. As we have noted, such courts are ephemeral, but this perhaps serves, if anything, to explain their turn to scientific patronage. Ever since the Abbasids, the presence of science in the court seems to be associated, above all, with the campaign of legitimization of the court's new authority. The scientific mastery of the scholars – to mark the sultan, himself, as a master. We can understand all of this, based on Ibn Khaldun.

The revolving door of ever-new sultan dynasties, throughout the history of Islamic societies, was thus accompanied by a steady accumulation of scientific knowledge. Is it a coincidence, then, that the significance of scientists at the court seems to diminish right at the moment that the sultans, finally, become stable? From the fifteenth century onward, Islamic states were consolidated into three major imperial powers – Ottomans, in the Islamic Mediterranean; Savafids, in Persia; Mughals, in India. For once, those powers *stayed* – crumbling, at the end, only under the pressure of European colonialism itself. Indeed, in the very same fifteenth century, we see the emergence of stable state power in Europe, as well. Global developments such as these often have a base in shared technology, and it is likely that the new stability of the state was based on new military technologies, above all, the creation of more effective guns. Perhaps there is less need for symbolic power when one is in possession of the real thing? Why have an entire retinue of astronomers in court to impress lesser rulers when you can simply cow them with your weapons? More artillery, less armillary. If so, we begin to hold the thread that might lead to an explanation for what would soon transpire in Europe: however much the science of Latin Europe did belong to the sphere of Islamic science, it did differ from it in its sociology, and in particular, it did not depend on the state. Let us begin to follow this thread – and visit Greek mathematics' final vanishing act, into modern science.

## The Renaissance to End All Renaissances

We start at the margin. There, in Bachet's edition of 1621, next to Diophantus's problem II.8 (which asked to split a square into two squares; so, for instance, given a number such as 25, finding its composition from 9 and 16) – Fermat had these famous words jotted down:

> But one cannot split a cube into two cubes, nor a quadratoquadrate into two quadratoquadrates, nor in general any power in infinitum beyond the square into two like powers. I have uncovered a marvelous demonstration indeed of this, but the narrowness of the margin will not contain it.

That we do not have Fermat's Last Theorem proved anywhere else is not surprising. Fermat tended not to divulge his proofs. Most of his claims were left as mere challenges. He would send out letters to his circle of correspondents, goading them to tackle problems that he claimed he, himself, had already solved. So, for instance: find – but through a generalizable method – a cube integer number so that, adding to it the sum of its divisors, you get a square! Or find a cube integer number so that, adding to it the sum of its divisors, you get a cube! (Both were sent out in the same letter of January 3, 1657.) Such challenges are highly reminiscent of Archimedes's letter to Conon, as, of course, is the more fundamental spirit of the mathematical tournament. Fermat was not exactly a marginal figure – in fact, he led the comfortable life of a successful lawyer – but he did not belong to any "School of Mathematics" in the city of Toulouse, where he spent most of his career. He belonged to a larger community but mostly through his correspondence, rather infrequently broken by face-to-face visits. The setting is indeed so reminiscent of Archimedes: the practice of science – as a network. For Archimedes, so much of our historical reconstruction is guesswork; for Fermat, we know more, and we can say this: he never did prove his "Last Theorem." He was often convinced of the truth of general claims based on some special, inspired solutions. The Last Theorem, in particular, was rigorously proved for the first time only in 1995, based on techniques that would have been inaccessible to Fermat many, many times over. Most likely, in my view, Fermat was simply assured of the truth of his generalization based on his discovery that a quadrato-quadrate number – as he called it – could not be divided into numbers. In our words, there are no integer solutions of

$$a^4 + b^4 = c^4.$$

Fermat's proof for that is, in fact, known: it is one of those that he did jot down, next to another of Diophantus's problems. The proof is based on a general technique – always a key concern for Fermat – that he was especially proud of and did discuss explicitly in his correspondence: infinite descent. Let me sketch the contours of the argument.

If we have $a^4 + b^4 = c^4$, it follows that we have a right-angled triangle such that each of its sides, themselves, are squares. Now, through the experience one develops in solving problems such as those proposed by Diophantus, it becomes clear that Pythagorean triplets can be generated from a couple of integers, $p$, $q$: specifically, where the right-angled triangle in integers has the hypotenuse equal to $p^2 + q^2$ and the sides equal, respectively, to $p^2 - q^2$ and $2pq$. Thus, when we have $a^4 + b^4 = c^4$, all three expressions ($p^2 + q^2$, $p^2 - q^2$, $2pq$) are, each, a square number. I will now skip the meat of the proof, which is more technical. The key idea is that it becomes possible to reduce those expressions and find factors, smaller in value from the $p$ and $q$ from which we started, so that those new values, once again, are integers producing another Pythagorean triplet, smaller than the first, once again with square integers.

This is all that Fermat needs. The assumption of a Pythagorean triplet where all sides are themselves squares gives rise to a smaller Pythagorean triplet where all sides are, once again, themselves squares. There is nothing wrong with that in and of itself, but a moment's reflection shows that this is not sustainable. For if we did it once, it is clear that we can do this again and so find a smaller triplet still, where all sides are themselves squares – and so on, *infinitely*. But obviously, we cannot go on infinitely because, you see, numbers are bounded on the way down: if you start with a finite integer such as, say, a trillion, and keep finding a smaller integer, sooner or later – in this case, after no more than a trillion moves – you will be out of integers.

Hence we prove, by *infinite descent*, the impossibility of there being

$$a^4 + b^4 = c^4$$

with integers $a$, $b$, and $c$ (which, I suspect, Fermat intuited – and was proven right, centuries later – must therefore hold for all other powers).

The method is brilliantly original, but its inspiration is Greek. In some ways, this is like Euclid's proof for the infinity of prime numbers, taken upside down. (Euclid proves that if a certain set of prime numbers exists, another, bigger one may be found; hence, there are infinitely many. Fermat proves the same, going *down* – but because the way down is limited, this proves not that there are infinitely many triplets but that there can be no such triplets at all.) In other ways, it harks right to the beginning – Archytas's argument that if there is an integer ratio between superparticulars, it must be reducible all the way down to their smallest versions. Fermat is what you get when you start proving results in arithmetic, again. It seems as if even the generation of Archimedes did not do that (arithmetic, for them, might have been understood mostly as music's handmaiden, hence was frowned upon?). Now, however, the range of the problems was provided by Diophantus, which ultimately went back to the practices of the Near Eastern classroom. In a sense, yet another synthesis of Greece and Babylon.

We might as well take this small example and generalize it: Fermat, we see, takes the transmitted text of Diophantus – based purely on the solution of arithmetic problems – and expands it many times over by applying to it Euclidean proof techniques. The arithmetic is transformed into number theory, and this happens not because Fermat escaped from the ancient legacy but because he inserted himself within it: applying Euclid-like proofs, living an Archimedes-like life. (To be clear, Fermat did much more besides arithmetic: this section is *very* selective.)

Bachet published his Diophantus in 1621; Commandino (of whom we will hear more later in the chapter) published Pappus's *Collection* in 1588. Here was a book that neither the Arabs nor the scholars of the Palaiologan renaissance ever knew – surviving in a single manuscript of the ninth century, possibly the direct copy (conceivably, the only medieval copy ever made?) of the book put together in Alexandria itself, five centuries earlier. The manuscript carries no annotations in later hands, and it may well have been – like the manuscripts of Archimedes – no more than an adornment to a library of Byzantium's first renaissance, gathering dust for centuries (and reaching Italy immediately following the Fourth Crusade). Now, printed! Clearly, there were many more readers now.

We focus on one of them, Descartes. A contemporary of Fermat and, like him, a gentleman of comfortable means, he, too, found his own path – comfortably – not far from the margins. He studied at university but did not seek an academic position; he set out for military adventure (of which the seventeenth century had many to offer) but never attained or perhaps never sought political or military glory. By 1629, at thirty-three years old, he ensconced himself in the Netherlands, by no means the boondocks but certainly at a remove from French society. So what? He had his own massive network of correspondence intercrossing that of Fermat, and unlike the lawyer of Toulouse, Descartes made good use of his access to Dutch printers, issuing an array of works that were decisive to the seventeenth century. His *Discourse on the Method*, published in 1637, was a key philosophical work, and an appendix to this discourse – *The Geometry* – took a problem in Pappus and turned it into a new mathematics.

Here is what intrigued Descartes so much. In Book VII, as we recall, Pappus expressed his displeasure with Apollonius for his – Apollonius's – failure to treat Euclid, the teacher of his teachers, with appropriate respect. What business is it of Apollonius to argue that Euclid did not solve the problem of three and four lines? What mattered to Descartes is that this report suggested *an open debate from antiquity*. This was an obvious challenge: How can I, Descartes, solve that which neither Euclid, nor Apollonius, not even Pappus did? Now, Descartes – like most mathematicians of the time – believed that the works transmitted from the Greeks were just the tip of the iceberg and that there was much more hidden from view. In particular, it was correctly observed that the extant Greek mathematical works made it hard to see how their results could be found. (I would say that many ancient mathematical authors, especially in the generation of Archimedes, sought maximal effect and surprise; hence, they made less effort to explain the routes leading to their results.) From there, it is but a small step to a suspicion, common to many of the mathematicians of Descartes's time, that the Greeks had possessed some secret methods, never divulged in their works. A certain historiographical paranoia (*they're hiding something!*) is typical of the seventeenth-century attitude toward the Greeks.

The project, then, almost determined itself: find a solution to Pappus's problem – by first discovering some kind of secret, more general method, with which Greek geometrical problems can universally be solved. Pappus's problem – known as the three- and four-line locus problem – is indeed a very general one and so suggests itself for such a treatment:[4]

> Having three, four, or more lines given in position, it is required to find a point from which the same number of lines can be drawn, each making a given angle with one of the given lines, so that the rectangle of two of the lines drawn bears a constant given ratio to the square of the third line (if there be only three), or to the rectangle of the other two (if there be four).

Not only is this a very general problem, but it is also quite obviously a quantitative one (what matters is not the resulting configuration but certain relations of quantity). This immediately suggests treating the geometrical elements in terms of the manipulation of quantity, which is precisely what Arabic mathematicians – and so, following them, Europeans – had been doing in their algebra. The very fact that there was no trace of this in any extant Greek mathematical writing dictated – by the force of historiographical paranoia – that this must have been *what the Greeks were hiding all along*. Algebra must have been the secret method.

Descartes begins his treatise with the statement that all problems of geometry can be seen to require the finding of the lengths of certain straight lines and that all geometrical manipulations, such as taking a line from a line, finding a line that is the fourth term in a proportion, and so forth, have analogues in arithmetic operations such as subtraction and multiplication, so that one can use such arithmetic operations (that is,

4 D. E. Smith and M. L. Latham, *The Geometry of René Descartes* (Mineola, NY: Dover, 1954), p. 22.

effectively, algebra) to solve any geometrical problem. The treatise ensues, leading to a solution of Pappus's problem – and to the invention of analytic geometry. This, then, is the vanishing act. Diophantus became number theory; Pappus became analytic geometry – all because Fermat and Descartes fully immersed themselves within the project of reviving a Greek science.

Let us pause to reflect.

I think of this period – the sixteenth and seventeenth centuries – as the scientific renaissance. It is most often referred to as the "scientific revolution" (although scholars today are usually coy about such terms, with their implication that history has a simple structure, indeed, even a direction). We sped through the history of Islamic mathematics, and I noted that, unfortunately, that history is yet to be written. The history of the scientific revolution is written many times over, and we do not need to rewrite it here. In this section, I merely consider a few vignettes and concentrate on just one question: What stands out in such vignettes, relative to the long history of Greek mathematics?

First, a rather obvious thing: this is the science of western and northern Europe. Consider Descartes in Leiden, even Fermat in Toulouse. Those spaces did not belong in any significant way to the civilization of Greek science. In a grand historical narrative, it is correct to think of the Middle Ages as a period of retreat: especially so *for the old centers*. If one could produce a time-lapse series of Rome, or of Constantinople, from 400 CE to 800 CE, the images would appear catastrophic. But if you would do the same time-lapse through the following centuries, and farther away, the picture would be very different. A time-lapse of Oxford, or of Samarkand, from 800 to 1300 would be very upbeat. Things were continuously improving! The Middle Ages, all told, are a period of expansion for Mediterranean culture, mostly in two directions: into Central Asia (spearheaded by Muslims) and into northern Europe (spearheaded by Christians). Many of the scientific advances of the Middle Ages came from that Central Asian region (this is the land of Omar Khayyam, Ibn Sinna, of the Maragha school, and of Ulugh Beg's observatory). It is unsurprising that northern Europe – with its coastline, rivers, and fertile plains – would ultimately be the more prosperous and populous and so should contribute even more.

Another thing that pops out of the evidence, compared to Islam, is the multiplicity of contexts for scientific practice. We will note a few more vignettes. Many authors of the scientific renaissance did rely on princely patronage. Descartes, as is well known, was lured, toward the end of his career, to Queen Christina's court in Sweden (where the severity of the weather, and of Christina's routine, promptly killed him off). But this is of a piece with the peripatetic theme of his life and underlines the multiplicity of contexts within which science in early modern Europe was pursued, even by a *single* person. (Should we even think of Descartes as a Frenchman – or as Dutch?) And quite simply, there were plenty of princes to choose from. Islam had its own political instability, but this was more in time than in space – Ibn-Khaldun's iteration of rises and falls. As emphasized by Scheidel, western Europe stands out, in historical comparison, in the degree of its political fragmentation.[5] This is especially significant, then, from our perspective, in comparison with Islam.

Political fragmentation was essential to the context in which Greek science began, and the relative consolidation of Hellenistic monarchies did not remove the dispersion and multiplicity of ancient science. It was not essentially produced for patronage, but rather, it was made for status – seeking the approval of an audience and of one's peers. It was produced from many places, from many contexts – which is perhaps related to its more open-ended creativity. Fermat's challenge is reminiscent of Archimedes's challenge, *for a reason*. So, early modern Europe, re-creating Greek political fragmentation and, with it, Greek authorial practices?

But there is also one further thing that stands out in historical comparison. Fermat and Descartes both passed through universities without, however, becoming academics. Which was the rule: most scientific authors in early modern Europe had a university education, and only a few settled down as professors (this final professionalization, with science becoming a profession pursued mostly in universities, happened only throughout the nineteenth century). As a context for one's overall career, the university was merely one option among many. But as a trigger, it was ubiquitous. When still young men, future scientists, in early modern Europe, traveled to a place of learning, where they met like-minded individuals, leafed through library collections, and heard lectures that at least touched on scientific topics. (Young "men": a masculine science, embedded now in gendered institutions.) There is hardly any precedent for that in antiquity, and although the university – like most aspects of intellectual life in medieval Europe – is technically an Islamic innovation, it does differ from its antecedent, the madrasa, in a crucial way. The Islamic madrasa was in a kind of binary opposition to the Islamic court. So far, it is comparable to

[5] W. Scheidel, *Escape from Rome: The Failure of Empire and the Road to Prosperity* (Princeton, NJ: Princeton University Press, 2019).

the church. However, as the Islamic court monopolized Greek learning, so the madrasa found its scope in Quranic learning: literary, grammarian, and above all, legal scholarship. The Islamic lands did build public institutions for medicine, separately, in state-funded hospitals, which often became centers for medical education – and this, too, was later imitated by Christians. But the exact sciences did not become part of any curriculum: one typically became a mathematician through private tutoring followed by service to the sultan.

Medieval European courts were far too weak to become the main site of culture, à la Baghdad (although not for lack of trying, as witnessed, for instance, by the ephemeral renaissance of Charlemagne). As we have seen in Byzantium: culture – all of it, religious, literary, scientific – remained the property of the church. And so, as the madrasa of Islamic legal law was adapted, in the twelfth and thirteenth centuries, to become the university of the Catholic Church, it naturally became the site of the *entirety* of all available cultural legacy. The exact sciences were still very far from the core of the mission of the university: in each individual university, science was at the margins. But there were plenty of universities (reflecting, once again, the massive political fragmentation). And so – plenty of margins for science to take hold in.

So, on to another vignette – now, at the very geographical edge. We move much earlier in time, to the first years of the sixteenth century, and farther north and east, to the Polish borderlands: obviously, time to look at Copernicus. He was born in 1473 in Torun, near present-day Gdansk, to a family well connected in the church. And so he naturally got a good education, studying at Krakow and then at Bologna and Padua. There were many universities, with their own local traditions: from this smorgasbord, one could fashion for oneself the education one wanted. His connections secured him the sinecure of a cathedral canon in Frombork, not far from his native Torun, and he took his work seriously, spending forty years in the uneventful pursuit of his duties to the church and his passion for learning. The passion was wide in scope. The only project Copernicus ever pursued by himself all the way to print was a Latin translation of an early Byzantine moralistic piece in epistolary form. He definitely knew his Greek! He was taught in the universities of the Renaissance, and he was eager to discover and revive ancient learning in its pristine form. Above all, ancient astronomy. Like many others before him – from Proclus through Al-Haytham and Al-Bitruji to the school of Maragha – he admired Ptolemy but was critical of him. Indeed, the continuity between Copernicus's techniques and those of the school of

Maragha is such that most scholars now assert (although no documents have yet been found to confirm this) that Copernicus learned some key technical ideas, including the Tusi couple, in some indirect way, from that school.

As noted earlier, there were usually two grounds for criticizing Ptolemy. One was philosophical – his model did not fit well within Aristotelian cosmology. The other was technical – Ptolemy's geometry contained ad hoc assumptions, especially the equant, which, among other things, seemed inconsistent with the goal (expressly stated by Ptolemy himself) of reducing all motions to simple circles. Now, Aristotle still reigned supreme in fifteenth-century Europe, as he did for centuries since his canonization in Late Antiquity, but the sheer plurality of intellectual sites, coupled with a quest for erudition, made it far more natural to approach Aristotle, as it were, from the margins. Copernicus translated, from the Greek, the obscure Theophylact Simocatta, and he also sought out, from his readings, the cosmologies of more minor figures, a Philolaus, an Aristarchus. He thus took the Maragha project of expunging the equant and sought to place it within a sound metaphysical project that, for once, did not necessarily mean that of Aristotle. Now, just as the Ptolemaic system was marred by the ad hoc use of the equant, there was a potential redundancy in that each planetary model had to possess, independently, its own basic motion around the zodiac, rectified by its own combination of epicycle, eccentric, and equant. By making the system heliocentric, the earth's single motion around the sun could, at a stroke, provide all the individual planets with their own basic rotation relative to the earth. Heliocentrism was metaphysically complicated, as long as you were Aristotelian, but it was mathematically simpler. Hence Copernicus's model: Maragha – but with obscure Greek erudition!

Before 1514, Copernicus sent around the *Commentariolus* – a brief summary of his system – as a letter to his friends. Print had been around for decades, although not so commonly for scientific works. Most scholarship was still done in the way it always has been, scholarly networks of correspondence. The cathedral canon himself might have been stuck in Frombork, but the work did travel; eventually, an eager young admirer, Rheticus, extracted Copernicus's magnum opus from his hands and had it published in 1543 – as Copernicus was nearly on his deathbed – *Six Books on the Revolutions of the Heavenly Spheres*. This was now a printed book, and a cause célèbre.

Most definitely not a *scandal. On the Revolutions* was the most ambitious work of astronomy to date in Latin and was immediately admired as such.

Like the best of the Islamic astronomers, it displayed the mastery of the author by covering the same ground as Ptolemy – the six books compress Ptolemy's thirteen, for instance, by having a single book cover all the planets, but the sequence is essentially the same. The absolute requirement for a genuine astronomer was, of course, to calculate detailed tables, which is why Copernicus took his time: this is, indeed, hard work for a mathematician working alone. (Already by 1551, tables based on Copernicus were issued separately as the *Prussian Tables* – a Zij! – recalculated by Erasmus Reinhold.)

Why was Copernicus successful? Modern readers are impressed especially by Copernicus's heliocentrism – because it is true and Ptolemy's geocentrism isn't – but this was, of course, not evident to readers in the sixteenth century. What they saw was, so to speak, not Aristarchus but Maragha. Copernicus's technical innovations got rid of the equant, and in this, they were clearly superior to Ptolemy. (We are less impressed because to us, the Tusi couple is just as erroneous as the equant – if anything, perhaps more distant from the truth of the ellipse! Which reminds us of the danger of anachronistically reading past science through the prism of what is now known to be true.) It is often remarked that Copernicus's system was not empirically superior to that of Ptolemy, but of course, this was so by design: Copernicus did not start from any worry concerning Ptolemy's empirical deficiency, and his task was, instead, to recover Ptolemy by alternative geometrical means. Empirical considerations were secondary.

Tycho Brahe was one of the finest – and richest – of Denmark's nobility. He had the connections, which mattered immensely within the burgeoning, many-centered scientific network. And he had the resources to pursue science. It was a pluralistic system: science could be found in many corners, pursued in many ways. It could also be pursued by offering distinct, alternative models, and Brahe proposed a rather obvious hybrid model. The elongation, from the sun, of two of the planets – Venus and Mercury – is bounded; that of the others is not. Brahe proposed that Venus and Mercury rotate around the sun while the other planets, like the sun itself – together with its satellites, Venus and Mercury – circle the earth. Unlike, say, Copernicus, Brahe had the means to set up his observatory, Oraniborg, on his private island not far from Copenhagen (partly funded by the Danish king, too – connections!). All this, in a sense, was nothing new – Islamic courts had been doing this literally for centuries – but a campaign of observations carried a radically new meaning once several rivaling astronomies were on offer. All of a sudden, it was back to

Hipparchus, criticizing rival astronomical models – with Oraniborg as the new Babylon. The observations came in fast and thick – or rather, fast and subtle (a genius observer and instrumenter, Brahe could determine positions down to the minute). By the end of the sixteenth century, Brahe – as noblemen of the era often did – got in trouble with the Danish court. He eventually decamped to Prague, where he was supposed to produce for the Habsburg emperor, Rudolph, a new publication based on all those observations – the *Rudolphine Tables*. (Of course! What else are astronomers for, if not a Zij?) An up-and-coming young scholar, Johannes Kepler, was recruited to help with the calculations. And then, when Brahe died in 1601, the thirty-year-old Kepler stayed on: someone had to do something with all those numbers.

In an era of many-sided geniuses, Kepler was more many-sided than most. His Neoplatonism was of his own making. He suspected that the precise details of mathematical astronomy were dictated by precise geometrical patterns – we are reminded of Ptolemy's *Harmonics* – and his first effort at an astronomical model fitted the solar system within the five regular solids (this is much more Platonist than your average philosophically motivated astronomy). He was indeed an erudite and brilliant mathematician: so, for instance, he was the first to make sense of the note in Pappus, according to which Archimedes discovered the thirteen semiregular solids. (Kepler did this by discovering, by himself, a proof – which is very difficult – for why only those thirteen were possible. Pappus never reported this. As with Descartes: original mathematics, motivated by Pappus's *omissions*!)

Starting out as a Platonist, Kepler was willing to consider the possibility of new astronomical systems and to evaluate them according to an array of considerations, mathematical and metaphysical. Now, one of the observations that made the heliocentric system attractive, from the metaphysical point of view, was that one then found that the planets assumed to be closer to the sun moved faster; the farther ones, slower. This suggested a certain metaphysics or even physics – the sun, as the *cause* of the planets' motion – and Kepler saw that this could generate the eccentric-like appearance of the motion (a new way, then, of reducing the irrationality of Ptolemy's devices). Suppose that the planets move not in a perfect circle but in some other trajectory, and this trajectory has the nature that the planet moves faster, closer to the sun, and slower, away from it. Just as the Maragha school and Copernicus got rid of the equant, so Kepler's more physical trajectory, in a sense, would get rid of both eccentrics and epicycles. It would not be a circle – but it would be a directly motivated curve. The obvious shape to correspond to this description was an *egg* – becoming wider in its slow base, elongated in its faster apex. A difficult curve, but Kepler, the widely read mathematician, knew how to approximate such curves. This was obviously rather similar to an ellipse, and Archimedes, in his *Conoids and Spheroids,* had directly shown how to measure a sector of an ellipse. One could therefore apply Archimedes and assume – on close analogy of Ptolemy's equant – that the planet covered equal sectors of the ellipse (the one approximating the egg) at equal times. Thus, a combination of techniques from Archimedes and from Ptolemy could be used to deduce planetary positions.

As Kepler went through the labor of producing the calculations and making them agree with Brahe's numbers, the realization gradually dawned on him: the scaffold was, all along, the building itself. The ellipses were best understood not as tools with which to calculate the real egg-like trajectory but as the actual curves: planets move, in fact, in *elliptic* trajectories. When the *Rudolphine Tables* was finally published in 1627, the tables were based on the assumption of elliptic trajectories and speeds corresponding to elliptic sectors. They were also decisively more precise than Ptolemy's tables (this was true for no previous Zij). For a good reason, too: Kepler has discovered the *truth*. This, then, is a good moment to pause again and reflect on the scientific revolution.

*The Structure of Scientific Revolutions* was Thomas Kuhn's second book. The first was titled *The Copernican Revolution.* The key example for Kuhn – the paradigm, if you will – of a "scientific revolution" was the transition from Ptolemaic to Copernican astronomy. Kuhn's model stemmed from a wide tendency, broadly Neo-Kantian, in the early twentieth century, to see the history of science as the unfolding of overarching conceptual systems. It was thus natural to look for "the conceptual system of Ptolemy" and "the conceptual system of Copernicus" and to consider the process through which the first conceptual system was transformed into the second. Kuhn's contribution – which made his work so captivating for a large audience in the 1960s – was the relativist implication of such an understanding of history. If indeed different scientific theories are each embedded within their own conceptual system (which is one of Kuhn's usages of the word *paradigm*), if important scientific changes – "revolutions" – involve an entire paradigm shift, and if, as seems philosophically plausible, theories are to be judged relative to their conceptual framework, then it becomes impossible, strictly speaking, to argue that one theory is better than another. Better, relative to which framework?

One may argue for or against this as a philosophy of science. Speaking as a historian and focusing just on the question of the legacy of Greek science,

I have two observations, one relatively minor (as far as Kuhn is concerned), the other more significant.

The minor observation is that the Ptolemaic–Aristotelian paradigm, overthrown by modern science, was much more recent than Kuhn imagined. Like most nonspecialists, Kuhn supposed that Aristotle was broadly canonical from the beginning and that although the ancients offered various astronomical variations, these had all to agree with the Aristotelian framework. Ptolemy, in this understanding, merely provided the final touch. This is wrong. In fact, Aristotle was not canonized throughout most of antiquity; Greek philosophers were in continuous, ever-shifting debate; the very practices of astronomy went through several stages in antiquity before they became stabilized through the ultimate canonization of Ptolemy – and of Aristotle – in Late Antiquity. This is significant because Kuhn, mistakenly, imagined a neat dichotomy between an ancient monolith and its modern replacement. That antiquity was not a monolith will matter in the discussion to come. But it also matters that the great Ptolemaic–Aristotelian continuity is to be found in a much narrower set of practices – those of commentators from Pappus onward but especially, and more creatively, in Islamic science.

This brings me to the second and more major point. Viewed against this tradition – of Islamic astronomy – Copernicus is seen not as a radical departure but as a brilliant culmination. He does exactly the same thing all creative astronomers were doing throughout the Middle Ages: he masters a particular book and re-creates it by eliminating difficulties having to do with Ptolemy's more ad hoc tools and by recalculating, and improving, Ptolemy's tables. Now, it is true that Copernicus, besides reinventing (or, more likely, applying) the Maragha ideas of replacing the equant, also had the idea of an extra simplification of removing one of the circles, in each planetary model, by reducing it to a single circle of the earth's rotation. And it is true that this idea, in Copernicus's own mind, was embedded within his brand of philosophical erudition, somehow related to his own version of Platonism or Pythagoreanism. But this was not a move, in Copernicus's mind, *away* from the Greeks. This was meant, if anything, as a virtuoso display of Greek erudition, akin to his translation of the difficult text of Theophylact. This is not the mindset of a revolutionary trying to overthrow the past. It is the mindset of someone trying to *reenact* the past – a person of the Renaissance. Which is, of course, what Copernicus was. Indeed, this was more than an aspiration: to an extent, it is fair to say that Copernicus's success was precisely in that he re-created a Greek experience.

What I mean is the following. Conceive of something akin to Copernicus's *On the Revolutions* – published in around 180 CE. I hasten to explain: Copernicus's work is so deeply Islamic that it would be absurd just to plop it into the second century CE. But we may conceive of a response to Ptolemy, proposed right at the end of that century, recalculating Ptolemy's positions without the equant, and with a heliocentric model instead. This would not have been a revolution at all. (And this is why my observation on the absence of a scientific consensus, prior to Late Antiquity, is significant.) This would have been ancient science as it always was, *zigzagging through debate*. Had there been a Hipparchus-like mind, not much later than Ptolemy, then of course, the thing to do would have been to attack Ptolemy and replace him! Now, Copernicus – in comparison to a typical ancient author – is extremely deferential to Ptolemy (although recall that Ptolemy, already, was remarkably collegial to his sources), and in particular, his close adherence to the basic structure of Ptolemy's approach would have been very surprising from an ancient author. In this adherence, Copernicus was still Islamic. But the major original move – the attitude that science should be made via a critique – is a *revival* of a Greek context and is of a piece of Copernicus's erudition as a humanist as, indeed, he reveals himself not only in his translation of Byzantine literature but in his famous introduction:[6]

> Some think that the earth remains at rest. But Philolaus the Pythagorean believes that, like the sun and moon, it revolves around the fire in an oblique circle. Heraclides of Pontus, and Ecphantus the Pythagorean make the earth move, not in a progressive motion, but like a wheel in a rotation from west to east about its own center.

The point is not that Copernicus had ancient antecedents for his heliocentrism. Those were, in fact, insignificant. The point is that Copernicus evoked the full cacophony of ancient debate as the immediate background for his own bold assertions. In this, he differed from, say, Ibn Al-Shatir, revealing that he was, in fact, made by the European Renaissance rather than just by global Islamic science.

The Maragha school did involve a number of astronomers, each proposing their own alternatives (if within fairly constrained parameters). Copernicus, recalling ancient cacophony, had revived it for Europe. The point is not that Copernicus's system gradually won but that, through the

[6] E. Rosen, *Nicolaus Copernicus: On the Revolutions* (Warsaw–Cracow: Polish Scientific Publishers, 1978), p. 5.

process of its victory, it unleashed a variety of alternatives in highly visible competition. So immediately there came Tycho Brahe – and his rival Ursus, who beat him to the publication of his own hybrid model – with new and improved techniques of computation invented along the way. (Ursus and Brahe quarreled not only over models of the sky but also about a certain technique of approximation in the construction of tables called *prosthaphairesis*. Another technique – that of logarithms, inspired by Archimedes's own astronomical calculations – was developed by several astronomers but finally published by Napier in 1615; this technique was crucial for Kepler.)

Thomas Kuhn, of course, was familiar with the sheer variety of astronomies and astronomical techniques on offer through the sixteenth and seventeenth centuries. As we will note, in some sense, this variety came to an end at the close of the seventeenth century, as Newtonianism became dominant. From Kuhn's perspective, then, this was a story of *crisis and resolution*. In this telling, the hitherto Ptolemaic system reached an impasse in the late fifteenth century; science was thrown into crisis with fierce debate and little agreement on the very basic terms of a world system, but of course, the revolution was over at the end, and order restored.

From our perspective, however, Copernicus was the one who achieved the restoration – by restoring, that is, the original *in*stability of Greek science.

The condition for it all is the sheer profusion of writing. This is clearly related to the rise of print, although the relation of cause and effect is complicated. After all, Gutenberg would not have printed his Bible had there not been demand for books, already, on a European scale. Print is made possible by the abundance of paper, at the very technical level that one could not have a print industry on parchment but also in that the massive production of paper manuscripts throughout the thirteenth and fourteenth centuries created the book culture that made print economically feasible. This, in turn, depended on the fragmented and multisited network of many princes, cathedrals, universities – and private patrons – each coveting their own library. Many of our handwritten manuscripts of Greek literature postdate even the Palaiologan renaissance and were produced in Latin-speaking Europe for Renaissance patrons throughout the fifteenth and sixteenth centuries. With one single exception – the Palimpsest – this is true, for instance, of *all* extant Archimedes manuscripts. Early in the fifteenth century, Bessarion had one copied for his collection – finally deposited in Venice; the Medici sponsored a copy for their own library in Florence, late in the same century (this is the misleading, quasi-facsimile imitating the early minuscule hand of its archetype, Codex A). Altogether, Heiberg counted eighteen Renaissance copies in Greek, but we should also add a Latin translation made in the middle of the fifteenth century and its own copies (one of those Latin copies – discovered in 2005 by James Banker – was copied in Piero Della Francesca's hand;[7] it seems clear that Leonardo da Vinci read it, as well). Copies of Moerbeke's translation were known, directly or indirectly, as well: in 1503 there came the first printing of works by Archimedes, produced by the Italian astronomer Luca Gaurico. (This was the *Quadrature of the Parabola* and the *Measurement of the Circle.*) Tartaglia published in 1543 – still in Latin – *On Balancing Planes*, as well as a very garbled version of *On Floating Bodies*, Book I; in the same year there came out a bilingual printing in Basel, with Regiomontanus's translation of the entirety of the contents of Codex A (so, without *On Floating Bodies*).

Scholars in Italy and beyond were bringing works from the past and acting out their roles as Greek scientists. Brunelleschi, the famous mathematician and architect of the dome of the cathedral in Florence, was "the Second Archimedes," a title sought by many. Maurolico was born in Syracuse (no less!) in 1494 – one of the best mathematicians of the century, he was recognized for his new studies in Archimedean topics, finding, for instance, the centers of gravity of solids, which Archimedes assumes in his writings without proof (new mathematics – as with the case of Descartes – produced by the *sense of loss*). Maurolico was – so Commandino – "so skilled in mathematics that in these times he can with justice be said to be another Archimedes." (This specific praise, we note, is entirely a cliché: to be a supreme mathematician meant to be an Archimedes.)

Commandino should have known. The texts issued by Gaurico, Tartaglia, and Regiomontanus were garbled, thwarted by their faulty manuscripts as well as by the sheer difficulty of so much of Archimedes's mathematics; Archimedes aimed, after all, at a certain enigmatic effect, and he was an enigma to his Renaissance readers. All the more tantalizing! Born in 1509, Federico Commandino would become master of the entire breadth of Greek mathematics. Throughout his career, he published superb Latin translations that may often serve, even now, as the guide for the edition of the Greek text itself, all thanks to his great skill in uncovering ancient authorial intentions. He produced the Pappus that

[7] J. R. Banker, "A Manuscript of the Works of Archimedes in the Hand of Piero della Francesca," *The Burlington Magazine* 147 (2005): 165–169.

impelled Descartes; Apollonius's *Conics*; and above all, the complete works of Archimedes, in 1558 (most of the works) and 1565 – *On Floating Bodies*.

Vincenzo Galilei was one of Florence's leading musicians, respected for his erudition in harmonic theory and historical lore concerning Greek music: he belongs to a circle of musicians at the time who tried to revive the music of ancient Greek tragedy and ended up inventing modern opera (ancient music: in yet another vanishing act). He never quite reached, himself, the top rung of Florentine society, never managed entirely to escape the low status of a mere musician-artisan. Hence, perhaps, the ruthless ambition of Vincenzo's son, born in 1564. This son – Galileo Galilei – would navigate the many facets of the culture of Renaissance Europe, a tightrope act conducted with great aplomb and with a single famous error.

As a mere youth, Galileo shone in elegant disputation and poetry, delighting Florence's high society. This was followed by a brilliant imitation of Archimedes, the *Bilancetta*, or *Small Balance* – the booklet in manuscript, with which we started this book. We recall: Galileo, in a display of his historical erudition, recounted Vitruvius's account of the problem of the crown; followed this by mathematical erudition, connecting this account to Archimedes's actual *On Floating Bodies*; and constructed his own balance so as to perform Archimedes's trick. If the goal was to become a new Archimedes, what better than to remake his supposed instruments? It was on the strength of such mathematical feats that Galileo got his university positions, first in Pisa, then in Padua. Galileo was alerted in 1609 to a new optical invention – the telescope. He knew best about mathematical instruments – and so he jumped right in, made his own telescope, directed it to the sky ... He was surrounded by the cacophony of competing theories incited by Copernicus, and this meant that such observations could gain a new meaning. Galileo supported Copernicanism, completely bypassing the technical questions of astronomical models (this entire Ptolemaic tradition hardly interested him, his implicit attitude being, so to speak, "equant schmequant"). Instead, Galileo concentrated directly on the metaphysical – and, literally, metaphysical – questions. The telescope had shown mountains and seas on the surface of the moon, moons surrounding Jupiter ... surely this all refuted Aristotle's philosophical account of a distinct astral realm made of pure spheres of ether. This was Galileo's point: he was mathematically trained – he understood the optics of refraction, the geometry of revolving moons – and so he was the one most capable of resolving philosophical questions, and all this just *because* he was mathematically trained. That's the metaphysics point of it all, the question of how to study physics – and *who* should study physics. As he would later write, in famous words, "The book of nature is written in the language of mathematics"! The study of nature – belonging, in Galileo's view, to the mathematician.

Archimedes and his generation developed an autonomous science because, in the third century BCE, mathematics and philosophy grew apart. But as Galileo – like so many before him – tried to revive Archimedes as an authority, seeking once again the autonomy of his own science, this was in a setting very different from that of third-century Alexandria. Mathematics and philosophy were woven together, and Galileo's remaking of Archimedean autonomy was of mathematics as superior to philosophy – not science, apart from philosophy, but science, *rather than philosophy*, as the route for learning the truth about nature. A bold claim – made by a shrewd author who knew his way around Italian courts. The new discoveries were published in a booklet addressed to the prince of Florence; even better, Jupiter's moons were called the Medici. (We recall Conon finding Berenice's lock of hair in the sky.) Soon, Galileo was installed as court philosopher in Florence, one of the most visible positions in all of European scholarship. The quest for status, finally complete!

Of course, his position might have become a little too visible. Copernicus had conceived of his system even before Martin Luther conceived of his theses, but by the beginning of the seventeenth century, the Catholic Church was in full fight, seeking to reestablish its authority. Even more worrisome than the idea of the motion of the earth was the temerity of mathematicians to dictate truth to philosophers. This could not please the masters of the church's doctrines, all trained Aristotelians. Galileo was nearby; unlucky events followed each other; two trials, in 1616 and then in 1633, gradually constrained his activities. The notion that there is some kind of inherent antagonism between science and the church rests almost entirely on the popular legend surrounding this episode. As for Galileo, he could not publish in Italy now, so his final and most important work – the culmination of his life project – was published in 1638 in the Netherlands, instead. (As ever in early modern Europe: fragmentary politics – benefitting the progress of science.)

*Discourses and Mathematical Demonstrations Concerning Two New Sciences* is as elegant a piece as any written by Galileo. Recalling the canonical works of Plato and his Latin imitator, Cicero, this is in dialogue form, and unlike most such imitations, it lives up to its models. The sense of place and persons is vivid, and their gradual learning builds up genuine drama. Indeed, the science itself builds up its surprises and climaxes. Galileo learned the craft of mathematical narrative from Archimedes himself. Let us describe one final proof from the "first day" (the conversation of the two new sciences is supposed to take place over four days). AFB is a semicircle around center C, contained in the rectangle ABED, with CD, CE joined and CF perpendicular to AB.

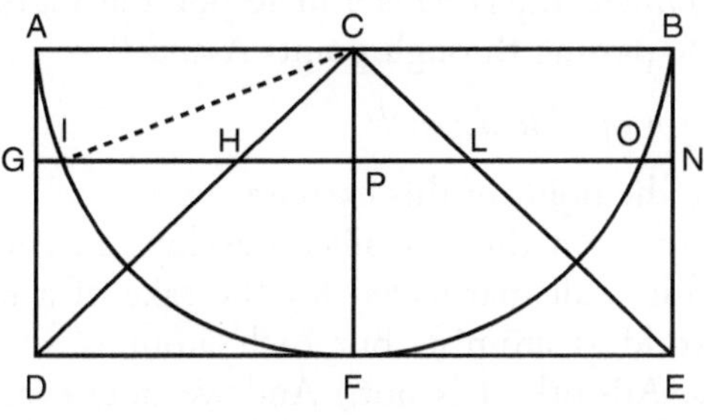


Figure 104

Have the entire thing rotate around CF, and you have a cylinder arising out of the rectangle ABED, a cone arising out of the triangle DCE, and a hemisphere arising out of the semicircle AFB. We are asked to imagine the cylinder with the hemisphere removed from it – the "bowl" ADFEB. Draw now an arbitrary parallel to DE, namely, GN, cutting the lines as shown in Figure 104. As the main speaker in the dialogue notes, from now on, "the demonstration is short and easy." Paraphrasing, we have the following:

$$IC^2 = IP^2 + PC^2 \text{(Pythagoras's theorem)},$$
$$\text{but} \quad IC = AC \text{ (radii)} = GP,$$
$$\text{and} \quad CP = PH \text{ (because CF} = \text{DF)}.$$

And so, bringing it all together:

$$GP^2 = IP^2 + PH^2, \text{ or}$$
$$GP^2 - IP^2 = PH^2.$$

Circles are to each other as the squares on their radii, so that, at the end:

> The circle associated with the cylinder, minus the circle associated with the hemisphere, is equal to the circle associated with the cone.

Now, "the circle associated with the cylinder, minus the circle associated with the hemisphere" is the ring associated with the "bowl." Hence, the ring associated with the bowl is equal to the circle associated with the cone, and – because both solids are made of all such planes – we conclude that the cone is equal to the bowl. This is true, to repeat, plane by plane – which is most evident "at the bottom." Around the line DE, the circle DE is the base of the bowl but also the base of the cone. They just coincide. But as we go upward, the cone becomes a smaller and narrower circle; the bowl becomes an ever-more-hollowed ring. They are still equal! Indeed, they must therefore be equal all the way to the top. Right at the topmost tip of the arrangement, the cone is a mere point at C; the bowl is a one-dimensional circle passing through points A and B –

*and the point is equal to the circle.*

Which is literally the point of this exercise.

Now that's quite something. Galileo develops an entire array of such paradoxes of infinity, all marshalled for the sake of a rather wild metaphysics. (The world is atomist, but each atom is an infinitesimal … Whatever this is, Aristotle it is not.) And we note everything, together: the very Archimedean mathematical proof – down to the very techniques of curvilinear measurement; the characteristic Galilean move of deploying Archimedes – as against Aristotle. Galileo's particular brand of atomism was not to last, but the mathematical technique was part and parcel of a much broader tendency in seventeenth-century mathematics. It was especially from the series of works sent to Dositheus that mathematicians learned of the technique of considering a curved figure as an approximation from a series of ever-closer rectilinear figures. They suspected, as ever, that there was a hidden method underlying it all, and incredibly, they were more or less right. (The "method" they imagined was in fact very much the same as that of the treatise of that name – its only extant copy, however, was then hidden under a prayer book in a monastery not far from Jerusalem.) Kepler wrote a magnificent *New Stereometry of Wine Barrels* (!), with curvilinear measurements of various strange figures, all achieved by their consideration slice by slice. Cavalieri, one of Galileo's pupils, published in 1635 a much more ponderous volume on the same topic. This book, *Geometry of the Indivisibles*, is almost seven hundred pages long (for his pains, however, Cavalieri was dubbed by Galileo "a new Archimedes": of course!). What Cavalieri did achieve was a more formal statement of a general method with which such measurements, based on indivisibles, could be obtained. And in this, he was not Archimedes: he went beyond Archimedes. As ever, trying to recover the lost past, mathematicians quickly sought methods and generalizations, and this, in and of itself, constituted a new science. By the mid-seventeenth century, a standard research area in mathematics was the quest for general methods for finding the areas contained by curves. Archimedes, in *Spiral Lines*, had obtained such an area – and had also shown how the circumference of the circle was related to the tangent of the spiral. Inspired by Apollonius, especially, mathematicians were also looking for general methods of finding tangents and other lines defined by curves. What else did the ancients discover? (It is said that one of the reasons attracting Descartes to Leiden was the knowledge that Golius – a Renaissance scholar of Arabic – had there a manuscript with Apollonius translated into Arabic, with more of the *Conics* than were known in Greek.) Looking for general methods for finding areas, general methods for finding tangents ... It was evident to many that the two were connected, and in 1670 Barrow showed – amazingly, Archimedeanly! – that if one curve is area-finding to another curve, then that other curve is tangent-finding to the first. Amazing, Archimedean ... But once again, not quite: such proofs in the latter half of the seventeenth century repeatedly refer to "general curves," a creature Archimedes never knew. One seeks the ancient, lost methods – unsurprisingly, erring to the side of generality.

But we are pushing too far ahead. We are still in 1634 – we should peer ahead merely from day one to days three and four of Galileo's dialogue.

Galileo was impelled onto his scientific trajectory by Archimedes's *On Floating Bodies*. It all began, after all, from the *Small Balance*! His first attempt at a major study was tentatively called "De Motu," and what Galileo was trying to achieve in this manuscript was a theory whereby all falling bodies were ruled by considerations similar to Archimedes's law of buoyancy. In other words, Galileo's first effort was to generalize *On Floating Bodies* from water to air. This is not a silly thought – aerodynamics is essentially an extension of hydrodynamics – but of course, this cannot work without a better understanding of the behavior of falling bodies. The manuscript "De Motu" was aborted, and Galileo put it aside, trying out alternative approaches. But then again, Galileo knew – as probably Ptolemy did but surely as Philoponus and many later medieval philosophers knew before him – that Aristotle's account wouldn't do, either. The task was to remake Aristotle's physics on mathematical – that is, Archimedean – foundations. Aristotle – not a geometer but heir to the mathematics of the generation of Archytas – assumed proportions everywhere: so, he assumed that the heavier an object was, the faster it fell. Whether or not Galileo checked this experimentally, he had good reasons to doubt this, and he had the good physical intuition: the proportion, Galileo realized, must hold for a different set of relations.

And so, Galileo found a different way of using proportions. Correlated with the speeds were not the weights but the *times*. Falling objects accelerated at a uniform rate: their speeds at a given time could be described geometrically as growing lines. The distance traversed in the course of a given time could then be described as the area of a triangle (we are looking for infinitesimal lines, composing areas: the habit of mathematicians, trying to re-create lost Archimedean insights). This is day three of the dialogue (Figure 105).

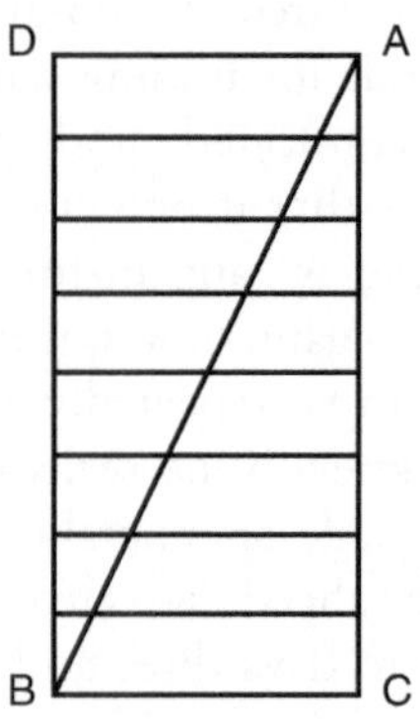


Figure 105

But more than this: suppose a body is pushed sideways; it will tend to continue horizontally at the constant speed in which it was pushed but would also be pulled down, vertically, in a uniformly accelerated motion. As the projectile progresses, the horizontal component grows linearly while the vertical component grows in the ratio of squares. A curve, where the linear in one dimension is correlated with the squared on the orthogonal – well, of course, this is the parabola, the most Archimedean of all geometrical figures. And this is day four of the dialogue (Figure 106).

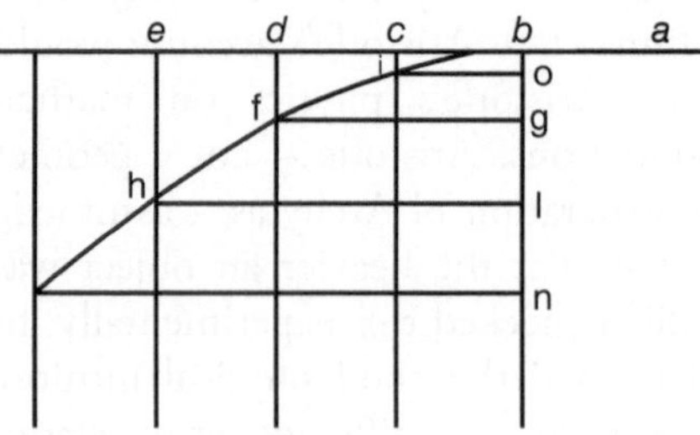


Figure 106

The rules governing falling bodies are found by Archimedean, not Aristotelian, notions; the book of nature is written in the language of mathematics . . .

The denouement is nearby. Falling bodies trace a parabola; planets trace an ellipse; new geometrical techniques – due to Descartes's analytic geometry and due to infinitesimal studies by Cavalieri, Barrow, and many others – make it natural to generalize across curves as the expression of more abstract, algebraic relations. In 1669 Barrow relinquished his chair of mathematics at Cambridge to his pupil, Newton. By that time, Newton was already in possession of his physical system, a synthesis of Kepler and Galileo, to some extent supported by Newton's own general method of infinitesimals, the calculus. Like Archimedes, Newton, too, was an enigmatic author – although of a more paranoid style – and, like Fermat, he conducted much of his career by throwing out hints and promises. He therefore took his time, but finally, in 1687, he published *Mathematical Principles of Natural Philosophy* – the *Principia*. As the title proclaims, Galileo was right; mathematics did triumph over philosophy. The system of the world is given by simple geometrical laws. The law of gravity is quadratic, and for this reason, it ensues in conic sections: hence the ellipses of planets and the parabolas of falling bodies and the greatest synthesis, since Ptolemy, of all science. The *Principia* was the final vanishing act of Greek mathematics, and we might as well end there.

Let me conclude this survey with two observations, one more contingent, the other more structural.

The contingent point is that without the towering synthesis of the *Principia*, there would have been no Newtonianism to define the eighteenth and nineteenth centuries, arguably no Enlightenment, and a very different trajectory to modern history. But working backward, without Galileo and Kepler, there would have been no *Principia*, and at a very basic technical level, both Kepler and Galileo would have been strictly impossible without conic sections. (I return to this in the epilogue.)

Well, you might say: had there not been conic sections, one ought to have invented them! This is all well and good. As a logical possibility, Kepler and Galileo, had they been forced to, could, of course, have invented conic sections from scratch. But to state this logical possibility is to miss the key historical realities. One is that Kepler used conic sections not because he needed to invent them but, on the contrary – he did so as a mere scaffold, simply because this was the only tool he had available *with which to measure eggs*. Another is that Galileo noted and emphasized the shape of the curve resulting from the fall of a projectile not because this was a necessary corollary but because this was an especially valuable feature – because it provided Galileo's science with an Archimedean respectability.

The same realities could be repeated for all mathematicians of the seventeenth century. Kepler and Galileo, and their entire generation, turned to conic sections because they had Archimedes. Had they not had him, they would have not. Which is indeed obvious, given that conic sections, as the center of a mathematical research agenda, emerged exactly once in history – as the parting shot of the generation of Archytas and as the central theme of the generation of Archimedes. Take away these two generations, and you take away the tools with which to make a Newton.

This, then, is a contingent fact, a coincidence, if you will. The more structural point concerning the scientific Renaissance has to do, more generally, with its attitude toward the Greeks.

We are reminded, once again, of Descartes, relocating to Leiden so as to see the manuscript (even if in translation!) of Apollonius's lost *Conics*; we are reminded of Commandino, recovering the text of so many diverse authors, cleaning them up, and finding their mathematical significance; we are reminded of Copernicus, translating the absolutely obscure letters of Theophylact ... One feature of the Renaissance of early modern Europe was its *philological* ambition. The goal was, quite simply, to find out everything knowable about the ancients. This urge, indeed, is noticeable already with the Palaiologan renaissance, and it becomes systematic and even professional by the fifteenth and sixteenth centuries. Incredibly, this project was, on the most part, achieved. Throughout the seventeenth century at the latest, nearly all the manuscripts were identified, nearly all first editions printed. (The eighteenth and nineteenth centuries could dedicate themselves to the task of professional textual criticism based on the manuscripts and texts that were already known; the twentieth and twenty-first centuries, left with the thankless task of *interpretation*.)

Contrast this with the Arabic reception of Greek culture. Here the impetus was much more restricted and short-lived. There was an initial engagement with a project of translation, much of it mediated by middlemen (as it were, it was Syriac readers pushing Greek into Arabic, no less than Arabs seeking to absorb Greek). Once a certain corpus was brought into Arabic – and once there grew, surrounding that translated corpus, an original, Arabic library tradition of interpreting and using that selection of the Greek past – the impetus was gone, and new translations, largely speaking, were no longer sought. We notice Ibn Al-Haytham, wondering about the contents of the final book of Apollonius's *Conics*, trying to reconstruct it, perhaps giving up on the project because he was not sure that this was the correct reconstruction ... Al-Haytham was born in Basra and died in Cairo. Was it that hard to travel, even once, the same distance so as to visit Constantinople, to seek that which was lost? Of course, Apollonius could not be found there. But, in the eleventh century, Pappus was still available – and just what could an Islamic mathematician do with *that*? One can imagine a whole new Islamic renaissance out of such what-ifs. The thing is that they remain as counterfactuals. Al-Haytham did not cross the Mediterranean. Nor is this some kind of peculiar Islamic deficiency: an exactly analogous story can be told of medieval Latin Europe, absorbing (once again, mostly via middlemen) many Islamic works but then showing almost no curiosity for finding out what more was there in Arabic. And once we put the two indifferences side by side, the Islamic and the Latin, they make sense: in both cases, the absorbing culture experienced itself as fundamentally alien from the source culture. Islamic authors highly valued their Greek scholars, but to them, they were ciphers – as in Borges's famous image of Ibn-Rushd, trying in vain to understand what Aristotle's distinction between tragedy and comedy even meant ... Greek history was known as vague legends, and its main figures were only very tentatively put in particular spaces and times. There was no clear map of influences, proximities, oppositions: and so, no urge to fill in the gaps because there was no mental map in which such gaps could present themselves. Similarly, of course, for Europe's reception of Islam. (How many Europeans even realized that, say, the two towering figures of Avicenna and Averroes – Ibn-Sina and Ibn-Rushd – came from the opposite geographical ends of the Islamic world?)

But the reception of classical antiquity in early modern Europe was very much unlike that. Vincenzo, the father, recovering Athenian tragedy – followed by Galileo, the son, recovering Syracusan geometry – and the two are of a piece, part of a similar project where antiquity, as an entire world of meaning, is vividly present. This explains why Europeans sought to know more and more of this past antiquity but, even more significantly, why they felt themselves *at home* in the past. This is the constant theme of the quest, to find and learn all the ancient works – but also, to reconstruct for oneself the secret methods of the Greeks, now no longer available. Authors in early modern Europe naturally slid into a reenactment of the ancient sources. This quality – the *live-action role-playing* of classical antiquity – gave the Renaissance of early modern Europe a distinct quality and meant that, for once, one did not merely survey and master past achievements, but instead, one sought to insert oneself within the past; one sought to become the "new Archimedes." The new Archimedes they became; and then, the vanishing trick – and Archimedes was gone. From now on, he would belong, strictly, to the pages of "A History of Greek Mathematics."

## Suggestions for Further Reading

It is very clear what you should read for book culture in Byzantium: this is N. Wilson, *Scholars of Byzantium* (London: Duckworth, 1983), an erudite and readable introduction. There is nothing comparable specifically on the tradition of mathematics in Byzantium. The recent edition, R. Netz, W. Noel, N. Tchernetska, and N. Wilson, *The Archimedes Palimpsest* (Cambridge: Cambridge University Press, 2011), adds much more on book culture and specifically on the Archimedean tradition.

The field of Arab language mathematics is vast, and as yet, there are no ready syntheses. G. Saliba, *A History of Arabic Astronomy* (New York: New York University Press, 1995), is essential for the Islamic route to non-Ptolemaic astronomy. On pure mathematics specifically, J. L. Berggren, *Episodes in the Mathematics of Medieval Islam*, 2nd ed. (New York: Springer, 2016), is excellent, although technical and also – by design – highly selective.

Not enough has been written on Islamic science. There is, on the other hand, quite a lot to read on science in early modern Europe! You should read, of course, T. S. Kuhn, *The Structure of Scientific Revolutions* (Chicago: University of Chicago Press, 1962). Its picture of a normal Ptolemaic–Aristotelian science thrown into crisis through empirical difficulties, ultimately ensuing in a paradigm shift into Newtonian science, is to be qualified in several ways. The actual practices of the new science are the subject of another classic, S. Shapin and S. Schäfer, *Leviathan and the Air-Pump* (Princeton, NJ: Princeton University Press, 1986) – which is essential for the social realities of science in the era (on which I touch only very lightly). P. L. Rose, *The Italian Renaissance of Mathematics* (Geneva: Librairie Droz, 1975), remains crucial for the origins of the so-called scientific revolution in the Renaissance project of restoration of the past.

For specific trends and authors, H. J. M. Bos, *Redefining Geometrical Exactness* (New York: Springer, 2001), is essential for the mathematics of Descartes and his generation, understood on its own terms; for the route leading to the new astronomy, two books are outstanding in different ways. O. Gingerich, *The Book Nobody Read* (London: Penguin, 2004) (warning: an ironic title!), an account of the legacy of Copernicus's *De*

*Revolutionibus*, is extraordinarily fun for such an intricate and precise book; J. Voelkel, *The Composition of Kepler's* Astronomia Nova (Princeton, NJ: Princeton University Press, 2001), is more technical but still an accessible account of the process leading Kepler to his major astronomical model. There are so many books to recommend on Galileo and Newton, one hardly knows where to begin, and perhaps it is best to read a good biography: J. L. Heilbron, *Galileo* (Oxford: Oxford University Press, 2010), and, of course, R. Westfall, *Never at Rest* (Cambridge: Cambridge University Press, 1983).

# *Epilogue: Bringing to the Boil*

In a world of ambitious mathematicians, each vying to become the new Archimedes, Evangelista Torricelli, born in Faenza, Italy, in 1608, was determined to out-Archimedes them all. The one book he published in his short life (he died at age thirty-nine) is the *Geometrical Works*, a series of mathematical marvels, some reproducing results from Archimedes, some extending the spirit of Archimedes by measuring new curved figures. Torricelli may have been the first to measure the area of the cycloid, a figure originally proposed by Galileo. (The seventeenth century was full of Archimedes-like challenges; unlike in the third century BCE, such challenges tended to generate many responses, with many ensuing fights over priority.) He invented and measured his own – paradoxical – figure. Consider a hyperbola, one of its asymptotes, and some perpendicular to that asymptote cutting the hyperbola. The three taken together define an infinitely long figure that is widest at its base (the perpendicular) and keeps narrowing – without ever reaching the zero width of a point – as it moves away from that base. Rotate this figure around the asymptote, and you get an infinitely long solid, shaped rather like an infinitely long, curved funnel. Torricelli measured the volume of this infinite solid – which turned out to be, incredibly, finite. Infinite surface, finite volume! A new Archimedes, right there!

When they met in 1641, Galileo took an immediate liking to Torricelli, who stayed on as a member of Galileo's household in the master's last days. Indeed, when Galileo died the year after, Torricelli was asked to stay in Florence as the new court mathematician. Two years later, *Geometrical Works* was published, and for the remaining three years of his life, Torricelli was recognized as one of Europe's leading mathematicians. Only a single book published, but this was still the age of the intellectual tournament. What mattered most was the letters, exchanged between mathematicians, copied and recopied by hand. Archimedes changed history through a series of letters sent from Syracuse to Alexandria; Torricelli would change history through a single letter sent from Florence to Rome.

We keep going back to Florence, to Galileo. And here is yet another puzzle, another challenge: Why does the water not budge? Pumps, after all, could deliver water out of wells. They did so by creating a column of water rising above the surface of the well, and when the top of the column was tall enough, one could reach it and let the water spill sideways and so collect it. But here's the problem: engineers in Florence noted that such columns of water were limited in height to about ten yards. Try any taller, and the water would not budge, or the column might "break." Thus, wells could be used only up to a certain depth. A practical problem, posed to Galileo by practical engineers, but also a good Archimedean question because it involves the behavior of liquids. Surely an Archimedes mathematician – freed of Aristotle's false philosophy – could solve this?

Galileo's first efforts were more anti-Aristotelian than Archimedean. (His core idea had merit but was slightly beside the point: Aristotle, so Galileo, was wrong in claiming that vacuum does not exist – so far, so good – and vacuum, so again Galileo, exerts a certain force that is capable of attracting water, the force limited to the height of ten meters. This is no longer so good!) Torricelli had a better and more Archimedean account, which he sent in a letter to his colleague in Rome, Michelangelo Ricci (who then proceeded to copy and share the letter), on June 11, 1644:[1]

> We live immersed at the bottom of a sea of elemental air, which by experiment undoubtedly has weight, and so much weight that the densest air in the neighborhood of the surface of the earth weighs about one four-hundredth part of the weight of water.

The physical breakthrough is the recognition that air, in its place, has positive weight (this would not be evident because one's first observation – which even Galileo could not bring himself to deny – is that air's tendency is to move *upward* or, at the very least, to stay still, suggesting a negative or zero weight). Once this recognition is achieved, air becomes akin to water, and the mechanism of *On Floating Bodies* directly applies, which Torricelli immediately confirms by sharing a marvel (what we would think of as an experiment). Take a narrow but sturdy glass tube with a single opening – preferably, at least a meter long – and fill it up with the heaviest liquid you can get, which is mercury; have a big bowl, also filled with mercury, at hand. (To have access to all of this, better be a court mathematician; Archimedes, we recall – or so it was told – weighed silver and gold.) With your finger pressed on the open end of the tube, immerse it in the bowl and turn it so

[1] W. F. Magie, *A Source-Book in Physics* (Cambridge, MA: Harvard University Press, 1935), p. 71.

that it now stands upside down, with your finger at the bottom (prepare to feel a good amount of pressure on your finger; it's perhaps best to ask a friend to do this for you). Now remove your finger.

The mercury level at the tube goes down, obviously: this is a liquid, after all, and we know from Archimedes that liquids become stable only when they are level. And so, we should expect the mercury to descend entirely into the bowl. Not so! It goes down a little and then budges no further – a stable column of mercury (we would say, seventy-six centimeters tall) that prefers to stay inside the tube, even while it is completely free to descend. The liquid achieves stability in an *un*even position!

Think of it in the terms of Archimedes's *On Floating Bodies*, and the explanation is obvious: the mercury achieves stability when its adjacent surfaces are equally pressured from above. Once we realize that the air has weight and so exerts pressure, it becomes obvious that the column of mercury should remain precisely at the height at which the weight of the mercury is equal to the weight of a column of air of the same width. (It does remain significant, then, that the top of the tube, in this marvel, holds a weightless vacuum, but it is not the vacuum that keeps the mercury up, but rather – with a nice paradoxical twist – it is the surrounding air.)

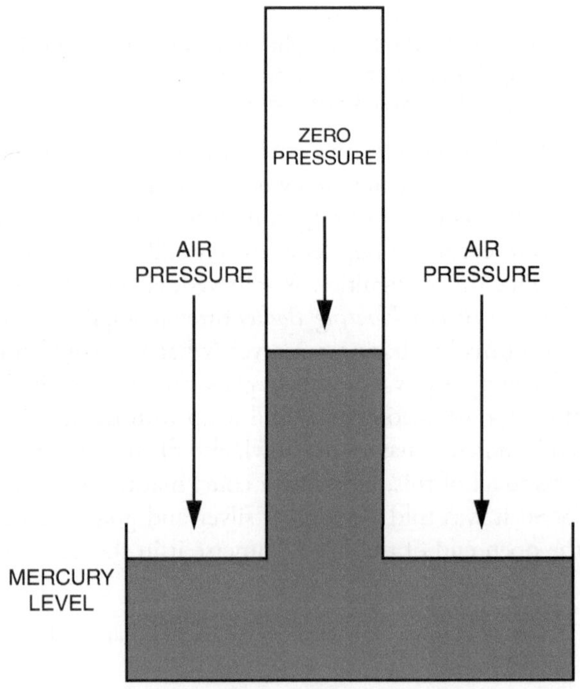


Figure 107

The logic of the water pump is now self-evident. The weight of ambient air is about the same as a column of seventy-six centimeters of mercury, and this is also the weight of about ten meters of water. The only function of the mercury, in fact, is to allow one to do everything within easy human grasp: the mercury does nothing that the water in the pump did not do already. But exercised within human grasp, the marvel is more impressive, and it spread across Europe like quicksilver. The thought quickly suggested itself that if air was rare on mountaintops (as was long suspected), the level of the mercury found there should be lower. Pascal, the French mathematician, repeated the marvel at different elevations: the level of the mercury fell down as one kept climbing up. Perhaps the word *marvel* is no longer fitting: the ancient-like marvel, replicated, becomes the modern experiment, in yet another of those vanishing tricks where repeating the past somehow changes its meaning. Pascal, then, invented the barometer. That the atmosphere has weight – and that Archimedes's hydrostatics applied to it – was no longer in any real doubt. It was 1648, Torricelli was already dead, and he had already set in motion the main trajectory of modern history.

What follows is well known. Armed with the double realization that (1) gases have weight and exert pressures, and (2) the production of void (or, perhaps, near-void) is a practical possibility, many authors extend the theory, and practice, of the pressure of gases. Several variations on a particular design were even found, beginning in the late seventeenth century, to be of practical use. Boil water and let the steam fill a chamber whose top is a movable piston. Allow a little cold water into the same chamber, and as the steam quickly condenses, the chamber becomes nearly empty. This means that the air above now exerts an enormous pressure down on the piston that, as a consequence, moves down violently. Engage that piston to a beam, and you can use the violent downward motion of the piston to lift up a weight. (This is all rather like an upside-down trebuchet.) Eventually this was known as the "Newcomen engine" (Figure 108), first put to use in a British coal mine in 1712 (although, to repeat, designs of this type were circulating between authors and engineers for decades, now). The weight lifted by this first Newcomen engine – in this first application and for many decades to come – was water, pumped out to drain the mine. The machine, at this stage, still belongs in the field from which it emerged, of water technology.

Newcomen's design was crude, as it were, the most direct application of Torricelli's marvel. There are many inefficiencies in the design, particularly because of the rapid alternation of heat and cold in its main chamber.

Improvements were gradually proposed and made, and ultimately, a mathematically trained Scottish engineer, James Watt, introduced a variation on the same design: keep the main chamber at a fixed temperature and create a mechanism that makes the steam escape regularly – through the same forces of gas pressure – into a separate condenser. In retrospect, this was the way to do this. The new machine – what we now know simply as the "steam engine" – produced considerably more power, and it was quickly recognized that this power could be harnessed to any device requiring force. There was no need to keep on just pumping mines. And so the steam engine – no longer a hydraulic tool but, instead, a universal source of energy – launched Britain, and then the entire planet, on the course to exponential, technologically driven growth (unfortunately, at least for its first quarter millennium, one based on fossil fuels).

This brings us to the question of the place of Greek mathematics in world history. The industrial revolution is the pivot separating the way things were, in the past, from the way they are now. So it is significant to note the very real through line: from Archimedes to Torricelli, to Newcomen, to Watt. It is tempting to propose a counterfactual history where Archimedes dies not as an old man, in the Second Punic War, but as a young man, in the first. He never writes anything; the second generation of Greek mathematics never takes off; there is no *On Floating Bodies* for Commandino to republish in Latin, to inspire Galileo and Torricelli; no letter to Ricci; Pascal gets excited by other things. There is no burst of atmospheric-pressure designs in the late seventeenth century and, of course, no application to the mines and so nothing to provoke Watt's curiosity. And then, our counterfactual continues: Britain, in the nineteenth century, is largely similar to what it has been in the eighteenth and what the Netherlands were in the seventeenth – a remarkably sophisticated and effective commercial state, maximizing the potential of the premodern economies without making the decisive break into modern technologies.

As such counterfactuals go, I would contend that this is a fairly reasonable one, but its one main drawback is its emphasis on the specific route leading from Archimedes's *On Floating Bodies*, via Commandino and Galileo, to the particular event of Torricelli's letter to Ricci. This letter is as important of a single event as one can think of in the history of science, but it is still just a single event, and there were plenty of others throughout the seventeenth century. There were even independent reflections and experiments concerning the atmosphere and the void, and in general, there was, throughout the century, lots of theoretical and experimental reflection upon nature, of a scale unprecedented in previous history.

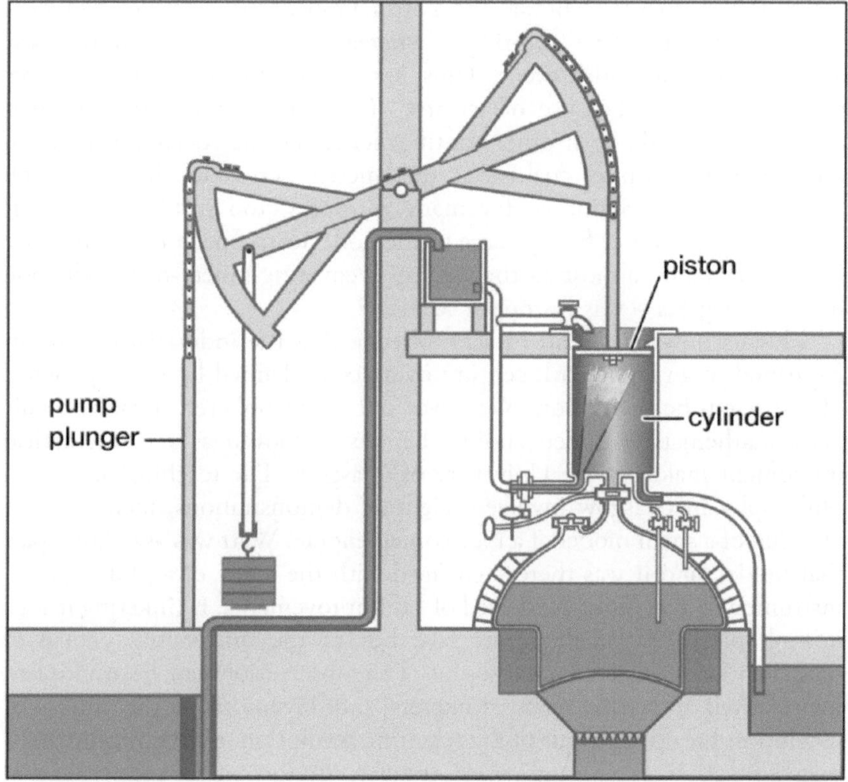


Figure 108 Newcomen steam engine. Encyclopaedia Britannica/Universal Images Group/Getty Images.

A ferment of creativity, not an isolated line of transmission. Indeed, the steam engine does not depend on any specific result obtained by Greek mathematicians (my argument is that it is ultimately inspired by an *analogy* to an Archimedean result; beyond this analogy, all it takes, to make the engine, is ingenuity). Surely other paths can be envisaged? There used to be a time when British scholars, in particular, emphasized Watt's identity as an engineer – rather than a scientist – and saw in the industrial revolution the reflection of a very wide rise in craft creativity and industrial organization, specific to Britain of the time. In this telling of the industrial revolution, science is almost irrelevant. Such an account is especially attractive if one is interested primarily in the question, Why Britain? Because, after all, the scientific revolution was a European event – taking place in the network encompassing Torun, Copenhagen, Prague, Florence, Leiden, Paris, and also Cambridge – whereas the industrial revolution was almost entirely a British event. Thus, asking, Why Britain? almost forces one to say: "Not because of science." (And surely the answer to *that* question has to do with other contingencies: specific social and political arrangements, empire, coal.) But this merely serves to show that the question, Why Britain? is, for many purposes, too insular. It was in Britain that the water finally came to the boil, steam finally rose. But what heated the water, almost all the way up – covering ninety-five of the one hundred degrees – was European science.

Scholars nowadays tend to acknowledge that the industrial revolution happened in an eighteenth century completely defined by the experience of the scientific revolution. Watt was not some tinkerer. Thanks to his good mathematical understanding, he was employed as a mathematical instrument maker at the University of Glasgow. The teaching of natural philosophy in Glasgow involved scientific demonstrations, including the working of a small model of a Newcomen engine. Watt was asked to repair that model, and it was there – engaged with the study of a philosophical instrument – that Watt conceived of his improvements. If tinkering it was, it took place not at the mine face but at the university's course of mathematically inspired philosophy. This observation can be multiplied many times over: the many tinkerers and inventors of the industrial revolution breathed the air of the scientific revolution, either through their own education or through their contacts in the wider society of the scientific enlightenment. And in this way, the scientific revolution was, after all, specifically British. It was Newton's land; throughout the Enlightenment, Europe as a whole would become Newtonian. Britain, more than most.

And so, it bears repeating – as noted in the final chapter of this book – that the same counterfactual, told for Newton's *Principia*, is quite robust. Remove Greek mathematics. All it would take is a set of fires, strategically spread throughout the seventh century CE, burning perhaps a hundred manuscripts or so (all that one then had of Greek mathematics?) – before Arab Sultans began to seek their translations, before Byzantine scholars began to restock their libraries.

Muslims and Christians, throughout the Middle Ages, would still recover Greek philosophy (and might be intrigued by the tiny reports of Greek science preserved there). But where would Newton's *Principia* come from? You cannot have a universal law of gravity without the combination of Galileo's parabola and Kepler's ellipse, and you will have neither without the conic sections, which the Middle Ages would never have come up with on their own. We know this because *no one else* ever came up with conic sections. And at an even more basic level, although there would be lots of cosmological reflection and metaphysics, there would be no Copernicus. No one would have constructed geometrical models and calculated tables based on them without the particular Ptolemaic synthesis of Babylon and Greece. We know this because *no one else* ever came up with anything like this independently.

Perhaps, if all you do is remove *On Floating Bodies*, you might still end up with Watt. But if you remove the entirety of Greek mathematics, you certainly cannot get Newton's *Principia*. It is extremely hard to conceive of what form the intellectual ferment of Europe in the seventeenth century then takes. But it should be borne in mind that throughout the same generations as those of the scientific revolution, European minds were fixed upon such questions as the nature of the Eucharist and of predestination, no less and even more than they were fixed on the nature of gases and projectiles. A more purely metaphysical and theological seventeenth century is completely conceivable. Without Greek mathematics, no one would have been particularly interested in reading the book of nature in the language of mathematics. Europeans might well have been content to go on reading the book of God, instead. Horses, harnessed to water-lifting machines, would continue to march in circles, right at the mouth of the coal mine, no one suspecting that a great secret – the one that could have spared all those horses – was lost forever, a thousand years before, by the fire that consumed Archimedes's codex.

Historical causation is not a matter of absolutes. Perhaps, had there been no Greek mathematics, aliens could have intervened to provide humans with the steam engine. Or perhaps, had there been no Greek mathematics, someone else in the centuries between Archimedes and Copernicus would have invented something analogous to it instead. Historical causation, to repeat, is not a matter of absolutes but of probabilities. The probability of aliens handing humans the steam engine is very small indeed. The probability of someone else, in the centuries between Archimedes and Copernicus, inventing something analogous to Greek mathematics is definitely higher. But I think that the narrative of this book makes clear that this, too, would not be very likely. Greek mathematics emerged under quite specific conditions – Greek democracy and its debates and emphasis on persuasion and proof; the encounter of science and philosophy in the generation of Archytas; the moment of autonomous science in the generation of Archimedes. The historical combination giving rise to the invention of Greek mathematics is precarious, and it is for a reason that this particular invention is not replicated elsewhere. The probability, then, is that without the seeds of Greek mathematics – the seeds sown by Archytas and Archimedes – the scientific revolution, hence the industrial revolution, would have been, at the very least, significantly delayed.

There are many legitimate questions to ask about why the industrial revolution happened. As noted, many scholars in the past were especially interested in the specific question of why it took place in Britain: this particular contingency shaped much of the history of the long nineteenth century and beyond. There are plenty of other questions to be asked, at different levels of specificity, looking at different historical parameters. For certain purposes, it is interesting to understand even more specific questions – for instance, why did Boulton and Watt end up setting shop in *Birmingham* of all places (this, too, has important consequences for the history of the British Isles). The causal thread pursued here, looking back from the industrial revolution to the scientific revolution and then to Greek mathematics, is in part an answer to a much wider question: why the industrial revolution took place in the Mediterranean region of the world. This means, in practice, that I suggest a rather obvious answer to Needham's question (noted on p. 474). Europe, rather than India or China, produced the scientific revolution because, unlike the other major civilizations, Europe had the resources of Greek mathematics. This seems as safe of a causal statement as one could ever hope to make in intellectual history.

This, however, does not touch on the more difficult question as to why the scientific revolution took place in Europe and not in the Islamic world. For this, perhaps two observations – both noted earlier – matter most. (1) The European scientific renaissance was part of a broader literary renaissance, and it was much more natural in a Latin-speaking world, whose literary canon was formed in Greco-Roman antiquity, than in the Arab- or Persian-speaking worlds, whose literary canons were distinct. (2) Europe, for reasons explained best by Scheidel, always possessed a more pluralistic system, with massive consequences for the possibility of a renaissance to begin and then take hold. (In particular, it was less dependent on the institution of the court, which, at the crucial moment, in the Islamic world, seems almost to have lost interest in science.) But to be fair, this particular question – why Europe and not Islam – is not easily answered by a "History of Greek Mathematics." It was here that Europe and Islam most neatly converged, after all.

In fact, it is perhaps better to concentrate on a different question altogether. When we note that Watt's steam engine became functional in Birmingham in 1776, this may give rise, as noted, to a series of concentric questions, each legitimate in its own way: Why Birmingham? Why Britain? Why Europe? Why the Greater Mediterranean? And the narrative of this book here answers best, I suggest, the last and widest one. But to conclude, I want to recall another, no less interesting question: Why 1776? At first glance, this appears like a tough question for my book: 1776 is a full two millennia later than Archimedes! But in fact, it is this question that the narrative of this epilogue, specifically, answers best. Watt's engine became functional in 1776 because Newcomen's engine was deployed in 1712, in turn because Torricelli's letter directed scientists to look for atmospheric effects, starting in 1644, finally because Archimedes's *On Floating Bodies* became available through Commandino's translation in 1565. And more broadly: there was a scientific renaissance, in the strict sense, throughout the sixteenth century, as the ancient works became available again. This gave rise, throughout the seventeenth century, to what is known as the scientific revolution, which, in the eighteenth century, made the industrial revolution possible. Once again: perhaps a different route might have presented itself, otherwise. But as it were, it was the sixteenth century that started us on our route to where we are now. And we can see, at the end of this history, that it did so, above all, through its rediscovery of Archimedes.

# Index